THE
CRC HANDBOOK
OF
MODERN
TELECOMMUNICATIONS

THE
CRC HANDBOOK
OF
MODERN
TELECOMMUNICATIONS

EDITORS-IN-CHIEF

Patricia Morreale
Kornel Terplan

CRC Press

Boca Raton   London   New York   Washington, D.C.

Library of Congress Cataloging-in-Publication Data

The CRC handbook of modern telecommunications / editors-in-chief, Patricia Morreale and Kornel Terplan.
    p. cm.
  Includes bibliographical references and index.
  ISBN 0-8493-3337-7 (alk. paper)
    1. Telecommunication--Handbooks, manuals, etc. I. Morreale, Patricia. II. Terplan, Kornel.

TK5101 .C72 2000
621.382—dc21

00-062155
CIP

Visit the CRC Press Web site at www.crcpress.com

Acknowledgments

The Editors-in-Chief would like to thank all their contributors for their excellent, timely work. Special thanks are due to our Associate Editors, Teresa Piliouras and James Anderson. Without their help, we would not have been able to submit this manuscript on time. We thank Mihaela Bucut, our Ph.D. student at Stevens Institute of Technology for her valuable help with voice and data communications.

We are particularly grateful to Dawn Mesa, who has supported our editorial work by providing significant administrative help from CRC Press. We would also like to thank Ramila Saldana, who greatly assisted the co-editors with the care and attention she provided to many details of the book.

Special thanks is due to Felicia Shapiro who particularly managed the production and Steve Menke for his excellent project editing work.

Foreword

In the preparation of this book, our objective was to provide an advanced understanding of emerging telecommunications systems, their significance, and the anticipated role these systems will play in the future. With the help of our talented associated editors and contributors, we believe we have accomplished this. By addressing voice, Internet, traffic management, and future trends, we feel our readers will be knowledgeable about current and future telecommunications systems.

In Section 1, the techniques of voice communication systems are outlined, with attention paid to both basic and advanced systems. Advanced intelligent networks (AIN) and computer telephony integrated (CTI) are key building blocks for future voice systems. Finally, voice over IP, and the anticipated integration of voice and IP data is closely examined. The second part of this section concentrates on state-of-the-art solutions for local area networks. In addition to data communication capabilities, multimedia attributes of LANs are also addressed.

Section 2 provides a detailed explanation of the Internet, including elements of its structure and consideration of how future services will be handled on the Internet. Internet management and security are discussed. A detailed discussion of virtual private networks (VPNs) is provided, as well as presentation of web design and data warehousing concepts. Electronic commerce and Internet protocols are presented in detail, permitting the reader to understand and select with insight from the available web-based technology choices.

Section 3 continues the exploration of advanced telecommunications concepts, focusing on network management and administration. As the services and features provided the network become larger in scale and scope, network management will become even more crucial and important than it is today. Telecommunications network management (TNM) and Telecommunications Information Networking Architecture (TINA) are presented. The telecommunications support process is outlined, including management frameworks and customer network management. A detailed consideration of outsourcing options, which will become even more frequent, is presented. The performance impact of network management is detailed.

Finally, in Section 4, future trends and directions are considered, with a view toward satisfying user needs in parallel with application trends, which will require system and service integration. While we know the future will hold new products and services, accounting for these services is a challenge, and an examination of telecommunications tariffing is also provided.

We hope our readers find this book an excellent guide to emerging telecommunications trends.

Patricia Morreale
Advanced Telecommunications Institute
Stevens Institute of Technology
Hoboken, NJ

Editors-in-Chief

Patricia Morreale, Ph.D., is Director of the Advanced Telecommunications Institute (ATI) and an Associate Professor in the School of Applied Sciences and Liberal Arts at Stevens Institute of Technology. Since joining Stevens in 1995, she has established the Multimedia Laboratory at ATI and continued the work of the Interoperable Networks Lab in network management and performance, wireless systems design, and mobile agents.

Dr. Morreale holds a B.S. from Northwestern University, a M.S. from the University of Missouri, and a Ph.D. from the Illinois Institute of Technology, all in Computer Science. She holds a patent in the design of real-time database systems and has numerous journal and conference publications. With Dr. Terplan, she co-authored *The Telecommunications Handbook*, published by CRC Press.

Prior to joining Stevens, she was in industry, working in network management and performance. She has been a consultant on a number of government and industrial projects.

Dr. Morreale's research has been funded by the National Science Foundation (NSF), U.S. Navy, U.S. Air Force, Allied Signal, AT&T, Lucent, Panasonic, Bell Atlantic, and the New Jersey Commission on Science and Technology (NJCST). She is a member of the Association for Computing Machinery (ACM) and a senior member of the Institute of Electrical and Electronic Engineers (IEEE). She has served as guest editor for *IEEE Communications* magazine, special issue on active, programmable, and mobile code networking. In addition, she is an editorial board member of the *Journal of Multimedia Tools and Applications* (Kluwer Academic).

Kornel Terplan, Ph.D., is a telecommunications expert with more than 25 years of highly successful multinational consulting experience. His book, *Communication Network Management*, published by Prentice-Hall (now in its second edition), and his book, *Effective Management of Local Area Networks*, published by McGraw-Hill (now in its second edition), are viewed as the state-of-the-art compendium throughout the community of international corporate users. He has provided consulting, training, and product development services to over 75 national and multinational corporations on four continents, following a scholarly career that combined some 140 articles, 19 books, and 115 papers with editorial board services.

Over the last 10 years, he has designed five network management-related seminars and given some 55 seminar presentations in 15 countries. He received his doctoral degree at the University of Dresden and completed advanced studies, researched, and lectured at Berkeley, Stanford University, University of California at Los Angeles, and Rensselaer Polytechnic Institute.

His consulting work concentrates on network management products and services, operations support systems for the telecommunications industry, outsourcing, central administration of a very large number of LANs, strategy of network management integration, implementation of network design and planning guidelines, products comparison, selection, benchmarking systems, and network management solutions.

His most important clients include AT&T, AT&T Solutions, Georgia Pacific Corporation, GTE, Walt Disney World, Boole and Babbage, Salomon Brothers, Kaiser Permanente, BMW, Siemens AG, France Telecom, Bank of Ireland, Dresdner Bank, Commerzbank, German Telecom, Unisource, Hungarian Telecommunication Company, Union Bank of Switzerland, Creditanstalt Austria, and the State of Washington.

He is Industry Professor at Brooklyn Polytechnic University and at Stevens Institute of Technology in Hoboken, NJ.

Contributors

John Amoss
Lucent Technologies
Holundel, New Jersey

James Anderson
Alcatel
Richardson, Texas

John Braun
Weston, Connecticut

Karen M. Freundlich
TCR, Inc.
Princeton, New Jersey

Joe Ghetie
Telcordia
Piscataway, New Jersey

Michel Gilbert
Hill Associates, Inc.
Colchester, Vermont

Takeo Hamada
Fujitsu Laboratories America
Sunnyvale, California

Stephanie Hogg
Telsta Research
Victoria, Australia

Hiroshi Kamata
OKI Electric
Red Bank, New Jersey

Matthew Kolon
Hill Associates, Inc.
Colchester, Vermont

Carel Marsman
CMG
The Netherlands

Patricia Morreale
Stevens Institute of Technology
Hoboken, New Jersey

Dermot Murray
Iona College
New Rochelle, New York

Mihir Parikh
Polytechnic University
Brooklyn, New York

Teresa Piliouras
TCR, Inc.
Weston, Connecticut

Andrew Resnick
Citicorp
New York, New York

Endre Sara
Goldman, Sachs & Co.
New York, New York

Endre Szebenyi
Industry Consultant
Budapest, Hungary

Kornel Terplan
Industry Consultant and Professor
Hackensack, New Jersey

Contents

1

Voice and Data Communications

Patricia Morreale
Stevens Institute of Technology

Michel Gilbert
Hill Associates, Inc.

Matthew Kolon
Hill Associates, Inc.

John Amoss
Lucent Technologies

1.1   Advanced Intelligent Networks (AIN)

Patricia Morreale

1.1.1   Definition

Intelligent network (IN) is a telephone network architecture originated by Bell Communications Research (Bellcore) in which the service logic for a call is located separately from the switching facilities, allowing services to be added or changed without having to redesign switching equipment. According to Bell Atlantic, IN is a service-specific architecture. That is, a certain portion of a dialed phone number, such as 800 or 900, triggers a request for a specific service. A later version of IN called advanced intelligent network (AIN) introduces the idea of a service-independent architecture in which a given part of a telephone number can be interpreted differently by various services depending on factors such as time of day, caller identity, and type of call. AIN makes it easy to add new services without having to install new phone equipment.

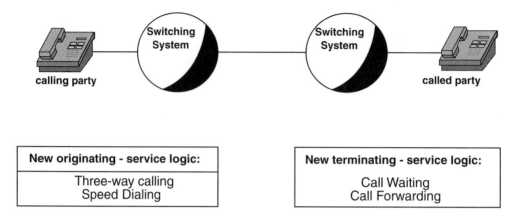

FIGURE 1.1.1   Plain old telephone service (POTS).

1.1.2   Overview

This chapter discusses how the network has evolved from one in which switch-based service logic provides services to one in which service-independent AIN capabilities allow for service creation and deployment.

As the IN evolves, service providers will be faced with many opportunities and challenges. While the IN provides a network capability to meet the ever-changing needs of customers, network intelligence is becoming increasingly distributed and complicated. For example, third-party service providers will be interconnecting with traditional operating company networks. Local number portability (LNP) presents many issues that can only be resolved in an IN environment to meet government mandates. Also, as competition grows with companies offering telephone services previously denied to them, the IN provides a solution to meet the challenge.

1.1.3   Network Evolution

1.1.3.1   Plain Old Telephone Service (POTS)

Prior to the mid-1960s, the service logic (Figure 1.1.1) was hard-wired in switching systems. Typically, network operators met with switch vendors, discussed the types of services customers required, negotiated the switching features that provided the services, and finally agreed upon a generic release date for feature availability. After this, the network operator planned for the deployment of the generic feature/service in the switching network fabric.

This process was compounded for the network operator with switching systems from multiple vendors. As a result, services were not offered ubiquitously across an operator's serving area. So, a customer in one end of a city, county, or state may not have had the same service offerings as a person in another part of the area.

Also, once services were implemented, they were not easily modified to meet individual customer's requirements. Often, the network operator negotiated the change with the switch vendor. As a result of this process, it took years to plan and implement services. This approach to new service deployment required detailed management of calling patterns, and providing new trunk groups to handle calling patterns. As customer calling habits changed — such as longer call lengths, larger calling areas, and multiple lines in businesses and residences — the demand on network operators increased.

1.1.3.2   Stored Program Control (SPC)

In the mid-1960s, stored program control (SPC) switching systems were introduced. SPC was a major step forward because now service logic was programmable where, in the past, the service logic was hard wired. As a result, it was now easier to introduce new services. Nevertheless, this service logic concept was not modular. It became increasingly more complicated to add new services because of the dependency between the service and the service-specific logic. Essentially, service logic that was used for one service

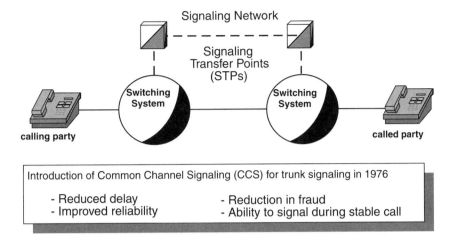

FIGURE 1.1.2   Common channel signaling (CCS).

could not be used for another. As a result, if customers were not served by a SPC switching system, new services were not available to them.

1.1.3.3   Common Channel Signaling Network (CCSN)

Another aspect of the traditional service offerings was the call setup information — the signaling and call supervision that took place between switching systems and the actual call. When a call was set up, a signal and talk path used the same common trunk from the originating switching system to the terminating switching system. Often there were multiple offices involved in the routing of a call. This process seized the trunks in all of the switching systems involved. Hence, if the terminating end was busy, all of the trunks were set up unnecessarily.

The network took a major leap forward in the mid-1970s with the introduction of the common channel signaling network (CCSN), or SS7 network for short. Signaling system number 7 (SS7) is the protocol that runs over the CCSN. The SS7 network consists of packet data links and packet data switching systems called signaling transfer points (STPs).

The SS7 network (Figure 1.1.2) separates the call setup information and talk path from the common trunks that run between switching systems. The call setup information travels outside the common trunk path over the SS7 network. The type of information transferred includes permission for the call setup, whether or not the called party is busy.

SS7 technology frees up trunk circuits between switching systems for the actual calls. The SS7 network enabled the introduction of new services, such as caller ID. Caller ID provides the calling party's telephone number, which is transmitted over the SS7 network. The SS7 network was designed before the IN concept was introduced. However, telephone operators realized that there were many advantages to implementing and using SS7 network capabilities.

1.1.4   Introduction of IN

During the mid-1980s, regional Bell operating companies (RBOCs) began requesting features that met the following objectives:

- Rapid deployment of services in the network
- Vendor independence and standard interfaces
- Opportunities for non-RBOCs to offer services for increased network usage

Bell Communications Research (Bellcore) responded to this request and developed the concept of Intelligent Network 1 (IN/1, Figure 1.1.3).

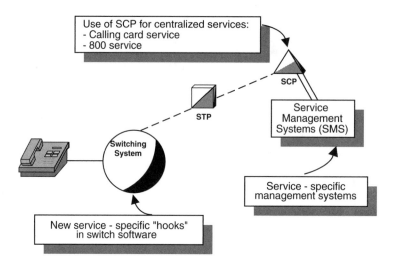

FIGURE 1.1.3   Intelligent Network (IN/1).

The introduction of the IN/1 marked the first time that service logic was external to switching systems and located in databases called service control points (SCPs). Two services evolved that required IN/1 service logic — the 800 (or Freephone) service and the calling card verification (or alternate billing service, ABS) service. Because of the service-specific nature of the technology, these services required two separate SCPs. In order to communicate with the associated service logic, software was deployed in switching systems. This switching system software enabled the switching system to recognize when it was necessary to communicate with a SCP via the SS7 network. With the introduction of the SCP concept, new operations and management systems became necessary to support service creation, testing, and provisioning. In Figure 1.1.3, note the term "service-specific management systems" under the box labeled "service management system." This means that the software-defined "hooks" or triggers are specific to the associated service. For example, an 800 service has an 800-type trigger at the switching system, an 800-service database at the SCP, and an 800-service management system to support the 800 SCP. In this service-specific environment, the 800-service set of capabilities cannot be used for other services (e.g., 900 service). Although the service logic is external to the switching system, it is still service-specific.

At first glance, Figure 1.1.4 looks similar to the previous diagram. However, there is one fundamental difference. Notice the wording "service-independent management systems" under the box labeled "service management system." Now, following the IN/1 800 service-specific example, the AIN service-independent software has a three-digit trigger capability that can be used to provide a range of three-digit services (800, 900, XXX, etc.) as opposed to 800 service-specific logic. Likewise, the SCP service logic and the service management system are service independent, not service specific. AIN is a service-independent network capability!

1.1.5   Benefits of INs

The main benefit of INs is the ability to improve existing services and develop new sources of revenue. To meet these objectives, providers require the ability to:

Introduce New Services Rapidly
IN provides the capability to provision new services or modify existing services throughout the network with physical intervention.

Provide Service Customization
Service providers require the ability to change the service logic rapidly and efficiently. Customers are also demanding control of their own services to meet their individual needs.

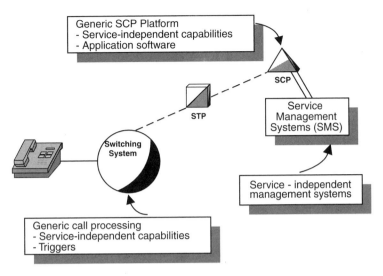

FIGURE 1.1.4   Advanced intelligent network (AIN) architecture.

Establish Vendor Independence

A major criteria for service providers is that the software must be developed quickly and inexpensively. To accomplish this, suppliers have to integrate commercially available software to create the applications required by service providers.

Create Open Interfaces

Open interfaces allow service providers to introduce network elements quickly for individualized customer services. The software must interface with other vendors' products while still maintaining stringent network operations standards. Service providers are no longer relying on one or two vendors to provide equipment and software to meet customer requirements.

AIN technology uses the embedded base of stored program-controlled switching systems and the SS7 network. The AIN technology also allows for the separation of service-specific functions and data from other network resources. This feature reduces the dependency on switching system vendors for software development and delivery schedules. Service providers have more freedom to create and customize services.

The SCP contains programmable service-independent capabilities (or service logic) that are under the control of service providers. The SCP also contains service-specific data that allows service providers and their customers to customize services. With the IN, there is no such thing as one size fits all — services are customized to meet individual needs.

Since service logic is under the service provider's control, it is easier to create services in a cost-effective manner. Network providers can offer market-focused service trials by loading service logic in a SCP and triggering capabilities in one or more switching systems.

Accepted standards and open, well-documented interfaces provide a standard way of communicating between switching systems and SCPs, especially in a multi-vendor environment.

1.1.6   Local Number Portability

The Telecommunications Act of 1996 is having a profound impact on the U.S. telecommunications industry. One area of impact that is being felt by everyone is Local Number Portability (LNP). For LNP, the Federal Communications Commission (FCC) requires the nation's local exchange carriers (LECs) to allow customers to keep their telephone numbers if they switch local carriers. The LECs must continue to maintain the quality of service and network reliability that the customer has always received.

The rules required that all LECs begin a phased deployment of a long-term service provider portability solution no later than October 1, 1997 in the nation's largest metropolitan statistical areas.

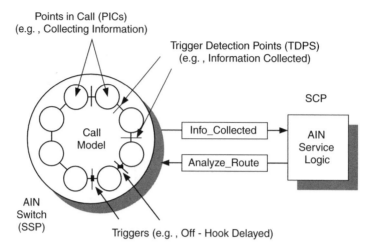

FIGURE 1.1.5   The call model: basic concept.

Wireless carriers are also affected by LNP. December 31, 1998 was the deadline date that wireless carriers had to be able to complete a call to a ported wire–line number. By June 30, 1999, the Act called for full portability between wireless and wireline, including roaming capabilities.

AIN is a logical technology to help service providers meet this mandate. Many providers are looking to AIN LNP solutions because of the flexibility that AIN provides without the burden of costly network additions.

1.1.7   The Call Model

The call model is a generic representation of service switching point (SSP) call processing activities required to establish, maintain, and clear a basic call. The call model consists of Point in Calls (or PICs), Detection Points (DPs), and triggers. These are depicted in Figure 1.1.5.

PICs represent the normal switching system activities or states that a call goes through from origination to termination. For example, the null state or the idle state is when the SSP is actually monitoring the customer's line. Other examples of states, or PICs, are off-hook (or origination attempt), collecting information, analyzing information, routing, alerting, etc.

Switching systems went through similar stages before AIN was developed. However, the advent of AIN introduced a formal call model that all switching systems must adhere to. In this new call model, trigger detection points (TDPs) were added between the PICs. SSPs check TDPs to see if there are any active triggers.

There are three types of triggers: subscribed or line-based triggers, group-based triggers, and office-based triggers. Subscribed triggers are provisioned to the customer's line, so that any calls originating from or terminating to that line would encounter the trigger. Group-based triggers are assigned to groups of subscribers, e.g., business or Centrex groups. Any member of a software-defined group will encounter the trigger. Office-based triggers are available to everyone connected to the telephone switching office or has access to the North American numbering plan. Office-based triggers are not assigned to individuals or groups.

If an active trigger is detected, normal switching system call processing is suspended until the SSP and SCP complete communications. For example, in Figure 1.1.5, suppose an AIN call has progressed through the null state or PIC, the off-hook PIC, and is currently at the collecting information PIC. Normal call processing is suspended at the information collected TDP because of an active off-hook delayed trigger. Before progressing to the next (analyzing information) PIC, the SSP assembles an information collected message and sends it to the SCP over the SS7 network. After SCP service logic acts on the message, the SCP sends an analyze route message that tells the SSP how to handle the call before going to the next PIC (analyzing information).

Essentially, when the SSP recognizes that a call has an associated AIN trigger, the SSP suspends the call processing while querying the SCP for call routing instructions. Once the SCP provides the instruction, the SSP continues the call model flow until completion of the call. This is basically how a call model works, and it is a very important part of AIN.

This concept differs from the pre-AIN switching concept in which calls were processed from origination state to the call termination state without call suspension.

1.1.8   AIN Releases

The demand for AIN services far exceeded the availability of network functionality. Service providers could not wait for all the features and functionality as described in AIN Release 1. AIN Release 1 defined all types of requirements, which made the capability sets too large to be adopted by the industry.

In North America, the industry agreed to develop subsets of AIN Release 1 that provided for a phased evolution to AIN Release 1. AIN 0.1 was the first subset targeted for use.

Bellcore developed functionality to address the FTS 2000 requirements set forth by the U.S. Government. The RBOCs AIN turn adopted these requirements to meet their customers' immediate needs. This effort resulted in AIN Release 0, which had a time frame before the availability of AIN 0.1.

Meanwhile, the global standards body, the International Telecommunications Union (ITU), embraced the concepts put forth in the AIN Release 1 requirements. The ITU developed an international IN standard called Capability Set 1, or CS-1. As with AIN Release 1 in North America, CS-1 was encompassing a rich functionality. To meet the market demand, the ITU formed a subgroup called European Telecommunications Standards Institute (ETSI) to focus on the immediate needs. This subgroup developed the Core INAP capabilities. Many Post Telegraph and Telecommunications (PTT) organizations and their switch vendors have adopted the ETSI Core INAP as the standard and are providing Core Intelligent Network Application Protocol (INAP) capabilities.

1.1.8.1   AIN Release 1 Architecture

Figure 1.1.6 shows the target AIN Release 1 architecture, as defined in Bellcore AIN Generic Requirements (GRs).

The SSP in this diagram is an AIN-capable switching system. In addition to providing end users with access to the network and performing any necessary switching functionality, the SSP allows access to the set of AIN capabilities. The SSP has the ability to detect requests for AIN-based services and establish communications with the AIN service logic located at the SCPs. The SSP is able to communicate with other network systems (e.g., intelligent peripherals) as defined by the individual services. The SCP

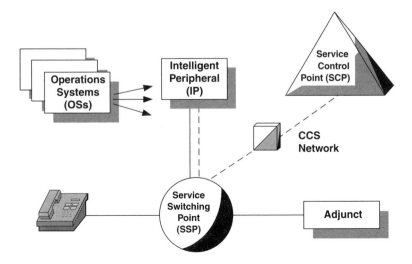

FIGURE 1.1.6   AIN Release 1.

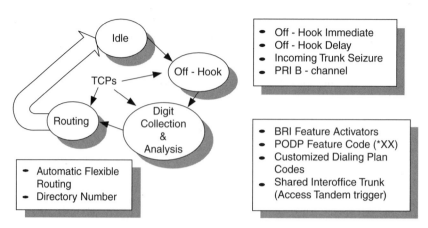

FIGURE 1.1.7 AIN Release 0 call model.

provides the service control. There are two basic parts to a SCP. One part is the application functionality in which the service logic is installed after the services have been created. This application functionality sits on top of the second basic SCP part: a set of generic platform functionalities that are developed by SCP vendors. This platform functionality is shared among the service logic application programs in the application functionality. The platform functionality also provides the SS7 interface to switching systems. As shown in Figure 1.1.6, the SCP is connected to SSPs by the SS7 network.

The intelligent peripheral (IP) provides resources such as customized and concatenated voice announcements, voice recognition, and dual tone multi-frequencies (DTMF) digit collection. The IP contains a switching matrix to connect users to these resources. In addition, the IP supports flexible information interactions between an end user and the network. It has the resource management capabilities to search for idle resources, initiate those resources, and then return them to their idle state.

The interface between the SSP and the IP is an integrated services digital network (ISDN), primary rate interface (PRI) and/or basic rate interface (BRI). The IP has the switching functionality that provides the ISDN interface to the switching system. The adjunct shown in Figure 1.1.6 is functionally equivalent to a SCP, but it is connected directly to a SSP. A high-speed interface supports the communications between an adjunct and a SSP. The application-layer messages are identical in content to those carried by the SS7 network between the SSP and SCP.

1.1.8.2   AIN Release 0

The AIN Release 0 call model has three trigger checkpoints (TCPs). At each TCP there are one or more triggers. For example, the off-hook TCP includes the off-hook immediate trigger. If a subscriber's line is equipped with this trigger, communications with the SCP will occur if the switching system detects an off-hook condition. For an off-hook delayed trigger, one or more digits are dialed before triggering to the SCP. At the digit collection and analysis TCP, collected digits are analyzed before triggering. Triggering may also occur at the routing stage of a call. This call model is shown in Figure 1.1.7.

When a switching system recognizes that a call needs AIN involvement, it checks for overload conditions before communicating with the SCP. This process is called code gapping. Code gapping allows the SCP to notify the switching system to throttle back messages for certain NPAs or NPA-NXXs. When code gapping is in effect, some calls may receive final treatment. For others, a provide instruction message is sent to the SCP. Depending on the SCP service logic, it will respond to the switching system with any of the call processing instructions shown in Figure 1.1.8.

AIN Release 0 provided 75 announcements at the switching system. Release 0 was based on American National Standards Industry (ANSI) Transaction Capability Application Part (TCAP) issue 1. TCAP is at layer 7 of the SS7 protocol stack. This means that there is only one message sent from the SSP to the SCP, no matter what trigger is hit at any of the three TCPs.

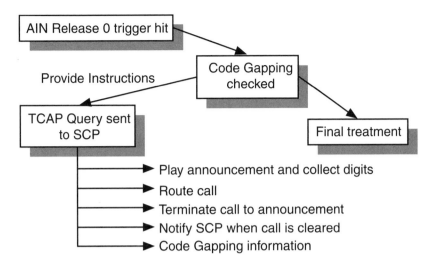

FIGURE 1.1.8 AIN Release 0 functions.

1.1.8.3 AIN Release 0.1

AIN 0.1 is the first subset of AIN Release 1. There are two fundamental differences between AIN Release 0 and AIN 0.1 The first is a formal call model and the second is the messaging sets between the switching system and the SCP. The formal call model is separated into the originating call model (originating half call) and the terminating call model (terminating half call). The AIN Release 0 call model did not distinguish between originating and terminating. A standard or formal call model is necessary as we evolve to the Target AIN Release 1 capability, because the capabilities will have more PICs and TDPs. Also, there will be multiple switch types and network elements involved. Therefore, the service logic will need to interact with every element that will be required in the network.

AIN 0.1 includes several other major features. There are 254 announcements at the switching system, which provide more flexible messages available to customers. There are additional call-related and non-call-related functions as well as three additional triggers — the N11 trigger, the 3–6–10-digit trigger, and the termination attempt trigger. More triggers provide additional opportunities for SCP service logic to influence call processing. (Note: TCP was an AIN Release 0 term that changed to TDP in AIN 0.1). There are several AIN 0.1 non-call-related capabilities. The SCP has the ability to activate and deactivate subscribed triggers. The AIN 0.1 SCP can also monitor resources. In addition to sending a call routing message to the switching system, the SCP may request that the switching system monitor the busy/idle status of a particular line and report changes. AIN 0.1 also supports standard ISDN capabilities.

As mentioned previously, there is a distinction between the originating side and the terminating side of a service switching point. This means that both originating and terminating triggers and service logic could influence a single call. Figure 1.1.9 shows a portion of the AIN 0.1 originating call model. The AIN 0.1 originating call model includes four originating trigger detection points — origination attempt, information collected, information analyzed, and network busy.

The AIN 0.1 terminating call model includes one TDP — termination attempt, as depicted in the partial call model in Figure 1.1.10.

1.1.8.4 AIN 0.1: SSP–SCP Interface

The AIN 0.1, as shown in Figure 1.1.11, is based on ANSI TCAP issue 2, which means that the message set is different than the message set in ANSI TCAP issue 1. For example, in AIN Release 0, there is only one message sent from the SSP to the SCP no matter what trigger is hit at any of the three TCPs. In AIN 0.1, separate messages are sent for the four originating and one terminating TDP.

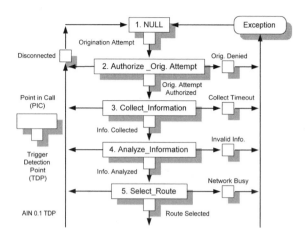

FIGURE 1.1.9   AIN 0.1 originating call model.

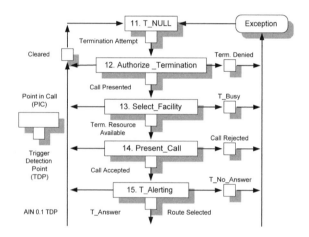

FIGURE 1.1.10   AIN 0.1 terminating call model.

FIGURE 1.1.11   AIN 0.1 SSP–SCP interface.

1.1.8.5   AIN Release 0.2

AIN 0.2 builds on AIN 0.1 with additional capabilities to support two service drivers — Phase 2 personal communication service (PCS) and voice activated dialing (VAD). While AIN 0.2 is focused on capabilities to support PCS and VAD, all requirements for these capabilities are defined in a service-independent manner. AIN 0.2 capabilities will include:

- ISDN-based SSP–IP interface
- Busy and no-answer triggers

- Next event lists processing
- Default routing, and
- Additional functions in all operations areas (e.g., network testing).

The two primary AIN 0.2 capabilities are the ISDN interface between a switching system and an ISDN-capable device (such as an IP) and the addition of busy and no-answer triggers.

Next event lists processing is another important capability. In addition to TDPs, AIN 0.2 includes event detection points (EDPs). With EDPs, the SCP will have the ability to send a next event list to the SSP. This next event list is used by the SSP to notify the SCP of events listed in the next event list. These events may include busy, no answer, terminating resource available, etc.

AIN 0.2 also includes default routing capabilities. This means that when calls encounter error conditions, they can be sent to a directory number, an announcement, etc., as opposed to sending it to final treatment, as is the case in AIN 0.1.

1.1.8.6   AIN 0.2 SSP–IP Interface

AIN Release 0 and AIN 0.1 assumed that the announcements were switch-based. With the introduction of AIN 0.2, announcements can reside in an external database, such as an IP. If the SCP sends a send-to-resource message to the switching system to have the IP play an announcement or collect digits, the switching system connects the customer to the IP via the SSP–IP ISDN interface. The end user exchanges information with the IP. The IP collects the information and sends it to the switching system. The switching system forwards the information to the SCP. One of the fundamental switching system capabilities is the interworking of SS7 (SCP) messages with ISDN messages (SSP–IP).

In addition the SSP may control IP resources without SCP involvement. VAD is an example. A VAD subscriber could be connected to the IP voice recognition capabilities upon going off-hook. The VAD subscriber says "call mom," and the IP returns mom's telephone number to the switching system. The switching system recognizes mom's number as if the subscriber had actually dialed the number.

1.1.9   AIN Service Creation Examples

The previous modules addressed the architecture and the theory of the AIN. This section will discuss various aspects of service creation — the tool that builds the representation of the call flow for each individual customer. Many AIN software vendors have paired service creation software with state-of-the-art computer graphics software to eliminate the need for traditional programming methods. Through the use of menu-driven software, services are created by inputting various service parameters.

1.1.9.1   Building Block Approach

Figure 1.1.12 provides an example of a building-block approach to creating AIN services. Play announcement, collect digits, call routing, and number translation building blocks are shown here. The SSP has the ability to play announcements and collect digits, as does the IP. Routing the call is a SSP function, and number translation is a SCP capability. By arranging these four capabilities or building blocks in various combinations, services such as 800 calling with interactive dialing, outgoing call screening, and area number calling can be created.

1.1.9.2   Service Creation Template

Figure 1.1.13 represents what a service creation template might look like. For an outgoing call screening service, the process begins with the customer's telephone number.

This example allows the customer to screen 900 numbers, while still having the ability to override 900 screening by entering a PIN. Except for 703-974-1234, all non-900 calls are processed without screening.

1.1.9.3   Digit Extension Dialing Service

A 5-digit extension dialing service is displayed in Figure 1.1.14. It allows for abbreviated dialing beyond central office boundaries. If an employee at location 1 wants to call an employee at location 2 by dialing the extension number 1111, 21111 would be dialed.

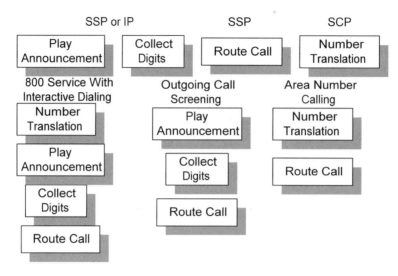

FIGURE 1.1.12   AIN service example: building block approach.

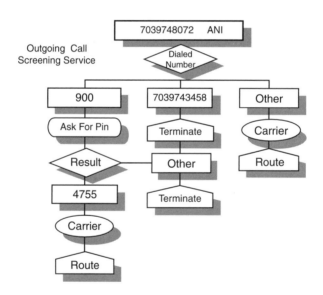

FIGURE 1.1.13   AIN service example: building block approach.

Although 21111 is not a number that a switching system can use to route the call, a customized dialing plan trigger is encountered after 21111 is dialed and a query is sent to the SCP. Service logic at the SCP uses the 21111 number to determine the "real" telephone number of the called party.

1.1.9.4   Disaster Recover Service

Figure 1.1.15 illustrates a disaster recovery service. This service allows businesses to have calls routed to one or more alternate locations based on customer service logic at the SCP. Calls come into the switching system served by the normal location. After triggering, communication with the SCP occurs. Based on the service logic, the call could be either routed to the normal business location or to one or more alternate business locations.

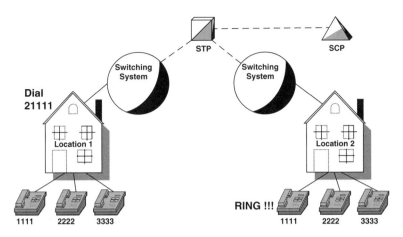

FIGURE 1.1.14 AIN service example: building block approach.

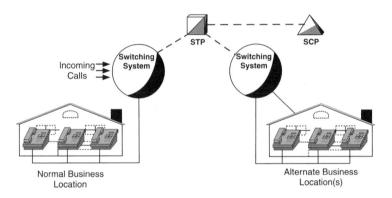

FIGURE 1.1.15 Disaster recovery service.

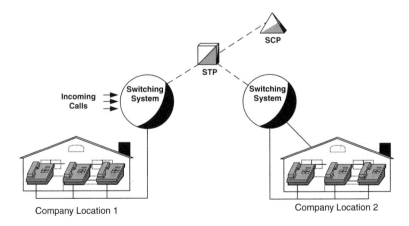

FIGURE 1.1.16 Area number calling (ANC) service.

1.1.9.5 Area Number Calling Service

An area number calling (ANC) service is shown in Figure 1.1.16. This service is useful for companies or businesses that want to advertise one telephone number but want their customer's calls routed to the nearest or most convenient business location. The SCP service logic and data (e.g., zip codes) are used

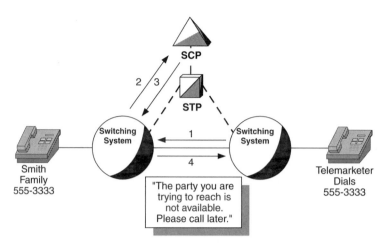

FIGURE 1.1.17 Do not disturb service.

to match the calling party's telephone number and their geographical location. The call is then routed to the company or business location that is closest to or most convenient for the calling party.

1.1.9.6   Do Not Disturb Service

Finally, a do not disturb service is displayed in Figure 1.1.17. This is a service in which the Smith family has terminating screening service logic at the SCP. Whenever someone calls them, the service logic determines whether the call should be routed to the Smith's telephone or play an announcement. In this particular case, a telemarketer calls the Smiths. The SCP tells the switching system to route the telemarketer to an announcement. The customer's SCP service logic may also contain a list of numbers that they want to get through while do not disturb is active. In that case, if the SCP finds a match between the calling party number and a number on the list, the call is routed to the Smiths.

1.1.10   Other AIN Services

The following list describes the services that companies have developed using AIN/IN technology. Some services are tariffed, deployed in the network, and generating revenues. Others are in market or technical trials, getting ready for deployment. There are other services that are either planned for deployment or were developed for demonstration purposes.

N11 access service: With this service, a unique code is used to access a service gateway to information service providers (ISPs), such as newspapers or libraries. The subscriber may either preselect an ISP for automatic routing or request block calls to ISPs.

Basic routing: Allows the subscriber to route calls to a single destination as defined in the system.

Single number service: Allows calls to have different call treatments based on the originating geographical area and the calling party identification.

Routing by day of week: Allows the service subscriber to apply variable call routings based on the day of the week that the call is placed.

Routing by time of day: Allows service subscribers to apply variable call routings based on the time of the day that the call is made.

Selective routing: Tied to the call forwarding feature generally offered as a switch-based feature. With the AIN, when a call to a selective routing customer is forwarded, the SCP determines where to route the forwarded call based on the caller's number.

Call allocator: Allows the service subscriber to specify the percentage of calls to be distributed randomly for up to five alternate call handling treatments.

Alternate destination on busy day: Allows the service subscriber to specify a sequence of destinations to which calls will be routed if the first destination is busy.

Command routing: A service subscriber predefines a set of alternate call treatments to handle traffic in cases of emergency, unanticipated or anticipated demand peaks, or for any other reason that warrants an alternate call treatment.

Call gate: This is a versatile outgoing call screening service. Call gate supports a personal identification number (PIN) and screening based on time of day and day of week.

Personal access: A type of "follow me" service. A virtual telephone number is assigned to the personal access service subscriber. When a caller dials this number, the software determines how to route the call.

Calling party pays: A service offered to cellular customers. It notifies the calling party that they are trying to reach a cellular number. If they choose to complete the call, they will incur the connect charge of the called party. If they elect not to incur the cost, the call may either be terminated or routed to the called party's voice mail.

Remote access to call forwarding (ultraforward): Allows remote access to call forwarding. Callers may, from any location in the world, call in remotely and activate and/or change their call forwarding number.

Portable number service: PNS Features enhanced call forwarding for large business subscribers. It provides subscribers with the ability to maintain a personal itinerary which includes time-of-day, day-of-week (TOD/DOW) schedules, call searching schedules, and call routing information. PNS subscribers also have the ability to override their schedules with default routing instructions. This service is intended for companies with employees who are in highly mobile environments, requiring immediate availability.

Enhanced 800 service: (Freephone) A customer's call to an 800 service subscriber can be routed to different destinations; instances of routing include the geographical location of the caller, the time and day the call is made, and the caller responses to prompts. The subscriber sets alternate routing parameters for the call if the destination is busy or unavailable, thereby redirecting and allowing for completion of the call.

Mass calling service: MCS A polling and information service that permits simultaneous calling by a large number of callers to one or more telephone numbers. MCS provides a variety of announcement-related services that connect a large number of callers (who dial an advertised number) to recorded announcement devices. Two types of offerings are mass announcements, such as time and weather, and televoting, which allows callers to register their opinions on a topic of general interest.

Automatic Route Selection/Least Cost Routing: Subscribers design a priority route for every telephone number dialed. The system either directs calls or blocks calls to restricted privilege users.

Work-at-home: Allows an individual to be reached at home by dialing an office number, as well as allowing the employee to dial an access code from home, make long distance calls, and have them billed and tracked to a business telephone number.

Inmate service: Routes prisoners' calls, tracks the call information, and offers call control features such as prompts for PINs, blocking certain called numbers, and time or day restrictions.

Holding room: Transportation companies' passengers use this service to inform families or business associates of transportation delays or cancellations.

Call prompter: Allows a service subscriber to provide an announcement that requests the caller to enter a digit or series of digits via a dual tone multi-frequency (DTMF) telephone. These digits provide information used for direct routing or as a security check during call processing.

Call counter: Increases a counter in the televoting (TV) counting application when a call is made to a televised number. The counts are managed in the SCP, which can accumulate and send the results during a specific time period.

500 access service: Allows personal communications service (PCS) providers the ability to route calls to subscribers who use a virtual 500 number.

PBX extend service: Provides a simple way for users to gain access to the Internet network.

Advertising effectiveness service: Collects information on incoming calls (for example, ANI, time, and date). This information is useful to advertisers to determine customer demographics.

Virtual foreign exchange service: Uses the public switched network to provide the same service as wired foreign exchange service.

ACNA originating line blocking: ACNA (Automated Customer Name and Address), with the ability to block their line from being accessed by the service.

AIN for the case teams: Allows technicians to dial from a customer premise location anywhere in the service region and connect to a service representative supported by an ACD. Through voice prompts, the technician is guided to the specific representative within a case team pool within seconds, with no toll charges to the customer.

Regional intercept: Instructs callers of new telephone numbers and locations of regional customers. This service also forwards calls to the new telephone number of the subscriber. Various levels of the service can be offered, based upon the customer's selection.

Work at home billing: A person who is working at home dials a 4-digit feature access code which prompts the system to track and record the billing information for the calls. Calls tracked in this manner are billed directly to the company rather than to the individual.

Inbound call restriction: Allows a customer to restrict certain calls from coming into the subscriber's location. This service is flexible enough to restrict calls either by area code, NNX, or particular telephone numbers. Restrictions may even be specified by day of week or time of day.

Outbound call restriction: Allows a customer to restrict certain calls from being completed from the subscriber's location. This service is flexible enough to restrict calls by either area code, NNX, or particular telephone numbers. Restrictions may even be specific to day of week or time of day.

Flexible hot line: Allows a customer to pick up a telephone handset and automatically connect to a merchant without dialing any digits. An example of this is a rent-a-car phone in an airport, which allows a customer to notify the rent-a-car company to pick them up at the terminal.

Acronyms

ABS	Alternative billing source
AIN	Advanced intelligent network
AMP	AIN maintenance parameter
API	Applications programming interface
ASE	Application service elements
BCSM	Basic call state model
BRI	Basic rate interface
BSTP	Broadband signaling transfer point
CCM	Call control module
CCSN	Common channel signaling network
CFM	Call failure message
CSM	Call segment model
DAA	Directory assistance automation
DAP	Data access point
DCN	Data communications network
DP	Detection point
DTMF	Dual tone multi-frequencies
EDP	Event detection point
EML	Element management layer
ETC	Event trapping capability
FE	Functional entity
GDI	Generic data interface
IF	Information flow
IN	Intelligent network
INAP	Intelligent network application protocol

IN/1	Intelligent Network 1
IP	Intelligent peripheral
IPC	Intelligent peripheral controller
IPI	Intelligent peripheral interface
ISCP	Integrated service control point
LEC	Local exchange carriers
LIDB	Line information database
LNP	Local number portability
MP	Mediation point
MSC	Message sequence chart
NAP	Network access point
NCAS	Non-call associated signaling
NCP	Network control point
NE	Network element
NEL	Next event list
NML	Network management layer
NNI	Network-to-network interface
OBCM	Originating basic call model
ONA	Open network architecture
OOP	Object-oriented programming
OPC	Originating point code
PCS	Personal communications service
PIC	Points in call
PP	Physical plane
RBOC	Regional bell operating companies
RDC	Routing determination check
RE	Resource element
RVT	Routing verification test
SCE	Service creation environment
SCMS	Service creation and maintenance system
SCP	Service control point
SDP	Service data point
SIBB	Service-independent building block
SLE	Service logic editor
SLEE	Service logic execution environment
SLI	Service logic interpreter
SLL	Service logic language
SLP	Service logic program
SM	Session management
SMS	Service management system
SN	Service node
SOP	Service order provisioning
SP	Service plane
SPC	Stored program control
SSP	Service switching point
STP	Signalling transfer point
TBCM	Terminating basic call model
TCP	Trigger check point
TCP	Test call parameter
TDP	Trigger detection point
TSC	Trigger status capability
WIN	Wireless intelligent network

References

1. *http://www.telecordia.com*
2. Uyless D. Black, *The Intelligent Network: Customizing Telecommunication Networks and Services*, Prentice Hall Series in Advanced Communications Technologies, 1998.

Further Readings

1. Uyless D. Black, *The Intelligent Network: Customizing Telecommunication Networks and Services*, Prentice Hall Series in Advanced Communications Technologies, 1998.
2. Bill Douskalis, *IP Telephony*, Hewlett Packard Professional Books, Prentice Hall PTR, 2000.
3. William Stallings, *High Speed Networks: TCP/IP and ATM Design Principles*, Prentice Hall, 1997.
4. Kornel Terplan, *Telecom Operations Management Solutions with NetExpert*, CRC Press, 1998.
5. Uyless Black, *ISDN and SS7: Architectures for Digital Signaling Networks*, Prentice Hall, 1997.
6. John G. van Bosse, *Signaling in Telecommunication Networks*, Wiley & Sons, 1997.
7. Paul Ferguson and Goeff Huston, *Quality of Service, Delivering QoS on the Internet and Corporate Networks*, Wiley & Sons, 1998.
8. Daniel Minoli and Emma Minoli, *Delivering Voice over IP Networks*, Wiley & Sons, 1998.

1.2   Computer Telephone Integrated (CTI)

Michel Gilbert

1.2.1   Abstract

In the universe of telecommunications, the worlds of voice and data have long been resistant to unification. The basic principles that underlie the two worlds have led to, at best, an uneasy truce. In recent times, however, integration has become the buzzword. The industry has seen the emergence of one technology after another that attempts to draw these two domains into closer proximity. Computer–telephone integration (CTI) is yet another arena in which data and voice encounter one another. In the CTI arena, however, voice and data appear to be on the cusp of a working relationship. This paper introduces and reviews the concepts that underlie the world of CTI, the elements that comprise a CTI application, and the standards that have emerged.

1.2.2   Basic Definitions

In a 1990 article titled "PBX/Host Interfaces: What's Real, What's Next" (Probe Research Conference Digest), Lois B. Levick of Digital Equipment Corporation defined CTI as, "A technology platform that merges voice and data services at the functional level to add tangible benefits to business applications." There are four key elements to this definition: 1) identifying CTI as a technology, 2) a focus on the integration of voice and data, 3) specifying a functional integration, and 4) the need to derive tangible benefits in a business environment.

First, some would dispute the notion that CTI is a new technology. They would suggest that CTI is actually a new application for pre-existing technologies. This is indeed the case. Not only is CTI simply a place to reuse existing technologies, it is also not (as we shall see) particularly new.

Second, the integration of voice and data is a key element in CTI, as the name itself implies. CTI builds on some remarkable convergence points in the evolution of computing and telephony. One of the earliest telephone exchanges was designed in 1889 by a frustrated funeral director! Almond B. Strowger was tired of seeing his competitor get the bulk of the funeral business by virtue of the fact that his competitor's spouse happened to operate the local telephone exchange. To deal with the problem, Strowger designed a telephone exchange that became generally known as a step-by-step (or stepper) exchange. Fifty-four years later, with funding from IBM, Howard Aiken created the Harvard Mark I. Both systems were entirely electromechanical, monstrous in size, and highly rigid in their design. Over the years, however, both computers and switches became entirely electronic and based on solid-state technologies.

Where early switches and computers tended to be hardwired, modern switches and computers are both stored-program machines and very flexible. The switch uses a stored-program model to handle call routing operations. The computer uses a variety of stored programs to support end-user applications. Both depend on a data communications infrastructure to exchange control information. Finally, the telephone network is rapidly converging to the digital communications model, which computers have used almost from the outset.

Telephone switches have become specialized computers designed to provide a switching function, and exchanging information via a complex digital data communications infrastructure.

The third major part of the definition, functional integration, requires a brief sidetrack to examine the anatomy of a phone call. A phone call can be divided into two logical activities, commonly referred to as call control and media processing. Call control is concerned with originating, maintaining, and terminating a call. It includes activities like going off-hook, dialing the phone, routing a call through a network, and terminating a call. Media processing is concerned with the purpose of the phone call. It deals with the type of information being conveyed across the call, and the format in which that information is presented.

Functional integration means the computer and switch collaborate in call control and/or media processing operations. They may actually interchange functions to meet the needs of an application. Data stored in the computer might be useful for routing incoming and/or outgoing calls. Perhaps the simplest example is an autocall application where the user can click on a name stored in a local application and the computer retrieves the associated phone number and dials the call automatically. Alternatively, call-related data can be used to trigger information retrieval from the computer. For example, automatic number identification (ANI) can provide the calling number, which can be used to key a database lookup to retrieve a particular customer's account information before the phone even rings. In both examples, the data of the computer and the routing of a call are bound together to do work.

Another form of functional integration is when computer and telephone peripherals begin to be used interchangeably. For example, computer peripherals can become alternative call control elements instrumental in call monitoring, and telephone network peripherals can become an alternative method for moving data between people and computers. There is even a degree of functional integration achieved when the computer and telephone system are managed from a single point.

The fourth and final element of Levick's definition concerns the benefits CTI brings to business applications. One of the obvious goals of any business application is to provide better service to customers. CTI can increase responsiveness, reduce on-hold waiting times, provide the customer with a single point of contact, and make it easier to provide a broader range of services.

CTI can also increase effectiveness by eliminating many of the mechanical tasks associated with telephony (e.g., dialing phones, looking up phone numbers, etc.), providing a better interface to the telephone system, and integrating control of the phone system into a familiar and regularly used computer interface (e.g., the familiar Windows desktop).

Perhaps the most telling benefit CTI brings to the corporate world (and the one most likely to garner the attention of the decision makers) is the potential for reductions in operating costs. Correctly applied, CTI can mean faster call handling, which translates to reduced call charges. Automation of call-related tasks means potentially fewer personnel, or greater capacity for business with existing personnel. Some CTI implementers have claimed 30% improvement in productivity.

1.2.3 A Brief History of CTI

Although CTI appears to be a recent introduction into the telecommunications arena, there were attempts to integrate voice and data into competitive business applications as early as the 1960s. In his book *Computer Telephone Integration* (ISBN 0-89006-660-4), Rob Walters describes an application put together by IBM for a German bookstore chain.

The bookstores were looking for a way to automate their ordering process. IBM produced a small, hand-held unit that each store manager could use to record the ISBN numbers of books they needed, together with the desired quantity of each. These small units were then left attached to the telephone at the end of the day. Overnight, an IBM 360 located at company headquarters would instruct the IBM 2570 PABX to dial each store in turn.

Once the connection was formed, the IBM mainframe would download the order and then instruct the PABX to release the connection and proceed to the next store. The link between the IBM 360 and the 2750 PABX was called teleprocessing line handling (TPLH). By the end of the night, the 360 would produce a set of shipping specifications for each store, the trucks would be loaded, and the books delivered.

In 1970, a Swedish manufacturer of ball bearings (SKF) replaced its data collection infrastructure with a CTI application that was also based on the IBM 360/2570 complex. Rather than using data collectors who would travel from shop to shop, local shop personnel provided the data directly. On a daily basis, they would dial a number that accessed the IBM 360/2750 complex at headquarters. Data was entered using push-button phones. The switch would pass an indicator of the numbers pressed to the 360 via the TPLH connection, and the computer would return an indication of acceptance or rejection of the data to the switch. The switch would, in turn, produce appropriate tones to notify the user of the status of the information exchange.

These two examples underscore the flexibility of this early system. Note that both outbound (IBM 360 initiates the calls) and inbound (users call the IBM 360) applications were supported. This system exhibited two classic hallmarks of a CTI application. First, the phone connection is used for media processing (i.e., the information being passed back and forth). Second, there is a linkage between the computer and the switch to exert call control.

Amazingly, after IBM's introduction of the 360/2570 applications, there was an attempt at a form of electromechanical CTI, albeit a short-lived one. In 1975, and largely in response to the IBM 360/2570 solution, the Plessey company designed a computer link to their crossbar PABX. Every line and every control register of the switch was wired to the computer so its status could be monitored and controlled. The computer could intercept dialed digits, make routing decisions, and instruct the switch to route a call in a particular fashion. Called the System 2150, only two were deployed before electronic switching rendered the technology obsolete.

At about the same time, a group of Bellcore researchers formed the Delphi Corporation to build a system for telephone answering bureaus. These bureaus were essentially answering services for multiple companies. At the end of the day, the company phones were essentially forwarded to these bureaus, where a person would answer the line and take a message. However, it was important for the person answering the phone to know what company was being called, and to be able to answer the phone as a representative of that company. Delphi 1, released in 1978, was the answer to the problem.

All calls were rerouted to a computer that could tell by the specific line being rung which company was being called. The computer would then retrieve the text for that company's standard greeting, as well as any special instructions for handling the call, and pass the call and instructions to an attendant. The answering bureaus saw a 30% increase in efficiency and the concept caught on quickly.

Through the 1980s, niche applications continued to appear, and new players entered the market. These included British Telecom (a telemarketing application), Aircall (paging), and the Telephone Broadcasting Systems (a predictive dialing system). Perhaps one of the best-known CTI applications to emerge in the 1980s was Storefinder™. The results of a collaboration between Domino's Pizza and AT&T, Storefinder™ used ANI to route a call to the Domino's Pizza nearest that customer. Before the phone in the store could ring, Storefinder™ provided the personnel at that store with the customer's order history, significantly enhancing the level of customer service.

Many early attempts to integrate computers and telephony focused on the media processing aspect of communication. This includes early versions of voice mail and interactive voice response (IVR) systems. These simple technologies did not need much more than specialized call receiving hardware in a computer system, and a hunt group. When a caller dialed in to the service, the telephone network switched the call to one of the access lines in the hunt group. The computer then proceeded to provide voice prompts to guide the user through the service. In the case of voice mail, the user was prompted to leave or retrieve recorded messages. In the case of IVR, the user was prompted to provide, by touch-tone or voice, the information necessary to perform a database lookup (e.g., current credit card balances, history of charges, mailing address, payment due dates, etc.).

Modern voice-mail and IVR systems, and more advanced CTI applications, include a strong call control component. They can transfer calls, provide outward dialing, and even paging. This requires a more complex physical and logical integration of the computer and telephony worlds. The two worlds must be physically connected, making it possible for data from the telephone network to be passed to the computer and call control information from the computer to be passed to the network. Logically, the integration of data from both the telephone network and the computer must be used to create new applications that give the corporation a competitive edge.

Today, the call center scenario dominates that CTI world. Resulting applications typically utilize the most advanced call control and media processing functions. CTI enables new call center models. A single call center can be logically partitioned to function as multiple smaller call centers, or multiple distributed call centers can be logically integrated to act as one. Modern CTI applications provide the knife, or the glue, to make these models possible.

1.2.4   Components and Models

The basic components of a CTI application are depicted in Figure 1.2.1. At the heart of the application lies the computer and the switch. The computer houses end-user data and hosts the end-user interface to the CTI application. The switch provides the ability to make and receive calls and hosts the network interface to the CTI application. The computer provides a set of peripherals (e.g., keyboard, screen, etc.) by which the user accesses the CTI application, and the switch provides the peripheral (e.g., telephone) by which the user communicates. Between the computer and switch there must exist a connection or link, the nature of which differs depending on the type of CTI application.

Consider the automated attendant application. A person needing to speak with someone within the company dials the company's published phone number. The switch routes the call to a computer that begins to play back a recorded message. The message prompts the caller to use the touch-tone buttons to select from an array of options. The caller can enter the extension of the person they wish to reach, in which case the computer directs the switch to reroute the call to that extension. The caller can use the keypad to enter the name of the person being reached. The computer has to translate each tone to the associated letter values, and determine if there is a match in the company personnel listing. If there is none, or if the match is ambiguous (e.g., "Sam" and "Pam" use the same key combination), the computer asks the caller to hold and transfers the call to an operator. If a single, unambiguous match is found, the

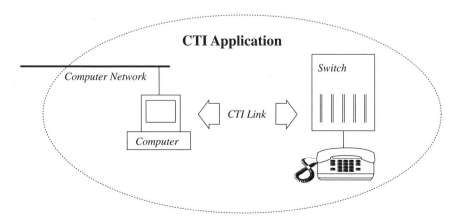

FIGURE 1.2.1   Basic components of a CTI application.

computer can ask the caller to confirm the match, retrieve the extension from the database, and direct the switch to transfer the call. At any point the caller can force the computer to transfer the call to an operator by pressing 0.

1.2.4.1 Media Processing

As has been noted, any phone call can be broken down into two broad activities: media processing and call control. CTI applications typically support both, albeit in different degrees of complexity and by using different strategies. However, a complete suite of CTI services requires both media processing and call control services.

Media processing is perhaps the easiest to understand. When a fax machine calls another fax machine, the transmission of the encoded image across the connection is media processing. When an end user uses their modem to dial in to the local Internet Service Provider (ISP), the exchange of data across the connection is also media processing.

In the CTI arena, the hardware required for media processing is relatively simple. It often takes the form of voice processing, speech digitization and playback, and fax circuitry. Many products integrate these functions into a single printed circuit board that can be installed in a desktop computer. Many of these integrated boards support multiple lines and hardwire the circuitry to each channel. This is sometimes referred to as dedicated media processing hardware (see Figure 1.2.2). Companies that provide such integrated boards include Dialogic Corporation (www.dialogic.com), Pika Technologies, Inc. (www.pika.ca), and Rhetorex (www.rhetorex.com). Rhetorex is now a subsidiary of Lucent Technologies (www.lucent.com).

This approach is appropriate for small-scale applications. For example, a company providing voice mail services in a small town might equip a standard desktop system with a four-line integrated board. A user dialing into the service would be switched by the network to one of the four lines. Based on the

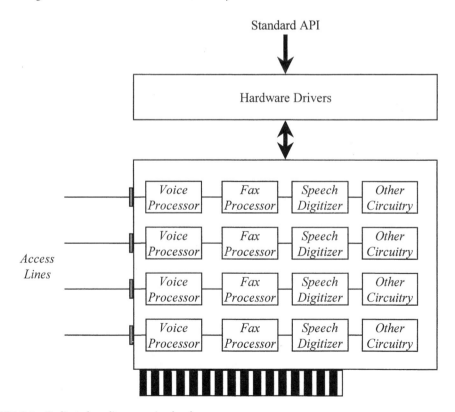

FIGURE 1.2.2 Dedicated media processing hardware.

tones provided by the user (e.g., "Please enter your mailbox number") or ANI information provided by the network, the user can retrieve recorded messages from the computer and play them back.

In these simple environments, standard application programming interfaces (API) are often adequate for controlling the resources. For example, the Microsoft Windows or Solaris APIs that are used to play sound files through a local speaker can also be used to send and receive multimedia content over a telephone connection.

Large-scale applications, however, are more complex. In these environments, sharing resources is more economically viable. A business person may be willing to purchase four complete sets of media processing circuitry, knowing that at any given time only a few components associated with any particular line are going to be used. However, equipping every line in a large application with all of the circuitry it might be called upon to use is not cost effective. For example, consider a large-scale application that implements a pool of four T1 circuit interfaces (96 voice channels). Usage patterns may show that this application needs 96 voice digitizers and playback units, but only 16 speech recognizers, 16 fax processing circuits, and 36 analog interfaces for headsets.

Assembling components at a more modular level is more cost effective and can scale more easily, but it also places new demands on the system. New APIs and standards are required for interconnecting, using, and managing these resources. There are two leading architectures for building such systems: the multi-vendor integration protocol (MVIP) and SCbus. In addition to describing the hardware architecture needed to interconnect telephony-related components, both GO-MVIP and SCSA define software APIs required to use and manage those resources (see Figure 1.2.3). The SCSA Telephony Application Objects (TAO) Framework™ is the API defined by the SCSA.

On the hardware side, both MVIP and SCbus describe a time-division bus for talk-path interconnection, and a separate communication mechanism for coordinating the subsystems. MVIP (www.mvip.org) is administered by the Global Organization for the MVIP (GO-MVIP). SCbus was originally developed by the Signal Computing System Architecture (SCSA™) working group (www.scsa.org). SCSA has since

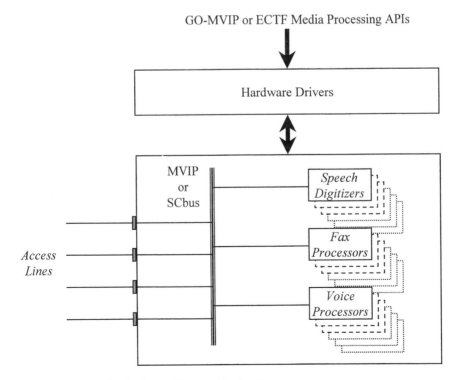

FIGURE 1.2.3 Architecture for sharing media processing hardware.

become part of the Enterprise Computer Telephony Forum (ECTF), a non-profit organization actively prompting the development of interoperability agreements for CTI applications (www.ectf.org). SCbus, announced in 1993, is now also an ANSI standard.

Both GO-MVIP and the ECTF also define a set of application program interfaces (API) for media processing.

1.2.4.2   Call Control

The other major activity a CTI application needs to support is call control. Call control is concerned with the successful establishment, maintenance, and termination of calls. To support these activities, the switching nodes in the telephone network must communicate with one another and with the end-user's terminal equipment. The process by which the switches do this is called signaling. Signaling can be done in-band or out-of-band. In-band signaling occurs on the same channel occupied by user information. This is common for terminal equipment (i.e., telephones), and has become less common within the network itself. Out-of-band signaling occurs on a separate channel from that occupied by user data. This approach is common within the telephone network, and less common between the user and the network (ISDN notwithstanding).

In addition to differentiating between in-band and out-of-band signaling, it is important to note that signaling between the network and the user is bidirectional. The user signals the network by going off-hook, dialing a phone number, and hanging up a phone. This signaling is well standardized. The most common standard today is dual tone multi-frequency (DTMF), the familiar tones we hear as we press buttons on a touch-tone phone. The network signals the user in-band by providing dial tone, busy signals, ringing tones, fast busy, and so forth. Each of these has a distinct meaning, but the sounds have not been well standardized internationally. This is a significant challenge for the CTI environment. Out-of-band network-to-user signaling is somewhat more standardized. Examples include the D-channel on an integrated services digital network (ISDN) interface, the proprietary interfaces defined by digital telephones, and dedicated CTI interfaces to private branch exchanges (PBX) and switches.

Perhaps the most challenging aspect of CTI applications is achieving accurate and reliable call control. In most applications, out-of-band signaling is preferred. Each option, however, has its scope, strengths, and weaknesses. In an ISDN environment, D-channel signaling can be used by the CTI application. One possible CTI application is a network-based automatic call distributor (ACD). Naturally the scope is limited to the domain for which the ISDN signaling is meaningful. For example, the ACD application may not be completely effective when calls cross some public network boundaries.

A CTI application could also leverage the proprietary signaling between a PBX and a digital telephone. Again, such an application may be limited to the scope of the PBX or a group of PBXs from the same manufacturer.

In the public network, the switch-to-switch signaling protocol is called Signaling System 7 (SS7). The domain for SS7 signaling can be as large as an entire public telephone network. Unfortunately, SS7 is usually not available to the CTI application. Closely associated with the internal operation of the public network, SS7 access is jealously guarded by most carriers. Where access is available to the corporate customer, a CTI application based on SS7 requires sophisticated customer premises equipment (CPE) that can handle the complexity of SS7. As a result, this signaling option is usually only appropriate for call centers handling large volumes of calls.

One of the most popular strategies for CTI applications is the dedicated CTI link implemented by many modern PBXs and some public exchange switches. The domain for a dedicated CTI link is a single telephone switch or a small number of tightly integrated switches or PBXs. These facilities are designed for CTI, and tend to offer the rage of signaling options best suited to this environment. These dedicated facilities can implement proprietary or standard call control strategies. Examples of proprietary strategies include Nortel's Meridian Link Protocol (MLP) and AT&T's ASAI Protocol.

Naturally, the industry is leaning strongly to standards-based strategies. The predominant standard is the Computer-Supported Telephony Application (CSTA) from the ECMA (formerly European Computer Manufacturers Association). Adopted in 1990, the CSTA protocol (www.ecma.ch) has now been implemented by

First-Party CTI

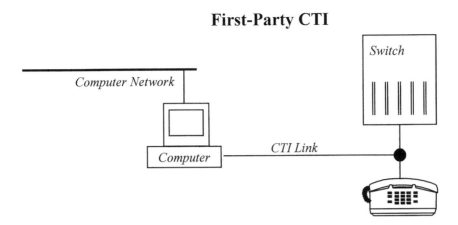

FIGURE 1.2.4   First-party CTI model.

such major players as Siemens ROLM, Ericsson, and Alcatel, to name a few. It is important to note that, although CSTA is a standard, the features any particular vendor elects to implement can vary. As a result, CSTA implementations from different vendors are not necessarily interoperable.

1.2.4.3 First-Party and Third-Party CTI

CTI applications can be broken into two broad classes based on the relationship between the computer and the switch. In first-party CTI, the computer is essentially on an extension to the line on which a call is being received. The computer can exert the same call control functions a human attendant could exert via a standard telephone set attached to the telephone system. This implies that call control is on a call-by-call basis. First-party CTI call control includes such activities as going off-hook, detecting dial tone, dialing a call, monitoring call status signals (e.g., ring, ring no-answer, answer, busy, and fast busy) conditions, and terminating the call.

In the first-party CTI model (Figure 1.2.4) the computer, the keyboard and screen, and the phone are all on the same line. The computer will tend to use the dedicated media processing hardware model, and tend to be a user end-system (as opposed to being a server). First-party CTI is further subdivided into basic and enhanced flavors. Essentially, basic systems use in-band signaling and have limited capability. Enhanced systems use out-of-band signaling, usually either ISDN or proprietary signaling to the PBX. While there are basic first-party CTI platforms on the market, the industry is more interested in enhanced first-party CTI systems.

The classic example of an inbound first-party CTI application is the voice mail system. In a voice mail application, an inbound call is received by the computer. The computer activates the local voice mail software to record and store, or retrieve and playback, voice mail. The simplest example of an outbound first-party CTI application is autocall.

APIs for first-party call control first appeared from the manufacturers of network access equipment (e.g., modems, fax boards, etc.). The only such API that achieved de facto standards status was the Hayes modem command set. Now universally understood by modem products, the Hayes command set defines basic commands for initiating and terminating calls, and altering the configuration of the modem.

Third-party CTI is the more sophisticated model. In third-party CTI, the computer exerts call control via a dedicated connection to the switch or PBX (Figure 1.2.5). This naturally implies out-of-band signaling. It also implies that call control can be exerted over several calls, or over the switch itself. The call control functions a third-party CTI application could exert are similar to those a human attendant could exert using a specialized telephone set with enhanced privileges, such as an operator's console.

In the third-party CTI application, the computer, the keyboard and screen, and the phone have no relationship to one another unless the computer establishes one. These environments tend to use the shared media processing hardware model, and tend to perform signaling via SS7 or (more commonly)

Third-Party CTI

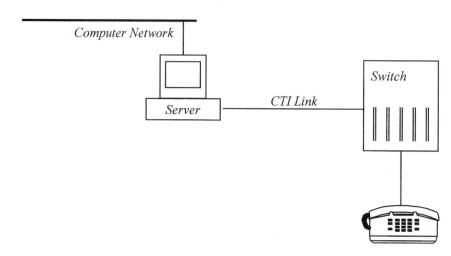

FIGURE 1.2.5   Third-party CTI model.

dedicated CTI links implementing the CSTA protocol. The CTI link typically terminates in a server rather than a specific application end-system.

There are three basic flavors of third-party CTI, which reflect the essential relationship between the computer and the switch. In the *compeer* model, the computer and switch are on equal terms. Each operates as the master of its own realm, passing information and receiving instructions from the other across a specialized interface. In the *dependent* model, the computer rules and the switch obeys. The switch has no innate call handling capability, and is actually incapable of processing calls without receiving instructions from the computer. Finally, the *primary* model is virtually identical to the compeer model, but the computer and switch do not share a specialized link. Rather, the computer attaches via a standard trunk or line port. Over the years, the dependent and primary models have seen diminishing emphasis as the market moves toward the compeer model. Unless explicitly identified as dependent or primary, third-party CTI is usually assumed to operate on the compeer model.

Automatic call routing applications are classic examples of third-party CTI. A server-based application is alerted, by the switch, to the arrival of a call. Based on ANI information, or the specific DNIS (i.e., called number), the computer directs the switch to divert the call to a specific line.

As with first-party CTI, the first third-party APIs were developed by manufacturers to support applications running on their own systems. Examples included the CallPath API from IBM, and the Computer-Integrated Telephony (CIT) API from Digital Equipment Corporation (DEC). Unlike the Hayes command set, however, none of these have achieved de facto standard status.

In the 1990s, three major APIs emerged, all strongly associated with a particular computing environment. Novell (www.novell.com) and Lucent collaborated to create the Telephony Services API (TSAPI). Novell's commercial product based on TSAPI is called NetWare Telephony Services, which links applications on remote clients with telephone system driver modules. TSAPI defines the boundary between CTI application software, and the drivers that control the links and signaling into the network.

Microsoft (www.microsoft.com) and Intel collaborated to create the Telephony API (TAPI). Like TSAPI, TAPI is concerned with call control. However, the TAPI architecture actually defines two distinct interfaces (see Figure 1.2.6). The first interface resides between CTI applications and the Windows operating system (OS). This interface, which unfortunately has the same name as the overall architecture, provides a standard means for CTI applications to access the telephony services provided by the Windows OS.

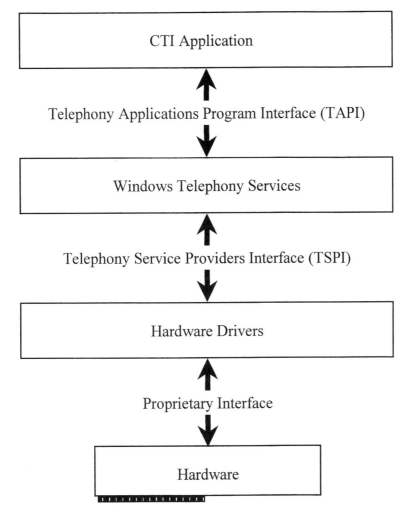

FIGURE 1.2.6 The TAPI architecture.

The second interface resides between the Windows OS and the CTI hardware drivers. Known as the telephony service providers interface (TSPI), this interface provides a standard mechanism for hardware vendors to write drivers that can support the telephony services provided by Windows. It is Microsoft's job to ensure that TAPI-compliant applications can access all of the resources provided by TSPI-compliant hardware drivers.

The third call control API is the more recent, introduced in October 1996, and brings CTI into the world of the Internet and the World Wide Web (WWW). Developed jointly by design teams from Sun, IBM, Intel, Lucent, Nortel, and Novell, the Java Telephony API (JTAPI) defines a call control interface for CTI applications running as Java applets. This opens the door to creating Web-based CTI applications. The Sun Microsystems product that implements this API is called JavaTel™.

Figure 1.2.7 integrates the various standards and concepts introduced in this paper into a single CTI model. A CTI application can be either first-party or third-party. First-party applications tend to use local, proprietary APIs (e.g., the Windows APIs) to access local call control and media processing services, and the Hayes command set to control dedicated telephony hardware.

Third-party CTI applications tend to use sophisticated call control APIs like TAPI, TSAPI, or JTAPI, and standardized media processing APIs like those defined by the ECTF. The link between the CTI server

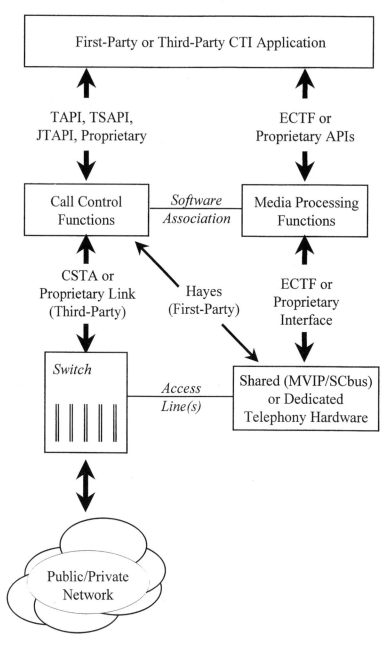

FIGURE 1.2.7   Combining the standards and components.

and the switch commonly implements the CSTA protocols. The server typically uses shared telephony hardware that is interconnected using the MVIP or SCbus architecture.

It is also possible to build a CTI server that supports several APIs and standards simultaneously. Such a product would have to map requests from all APIs into a single common function set. Dialogic's CT-Connect product takes this approach. It supports both the TAPI and TSAPI interfaces and includes built-in drivers for the ECMA CSTA link protocol and several other proprietary CTI link protocols.

1.2.5 CTI Applications and Trends

A few of the more common, and simpler, CTI applications have already been noted: voice mail, autocall, and automatic attendant. Each of these is commonly implemented as first-party CTI applications using dedicated media processing hardware. Digital dictation is another CTI application that is virtually identical to voice mail, but typically supports longer record times. The recorded dictation is usually retrieved and transcribed locally.

Many companies are beginning to provide interactive or on-demand fax services. For example, the real estate company could provide automated faxes of current properties for sale. In such a service, the user dials in and, using a touch-tone driven menu system, requests a particular fax or group of faxes and provides the number to which the fax is to be sent. The service retrieves the fax from a local file, initiates an outbound call to the specified number, and transmits the fax. As with the automated attendant application, interactive fax could be implemented as a first-party of third-party application.

Many pay-per-call applications are CTI applications. This is a common strategy for implementing fee-for-access Internet services. The user dials a 900 number and the PBX routes the call to the CTI application. The user is prompted to provide a code identifying the service they are trying to access. The CTI application provides an access code that permits the user to access the web site. The phone service bills the user for the 900 call and passes the majority of the fee to the pay-per-call service provider. The pay-per-call service provider takes an additional cut and passes the remainder of the fee to the company hosting the web service.

Perhaps the most common third-party CTI application is the inbound and outbound call center. Inbound call centers typically integrate an automatic attendant to collect initial customer information (i.e., credit card numbers, zip codes, pin numbers, etc.) and provide core services (e.g., account balances, mailing addresses, account histories, a list of service or product options, automated order taking, etc.). The caller always has the option, however, to abandon the automated system and speak to a person. In this case, the CTI application routes the call to an available attendant and provides all information the user has submitted. The application may also provide any call information provided by the phone network and any customer data retrieved from the computer's database.

The CTI market is showing clear signs of accelerated growth, fueled by a number of enabling factors in the industry. The pervasive deployment of LANs and internetworks provides the infrastructure over which many first-party and third-party CTI applications operate. The growth in digital communications and integrated networks that provide enhanced signaling capabilities (e.g., ISDN and digital telephones) create a rich set of network information on which CTI applications can be built.

The emergence of standard APIs in both the media processing and call control arenas has furthered equipment and service interoperability. Furthermore, the increasing maturity of voice processing technology makes interactive voice response (IVR) systems easier to deploy and use. Finally, the industry is seeing a broad array of CTI application development toolkits. Examples of these include OmniVox from Apex Voice Communications (www.apexvoice.com), Visual Voice from Artisoft (www.artisoft.com), MasterVox from Mastermind Technologies (www.mastermind-tech.com), and IVS Builder and IVS Server from Mediasoft Telecom (www.mediasoft.com).

1.2.6 Conclusion

The CTI market is a young one, but the technologies coming together into this application environment are relatively mature. As the CTI-related standards themselves mature, interoperability agreements emerge, and economies of scale begin to apply, CTI applications are likely to become pervasive. Furthermore, with the emergence of JTAPI and the increasing drive toward voice over IP (and hence over the Internet), CTI applications are finding a new niche in which to grow. The Internet is a significant niche indeed!

For further information, the reader is recommended to visit the various web sites identified in this chapter. There are also two periodical publications dedicated to CTI, both of which can be accessed via the Internet: *Computer Telephony* (www.computertelephony.com) and *CTI Magazine* (www.tmcnet.com).

1.3   Voice over IP

Matthew Kolon

1.3.1   The Coming Integration of Voice and IP Data

Companies in the U.S. spend $100B on long-distance and international telephony every year. Most of that money goes to the basic transit of voice and fax from one location to another. With the continued pervasiveness of intelligent peripheral (IP) networking, a new class of products and services has evolved to move some of that traffic from its traditional home on the public switched telephone network (PSTN) to a variety of packet-switched networks. While many of these new "voice" networks have not previously been considered telephony-class, they are nonetheless attractive because of their low cost.

The IP telephony scene has jumped from being a hobbyist's realm of custom solutions and cobbled-together software to a $400M per year industry hotly pursued by industry giants of hardware and software. Continued improvements in digital signal processor (DSP) technology, voice packetization techniques, and the networks that IP voice runs over have combined to make the start of the 21st century into the era that IP telephony begins the transition to a mainstream solution for business.

There are a number of reasons for the inevitability of this transformation, but all of them come back to the relief of high-cost long-distance telephone services. Reviewing a few comparative facts regarding the PSTN and voice over IP (VoIP) presents some compelling realities:

- **One can fit more voice on an IP network than one can on the PSTN.** The Bell System definition of a single voice channel as a 64kbps DS-0 has led to a long-standing institutional belief that 64k is necessary to carry a voice conversation. Thus a T-1 is commonly referred to supporting 23 "voice" channels over its 1.544 Mbps. Yet today's VoIP products can carry hundreds of voice conversations over that same amount of unchannelized bandwidth.

- **Packet networks are much better than they used to be.** Improvements in the quality of physical-layer packet networks over the past 30 years have resulted in a large general improvement in data integrity. The same forces that make simple frame relay an effective replacement for the robust X.25 protocol mean that even connectionless IP data — and voice — may be entrusted to today's connectionless networks and still have an excellent chance of getting through in a reasonable amount of time and with few errors (or little delay) of consequence.

- **Control of IP data networks rests largely in the hands of the customer.** As long as a minimum quality of service — particularly the establishment of maximum delay guidelines — is met, virtually every service available over IP is controllable from the sending and receiving stations. For example, packets may be routed over the Internet for free if tolerant of lower quality, over a private IP network if demanding of higher quality, or even over the PSTN if necessary — all at the discretion of the originating node.

These are just a few of the reasons why many network managers are examining the current possibilities for placing at least some of their voice traffic into IP networks.

1.3.2   Applications for Voice over IP (VoIP)

Of course, with long-distance services being the single most expensive portion of any company's telephony budget, the application of VoIP to the interexchange carrier (IEC) realm is taking the forefront when it comes to the immediate application of the technology. The basic design of such a network is rather simple: gateways within local calling areas connected by an IP network which spans the distance previously covered by the IEC.

While a company implementing VoIP for the purpose of saving charges on interoffice communications may have a design as simple as that in Figure 1.3.1, it is more likely that the IP network will connect multiple sites, each with its own gateway, each of which may then contact another dynamically when it

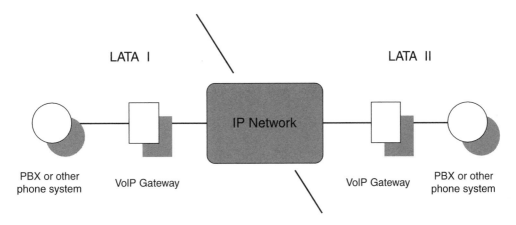

FIGURE 1.3.1   Business IEC replacement using VoIP.

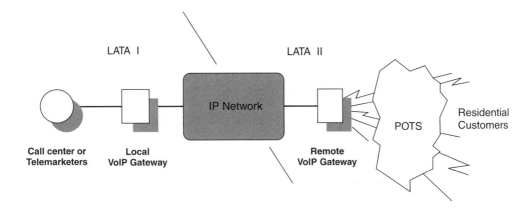

FIGURE 1.3.2   Business-to-residential VoIP network.

has a voice call destined for that site. The connectionless nature of IP ensures that new gateways may be added at will, with little need for reconfiguration at the other stations.

Many variations of this scheme are possible, depending upon the nature of the service one is trying to implement. For tie-line replacement and business-to-business calls, the simplest to exploit is that shown in Figure 1.3.1, that is, two or more gateways connected by an IP network. The reason that most pundits consider this setup to be the first area to exploit VoIP is because the difficult part — getting the voice to a few places where it can be digitized and packetized into IP — is already done. The private branch exchange (PBX) that currently connects via a leased line or IEC to another PBX can easily have that connection replaced by IP — with no changes in how users place calls.

Another application that is generating a large amount of industry interest is that of business-to residential telephony (Figure 1.3.2), to allow telemarketers or call centers to physically centralize while obtaining low-cost long-distance service via VoIP. In this scenario, residential customers are able to dial a local number and access a VoIP gateway which connects them to the implementer's customer support or sales office — wherever it may be. The customer makes a free call, and receives the same service had an 800 number been dialed, but the company avoids the cost of maintaining 800 service. It is also able to supply customers with a "local" number to call for service, which can enhance the company's image.

Reversing the above strategy — that is, using the remote gateway to *place* local calls rather than accept them — allows telemarketers access to large, yet distant, markets without the need to place large numbers of long-distance calls to get to them.

Yet another option exists for those eager to exploit the possibility of VoIP at their businesses or campus: replacing the PBX and its network with an IP network. Most businesses are already halfway there; they have local area networks (LANs), routers, and digital wide area network (WAN) facilities capable of handling IP traffic. New products, such as 100- and 1000-Mbps Ethernet, as well as the cost-effective speed of LAN switching, mean that network managers can build an enormous amount of capacity into their local and enterprise networks — capacity which might well be used to carry voice traffic. Traditional models for business traffic have always involved the creation and management of two separate networks, one for voice and one for data. The encapsulation of voice in IP packets means that the consolidation of voice into the data network is now possible, with the corresponding reduction in the need for equipment, data facilities, staffing, and expertise in several types of systems. Consolidation of voice traffic and data traffic into the same end-to-end network opens the door to true integration of messaging and telephony systems, such as integrated email and voice mail, and IP-based fax messaging.

The final area of interest for VoIP proponents is that of residential-to-residential connectivity, that is, friends and relatives speaking to each other from handsets or speakerphones integrated into Internet-connected PCs. While this is the application that "proved" the possibility of VoIP, it remains the most difficult to ensure acceptable quality for. The difficulty of obtaining quality voice this way has nothing to do with the equipment at the ends of the link, but rather with the lack of guaranteed, or even reliable, values for delay and delay variation over the Internet. Indeed, improvements in low-cost digitization hardware and "Internet telephony" software have made it possible to have a full-featured, high-quality VoIP gateway for the cost of a new PC. But even the best-quality digital voice will be unintelligible if only half of it arrives at the intended destination.

These are just the basic categories that some of the most obvious applications for VoIP fall into. But applications are as numerous as those for the telephone itself — perhaps even more so. The lower cost of VoIP means that some uses for telephony that were once deemed uneconomical may now be justified. And the integration of voice and data traffic over a single IP network may make some forms of integration possible that were unthinkable just a few years ago.

1.3.3   A Component-based Overview

What are the components of a successful IP telephony system? While there are of course a number of different approaches, there are a few basic ingredients that all systems must implement — although the use and location of parts changes with different network designs.

The VoIP Network: In the list of VoIP components (Figure 1.3.3), the IP network(s) over which the voice will travel is of primary importance. IP is first and fundamentally a connectionless protocol, with no guarantees concerning the traffic that it carries. It cannot ensure a maximum delay or variability of delay, cannot retransmit errored or lost packets, and does not even promise that its payload will arrive at all. The quality of service one receives from the PSTN, and that provided by even the most carefully managed and overbuilt IP network do not bear comparison. And for those thinking about using the Internet as the equivalent of their current expensive IEC service…well, suffice it to say that when a web page often takes 60 seconds to download, sending real-time voice traffic over that same series of links will be a challenge. Until the Internet infrastructure is managed under an agreement which includes concrete plans to provide some limited and predictable delay — in an interprovider fashion — voice traffic cannot travel the Internet and maintain the quality that business customers demand. It's worth mentioning that this agreement is nowhere in sight.

That does not mean that today's Internet has no place in the voice network, however. VoIP gateways can use the Internet to provide the non-real-time services that constitute much of today's "voice" traffic. The most obvious one of these is facsimile transmission. While fax machines thrive on the dedicated lines of the circuit-switched PSTN, there is no reason why their transmissions cannot be placed in IP for long-distance transit. Delay — the reason why interactive voice is so difficult over the Internet — doesn't affect fax transmissions at all, and transmission control protocol/Internet protocol (TCP/IP) can resend

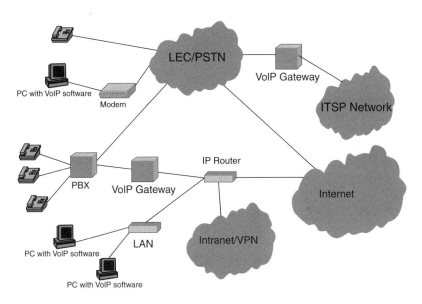

FIGURE 1.3.3 VoIP network components.

data until the network gets it right without bothering the receiver. The same could be said for voice mail messages.

The next step between the very public Internet and a completely private IP network is the ISP backbone itself, which is nothing more than a single provider's portion of the Internet. If this network extends close to the points where gateways will be placed, IP traffic between them may remain solely on that network. In almost all circumstances, this will result in less delay and better predictability for traffic of all types. But while the statistics for network performance may improve in a single-provider environment, the lack of user control over these fundamentally public networks may be unacceptable for the network manager who seeks to have some influence over the environment in which his traffic travels. Single Internet service provider (ISP) IP telephony, though, has the lowest cost of any of the non-Internet options, and therefore is attractive as long as acceptable quality can be achieved. This may be a matter of simply trialing a number of ISP networks and choosing the one with the best performance, or may actually involve a level of performance — with stated delay and throughput characteristics — to be specified in the user contract.

Luckily, the Internet and its constituent networks are not the only options for long-distance carriage of VoIP. Many of the larger ISPs offer, in addition to their public Internet network, access to a separate IP network designed for virtual private network (VPN), intranet, extranet, and other semiprivate usage. These networks are not any more remarkable in concept than an average ISP's network, except for their managed nature, that is, the knowledge the provider has of just how much traffic any one user is likely (or allowed) to subject the network to at any one time — something unheard of on an Internet access network. This knowledge allows the provider to predict and maintain a high level of quality, which can result in service level agreements in which end-to-end delay is specified to be well below 0.5 seconds — the point at which telephony starts becoming reasonable. In this environment, SLAs are becoming the rule rather than the exception.

The ultimate VoIP network, however, is the one where all aspects of IP traffic and performance can be managed by the users — a completely private intranet. Formed from private (leased) lines, with perhaps some links composed of frame relay or asynchronous traffic mode (ATM), the distinguishing characteristic of these networks is that they are completely under the control of the network managers who deploy and run them. Therefore, the amount of bandwidth reserved for voice traffic can be strictly controlled, as can the throughput of routers and other connectivity equipment. How those resources are actually apportioned may vary from protocol-based reservation systems like reservation protocol (RSVP) to completely manual intervention, but whatever the method, the manager has the ability to restrict the effect of data traffic that interferes with voice. While this sounds like — and in fact is — the ideal environment for packetized voice, it comes with a price. Completely private IP networks are by far the most expensive way to ship IP from one location to another. Whether the establishment of such a network is worth the ability to carry voice effectively depends on how much money can be saved by eliminating IEC charges from the IT budget.

If the number of options and the headaches of managing another network service are a serious disincentive, another possibility is to leave the network and its management to the specialists — that is, to contract with one of the growing number of Internet (or IP) telephony service providers (ITSPs). An ITSP functions as a plug-and-play replacement for a traditional IEC, by providing the gateway, network, and management needed to make VoIP successful. The tradeoff here, of course, is that since the ITSP does all the work, they also reap some of the rewards. Typically, ITSPs function like an IEC in terms of billing, with per-minute rates that range from one half to three quarters that of comparative IECs.

That level of discount may change before long, however. Much of the savings that ITSPs are able to pass on to their customers are possible because of a May 1997 FCC ruling that classifies ISPs and ITSPs as end users of the PSTN rather than as carriers. This classification currently makes it impossible for LECs to charge ITSPs the same access charges they demand from traditional IECs. Those access charges, when passed on to the IEC customer, can account for as much as one half of the average IEC bill. It is the lack of these charges, more than the technological benefits of VoIP, that allows ITSPs to sell services for so much less than their IEC counterparts.

While the level of savings on recurring charges is the least with the ITSP option, it may well be compensated for by the simplicity of setup and management, and the lack of gateway hardware or software costs. The users who benefit from the access charge loophole, however, may have some hard decisions to make if, as many believe will occur, the FCC reverses itself and decides to consider ITSPs as carriers. In that market, much of the price differential would disappear, and users would have to make their decisions based more on quality, service, and other points rather than price (Figure 1.3.4).

All of these networks can and will benefit from work currently underway to allow efficient prioritization of packets containing voice over those containing non-real-time data. Gigabit-speed routers, faster switches, better routing and path-reservation protocols, and the continued addition of cheap bandwidth are all reasons why VoIP quality will continue to increase.

In summary, there are a number of network options for VoIP. Which one best suits a particular need depends on a number of factors, primarily revolving around the level of expected quality. For those looking for a way to lower the cost of interoffice communications — an application where the "internal" aspect may allow slightly lower quality than that required for communications with customers — some of the lower-cost options like single-ISP VoIP networking may suffice. Those wishing to completely replace their IEC contract with an IP-based IEC solution are faced with replacing a complex network from the ground up, and will have to plan, and pay for, a much more robust service. And for the time

Network	Gateway	Cost	User Control	Performance
Internet	User-provided	Least	Least	Worst
Single ISP	User-provided			
Managed IP	User-provided			
Private IP	User-provided		Most	Best
ITSP	Included in contract	Most	N/A	N/A

FIGURE 1.3.4   VoIP network options compared.

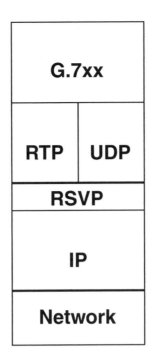

FIGURE 1.3.5 VoIP protocol components.

being, at least, voice over the public Internet remains in the realm of a hobby for those willing to tolerate indifferent and completely unpredictable voice quality.

Gateway Software and Hardware: The hard work of actually taking analog voice and sending it over an IP network, as well as receiving IP and converting it back into voice, is the job of the gateway. It is easiest to examine the issues related to this complex task if we break it down into its components (Figure 1.3.5):

Accept analog or digital voice: A gateway must have some connection to the non-IP world where the voice traffic originates, usually consisting of either a bank of dial-in plain old telephone service (POTS) ports or a digital connection to a PBX.

Prepare the voice signal: In order to use the available bandwidth as efficiently as possible, the voice signal must go through a number of transformations before it is ready to be digitized. First, it must be "cleaned up" by having as much noise and echo removed as possible. The techniques for doing this have been well established in the traditional telephony world for years, but the cooperation of the various systems and gateways through which voice may pass is essential. This means that calls traveling through a LEC on their way to the VoIP gateway may need to be treated differently than those coming directly from a PBX.

Second, it must be stripped of unnecessary silence, to avoid making the gateway send hundreds or thousands of packets per second carrying nothing. Most gateways have adjustable options for when silence suppression "closes off" and stops transmitting on behalf of a user, but the effectiveness of default settings may depend on usage characteristics that are themselves dependent on cultural factors. Some adjustment of this setting to achieve the best compromise between quality and throughput is usually necessary. Related to the subject of silence suppression is the modeling and regeneration (at the remote end) of background noise, without which users can become disconcerted.

Compress and digitize the voice signal: The standard compression and digitization of voice provided by traditional 64k PCM produces a stream of digital data that is enormous compared to that available by many newer codecs. While some vendors have achieved good results with proprietary schemes, most of the industry is settling down to the use of one or another International Telecommunications Union (ITU) G-series codecs, as specified in their H.323 standard. H.323 is a complex specification for point-to-point and multipoint teleconferencing, data sharing, and telephony over IP. While the full effect of this standard on "VoIP-only" products remains to be seen, the G.711, G.723, and G.729 codec specifications referenced by it are current favorites for coding voice.

These three standards differ primarily in the amount of work that the DSP must do in order to process the analog signal, and the number of bits that it takes to represent a given amount of voice. While recent advances in DSP design and manufacture have allowed vast improvement in these areas, there remains an inverse relationship between them, and also therefore a higher cost for greater efficiency. Nevertheless, the most aggressive of the standards — G.729 — can represent 10 msecs of voice with only 10 octets of IP data. The less intensive G.711 and G.723 trade higher traffic volume for higher quality. Many gateways can be configured to use whichever one of these standards provides the most acceptable trade-off between quality and traffic level.

Route the call: Once a gateway has a potential stream of packets ready to send, it must have some way to identify the address of the gateway it will send them to, and to inform that gateway of which local user it is destined for (or what local number to dial.) For simple point-to-point applications, IP address can be a manually configured variable, since there is only one destination possible. But in cases where a multipoint network means that packets may be simultaneously distributed among a number of destinations, there must be a process in which the called number is translated into an IP address.

Informing the destination gateway of the called phone number has its complications, too, because many of the codecs used in current gateways compress the analog signal so much that the dual-tone multi-frequency (DTMF) tones produced by phones become unreliable. Therefore, the calling gateway must be able to transform those DTMF tones into a code representing the called number and transmit them to the destination gateway for correct routing at the called end.

Packetize and send digital voice in IP datagrams: At first glance, this is the simple part. After all, IP stacks on end stations and routers have been performing this function since the late 1960s. Yet some of the characteristics of packet-switched networks with regard to real-time traffic are different than those regarded as common knowledge by those used to thinking of IP as data-only transport.

For example, the flexible size of an IP datagram, while an advantage in the transmission of data, complicates the problem of achieving low variability of delay, since IP routers handle packets of various sizes differently, and may tend to process smaller packets more quickly than larger ones. The destination gateway would then need to account for the tendency of larger packets to take longer, and thus delay reassembly. In practice, VoIP gateways by default transmit packets of a single size or small range of sizes in order to obviate this problem, but this is one area where the capabilities of the gateways and the network(s) over which they will transmit must be closely matched. Setting the maximum packet size of the gateway to any amount higher than the maximum transmission unit (MTU) of the underlying network will introduce latency as routers fragment datagrams that are too big to travel through networks attached to them.

Enabling routers to prioritize packets containing voice can enable voice and data to coexist on the same network more easily. Methods for doing this include enabling priority queuing based on transport layer port number, packet size, and source and destination addresses. RSVP can be used to reserve router bandwidth and processing capability, as well as network segment bandwidth, for packets that meet certain criteria, but implementing RSVP demands a network path in which all routers are RSVP-compliant, something that is not likely in a multiprovider (or even some single-provider) scenarios.

Receive, buffer, and decode the incoming stream of VoIP data: Again this is a well-understood process for data, which generally depends upon the IP suite's TCP protocol to retransmit lost data and reassemble segments in the proper sequence before it is passed to the application. VoIP software seldom makes use of TCP, largely because the services it provides introduce far too much latency into the transmission process for them to be useful (an exception to this rule is fax transmission, for which TCP makes sense given the lack of need for real-time treatment of data.) Instead, most gateways can use real time protocol (RTP) as the protocol in which voice data rides. While having no control over delay imposed by the network, RTP makes it possible to trade a small amount of additional delay for a reduction in the amount of delay variation. This is accomplished by transmitting each packet with a timestamp that can be read by the receiver and used to pass data to the upper layers of the VoIP software with something like the transmitted amount of inter-packet delay.

Alternatively, some gateways have the option to send digitized voice in user datagram protocol (UDP) packets, which travel in an unstructured stream, free of sequence numbers, timestamps, and acknowledgments — but also free of the delay imposed by processing these variables. Since the audio stream at the remote end must go on regardless of the actual receipt of data, large numbers of packets that are lost en route simply result in "holes" or "dropouts" in the audio signal. While this sounds as though it would spell the end for reproduction of any reasonable quality, in fact it takes the loss of a relatively large number of packets to create noticeable holes in outbound audio at anything but the highest compression levels. Whether the control and complexity of RTP or the simplicity and speed of UDP will prove to be the most effective way to carry datagram voice remains to be seen.

1.3.4   Keys to Successful Deployment

The large number of configurable variables and the many options within each make configuring VoIP networks a considerable challenge, especially since these networks' main role is to replace some of the most bulletproof networks in the world: those of the PSTN. Aside from performance issues, questions of interoperability abound, particularly for those users who wish to deploy distributed VoIP networks consisting of hardware and software from more than one vendor, and networks from more than one provider.

One thing is certain, though: IP telephony is here to stay. Despite the challenges that network managers face in order to reduce their IEC bills, in at least some applications the payoff is great enough to make the decision to at least trial the technology obvious. The astute manager, however, remembers a few things:

- Few, if any, of the products currently available for VoIP networking work well "out of the box." Nearly everyone who has implemented gateways on either a point-to-point or multipoint basis has a story to tell about the setup and configuration of their system, and the shakedown and subsequent adjustments, that had to occur before the network settled down. Almost as invariably, though, they can recount the time that things began to work well, and now can point to users who are happy with the price and performance of the VoIP network.

- All VoIP products aren't the same. Vendors are scrambling to improve quality and add features, and that translates into large variations in product lines — at least until the next revision is introduced.

The good news is that there are many positive signs for those considering putting their trust into VoIP. The current standards situation for components of VoIP products seems to be stabilizing. While any emerging technology — especially ones with such high visibility — generates a large number of proprietary solutions which get narrowed down by the market, VoIP is one example of how vendors can cooperate. Most of the standards for encoding (the ITU G-series) seem to be settling down for a long period of maturity.

With regard to the network technologies in use, a new generation of network designers and engineers feels more comfortable with IP than with any other technology — including voice traffic. The ubiquity of the Internet and of IP itself have created a large pool of experience from which managers can draw when deploying VoIP. As for the future, a knowledge of the workings of Internet protocols is commonplace among graduates of almost any technical program.

While the public telephone network has existed for years, fast public data networks have not existed until recently, and new data networks are being constructed at a staggering rate. Many of these networks will be suitable for voice traffic, and thus can extend the reach of VoIP networking. And the rapid pace of network improvement means that end-to-end latency will continue to drop, which can only mean good things for the quality, and success, of VoIP.

Acronyms

ATM — Asynchronous transfer mode
DSP — Digital signal processor
DTMF — Dual-tone multi-frequency
FCC — Federal communications commission
IEC — Interexchange carrier
IETF — Internet engineering task force
IP — Internet protocol
IP — Intelligent peripheral
ITSP — Internet (IP) telephony service provider
LAN — Local area network
LEC — Local exchange carrier
PBX — Private branch exchange
PSTN — Public switched telephone network
RSVP — Reservation protocol
RTP — Realtime protocol
UDP — User datagram protocol
VoIP — Voice over IP
WAN — Wide area network

1.4   Local Area Networks

John Amoss

1.4.1   Overview

1.4.1.1   Standards

The Institute of Electrical and Electronics Engineers (IEEE) 802 Local and Metropolitan Area Network Standards Committee has the basic charter to create, maintain, and encourage the use of standards for local and metropolitan networks. In the IEEE 802 Committee context the term "local" implies a campus-wide network and the term "metropolitan" implies intracity networks. The IEEE 802 Committee defines interface and protocol specifications for access methods for various Local Area Network (LAN) and Metropolitan Area Network (MAN) technologies and topologies. The project has had a significant impact on the size and structure of the LAN market.

The standards are jointly published by the IEEE, the International Organization for Standardization (ISO) and the International Electrotechnical Commission (IEC). An overview of the standards is published by these bodies. [1,2]

1.4.1.2   Reference Model

Figure 1.4.1 relates the specific protocol layers defined by the IEEE 802 Committee, which include Physical, Media Access Control (MAC) and Logical Link Control (LLC) layers, to the layers of the Open Systems Interconnection (OSI) Reference Model. [3] The protocol architecture shown in Figure 1.4.1, including the Physical, MAC and LLC layers, is generally referred to as the IEEE 802 Reference Model.

Working from the bottom up, the Physical layer of the IEEE 802 Reference Model corresponds to the Physical layer of the OSI Reference Model and includes the following functions.

- Encoding/decoding the signals to be transmitted in a manner appropriate for the particular medium, e.g., the use of Manchester or Non-return to Zero encoding schemes;
- Achievement of synchronization, e.g., by the addition of a preamble field at the beginning of a data frame;

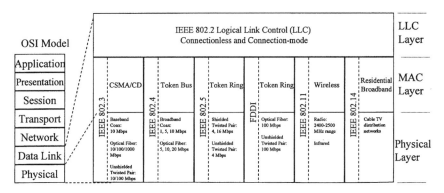

FIGURE 1.4.1   IEEE 802 reference model.

- Bit transmission and reception;
- Specification of the physical and electro/optical characteristics of the transmission media (e.g., fiber, twisted pair wire); and
- Network topology (e.g., bus, ring).

Above the Physical layer are functions concerned with providing the frame transmission service to LAN users. Such functions include the following.

- Governing access to the LAN transmission medium.
- Performing error detection (e.g., via addition of a Frame Check Sequence field);
- Assembling the frame for transmission; and
- Upon reception, performing address recognition.

These functions are collectively associated with a MAC sublayer, shown in Figure 1.4.1. As indicated in the figure, a number of MAC layers are defined within the IEEE 802 Reference Model including access control techniques such as Carrier Sense Multiple Access/Collision Detection (CSMA/CD) — also generally referred to as Ethernet — Token Bus and Token Ring.

Finally, the Logical Link Control (LLC) layer is responsible for providing services to the higher layers regardless of media type or access control method (such as those specified for CSMA/CD, Token Bus, Token Ring, and so on). The LLC layer provides a High-level Data Link Control (HDLC)-like interface to the higher layers and essentially hides the details of the many MAC schemes shown in Figure 1.4.1 from the higher layers. The LLC layer provides a multiplexing function, supporting multiple connections, each specified by an associated destination service access point (DSAP) and source service access point (SSAP), discussed later. As shown in Figure 1.4.1, the LLC layer provides both connectionless and connection-oriented services, depending on the needs of the higher layers.

1.4.1.3 Overview of the Major MAC Standards

Since its inception at Xerox Corporation in the early 1970s, the carrier sense multiple access with collision detection (CSMA/CD) method, also commonly termed Ethernet, has been the dominant LAN access control technique. The CSMA/CD method was the first to be specified by the IEEE, under the IEEE 802.3 working group, and was closely modeled after the earlier joint Digital/Intel/Xerox (DIX) Ethernet specification. [4] Ethernet has, by far, the highest number of installed ports and provides the greatest cost performance relative to other access methods such as Token Ring, Fiber Distributed Data Interface (FDDI) and the newer Asynchronous Transfer Mode (ATM) technology. Recent and in-progress extensions to Ethernet include Fast Ethernet, which, under the auspices of the IEEE 802.3u working group, increased Ethernet speed from 10 Mbps to 100 Mbps thereby providing a simple, cost-effective option for higher speed backbone and server connectivity, and Gigabit Ethernet, which under the auspices of the 802.3z working group increased the speed to 1000 Mbps.

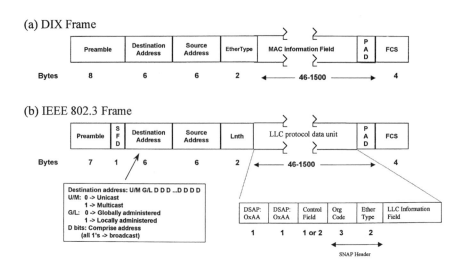

FIGURE 1.4.2   DIX and IEEE 802.3 frame formats.

The IEEE 802.4 Token Bus specifications were developed primarily in response to requirements for the deterministic performance of a token passing scheme, coupled with a bus-oriented topology. The use of a broadband technology option provided the additional benefits of increased bandwidth, geographic coverage, and number of terminations.

The IEEE 802.5 Token Ring specification was developed with major support from IBM and reflected IBM's perspective on local area networking. Improvements over the IEEE 802.3 scheme include deterministic performance and the specification of a priority mechanism.

As shown in Figure 1.4.1, work has been completed in several new technology areas including wireless LANs (IEEE 802.11) [5] and Cable Modems (IEEE 802.14). [6]

Due to their wide market acceptance, this section focuses on the details of the IEEE 802.3 (CSMA/CD) and 802.5 (Token Ring) specifications. The section also addresses the Logical Link Control layer and presents an overview of building wiring considerations which would ensure that the building cabling meets the requirements of the various LAN types.

1.4.2   IEEE 802.3 (CSMA/CD) Specifics

1.4.2.1   Frame Structure

As mentioned, the carrier sense multiple access with collision detection (CSMA/CD) method was the first to be specified by the IEEE and was closely modeled after the Digital/Intel/Xerox (DIX) Ethernet specification. Although there are differences between the Ethernet and the 802.3 specifications, manufacturers now typically produce hardware that can support both, so that effectively the two are compatible. Differences in the packet format are resolved in firmware for a particular implementation. We use the terms Ethernet and IEEE 802.3 CSMA/CD interchangeably.

The frame format in the original DIX specification is shown in Figure 1.4.2(a). Frame fields are as follows.

- Preamble — To allow synchronization by the receiving station and to indicate the start of frame, the frame starts with an eight byte sequence, the first seven of which have the format (10101010), and the eighth the format (10101011).
- Source and destination addresses are 48 bits each (a little-used option allows for 16 bits) and have the structure shown in Figure 1.4.2(b) except for a minor variation in the second bit of the address.
- EtherType — The EtherType field (16 bits) allows for the multiplexing of data streams from different higher level protocols and identifies the particular higher level protocol data steam carried

- Consider two stations (A and B) at the ends of an Ethernet network.
- Assume the maximum allowed frame size of 1518 bytes (12144 bits).
- At 10 Mbps, the resulting frame transmission time is 1214.4 μs.
- Assume a 500 m cable; propagation time is thus about 2.2 μs (using propagation speed of .77c, where c is the speed of light).
- This figure shows successful transmission of frame from A to B. Station A starts to send at t = 0 and completes transmission at t = 1214.4 μs; station B starts to receive at t = 2.2 μs and has received the entire frame at t = 1216.6 μs.

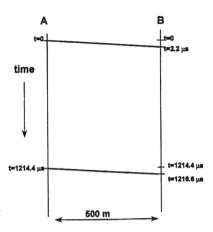

FIGURE 1.4.3   Example of successful frame transmission.

by this frame, e.g., an EtherType of Ox08-00[1] indicates a frame carrying an IP datagram. Values for the EtherType field can be found in [7].

- Data — The Data field carries the service data unit from the higher layer protocol entity and ranges in length from 46 (including an added PAD field if the service data unit is less than 46 bytes) to a maximum of 1500 bytes.
- Frame Check Sequence (FCS) — Finally, a four-byte FCS field is added for error detection purposes.

The IEEE 802.3 frame format is shown in Figure 1.4.2(b). The major difference in format arises from the need to accommodate other MAC specifications under the IEEE umbrella which may have no equivalent of the EtherType field. As a result, this multiplexing capability is included in the next higher layer of the IEEE 802 Reference Model, the LLC layer (see Figure 1.4.1). The method used to provide this additional protocol information is the Subnetwork Access Protocol (SNAP). A SNAP encapsulation is indicated by the LLC layer SSAP and DSAP fields both being set to OxAA. The SNAP header is five bytes long: the first three bytes consist of an organization code, which is assigned by the IEEE; the second two bytes use the EtherType value set from the Ethernet specifications. Using this scheme, the multiplexing service afforded by the EtherType field is available at the LLC layer, independent of the individual MAC layer capabilities. Note that several layers of multiplexing are available at the LLC layer; one provided by the LLC Destination Address/Source Address fields in Figure 1.4.2(b), and the other by the LLC/SNAP fields shown in the figure (which include the EtherType field). Again, when the length of MAC layer data field is less than 46 bytes, a PAD field is added to ensure a minimum data plus PAD field length of 46 bytes. The PAD field consists of an arbitrary array of bits.

1.4.2.2 Sample Frame Transmission

For a transmission media operating at a data rate of 10 Mbps, typical of many 802.3 specifications, Figure 1.4.3 shows the successful transmission of a frame between two stations at the ends of the cable, from station A (shown on the left) to station B (shown on the right). Cable length is assumed to be 500 meters, the approximate maximum length for a number of IEEE 802.3 configurations (per Section 13 of [8]). A frame size of 1518 bytes is assumed, also the maximum as per the IEEE 802.3 specification. From the figure, station A begins transmitting at time t = 0 and some time later the leading edge of the signal

---

[1]This notation indicates a string of bytes (groups of eight bits) with the values of the bytes given in hexadecimal form; thus Ox08-00 represents the two bytes 00001000–00000000.

TABLE 1.4.1 Minimum Propagation Speeds
for Sample Media

Media Type	Minimum Propagation Speed
Coax (10BASE5)	0.77 c
Coax (10BASE2)	0.65 c
Twisted Pair (10BASE-T)	0.585 c

appears at station B. This time is determined by the propagation speed of the signal on the particular media, with the speeds for a number of media shown in Table 1.4.1. Assuming a propagation speed of .77c, where c is the speed of light (3×10^8 m/s), yields a propagation delay of about 2.2 µs for the example in Figure 1.4.3.

The total signal transmission time, neglecting a short initial synchronization period when the preamble and start of frame delimiter are transmitted is

$$\left(1518 \text{ bytes}\right) \times \left(8 \text{ bits/bytes}\right) / 10 \text{ Mbps} = 1214.4 \text{ µs}$$

Thus station A completes transmitting the signal at t = 1214.4 µs and station B begins receiving the signal at t = 2.2 µs and receives the entire signal at time t = 1216.6 µs.

After a brief delay period to allow recovery time for other stations and the physical medium, termed the interframe gap, another frame can be transmitted if available. An interframe gap of 9.6 µs or 96 bit times for a 10 Mbps implementation is specified by the standard. This value is chosen to account for variability in the gap as frames travel over the media and connecting repeaters (discussed below). This variability occurs because two successive frames may experience different bit loss in their preambles. If the first packet experiences greater bit loss than the second, the gap will shrink as the repeater reconstructs the preamble and therefore introduces delay. If the second frame experiences greater bit loss, the gap will expand.

1.4.2.3 Carrier Sense Multiple Access

A simple addition to the above scheme is to require each station to "listen before talking," i.e., require a station to sense the medium to determine if another station's signal is present and defer transmission if this is the case. This situation is shown in Figure 1.4.4 where a third station at the middle of the cable

- **With Carrier Sense Multiple Access (CSMA), Station A would check that the media is idle before sending. If idle, it would generally send (as in last example) and if busy, it would perform a backoff algorithm (persistent or non-persistent).**
- **For example, suppose station C (in middle of network) began transmitting a 1518 byte packet at t = 0. If Station A received a frame to send at t = 300 µs and B at t = 600 µs, both would sense the media busy and perform a backoff algorithm.**

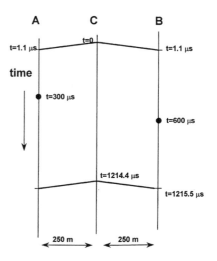

FIGURE 1.4.4   Use of carrier sense multiple access (CSMA).

TABLE 1.4.2   Typical Persistency Algorithms

Persistency Scheme	Description
Non-persistent	• idle $\Rightarrow$ transmit • busy $\Rightarrow$ wait random time and repeat
1-persistent	• idle $\Rightarrow$ transmit • busy $\Rightarrow$ wait until idle then transmit immediately (Note that if 2 or more stations are waiting to transmit, a collision is guaranteed)
p-persistent*	• idle $\Rightarrow$ transmit with probability p and delay one time unit with probability 1-p; time unit is typically the maximum propagation delay • busy $\Rightarrow$ continue to listen until channel is idle and repeat above for idle • delayed one time unit $\Rightarrow$ repeat above for idle

* Issue is choice of p
  • Need to avoid instability under heavy load.
  • If n stations are waiting to send, the expected number transmitting is np. np $> 1 \Rightarrow$ collision is likely.
  • New transmissions will also begin to compete with retries and network will collapse: all stations waiting to transmit, constant collisions, no throughput.
  • Thus np must be <1; but heavy load means p must be small and time will be wasted even on a lightly loaded line, e.g., p = 0.1 $\Rightarrow$ on average, will transmit in tenth interval on an idle line.

begins sending at time t = 0. Due to signal propagation delays, signal reception begins at both A and B at time t = 1.1 µs. In the figure, while sensing the presence of carrier from C, A and B both receive frames from higher layers to transmit but, adhering to the CSMA scheme, defer transmitting until some time after station C's transmission is completed.

For typical CSMA schemes, a number of strategies can be employed to determine when to begin transmitting after deferring to a signal already on the medium. These strategies typically involve invoking one of the persistency schemes shown in Table 1.4.2. The persistency parameter "p" relates to the probability that a station sends its frame immediately after the medium is sensed idle. To obtain maximum channel utilization, the choice of the persistency value, 0.1, 0.2, 0.3, ..., etc. is dependent on the traffic offered by the stations. A low level of traffic would operate best with a persistency value, p, near 1.0 (here, typically only a single station will be ready to send and thus should send immediately with high probability) and a high level of traffic would operate best with a lower value of p (here multiple stations will likely be ready to send and the lower value of p will make it more likely that only one station attempts to transmit). It should be noted that the above retransmission algorithm is not related to the binary exponential backoff algorithm discussed below, associated with resolving collisions. Also of note is that the IEEE 802.3 standard specifies the 1-persistent scheme, ensuring a collision if two or more stations are deferring to an ongoing transmission.

1.4.2.4   Adding Collision Detection

A problem with the CSMA scheme is depicted in Figure 1.4.5, where stations A and B both have something to send at t = 0 and, sensing the medium idle (no carrier) both begin transmission. For example, this case would occur if both have been deferring to another station transmitting on the medium and used the 1-persistent backoff scheme. At some short time later, the signals will collide at stations A and B (and at all other stations on the medium). In this case, no useful information is transferred for the entire transmission time of the frame, approximately 1200 µs for a frame of maximum length.

A solution to this problem is the addition of the collision detection mechanism depicted in Figure 1.4.6. The addition of such a mechanism reduces the wasted transmission time as both stations will stop transmitting upon detection of the collision. Here the stations have the added capability of detecting the occurrence of a "collision" of the two signals on the medium. With this added functionality, the stations can stop transmitting upon detecting collisions and immediately undertake a backoff scheme to allow one station to capture the medium.

- **Problem: Suppose both stations decide to transmit in the time frame t = 0 to t = 2.2 μs. With no collision detection, both will continue to transmit for the entire frame time.**
- **Considerable time will be wasted (over 1200 μs in this example).**
- **Also, a scheme is needed to determine who transmits when the media becomes idle or else, with probability 1, they will collide again, e.g., an n-persistent scheme (e.g., .1-persistent) could be used, i.e., a particular station transmits with probability n (e.g., probability of .1).**

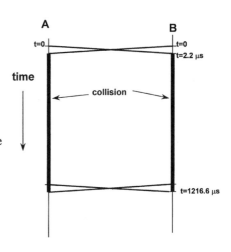

FIGURE 1.4.5   Wasted resources with CSMA.

- **Collision Detection (CD) solves this problem; the stations stop transmitting when they sense a collision. After they stop transmitting, they implement a *binary exponential backoff* scheme (random retransmission interval doubled each time a collision is detected).**
- **CSMA/CD also employs the 1-persistent scheme for stations accessing a media that just became idle. This can raise the chance of collision since multiple stations waiting for the media to become idle will collide but the collision detection scheme will minimize wasted resources.**

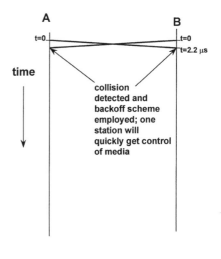

FIGURE 1.4.6   CSMA/CD.

1.4.2.4.1 Collision Backoff Scheme

Two stations, A and B, implementing the CSMA/CD media access control technique are shown in Figure 1.4.7. If, as shown in the figure, they both sense the media idle and begin to transmit at t = 0, a collision will occur and IEEE 802.3 specifies a "truncated binary exponential backoff" randomization scheme so that one of the stations can obtain control of the media. As shown in the figure, with this scheme, first one of two "slots" is chosen randomly by each station to attempt to capture the medium. The slot time is chosen based on factors that include the round trip transmission time between two stations at the ends of the medium, and the time required to detect a collision. It is specified in bit times; the IEEE 802.3 standard specifies a slot time of 512 bit times (51.2 μs for a 10 Mbps system). If a collision

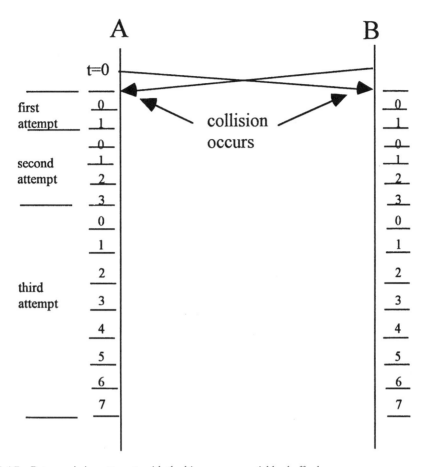

FIGURE 1.4.7 Retransmission attempts with the binary exponential backoff scheme.

occurs again (i.e., they both choose the same slot), one of four slots is chosen in the next attempt; then eight; then sixteen; etc. The number of slots grows in this manner to 2^{10} and truncates at this value. After a total of 16 retransmission attempts fail, this event is reported as an error.

With the CSMA/CD scheme, for reasonable traffic levels, a station should capture the medium in a rather short time, especially when compared to the CSMA scheme. For instance, consider the two stations in Figure 1.4.6 implementing a 1-persistent scheme (the recommended IEEE 802.3 scheme used after deferring to a transmitting station). Each will begin to transmit when station C stops transmitting and thus will suffer a collision on this first transmission. The binary exponential backoff scheme of Figure 1.4.7 will yield the following results.

For the two stations competing for the medium, the following outcomes are equally likely during the first retransmission.

(0,0), i.e., station A picks slot 0 and station B also picks slot 0,
(0,1), i.e., station A picks slot 0 and station B picks slot 1,
(1,0), i.e., station A picks slot 1 and station B picks slot 0,
(1,1), i.e., station A picks slot 1 and station B also picks slot 1.

Two of these outcomes, (0,1) and (1,0), will result in a station capturing the medium, station A in the first of these and station B in the latter. Two of the outcomes, (0,0) and (1,1) result in collisions in slots 0 and 1, respectively. Thus with a probability of _ = 0.5, one of the stations will capture the medium in this first retransmission period.

If there is a collision for the first retransmission, the second retransmission uses 4 slots chosen at random by the stations (numbered 0, 1, 2 and 3 in Figure 1.4.7) , resulting in 16 possible outcomes. Using similar notation as above, these outcomes are

(0,0), (0,1), (0,2), (0,3)
(1,0), (1,1), (1,2), (1,3)
(2,0), (2,1), (2,2), (2,3)
(3,0), (3,1), (3,2), (3,3)

and only four of these, (0,0), (1,1), (2,2), and (3,3), result in collisions yielding a probability of 12/16 or _ for a successful outcome. Thus the probability of exactly two retransmissions is (prob of collision on first retransmission) × (prob of success on second) = _ x _ = 3/8 = .375.

Similarly, the probability of exactly three retransmissions is

$$1/2 \times 1/4 \times 7/8 = 0.109,$$

and that for four is

$$1/2 \times 1/4 \times 1/8 \times 240/256 = 0.0146.$$

The likelihood of more than four transmissions is rather small.

Finally, the average number of retransmissions is

$$\sum_{i=1}^{\infty} i \times \left(\text{prob of i retransmission}\right) = \left(1 \times 0.5\right) + \left(2 \times 0.375\right) + \left(3 \times 0.109\right) + \left(4 \times 0.0146\right)\ldots$$

$$= 0.5 + 0.75 + 0.327 + 0.058 \ldots = 1.635$$

On average then, with two stations competing for the medium, one will capture the medium during the second retransmission attempt. Note that this saves medium resources when compared to Figure 1.4.5 where over 1200 μs are wasted due to the collision and additional time will be spent in some sort of backoff scheme.

It is interesting to note that for 3 stations competing, media capture will occur more quickly. Here, the three stations will collide on the first transmission attempt and the eight possible outcomes for the first retransmission are (0,0,0), (0,0,1), (0,1,0), (0,1,1), (1,0,0), (1,0,1), (1,1,0) and (1,1,1), with only (0,0,0) and (1,1,1) resulting in unsuccessful outcomes. For example, the (0,0,1) outcome will result in media capture: stations A and B will collide in slot 0 and station C will capture the medium in slot 1. Thus the probability of one station capturing the medium in the first retransmission is _, greater than the case with two stations competing. The average number of retransmissions for this case can be shown in the above manner to be on the order of 1.27. The reduced average number of retransmissions in the case of three stations competing is somewhat of an anomaly; for more stations competing, the average number of retransmissions steadily increases. Of course, the average waiting time for a particular station to capture the medium will increase with the number of stations competing for the medium.

1.4.2.5 CSMA/CD System Components

As mentioned, the IEEE 802.3 specifications support multiple media types, including coaxial cable, twisted pair, and fiber. Thus one of the component interfaces will of necessity vary with the media type. For example, the physical media dependent interface for twisted pair differs in a number of respects from that for fiber, including the physical connector (or plug), the electrical vs. optical nature of the interface

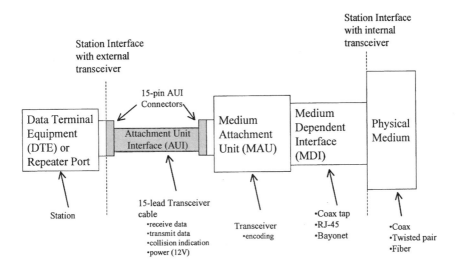

FIGURE 1.4.8 CSMA/CD system components.

and the encoding scheme (translating a logical sequence of bits to the electrical/optical signal on the medium), e.g., Manchester or Non-return to Zero (NRZ) schemes. In a number of IEEE implementations, this interface is remote from the station itself and the IEEE specification provides a media-independent manner of extending the station interface to the media-dependent interface.

This general situation is shown in Figure 1.4.8, where the various functional components that make up the CSMA/CD system are also shown. The Medium Dependent Interface (MDI) is shown on the right of the figure and provides the direct electrical/optical connection. Examples include an 8 pin connector (RJ-45 telephone style jack) for twisted pair, a coaxial cable clamp for coax medium and a spring-loaded bayonet connector (termed an ST connector) for fiber. The Medium Attachment Unit (MAU) shown in the figure, also commonly known as a transceiver, is also specific to the type of medium and performs signal encoding (e.g., Manchester or NRZ). Transceivers also contain a jabber protection circuit that protects the network from a station or transceiver that is transmitting frames whose length exceeds the maximum allowed.

The interconnection of the station to the remote transceiver is accomplished by the Attachment Unit Interface (AUI), also termed the transceiver cable. This 15-lead cable is media independent and carries data receive, data transmit and collision detection signals along with power from the station to the transceiver. The maximum length of the transceiver cable is 50 m. The associated 15 pin AIU connector is commonly found on Ethernet station cards.

It is important to note that many implementations include the transceiver as part of the Ethernet station card. For example, twisted pair implementations use this scheme. In this case, only the media-dependent interface in visible on the station Ethernet card; e.g., the RJ-45 type interface is commonly found on interface cards connecting to twisted pair medium.

1.4.2.6 Example Implementations

IEEE 802.3 standards are characterized by a shorthand notation to facilitate their description. The notation (e.g., 10BASE5) is composed of three elements as shown in Figure 1.4.9, indicating the transmission rate of the system in Mbps, the modulation scheme, baseband or broadband, and the maximum length of the segment in hundreds of meters. With standards adopted more recently, such as 10BASE-T, the IEEE has been more descriptive with its notation. For example, the "T" in the 10BASE-T notation is short for "twisted-pair wiring."

This section describes the most commonly implemented versions of the IEEE 802.3 specification.

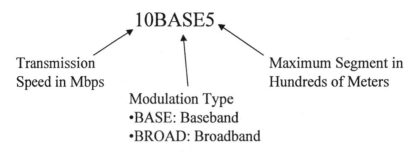

FIGURE 1.4.9   IEEE 802.3 nomenclature.

1.4.2.6.1   10BASE5

10BASE5 was the first version of the IEEE specification to be developed and it most closely resembles the earlier DIX versions 1 and 2. [4] The 10BASE5 specification employs a "thick Ethernet" 50-ohm coaxial cable. While this cable is difficult and relatively expensive to install, it provides advantages over other implementations in terms of distance and the number of terminations permitted for each segment.

This specification uses the standard outboard transceiver option discussed above: the station adapter board connects to the standard Attachment Unit Interface (AUI) cable; this in turn is connected to the transceiver which is connected to the Ethernet trunk cable via a "vampire" tap. Up to 100 devices can be placed on a 500-meter segment, with a maximum of 1024 devices on a multi-segment network, discussed below.

1.4.2.6.2   10BASE2

10BASE2 (also known as "thin Ethernet" or "Cheapernet") employs a thin flexible coaxial cable (RG-58). In earlier implementations, the transceiver functions were onboard the station and the connection to the station was by means of a media-dependent BNC "T" connector. To provide the flexibility to use the station board for either the 10BASE 5 or 10BASE2 systems, station boards have been developed which provide options for both external 10BASE5 "vampire" taps and 10BASE2 BNC connectors. Board manufacturers commonly provide boards with built-in transceivers that can be switched on or off for a particular application.

The standard 10BASE2 LAN can support only 30 terminations on each coaxial cable segment of 185 meters. While this may seem like a constraint, it is often adequate for most work area environments. Where a requirement exists for interconnecting multiple work areas, or work areas with multiple 10BASE2 segments, a backbone 10BASE5 segment can be employed to provide intersegment connectivity. Table 3 highlights the differences between the 10BASE5 and 10BASE2 systems.

1.4.2.6.3   1BASE5

This standard approach was contributed by AT&T to accommodate its earlier Starlan products. It operates at 1 Mbps, and as such is often most useful for small work areas or low-traffic environments. 1BASE5

TABLE 1.4.3   802.3 10BASE5/10BASE2 Comparison

	10BASE5	10BASE2
Common name	802.3 "Ethernet"	Cheapernet, THIN Ethernet, THINWIRE Ethernet, etc.
Type of cable	50 Ω Thick dual shield	50 Ω RG-58
Maximum segment length	500 m.	185 m.
Spacing of devices on cable	2.5 m. minimum	0.5 m. minimum
Maximum number of taps for a segment	100	30
Maximum number of full repeaters in a path between two stations	2	2
Type of taps	Vampire or coax	BNC "T" connector for "daisy chaining"

also employs inexpensive twisted-pair wire interconnected through a hierarchical system of concentrator hubs. The hubs emulate a bus configuration by broadcasting data and collision information on all ports.

1.4.2.6.4   10BASE-T

One of the most important developments in the IEEE 802.3 area was the specification of the 10 Mbps unshielded twisted-pair (UTP) Ethernet system, 10BASE-T. Virtually every vendor active in the Ethernet market now offers 10BASE-T products.

Like the 1BASE5 specification, this system uses a hub concentrator to interconnect multiple stations and emulate bus operation. These implementations are limited to 100-meter segments due to the greater attenuation and signaling difficulties of twisted pair. This does not present any unusual problems since these connections only reach to the communications closet. From there, fiber and coax segments can be used to concatenate and extend the LAN system. 10BASE-T systems use one twisted pair for transmitting data and a separate pair for receiving. "Collisions" are detected by sensing the simultaneous occurrence of a signal on both the transmit and receive pairs.

It is imperative, however, that organizations planning these networks have their existing twisted-pair wire certified for both attenuation and capacitance before making any assumptions on its reuseabilty.

1.4.2.6.5   10BASE-FL

The 10BASE-FL specification allows for the use of two fiber optic cables as the medium, one for signal transmission and one for reception. Such a medium allows advantages of greater distances, e.g., the standard allows segment lengths of up to 2000 meters, evolution to higher transmission speed, and isolation from electromagnetic radiation. The system components are identical to those shown in Figure 1.4.8 with the use of a fiber-optic Medium Attachment Unit (MAU).

1.4.2.6.6   10BROAD36

The 10BROAD36 implementation uses much of the same hardware as the baseband implementations. The specification enables an organization to use its existing workstation boards for connection to either a baseband or broadband system. The essential difference is the substitution of a broadband electronics unit and a passive broadband tap for the baseband MAU. The primary function of the broadband electronics unit is to create the frequency-derived data channels and to monitor for collisions. It also converts the signals from the baseband-coded signal on the AUI to the analog signal necessary on the broadband channel. Workstations can be placed up to 1800 meters from the "head-end" of the broadband cable plant. By placing the head-end in the center of the configuration, workstations can be installed up to 3600 meters from each other. In recent years, this standard has been less frequently used.

1.4.2.7   Topology Extensions

1.4.2.7.1   Repeaters

Repeaters regenerate the signals from one LAN segment for retransmission to all the others. The earliest repeaters were simple two-port devices that linked a couple of coaxial cable segments. Later, repeaters evolved to multiport devices deployed as the hub of a star topology. Since with repeaters, all segments are part of a unified LAN, the nature of the shared channel must be preserved by broadcasting all information to all attached devices. An aspect of these repeaters is that they must be capable of retransmitting collisions as well as data frames. In the case of IEEE 802.3 CSMA/CD LANs, physical LAN segment connection standards for repeaters are well developed and mature. The latest specifications for implementation of 10BASE-T repeaters are contained in the IEEE 802.3 specifications (see Section 9 of [8]).

In addition to the functions described above, repeaters can provide an optional "partitioning" feature between segments. This function is designed to address an abnormal situation such as a cable break or network card failure. Thus, if conditions on a given segment are causing an extensive proliferation of collisions, the rest of the LAN can be protected from this anomaly. The repeater will count the number of collisions from the source segment and when excessive, isolate these from transmission to the next segment.

1.4.2.7.2 Switched Hubs (Ethernet Switches)

In Ethernet switching, the interconnecting device (termed a switching hub) has intelligence to use the MAC-layer address of a received frame to determine the specific port on which the destination station is attached and transmit the frame on only that port. No other stations are aware of the frame. If frames arrive destined for a busy port, that port can momentarily holds them in its buffer; the size of port buffers differs by vendor from a few hundred to more than a thousand packets. When the busy port becomes free, frames are released from the buffer and sent to the port. This mechanism works well unless the buffer overflows, in which case packets are lost. To avoid this, some vendors offer a throttling capability; when a port's buffer begins to fill up, the hub begins to transmit packets back to the workstations. This effectively stops the stations from transmitting and relieves the congested state.

Some LAN switching products offer a choice of packet-switching modes, e.g., fast forward and "store and forward." These modes affect the amount of packet latency, the time from which the first byte of a packet is received until that byte is forwarded. Each mode reads a certain number of bytes of a packet before forwarding. This creates a trade-off among latency, collision, and error detection. The greater number of bytes read, the greater the latency, but the fewer errored or collision terminated frames propagated through the network. The fast-forward mode passes packets shortly after receiving the Destination Address portion of the Ethernet frame (see Figure 1.4.2) and, based on this field, determines the appropriate destination port. In this mode, typical latencies are on the order of tens of μs for a 10 Mbps Ethernet system. Store-and-forward mode receives entire frames and performs error detection via the FCS field (see Figure 1.4.2) before forwarding, resulting in increased latency but maximizing frame error detection. Here, for a maximum length frame, latency will be greater than 1200 μs. Some LAN switching devices support only 1 address per port, while others support 1500 or more. Some devices are capable of dynamically learning port addresses and allowing or disallowing new port addresses. Disallowing new port addresses enhances hub security; in ignoring new port addresses, the corresponding port is disabled, preventing unauthorized access. These and other techniques used in conjunction with port switching enhance overall network performance by eliminating the contention problem that occurs in shared Ethernet networks.

1.4.2.7.3 Multi-segment Guidelines

The IEEE 802.3 specifications provide guidelines for the number and types of segments that can be interconnected via repeaters and switch hubs. A number of example configurations are presented which would be typical of many implementations. For example, one possible configuration that taxes the CSMA/CD configuration guidelines has a distance of about 1800 meters between stations, with three 500 meter segments and two shorter segments interconnected by four repeaters. For more complex cases, the specifications provide guidelines for calculating system parameters such as the round-trip delay time and interframe gap shrinkage to ensure that these parameters are within allowed limits.

1.4.2.8 Higher Speed Extensions (100 Mbps and 1000 Mbps)

1.4.2.8.1 Fast Ethernet (100 Mbps) Overview

100 Mbps CSMA/CD operation, also termed Fast Ethernet, is specified in the IEEE 802.3u supplement to the standard. While attaining a ten-fold increase in transmission speed, other important aspects, including the frame format and the CSMA/CD access control scheme, remain unchanged from a 10 Mbps system, making the transition to higher speeds straightforward for network managers. The 100 Mbps specification uses Ethernet's traditional CSMA/CD protocol and is designed to work with existing medium types, including Category 3 and Category 5 twisted-pair and fiber media. In addition, Fast Ethernet will look identical to lower-speed Ethernet from the LLC layer upward. Since Fast Ethernet significantly leverages existing Ethernet technology, network managers will be able to use their existing knowledge base to manage and maintain Fast Ethernet networks.

The Fast Ethernet specifications include mechanisms for Auto-Negotiation of the media speed. This makes it possible to provide dual-speed Ethernet interfaces that can be installed and run automatically at either 10 Mbps or 100 Mbps.

There are three media varieties that have been specified for transmitting 100-Mbps Ethernet signals: 100BASE-T4, 100BASE-TX, and 100BASE-FX. The third part of the identifier provides an indication of the segment type.

"T4" is a twisted-pair segment that uses four pairs of telephone-grade twisted-pair wire.
"TX" segment type is a twisted-pair segment that uses two pairs of wires and is based on data-grade twisted-pair physical medium standards, covered later in this section.
"FX" segment type is a fiber-optic link segment that uses two strands of fiber cable and is based on the fiber optic physical medium standard developed by ANSI.

The TX and FX medium standards are collectively known as 100BASE-X. The 100BASE-TX and 100BASE-FX media standards used in Fast Ethernet are both adopted from physical media standards first developed by ANSI for the Fiber Distributed Data Interface (FDDI) LAN standard (ANSI standard X3T9.5), and are widely used in FDDI LANs. Rather than "re-inventing the wheel" when it came to signaling at 100 Mbps, the Fast Ethernet standard adapted these two ANSI media standards for use in the new Fast Ethernet medium specifications. The T4 standard was also provided to make it possible to use lower-quality twisted-pair wire for 100 Mbps Ethernet signals.

1.4.2.8.2   *Gigabit Ethernet (1000 Mbps) Overview*
Gigabit Ethernet, under the auspices of the IEEE 802.z working group, builds on the CSMA/CD MAC scheme and increases the transmission speed to 1000 Mbps. A key feature of Fast Ethernet implementations is the autoconfiguration capability, and Gigabit Ethernet solutions providing 10/100/1000 Mbps operation allow comparable features.

It should be noted that, as in the case for Fast Ethernet, a number of challenges involved in achieving rapid time to market for Gigabit Ethernet were resolved by merging existing technologies:

1. IEEE 802.3 CSMA/CD and
2. ANSI X3T11 Fibre Channel; Fibre Channel encoding/decoding integrated circuits (ICs) and optical components were readily available and optimized for high performance at relatively low cost.

Leveraging these two technologies meant that the Gigabit Ethernet standard could take advantage of the existing, proven high-speed physical interface technology of Fibre Channel while maintaining the IEEE 802.3 Ethernet frame format, backward compatibility for installed media, and use of CSMA/CD. This strategy helped minimize complexity and resulted in a technology that could be quickly standardized. Figure 1.4.10 shows how key components from each technology have been leveraged to form Gigabit Ethernet. As a result the Gigabit Ethernet standard based on fiber optics for the MAC and Physical layers has progressed rapidly. Unshielded twisted-pair media did not have the advantage of a proven, existing technology base and the standards remained in further development; this work is under the auspices of the IEEE 802.3 Working Group and referred to as 1000BASE-T.

Physical Layer Characteristics of Gigabit Ethernet
The initial Gigabit Ethernet specification from the IEEE 802.3z working group calls for three transmission media: single-mode and multimode fiber and balanced shielded 150-ohm copper cable. There are two supported types of multimode fiber: 62.5-micron and 50-micron diameter fibers. The IEEE 802.3ab committee is examining the use of unshielded twisted pair (UTP) cable for Gigabit Ethernet transmission (1000BASE-T). The distances for the media supported under the IEEE 802.3z standard and those projected for the IEEE 802.3ab are summarized in Figure 1.4.11.

Fiber Optic Media (1000Base-SX and 1000Base-LX)
As mentioned, the Fibre Channel physical medium dependent specification was employed for Gigabit Ethernet to speed standardization. This standard provides 1.062 gigabaud in full duplex mode and Gigabit Ethernet will increase this rate to 1.25 gigabaud with an 8B/10B encoding scheme allowing a data transmission rate of 1000 Mbps. In addition, the connector type for Fibre Channel was also specified for both single-mode and multimode fiber.

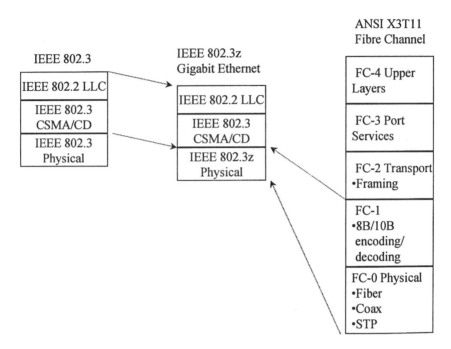

FIGURE 1.4.10   Gigabit Ethernet and the ANSI Fibre Channel Standard.

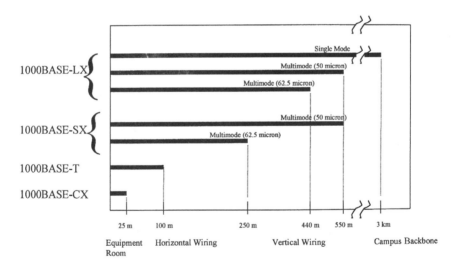

FIGURE 1.4.11   Distance specifications for gigabit ethernet media.

The standard supports two laser types, a short-wave laser type (termed 1000Base-SX) and a long-wave laser type (termed 1000Base-LX). Both short-wave and long-wave lasers are supported over multimode fiber. There is no support for short-wave laser over single-mode fiber. The key issues between the use of long-wave and short-wave laser technologies are cost and distance. Short-wave lasers are readily available since variations of these lasers are used in compact-disc technology. Long-wave lasers take advantage of attenuation dips at longer wavelengths in the cable and suffer lower attenuation. The net result is that short-wave lasers will cost less, but traverse a shorter distance. In contrast, long-wave lasers will be more expensive but will traverse longer distances.

The 62.5-micron fiber is typically seen in vertical campus and building cable plants and has been used for Ethernet, Fast Ethernet, and FDDI backbone traffic. However, this type of fiber has a lower modal bandwidth (the ability of the cable to transmit light), especially with short-wave lasers. This means that short-wave lasers over 62.5-micron will generally traverse shorter distances. The 50-micron fiber has significantly better modal bandwidth characteristics and will be able to traverse longer distances with short-wave lasers relative to 62.5-micron fiber.

150-Ohm Balanced Shielded Copper Cable (1000Base-CX)

For shorter cable runs (of 25 meters or less), Gigabit Ethernet will allow transmission over a new type of special balanced shielded 150-ohm cable (termed 1000Base-CX). Because of a distance limitation of 25 meters, this cable will likely have limited use.

1.4.2.9 Full Duplex Operation

The previous discussion has focused on half-duplex operation, where only a single communications channel was available and thus data could be transmitted in only one direction at a time. The CSMA/CD Media Access Control mechanism was required to determine which station could use the single channel. Full-duplex is an optional point-to-point mode of operation between a pair of devices allowing simultaneous communication between the devices and thus a doubling of aggregate capacity. Two separate communications channels are needed in this case to allow both stations to simultaneously transmit and receive. Thus the allowed physical media for this operation are only those with the capability of supporting two simultaneous channels, e.g., 10BASE-T provides independent transmit and receive data paths that can be simultaneously active. Full duplex operation cannot be supported on coaxial cable systems since they do not provide independent transmit and receive data paths. The optional full duplex mode of operation is specified by the 802.3x supplement to the standards.

1.4.3 IEEE 802.2 Logical Link Control Layer

The IEEE 802.2 Logical Link Control (LLC) layer specifications [9] include those Data Link Layer functions that are common to all 802 LAN MAC sublayer alternatives. The LLC frame format is shown in Figure 1.4.2.

Three basic types of service are defined in the standard.

Type 1 (Connectionless)

This service provides a best-effort delivery mechanism between origin and destination nodes. No call or logical circuit establishment procedures are invoked. Each frame is treated as an independent entity by the network. The type of frame used to provide this service is the unnumbered type; no flow control or acknowledgments are provided with this service. If the packet does not arrive at the destination, it is the responsibility of higher layers to resolve the problem through time-outs and retransmission. This type of service would be provided to the IP network layer protocol.

Type 2 (Connection Oriented)

Many wide area network protocols require that a logical circuit or call be established for the duration of the exchange between the origin and destination nodes. Packets usually travel in sequence over this logical circuit and are not routed as independent entities. LLC Type 2 provides this type of service. The service involves a number of control frames to manage the logical circuit (establishment, disconnection) and numbered frames for information transfer. Positive acknowledgments and flow control mechanisms based on this frame numbering are an integral part of this service. LLC Type 2 is commonly found in implementations of IBM's Systems Network Architecture (SNA).

Type 3 (Acknowledged Connectionless)

No circuit is established in this service variation, but acknowledgments are required from the destination node. This type of service adds additional reliability to Type 1, but without the overhead of Type 2.

These LLC alternatives are summarized in Table 1.4.4.

TABLE 1.4.4   Summary of Logical Link Control Alternatives

Service Type	Type 1	Type 2	Type 3
Description	Connectionless	Connection	Acknowledged Connectionless
Acknowledgments	No	Yes	Yes
Error recovery	No	Yes	Yes
Flow control	No	Yes	No

1.4.4   Building Cabling Specifications

The major components of a building cabling architecture that apply to the implementation of LANs are shown in Figure 1.4.12 and include the following.

- Equipment room — This location houses major data center processing and communications equipment for the building including servers, routers, and LAN switches. For a campus environment involving a number of buildings, one such location would serve as a data processing and communications center for the campus; other equipment rooms on the campus would serve specific buildings.
- Telecommunications closet — This is an area, typically located on each floor of a building, that houses data and telecommunications equipment providing wiring concentration, cross-connect and hubbing functions.
- Backbone cabling — This cabling provides connectivity between equipment in the equipment room and the telecommunications closets. It includes vertical connections between the floors and connections between buildings.
- Horizontal cabling — This cabling extends from the telecommunications closet to the individual work areas on the building floors.

The American National Standards Institute (ANSI), the Electronics Industry Association (EIA) and the Telecommunications Industry Association (TIA) develop specifications for commercial building

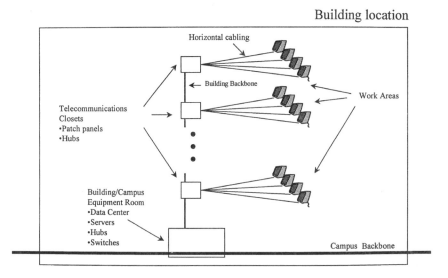

FIGURE 1.4.12   Architecture for building/campus telecommunications cabling.

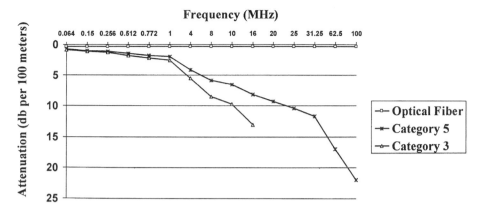

FIGURE 1.4.13 Attenuation of various media — EIA 568A.

cabling standards. This set of standards, referred to as ANSI/EIA/TIA 568A, defines the installation practices, certification, and physical, electrical, and optical characteristics for various physical media such as unshielded twisted-pair and fiber-optic cable [10]. The intent of the standard is to provide a guideline by which a cabling system can be designed and implemented as part of the overall design of a new building, even if the systems that the cabling must support are not yet defined. It guides the user toward the selection of cabling that will support current and future communications needs. The 568A standard and other EIA standards are recognized by architectural and engineering firms as definitive guidelines to use during the design phase of a building.

The 568A standard defines the requirements of a cabling system on a generic level that is appropriate to a commercial environment. The standard allows certain options, such as the use of various cabling media, including the following.

- 100-ohm unshielded twisted-pair cable in a four-pair configuration;
- 150-ohm shielded twisted-pair cable in a two-pair configuration;
- 50-ohm coaxial cable (not recommended for new installations); and
- 62.5 micron optical fiber cable in a two-pair configuration.

These cables exhibit performance that varies greatly depending on the frequency of the signal that is carried. For example, at Mbps speeds, a signal on a twisted-pair cable deteriorates in quality over a fairly short distance. The 568A standard provides performance criteria for the above cabling which must be met to be classified as 568A compliant. A summary of the attenuation limits specified by 568A for certain cable types is shown in Figure 1.4.13.

References

1. ISO/IEC TR 8802-1, Overview of LAN/MAN Standards.
2. IEEE Std. 802, Overview and Architecture.
3. ISO/IEC 7498-1: 1994, Open Systems Interconnection Basic Reference Model.
4. *The Ethernet, A Local Area Network, Data Link Layer and Physical Layer Specifications,* Digital Equipment Corp., Maynard, MA; Intel Corp., Santa Clara, CA; Xerox Corp., Stamford, CT; Version 1.0, Sept. 30, 1980, and Version 2.0, Nov. 1982.
5. IEEE 802.11, Wireless LAN Medium Access Control (MAC) Sublayer and Physical Layer Specifications.
6. IEEE 802.14, Standard Protocol for Cable-TV Based Broadband Communication Network.
7. RFC 1700, *ASSIGNED NUMBERS,* J. Reynolds, J. Postel. Oct. 1994.

8. International Standard ISO/IEC 8802-3: 1996(E), ANSI/IEEE Std 802.3, 1996 Edition, *Part 3: Carrier sense multiple access with collision detection (CSMA/CD) access method and physical layer specifications.*
9. ANSI/IEEE Std 802.2 [ISO/IEC 8802-3], Logical Link Control.
10. ANSI/EIA/TIA 568A, Commercial Building Telecommunications Cabling.

1.5 Token Ring Specifics

John Amoss

The IEEE token-passing ring local area network architecture, first approved in 1985 by the IEEE 802.5 Working Committee, has undergone a variety of additions and modifications and its essential specifications are complete. The token ring architecture specification addresses the media access control (MAC) layer and the physical layers shown in Figure 1.5.1 and the current version is defined in [1].

1.5.1 Topology

An IEEE 802.5 token ring consists of a star wired system of stations with each station connected by lobe cabling to a trunk coupling unit (TCU) on a concentrator. Figure 1.5.1 shows a sample token ring configuration consisting of two concentrators with three stations attached to each concentrator. The TCU provides a mechanism for insertion of a station into the ring and removal of the station from the ring. For example, note in the figure that station 5 is in bypass mode and is not participating in the ring operation. Concentrators can support multiple TCUs and are in turn serially connected via a trunk cable between ring in and ring out ports.

In addition to the architecture shown in the figure, a supplement to the IEEE 802.5 standard [2], defines a dual ring architecture intended for applications that require very high availability and recovery from media and station failures. This architecture uses two separate counter-rotating token passing rings; counter-rotating implies that information flow is in opposite directions on the two rings. Thus in addition to the ring shown in Figure 1.5.1, there is a separate independent secondary ring. The primary ring is normally the operational ring, with the secondary ring becoming operational in case of ring element

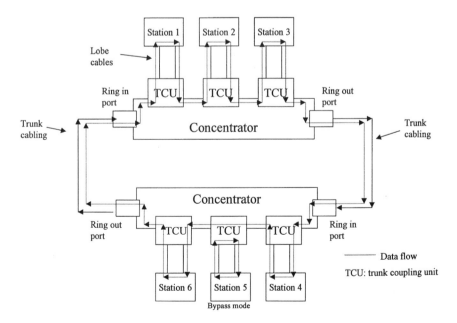

FIGURE 1.5.1 Examples of token passing rings confirmation.

failures. A typical reconfiguration to account for a failure condition involves wrapping the signal around from the primary ring to the secondary ring.

1.5.2   Station Attachment

As mentioned, each TCU provides insertion to or bypass from the ring for the station. Prior to insertion, while in the bypass mode, a station can perform self tests of the attaching lobe cabling and its station circuitry to assure proper operation. To attach to the ring, the station sends a signal over the lobe cable and the TCU switches from bypass to insert mode.

1.5.3   Token Ring Operation

Unlike CSMA/CD operation, each station regenerates and repeats each bit. Information on the token ring is transferred sequentially from one inserted station to the next. A given station transfers information onto the ring, where the information circulates from one station to the next. The addressed destination station copies the information and the station that transmitted the information removes it from the ring. Note also that the concentrators shown in Figure 1.5.1 may be passive (with no active elements) or active (performing a repeater function).

   A station gains the right to transmit its information onto the medium when it detects a "token," a special media access control signal, passing on the medium. The token format is discussed in a later section. Any station may "capture" the token by modifying it slightly and appending additional fields including information to be transferred to a destination station. At the completion of the information transfer, the station initiates a new token, which provides other stations with access to the media.

   The current specification includes an "Early Token Release" feature intended to make more efficient use of the available bandwidth on physically large rings operating with particularly small frames. In earlier versions of the token-passing protocol, a new free token could not be released by the sending station until it recognized the address in its own frame coming back around the ring to itself. If the frame was small, and the ring was large, there was a great deal of wasted time on the medium. Using Early Token Release, a sending station can release the free token immediately upon completing its transmission. The unused capacity on the ring can now be used by other stations. When coupled with the 16 Mbps ring operation, this new feature has significant advantages in terms of performance.

1.5.4   Priority Feature

An important feature of the IEEE token ring architecture, not found in the CSMA/CD architecture, is the support of multiple levels of priority. These priority levels are available for use by applications with varying classes of service needs. For example, the standard mentions real-time voice and network management as potentially high-priority applications. The priority mechanism operates in such a way that fairness is obtained for all stations within a priority level. This is accomplished by having the same station that raised the service priority level of the ring return the ring to the original service priority.

   As mentioned, station access to the physical medium is controlled by passing a token around the ring. Operation of the priority feature is outlined in Figure 1.5.2 and is governed by the following parameters.

- P_{msg} = Priority of the frame to be transmitted by the station
- P_{rcvd} = Priority of the received token or frame (contained in the header)
- R_{rcvd} = Reservation level of the received token or frame (contained in the header)

   As a token passes a station it has the opportunity to transmit one or more frames or place a request for a token of the appropriate priority. If a frame from another station passes the station, it may also place a request for a token of the appropriate priority. As shown in the figure, the station will send a frame if a token is received and the priority of the waiting frame is greater than or equal to the priority of the token. If the token has higher priority than the frame or if it is a frame from another station that is received, the station will attempt a reservation.

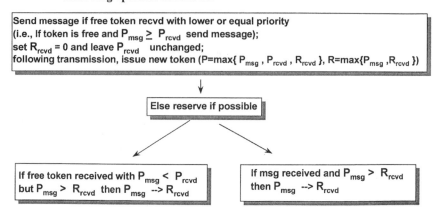

Let P_{msg} = priority of msg to be transmitted by station
P_{rcvd} = priority of rcvd token or message
R_{rcvd} = reservation level of rcvd token or message
Then ring operates as follows:

Send message if free token recvd with lower or equal priority
(i.e., If token is free and $P_{msg} \geq P_{rcvd}$ send message);
set $R_{rcvd} = 0$ and leave P_{rcvd} unchanged;
following transmission, issue new token (P=max{ P_{msg} , P_{rcvd} , R_{rcvd} }, R=max{P_{msg} ,R_{rcvd} })

Else reserve if possible

If free token received with $P_{msg} < P_{rcvd}$
but $P_{msg} > R_{rcvd}$ then P_{msg} --> R_{rcvd}

If msg received and $P_{msg} > R_{rcvd}$
then P_{msg} --> R_{rcvd}

FIGURE 1.5.2   802.5 Token passing ring — priority and reservation scheme.

1.5.5   Management

Special stations are defined on the ring for system management purposes. Termed server stations, these stations act as data collection points on the ring, gathering information from the ring stations on any errors encountered, such as lost tokens, and lost or errored frames.

The server stations interact with a ring management system to provide the information necessary to manage the ring.

1.5.6   Physical Attributes

Distances covered by a token ring network are specified in terms of what is termed ring segment length, which is the transmission path between repeaters (i.e., stations or active concentrators). Since an active concentrator performs a repeater function, it can be seen from Figure 1.5.1 that allowable lobe cable lengths are longer for active than passive concentrators. For example, the token ring specification (Annex B) presents sample designs with lobe lengths on the order of 65 meters for passive concentrators and 200 meters for active concentrators.

The initial version of the IEEE token-passing ring was a 4 Mbps implementation which ran on shielded twisted-pair (STP) wire. The current token ring specification supports transmission rates of 4 and 16 Mbps and both STP and unshielded twisted pair (UTP) media.

Based on a number of factors such as the media type, data rate, and concentrator type, the number of station on a ring may be up to 250.

1.5.7   Formats

1.5.7.1   Tokens

The token, or control signal, has the format shown in Figure 1.5.3. This signal is the means of passing the right to transmit from station to station on the ring. As shown in the figure, the token consists of three octets.

The starting delimiter (SD) and ending delimiter (ED) each consist of a fixed sequence of symbols, some of which deliberately violate the differential Manchester encoding scheme (see Figure 1.5.4). Because of these violations, these delimiters cannot be part of the fill sequence, defined as a stream of validly encoded symbols, sent on the media between frames. Thus false SD and ED indications will not occur.

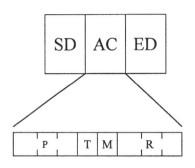

SD = Starting Delimiter (1 octet)
AC = Access Control (1 octet)
ED = Ending Delimiter (1 octet)

P = priority bits (lowest 000 to highest 111)
T = token bit (0 in a token and 1 in a frame)
M = monitor bit (prevent token or frames from
    continuous circulation)
R = reservation bits (for reservation requests -
    lowest 000 to highest 111)

FIGURE 1.5.3 Token format.

- Each half of the bit cycle is termed a signal element.
- Value to be transmitted, termed a symbol, is encoded as follows.
  - Data-zero: polarity of the leading signal element is opposite to the trailing signal element of preceding symbol and transition occurs at mid-symbol (thus transitions at symbol boundary as well as mid-symbol)
  - Data-one: polarity of the leading signal element is the same as the trailing signal element of preceding symbol and transition occurs at mid-symbol (thus no transitions at symbol boundary)
- IEEE 802.5 uses two code violations (no mid-symbol transition) to encode special symbols, termed J and K. These are used for special purposes, e.g., Ending and Starting Delimiters.

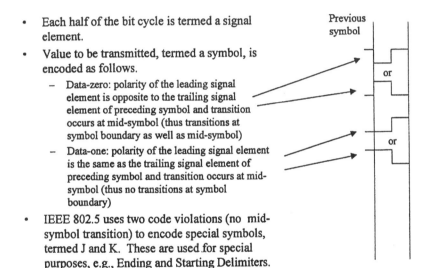

FIGURE 1.5.4 Differential Manchester encoding.

The access control field is a key field associated with many of the services provided by a token ring network.

- A 3-bit priority field indicates the priority of the token. When frames are eventually sent on the media, they carry the same priority as the token by which they gained access to the network.
- A 1-bit field indicates whether this is a token (value of 0), or the beginning of a frame transmitted by a station (value of 1).
- A 1-bit monitor field, set initially to 0 and set to 1 by a monitoring station, is used to prevent certain tokens or data frames from continuously circulating around the ring.
- A 3-bit reservation field allows stations with higher priority protocol data units to gain quicker access to the media.

1.5.7.2 Frames

Frames are the resulting messages sent between stations. It is important to note that two very different types of frames may be sent. One carries LLC messages, or user data, containing higher level protocol

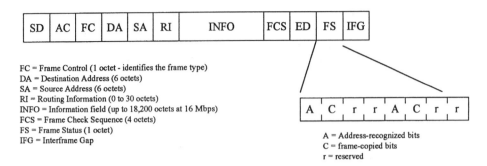

FC = Frame Control (1 octet - identifies the frame type)
DA = Destination Address (6 octets)
SA = Source Address (6 octets)
RI = Routing Information (0 to 30 octets)
INFO = Information field (up to 18,200 octets at 16 Mbps)
FCS = Frame Check Sequence (4 octets)
FS = Frame Status (1 octet)
IFG = Interframe Gap

A = Address-recognized bits
C = frame-copied bits
r = reserved

FIGURE 1.5.5   Frame format.

exchanges as described in the MAC section. A second type of frame, described in detail in the IEEE 802.5 standard carries MAC messages used to manage the MAC layer.

The format of these frames is shown in Figure 1.5.5.

In addition to those fields contained in a token, these frames contain additional fields, including the following.

- A Frame Control (FC) field indicates the frame type – MAC (either a MAC protocol frame or a MAC management frame) or LLC.
- A 48-bit source address (SA) field and a 48-bit destination address (DA) field identify the sending station and the receiving station. The destination address can represent an individual station, a group of stations or all stations (broadcast). The source address field contains 1 bit which is used to indicate whether routing information is included in the frame, as would be the case for a source-routed network.
- A routing information (RI) field is used in the case of source routing to specify the route for the frame through a bridged network.
- The information field contains octets destined for the MAC Protocol or Management entity or the LLC entity. A 32-bit frame check sequence field is used for error detection.
- A frame status (FS) field is used to provide an indication as to whether or not a frame reached its destination and whether it was successfully read from the media. "A" bits are set by a destination station to indicate that it has recognized its address in the DA field and "C" bits are set to indicate that it has successfully copied the frame.
- Finally, an interframe gap (IFG) field is added to account for variability in the gap between frames.

References

1. *Token ring access method and physical layer specifications*, ISO/IEC 8802-5 : 1998, ANSI/IEEE Standard 802.5, 1998 Edition.
2. *Recommended practice for dual ring operation with wrapback reconfiguration*, IEEE Standard 802.5c-1991, Supplement to 802.5

1.6   Summary

This section presented the current trends in voice and data networks regarding voice services; the evolution from the connection oriented networks to the connectionless networks, and unification between voice and data networks.

An Intelligent Network (IN) is a telephone network with a "service specific" architecture. It evolved in an Advanced Intelligent Network (AIN), which has a "service independent" architecture. This means that a given part of a phone number won't trigger a request for a specific service, but it can be interpreted differently by various services, depending on different factors such as time of day, caller identity, and type of call.

CTI is a technology platform that merges voice and data services at the functional level to add tangible benefits to business applications. In fact, CTI is a new application for pre-existing technologies. The most important benefit CTI brings to the corporate world is the potential to reduce operating expenses.

The low cost of the public Internet resulted in a great increase of data traffic in the last couple of years. Besides this, the quality of service (QoS) has been considerably improved, and offered acceptable delays for traffic less tolerant to delays. This raised a considerable interest in transporting packetized voice (not tolerant to delays) over data, in particular, IP networks. The support for voice communications over the Internet Protocol is usually called Voice over IP (VoIP).

Local area networks don't lose their importance for users independently of how much bandwidth is available in the wide area. They serve as local connectors between end-user-devices, server farms, storage farms, and the ingress/engress nodes of wide area networks. Their limitation to data is changing by supporting multimedia applications up to the desktop. Step-by-step, throughput capabilities could be improved; in this respect, Giga Ethernet seems to be the winner for high-speed LAN technology.

2

Intranets

Teresa Piliouras
TCR, Inc.

Andrew Resnick
Citicorp

John Braun

Endre Sara
Goldman, Sachs & Co.

Karen M. Freundlich
TCR, Inc.

Dermot Murray
Iona College

Mihir Parikh
Polytechnic University

Introduction

Teresa Piliouras

The Internet started as a technological revolution, designed to protect national interests by ensuring redundancy and resiliency in governmental networks, particularly in time of war. It has spawned world-wide cultural revolution, fostering universal communication exchange with limitless geographic, time, and subject matter boundaries. The extent and ease of the Internet's adoption has had profound implications on all — including personal, business, and governmental — aspects of life. There is no place on earth that cannot be reached by the Internet.

In this chapter, we review the basic technological underpinnings of the Internet and discuss why it is so flexible. As we explore the evolution of the Internet, which continues at an ever-increasing pace, we also examine corresponding effects on communication paradigms, particularly in a business context.

Once caution was the order of the day. Now, businesses small and large alike are racing to join the Internet bandwagon and to have a "Web" presence.

In a dynamic environment that changes faster than words can be put into print, we can only hope to scratch the surface on the Internet's development and major trends in this Chapter.

2.1   Internet and Intranet Management Concepts

Teresa Piliouras

2.1.1   Management Overview of the Internet and Intranets

An Intranet is a company specific, private network based on Internet technology, and as such, it is a form of local area network (LAN). However, one of the major distinctions between traditional LANs and Intranets is the reliance of the latter on TCP/IP, packet switching, and Internet technologies. In the case of the Internet, the technology is deployed over a public network, while in the case of Intranets, the technology is deployed within a private network.

According to George Eckel, author of *Intranet Working*, one of the important benefits of Intranets is that they provide a cost-effective vehicle for communication, since the expense of reaching one person or one million people is essentially the same. Intranets are becoming the corporate world's equivalent of a town hall where people can meet, chat, and exchange information.

The emergence of Intranets promises to change the way companies communicate with their employees and how they conduct their business. For example, after years of using satellite feeds to disseminate information to its 208 network affiliates, CBS News now uses an Intranet to provide affiliates with point-and-click access to information on upcoming news stories. Access to this information is provided through the CBS Newspath World Wide Web home page.

2.1.1.1   Benefits of Intranets

Intranets offer many potential benefits, including:

- Reduced operating costs
- Improved employee productivity
- Streamlined processing flows
- Improved internal and external communication
- New and improved customer service
- Cross-platform capability

We will discuss some of the ways these benefits can be achieved.

2.1.1.1.1   *The Paper-less Office*
Many companies find that Intranets simplify corporate-wide communications, and reduce printed material costs by eliminating the need for many paper-based processes. For example, some organizations offer complete manuals on their corporate Web site in electronic form, instead of distributing the information in printed form. Companies can benefit immediately from an Intranet by replacing their printed materials, little by little, with electronic versions. Electronic media is cheaper to produce, update, and distribute than printed material. Often times, printed material is out of date by the time it is distributed. Electronic documents, however, can be easily modified and updated as the need arises.

2.1.1.1.2   *Improved Customer Service*
For many organizations, having the right information at the right time can make a significant difference in their ability to close a sale or meet a deadline. In today's competitive business environment, companies are also under constant pressure to improve productivity while reducing costs. To achieve these productivity

gains, companies must constantly improve their relationships with employees, customers, vendors, and suppliers. Intranets provide an important avenue for making these advancements.

Using an Intranet, vendors, employees, and customers can access information as it is needed, alleviating delays associated with mailing or distributing printed materials. For example, Intranets have been used to:

- Distribute software updates to customers, reducing the need to send printed materials and floppy diskettes or CD-ROMs
- Handle customer orders on-line
- Process and respond to customer inquiries and questions about products and services
- Collect customer and survey data

Using an Intranet, all these activities can be completed electronically in a matter of minutes.

2.1.1.1.3 Improved Help Desks

Intranets have been used to augment help desk services. For example, when someone in the organization learns about a new technology or how to perform a new task (for example, running virus software), he/she can put information and instructions for others on a personal Web page. Others within the organization, including help desk staff, can then access this information as needed. In an organization empowered by an Intranet, all employees can leave the imprints of their expertise.

2.1.1.1.4 Improved Corporate Culture

Intranets help to cultivate a corporate culture that encourages the free flow of information. Intranets place information directly into the hands of employees, promoting a more democratic company structure. The danger of "information democracy" is that once it is in place and taken for granted, management can not easily revert to older, more controlled forms of communication without seriously damaging employee morale and cooperation. Every individual in an Intranet environment is empowered to access and distribute information, both good and bad, on a scale heretofore unknown in the corporate realm.

Intranets dissolve barriers to communication created by departmental walls, geographical location, and decentralized organizations. Placing information directly in the hands of those who need it allows organizations to decentralize and flatten decision making and organizational processes, while maintaining control over the information exchange. Individuals and groups can distribute ideas freely, without having to observe traditional channels of information (i.e., an individual, a printed document, etc.) that are far less effective in reaching geographically dispersed individuals.

2.1.1.1.5 Cross-platform Compatibility

Since the early 1980s, organizations with private networks have struggled with connecting and disseminating information between different types of computers — such as PCs, Macintoshes, and Unix-based machines. To help manage potential barriers to electronic communication posed by hardware and software incompatibilities, many companies have instituted strict standards limiting corporate users to specific hardware and software platforms. Even today, if a company uses PCs, Macs, and Unix-based machines, sharing a simple text document can be a challenge.

Intranets provide a means to overcome many of these software and hardware incompatibilities, since Internet technologies (such as TCP/IP) are platform independent. Thus, companies using Intranets no longer need to settle on one operating system, since users working with Macintosh, PC, or Unix-based computers can freely share and distribute information. In the sections that follow, we will explain why this is so.

2.1.2   Intranet Planning and Management

To implement an Intranet, a company needs a dedicated Web server, communications links to the Intranet, and browser software. *Unfortunately, Intranets do not come prepackaged and fully assembled.* They require careful planning and construction, if they are to be effective in meeting the needs of the organization. In the sections that follow, we discuss recommendations for planning and implementing an Intranet.

2.1.2.1 Gaining Support

The first step toward a successful Intranet implementation is to obtain company-wide support for the project, including endorsement from upper management. A quality presentation should be made to both management and staff to explain the benefits of the Intranet project. Some of these are tangible and easy to measure, while others are intangible and difficult to measure. To gain widespread support for the Intranet project, decision makers must be shown what an Intranet is and how it will benefit the organization. There are many resources (including complete presentations) available on the World Wide Web to help promote the Intranet in a corporate environment.

2.1.2.2 Planning the Intranet Strategy

After selling upper management on the idea of an Intranet, the next step is to define the goals, purpose, and objectives for the Intranet. This is an essential part of the Intranet project planning.

The Intranet project plan should include an overview of the organizational structure and its technical capabilities. The current communication model used to control information flows within the organization should be examined with respect to its strengths and weaknesses in supporting workflow processes, document management, training needs, and other key business requirements. It is important to understand and document existing systems within the organization before implementing the Intranet.

The Intranet plan should clearly define the business objectives to be achieved. The objectives should reflect the needs of the Intranet's potential users. Conducting interviews with employees and managers can help identify these needs. For example, the human resource department may wish to use the Intranet to display job opportunities available within the organization. If this need is to be satisfied, the Intranet should be designed to display job information and job application forms on a Web server, so applicants can apply for positions electronically. The human resource department might also wish to offer employees the ability to change their 401K information by using the Intranet. Each identified goal shapes and defines the functionality that the Intranet must support. An employee survey is also an excellent way to collect ideas on how to employ the Intranet within the organization.

In summary, the following questions are helpful in defining the requirements of the Intranet project:

- Will Intranet users need to access existing (legacy) databases?
- What type of training and support will Intranet users require?
- Who will manage, create, and update the content made available through the Intranet?
- Will individual departments create their own Web pages autonomously?
- Will there be a central authority that manages changes to the content offered on the Intranet?
- Do users need remote access to the Intranet?
- Will the Intranet need to restrict access to certain users and content?
- Will a Webmaster or a team of technicians/managers be assigned to coordinate and manage the maintenance of the Intranet?
- Will the Intranet be managed internally or will it be outsourced?

2.1.2.3 Selecting the Implementation Team

After the Intranet project plan has been developed and approved, the implementation team should be assembled. If the organization does not have an infrastructure in place that is capable of implementing the Intranet, additional staff and resources will need to be hired or the project will need to be outsourced to a qualified vendor.

It is important that the Intranet team has the requisite skills to successfully execute the project plan. A number of skill assessment checklists are provided below to help evaluate the resources available within an organization and their abilities to successfully support the Intranet implementation.

Technical Support Skills Checklist

The Intranet project will require staff with the technical skills needed to solve network problems, understand network design, troubleshoot hardware and software compatibility problems, and implement client-server solutions (such as integrating network databases). Thus, the following skills are required to support an Intranet:

- Knowledge of network hardware and software
- Understanding of TCP/IP and related protocols
- Experience implementing network security
- Awareness of client-server operations
- Practice with custom programming
- Abilities of database management

Content Development and Design Checklist
A typical organization has many sources of information: human resource manuals, corporate statements, telephone directories, departmental information, work instructions, procedures, employee records, and much more. To simplify the collection of information that will be made available through the Intranet, it is advisable to involve people familiar with the original documentation and also those who can author content for Intranet Web pages. If possible, the original authors of the printed material should work closely with the Intranet content developers to ensure that nothing is lost in translation.

The following technical skills are needed to organize and present information (content) in browser-readable format:

- Experience in graphic design and content presentation
- Basic understanding of copyright law
- Knowledge of document conversion techniques (to convert spreadsheet data, for example, into a text document for HTML editing)
- Experience in page layout and design
- Experience with Web browsers and HTML document creation
- Knowledge of image-conversion techniques and related software
- Knowledge of programming languages and programming skills
- CGI programming and server interaction

Management Support Skills Checklist
As previously discussed, the company's management should be involved in the planning and implementation of the Intranet. Ideally, management should have a good understanding of the Intranet benefits, and the expected costs and time frames needed for the project completion. Managers with skills relating to quality-control techniques, process-management approaches, and effective communication are highly desirable. Thus, the following management skills are recommended:

- Understanding of the organization's document flow
- Experience with the re-engineering process
- Knowledge of quality-control techniques
- Knowledge of the company's informal flow of information
- Experience with training and project coordination

2.1.2.4 Funding Growth

The initial cost of setting up a simple Intranet is often quite low and may not require top management's approval. However, when complex document management systems are needed to integrate database access, automate workflow systems, implement interactive training, and other advanced features, the Intranet should be funded with the approval of top management. To gain approval for the project, upper management must be convinced that the Intranet is an integral part of the company's total information–technology deployment strategy. This involves quantifying the tangible benefits of the Intranet to the organization. Management also needs to understand how the Intranet will change the way people work and communicate.

2.1.2.5 Total Quality Management (TQM)

Effective deployment of an Intranet often involves re-engineering current process flows within the organization. Employees are usually most receptive to changes that make their jobs easier. To avoid perceptions that the Intranet is an intimidating intrusion of yet another technology, it is advisable to involve staff as early on as possible in the deployment planning. This will facilitate the transition to the Intranet, and encourage employee participation in the Intranet's success.

After migrating the company's work processes to the Intranet, it is up to managers and employees to adhere to the procedures that have been put in place to improve productivity and teamwork. Management should not assume that because employees have a new tool — the Intranet — that this alone is sufficient to ensure that the desired attitudes and service levels will be attained. Instead, managers should view the Intranet as one aspect of their quest for total quality management (TQM).

TQM involves creating systems and workflows that promote superior products and services. TQM also involves instilling a respect for quality throughout the organization. TQM and the successful deployment of Intranets represent a large-scale organizational commitment, which upper management must support.

2.1.2.6 Training Employees

If employees are expected to contribute content to the Intranet, they will need to be given tools and training so they can author HTML and XML documents. In general, it is a good idea to encourage employees to contribute to the content on display through the Intranet. To do otherwise means that the organization may have to depend on only a few people to create HTML and XML documents.

After initial training, users should be surveyed to determine if the tools they have been provided satisfy their needs. Many users find that creating HTML/XML documents is difficult. If so, then they may also need special training. In corporations, this training is often provided by one person in each department who has been given responsibility for training the rest of the department.

In summary, the following actions are recommended to help develop an effective program for training employees to author high-quality HTML/XML documents:

- Conduct a survey to assess user training needs and wants
- Train users how to develop HTML/XML content
- Provide users with HTML/XML authoring tools that complement what they already know (for example, the Internet Assistant for Microsoft Word is a good choice for users already familiar with Microsoft Word)
- Review the design and flow of material that will be "published" on the Intranet
- Give feedback to HTML/XML authors on ways to improve the site appearance and ease of use

2.1.2.7 Organizational Challenges

In addition to technological challenges, companies may also face the following organizational challenges after the initial release of an Intranet:

- Marketing the Intranet within the organization so that all employees will support its growth and continued use
- Obtaining additional funding on an ongoing basis to implement new capabilities
- Encouraging an information-sharing culture within the company so that all employees will contribute toward building a learning organization
- Merging a paper-based culture with the new culture of electronic documentation
- Ensuring that the content on the Intranet is updated on a regular basis
- Preventing one person or group from controlling (monopolizing) the content on the Intranet
- Instructing employees to author HTML/XML content so they can contribute material to the Intranet
- Informing employees on Intranet etiquette, thereby facilitating courteous on-line discussion forums and other forms of user interaction on the Intranet

- Using the Intranet as an integral part of working with customers and vendors
- Measuring the Intranet's overall effectiveness and contribution to the organization

As is the case when introducing any new information technology to an enterprise, Intranet deployment requires careful planning, effective implementation, and employee training. In the short term, most of the organizational focus is usually on the technical aspects of the Intranet deployment. But as time goes on, organizational issues relating to how the Intranet is used within the organization must be managed. When an organization actively examines and works toward resolving these issues, they are better able to achieve a culture of teamwork and collaboration.

2.1.2.8 Management Summary

The following list summarizes key points surrounding the use of Intranets:

- An Intranet is a company-based version of the Internet. Intranets provide an inexpensive solution for information sharing and user communication.
- An Intranet provides an easy way for users to communicate and share common documents, even if they are using different machines, such as IBM compatible and Macintosh personal computers.
- Some organizations have expanded their Intranet to allow customers to access internal databases and documents.
- Many companies can establish a functional Intranet using in-house personnel with a minimal amount of new equipment.

Internet technology adheres to open standards that are well documented. This, in turn, encourages the development of cost-effective and easy-to-implement Intranet solutions. As the popularity of Intranets has increased, so has the demand for new tools and Web-based solutions. This demand has fueled competition among software manufacturers which, in turn, has resulted in better and less expensive Intranet products.

In summary, Intranets can be used to improve productivity, simplify workflows, and gain a competitive advantage over those who have yet to learn how to capitalize on the benefits of Intranets.

2.1.3   Technical Overview

2.1.3.1   Internet Basics

2.1.3.1.1   *Packet Switching*

Packet switching was introduced in the late 1960s. In a packet-switched network, programs break data into pieces, called packets, which are transmitted between computers. Each packet contains the sender's address, the destination address, and a portion of the data to be transmitted. For example, when an e-mail message is sent over a packet-switched network, the e-mail is first split into packets. Each packet intermingles with other packets sent by other computers on the network. Network switches examine the destination address contained in each packet, and route the packets to the appropriate recipient. Upon reaching their destination, the packets are collected and reassembled to reconstitute the e-mail message.

2.1.3.1.2   *TCP/IP*

The U.S. Advanced Research Projects Agency (ARPA) was a major driving force in the development and adoption of packet-switched networking. The earliest packet-switched network was called the ARPAnet. The ARPAnet was the progenitor to today's Internet. By the early 1980s, ARPA needed a better protocol for handling the packets produced and sent by various network types. The original ARPAnet was based on the Network Control Protocol (NCP). In January 1983, NCP was replaced by the Transport Control Protocol/Internet Protocol (TCP/IP). TCP/IP specifies the rules for the exchange of information within the Internet or an Intranet, allowing packets from many different types of networks to be sent over the same network.

2.1.3.1.3   *Connecting to the Internet*

One way to connect to the Internet is to install a link from the company network to the closest computer already connected to the Internet. When this method is chosen, the company must pay to install and maintain the communications link (which might consist of a copper wire, a satellite connection, or a fiber optic cable) to the Internet. This method was very popular with early adopters of the Internet, which included universities, large companies, and government agencies. However, the costs to install and maintain the communications link to the Internet can be prohibitive for smaller companies.

Fortunately, specialized companies — called Internet Service Providers (ISPs) — are available to provide a low-cost solution for accessing the Internet. ISPs pay for an (expensive) connection to the Internet, which they make accessible to others through the installation of high-performance servers, data lines, and modems. Acting as middlemen, the ISPs rent time to other users who want to access the Internet.

Two important decisions must be made when deciding what type of Internet connection is the most appropriate. The first decision is the company budget allocated for Internet connectivity, and the second is the Internet connection speed needed to support the business requirements. Both decisions are inter-related. ISPs offer a variety of options for connecting to the Internet, ranging from a simple dial-up account over phone wires to high-speed leased lines from the company to the ISP. Dial-up accounts are typically available for a low, flat monthly fee, and are generally much cheaper than a leased line connection. However, the leased line connection is usually much faster than the dial-up connection.

When a dial-up account is used, a modem and a phone line are used to call and log into the ISP server (or computer), which, in turn, acts as the doorway to the Internet. The transmission speed of the connection is limited by the speed of the modems employed by the user and the ISP. A modem is unnecessary when a leased line connection is available to the ISP. Leased lines are offered in many different configurations with a variety of options. The most common link types are ISDN (which support trans-mission speeds from 56 Kbps to 128 Kbps), T1 (transmitting at speeds up to 1.54 Mbps), and T3 (transmitting at speeds up to 45 Mbps).

If a company only needs to make an occasional connection to the Internet — for example, less than 20 to 50 hours per month for all users — a dial-up account should be sufficient. However, if a company needs faster data transfer speeds or has several users who must access the Internet for substantial periods of time over the course of a month, a leased line connection should be considered.

The fastest growing segment of Internet users are those who connect to the Internet through an ISP via an ordinary telephone connection. There are two major protocols for connecting to the Internet in this way: Serial Line Internet Protocol (SLIP) and Point-to-Point Protocol (PPP). SLIP is the older protocol and is available in many communications packages. The faster PPP is newer and therefore it is not as widely supported.

Principles of queuing analysis can be applied to the problem of sizing the links needed to support the Internet access, whether or not that access is to an ISP or to a direct Internet connection. The reader is referred to Chapter 2, *Network Design: Management and Technical Perspectives* by Mann-Rubinson and Terplan, for specific techniques on how to estimate the throughput and performance characteristics associated with using different size link capacities. This analysis can be used to determine whether or not a dial-up or leased line connection is sufficient to support the bandwidth requirements with tolerable transmission delays.

2.1.3.1.4 Basic Terminology

In this section, we define commonly used Internet and Intranet terminology.

The World Wide Web

The World Wide Web — or Web — is a collection of seamlessly interlinked documents that reside on Internet servers. The Web is so named because it links documents to form a web of information across computers worldwide. The documents available through the Web can support text, pictures, sounds, and animation. The Web makes it very easy for users to locate and access information contained within multiple documents and computers. "Surfing" is the term used to describe accessing (through a Web browser) a chain of documents through a series of links on the Web.

Web Browsers

To access and fully utilize all the features of the Web, special software — called a Web browser — is necessary. Its main function is to allow the user to traverse and view documents on the Web. Browser software is widely available for free, either through a download from the Internet or from ISPs. Commercial on-line services — such as America Online and Prodigy — also supply browsers as part of their subscription products. The two most commonly used browsers are Netscape Navigator and Microsoft Internet Explorer. Some of the common tasks which both support include:

- Viewing documents created on a variety of platforms
- Creating and revising content
- Participating in threaded discussions and news groups
- Watching and interacting with multimedia presentations
- Interfacing with existing legacy data (non-HTML based data) and applications
- Gaining seamless access to the Internet

It should be noted that the same Web browser software used for accessing the Internet is also used for accessing documents within an Intranet.

Uniform Resource Locator (URL)

The Web consists of millions of documents that are distinguished by a unique name called a URL (Uniform Resource Locator), or more simply, a Web address. The URL is used by Web browsers use to access Internet information. Examples of URLs include:

http://www.netscape.com
ftp://ftp.microsoft.com

A URL consists of three main parts:

1. A service identifier (such as http)
2. A domain name (such as www.ups.com)
3. A path name (such as www.ups.com/tracking)

The first part of the URL, the service identifier, tells the browser software which protocol to use to access the file requested. The service identifier can take one of the following forms:

- http:// — This service identifier indicates that the connection will use the hypertext transport protocol (HTTP). HTTP defines the rules that software programs must follow to exchange information across the Web. This is the most common type of connection. Thus, when Web addresses start with the letters "http" it indicates that the documents are retrieved according to the conventions of the HTTP protocol (hypertext transport protocol).
- ftp:// — This service identifier indicates that the connection will use the file transfer protocol (FTP). This service identifier is typically used to download and copy files from one computer to another.
- gopher:// — This service identifier indicates that the connection will utilize a gopher server to provide a graphical list of accessible files.
- telnet:// — This service identifier indicates that a telnet session will be used to run programs from a remote computer.

The second part of the URL, the domain name, specifies which computer is to be accessed when running server software. An example of a domain name is: www.tcrinc.com.

The final part of the URL, the path name, specifies the directory path to the specific file to be accessed. If the path name is missing from the URL, the server assumes that the default page (typically, the homepage) should be accessed. Large, multi-page Web sites can have fairly long path names. For example, these URLs request specific pages within a given Web site:

- http://www.apple.com/documents/productsupport.html
- http://www.bmwusa.com/ultimate/5series/5series.html
- ftp://ftp.ncsa.uiuc.edu/Mac/Mosaic
- http://www.microsoft.com/Misc/WhatsNew.htm

Home Pages

Companies, individuals, and governments that publish information on the Internet usually organize that information into "pages," much like the pages of a book or a sales brochure. The first page that people see in a sales brochure is the cover page, which may contain an index and summary of the brochure contents. Similarly, a home page is the first page that users see when they access a particular Web site. The home page is to the Web site what the cover page is to a sales brochure. Both must be appealing, concise, informative, and well organized to succeed in maintaining the reader's interest. The home page is usually used to convey basic information about the company and what it is offering in the way of products and/or services.

Many companies publish the Internet address (or URL) of their home page on business cards, television, magazines, and radio. To access a Web site, a user has merely to type the URL into the appropriate area on the Web browser screen.

Client Programs and Browsers

Across the Internet, information (i.e., programs and data) is stored on the hard disks of thousands of computers called servers. These are so named because, upon request, they serve (or provide) users with information. A server is a remote computer that may be configured to run several different types of server programs (such as Web server, mail server, and ftp server programs).

A client program is used to initiate a session with a server. Client programs are so named because they ask the server for service. In the case of the Web, the client program is the Web browser. All client–server interactions take the same form. To start, the client connects to the server and asks the server for information. The server, in turn, examines the request and then provides (serves) the client with the requested information. The client and server may perform many request–response interactions in a typical session.

Software programs — such as a browser — use HTTP commands to request services from an HTTP server. An HTTP transaction consists of four parts: a connection, a request, a response, and a close.

Where Web Documents Reside

When users publish Web pages, they actually store the pages as files that are accessible through a file server. Typically, Web pages reside on the same computer on which the server program is running, but this is not necessarily true. For security reasons, it may be necessary to limit accessibility to various files on the Web server. Obviously, it might be disastrous if internal documents and data were made available to competitors. To prevent this type of security risk, a WebMaster (or Systems Administrator) can configure the Web server so it only allows specific clients to access confidential information, based on a need-to-know basis. The WebMaster can control access to the server by requiring users to log-in with a username and password that has predetermined access privileges.

HTML — The Language of the World Wide Web

The European Particle Physics Laboratory at CERN, in Geneva, Switzerland, developed Hypertext Markup Language (HTML) in the late 1980s and early 1990s. HTML is the language of the World Wide Web. Every site on the Web uses HTML to display information.

Each Web document contains a set of HTML instructions that tell the browser program (e.g., Netscape Navigator or Microsoft Internet Explorer) how to display the Web page. When you connect to a Web page using a browser, the Web server sends the HTML document to your browser across the Internet. Any computer running a browser program can read and display HTML, regardless of whether that computer is a personal computer running Windows, a Unix-based system, or a Mac.

If word processor formatted files — such as Microsoft Word — were used to create Web pages, only users with access to Microsoft Word would be able to view the Web page. HTML was designed to overcome this potential source of incompatibility. All users can access Web pages from their browser since all Web pages conform to HTML standards. A HTML Web page is a plain text file (i.e., an ASCII text file) that can be created and read by any text editor. There are many software programs available to convert document files to HTML equivalents. In addition, many standard presentation and word processing packages offer built-in routines to convert a standard document into a Web-ready HTML file. This type of conversion might be helpful, for example, if you wanted to convert a Microsoft PowerPoint presentation into a set of HTML files for display on the Web.

After HTML files are transferred to a Web site, anyone with a browser can view them. HTML provides the browser with two types of information:

1. "Mark-up" information that controls the text display characteristics and specifies Web links to other documents.
2. "Content" information consisting of the text, graphics, and sounds that the browser displays.

Hypertext and Hyperlinks

Documents on the Web can be interconnected by specifying links (called hyperlinks) that allow the user to jump from one document to another. The HTML code, which drives all Web pages, supports hypertext. Hypertext, in turn, supports the creation of multimedia documents (containing pictures, text, animation, sound, and links) on the Web.

Hyperlinks (or simply, links) are visually displayed on the Web pages as pictures or underlined text. When a user clicks on a hyperlink displayed on their browser screen, the browser responds by searching for and then loading the document specified by the hyperlink. The document specified in the hyperlink may reside on the same computer as the Web page on display or it may reside on a different computer on the other side of the world. Much of the Web's success has been attributed to the simplicity of the hyperlink point-and-click user interface.

There are four basic layouts for linking Web pages with hyperlinks: linear, hierarchical, Web, and combination. Which layout is the most appropriate depends on the type of information that is being presented and the intended audience.

FTP — The File Transfer Protocol

The FTP (file transfer protocol) is a standard protocol for transferring and copying files from one computer to another. Depending on the configuration of the FTP server program, you may or may not need an account on the remote machine to access system files. In many cases, you can access a remote computer with FTP by logging on with a username of "anonymous," and by entering your e-mail address as the password. This type of connection is referred to as "anonymous FTP session."

After logging in to the remote FTP server, it is possible to list a directory of the files that are available for viewing and/or copying. The systems administrator determines which files can be accessed on the remote server, and who has access privileges. When system security is a major concern, the system administrator may require a specific username and password (as opposed to allowing an anonymous log-on procedure) to gain access to system files.

FTP is very useful in accessing the millions of files available on the World Wide Web. Most browsers have built-in FTP capabilities to facilitate downloading files stored at FTP sites. To access a FTP site using your browser, you type in the FTP site address, much like entering a Web address. For example, to access the Microsoft FTP site, the address "ftp://ftp.microsoft.com" would be entered into the browser address window.

Java

Java is a new programming language released by Sun Microsystems that closely resembles C++. Java is designed for creating animated Web sites. Java can be used to create small application programs, called applets, which browsers download and execute. For example, a company might develop a Java applet for their Web site to spin the company's logo, to play music or audio clips, or to provide other forms of animation to improve the appeal and effectiveness of the Web page.

Network Computers

Although personal computers (PCs) are becoming more and more common, the number of households with a PC is still only about one third the number of households with a television. The main reason is that PCs are still too expensive for the masses. The network computer is a scaled-down, cheaper version (under $500) of the PC. A network computer is designed to operate exclusively with the Internet and Java applets.

2.1.4   Intranet Components

This section will provide an overview of the components necessary to create an Intranet. The final selection of the Intranet components depends upon on the company's size, level of expertise, user needs, and future Intranet expansion plans. In addition, we also examine some of the costs associated with the various Intranet components.

An Intranet requires the same basic components found on the Internet, including:

1. A computer network for resource sharing.
2. A network operating system that supports the TCP/IP protocol.
3. A server computer that can run Internet server software.
4. Server software that supports hypertext transport protocol (HTTP) requests from browsers (clients).
5. Desktop client computers equipped with network software capable of sending and receiving TCP/IP packet data.
6. Browser software installed on each client computer.

It should be noted that if a company does *not* want to use an internal server, an ISP can be used to support the Intranet. It is very common for organizations to use an ISP, especially when there is little information content or interest in maintaining a corporate-operated Intranet server. ISPs are also used when the organizational facilities can not support the housing of an Intranet server.

In addition to the software and hardware components listed above, HTML/XML documents must be prepared to provide information displays on the Intranet. The creation and conversion of documents to HTML/XML format is very easy using commercial software packages, such as Microsoft's FrontPage. Third-party sources are also available to provide this service at a reasonable cost.

2.1.4.1   Network Requirements

The first requirement for an Intranet is a computer network. For the purpose of this discussion, we assume that a basic computer network is in place. We now focus on the hardware and software modifications needed to support an Intranet.

Most computer networks are local-area networks (LANs). LANs are based on a client–server computing model that uses a central, dedicated computer — called the server — to fulfill client requests. The client–server computing model divides the network communication into two sides: a client side and a server side. By definition, the client requests information or services from the server. The server, in turn, responds to the client's requests. In many cases, each side of a client–server connection can perform both client and server functions.

Network servers are commonly used to send and receive e-mail, and to allow printers and files to be shared by multiple users. In addition, network servers normally have a storage area for server programs and to backup file copies. Server applications provide specific services. For example, a corporate-wide e-mail system typically uses a server process that is accessible from any computer within the company's network.

A server application (or server process) usually initializes itself and then goes to sleep, spending much of its time simply waiting for a request from a client application. Typically, a client process will transmit a request (across the network) for a connection to the server, and then it will request some type of service through the connection. The server can be located at either a local or remote site.

Every computer network has a physical topology by which it is connected. The most common topologies used to connect computers are the star, token ring, and bus topologies.

A network-interface card (NIC) is needed to physically connect a computer to the network. The network-interface card resides in the computer and provides a connector to plug into the network. Depending on the network, twisted-pair wiring, fiber optic, or coaxial cable may be used to physically connect the network components. The network-interface card must be compatible with the underlying network technology employed in the network (e.g., Ethernet or Token Ring).

2.1.4.2 Network Operating Systems

The Internet supports connectivity between various hardware platforms running various operating systems. In theory, there is no reason why an organization must stay with one type of machine or operating system when implementing an Intranet. However, in practice, many organizations use only one network operating system to simplify the task of managing the network.

The primary choices for network operating systems are: UNIX, Windows NT, and Novell NetWare. We now discuss each of these operating systems and important considerations surrounding their use.

UNIX

Many larger companies use UNIX-based machines as their primary business application server platform. UNIX is a proven operating system that is well suited for the Internet's open system model. Unfortunately, learning how to use UNIX is not easy. Also, using a UNIX-based machine limits the choices available for developing interactive Intranets and other software applications. Many programmers, for example, prefer to develop applications using Windows-based machines and programming languages (such as Microsoft's Visual Basic or Borland's Delphi).

Windows NT

Many companies choose Windows NT over UNIX because NT is easy to install, maintain, and administer. Windows NT, like UNIX and OS/2, provides a high-performance, multi-tasking workstation operating system. It also supports advanced server functions (including HTTP, FTP, and Gopher) and communications with clients running under MS-DOS, Windows 3.1, Windows 95, Windows for Workgroups, Windows NT Workstation, UNIX, or Macintosh operating systems. The latest version of Windows NT Server includes a free Internet Information Server (IIS) and a free Web browser (Internet Explorer). Microsoft designed the IIS so that it can be installed and up and running on a Windows NT workstation in less than 10 minutes. The Windows NT Server comes with a built-in remote access services feature that supports remote access to the Intranet through a dial-up phone connection.

Novell NetWare (IPX/SPX)

The NetWare operating system provides network-wide file and printer sharing for Ethernet or token ring networks. It runs on all major computer platforms, including UNIX, DOS, Macintosh, and Windows. However, behind the scenes, NetWare sends and receives data packets based on the Internetwork Packet Exchange/Sequenced Packet Exchange (IPX/SPX) protocol. Like TCP/IP, the IPX/SPX protocol defines a set of rules for coordinating network communication between network components.

Many companies use Novell network products within LANs to operate file and print servers. Therefore, it is important to understand how Novell's NetWare products can be used in an Intranet implementation strategy. If a company has an existing NetWare network, it might choose not to provide TCP/IP software to its network clients. Instead, the local-area network features provided by the NetWare software can be used.

A true Intranet uses Internet technology. This implies that an Internet Protocol (IP) address is assigned to each network computer (actually to each network-interface card), and that the TCP/IP protocol is used in the network. However, it is possible to run an Intranet on top of a NetWare LAN using various software products that translate IPX to IP. Many of these software packages provide IPX to IP translation while leaving the existing LAN infrastructure unchanged.

For example, the Novell product, IntranetWare, does not require assignment of an IP address to each client on a network. Instead, an IP address is only assigned to the NetWare Web server. The software performs IPX to IP translation on the client side, translating the TCP/IP protocols used by a Web browser to IPX protocols. After the protocol translation on the client side, the messages travel across the network until they reach the NetWare Web Server. At this point, the IntranetWare Server running on the NetWare Web server translates the IPX messages back into TCP/IP so they can be sent on to other servers on the network.

Another way to create an Intranet using a NetWare-based LAN is to use the Internet gateway product Inetix from Micro Computer Systems, Inc. Inetix runs on NetWare, Windows NT, or UNIX servers. Inetix does not require that the NetWare client machines support TCP/IP, only that a single IP address be assigned to the Intranet Web server. The Inetix client software allows a Web browser to execute on the client side, even though the client machine does not use TCP/IP-based software.

In a Mac-based network, NetWare for Macs, or AppleTalk[1] can be used to support the Intranet.

2.1.4.3 Server Hardware

Server machines run the network operating system and control how network computers share server resources. Large businesses with thousands of users typically use high-speed Unix-based machines for their servers. Small and medium-sized companies normally use less expensive Intel-based machines. The load (i.e., the number of users and the amount of network traffic) on the Intranet server machine will influence the selection of a specific processor type.

There is considerable debate in the industry as to which machine makes a better Intranet server: a Unix workstation, an Intel-based machine, or a PowerPC-based system. In general, the server choice depends on the plans for the Intranet and the level of familiarity the company has with each of these platforms. The server hardware selection also depends on the network operating system in use.

If a Unix-based server is chosen, a company will pay more for an equivalent amount of computing power provided by an Intel-based machine. Unix machines still carry a price premium over PCs, because they are made from custom parts, while Intel-based machines are made from commodity components available from many hardware vendors and suppliers. For example, a high-end Pentium machine with the capacity to serve over 1000 client machines can be bought for about one tenth of the cost of a comparable Unix server. Macintosh-based systems are more expensive than comparable Intel-based machines, but they are still much less expensive than Unix-based machines.

The decision to use a Unix-based machine vs. an Intel-based machine as the Intranet server is also influenced by maintenance costs. Maintaining a Unix-based machine requires more resources than maintaining an Intel-based machine. Hardware upgrades for Intel-based machines are also cheaper than hardware upgrades for Unix workstations. A Macintosh server will cost more to upgrade than an Intel-based machine. However, these costs are still lower than a comparable upgrade on a Unix-based machine.

The debate over Unix- and Intel-based machines focuses primarily on their performance in supporting business application servers. For example, companies that use large accounting and financial software packages often use Unix servers. On the other hand, companies that do not want to pay the price premium for Unix machines and/or are not familiar with Unix machines often select Intel-based machines as their business application servers.

A Pentium-class machine supports a vast array of software applications and server software. Many industry experts believe that Pentium-class machines will take a significant amount of the market share away from Unix workstations. This, in turn, means that more and more Intranet-based applications will use Pentium-class machines in the future.

2.1.4.4 Web Server Software

A working Intranet requires server software that can handle requests from browsers. In addition, server software is needed to retrieve files and to run application programs (which might, for instance, be used to search a database or to process a form containing user-supplied information).

For the most part, selecting a Web server for an Intranet is similar to selecting a Web server for an Internet site. However, Internet servers must generally handle larger numbers of requests and must deal with more difficult security issues. The performance of the Web server has a major impact on the overall

---

[1]AppleTalk — AppleTalk is Apple's proprietary network operating system for supporting a LAN. Unlike AppleTalk, TCP/IP supports both LANs and WANs.

performance of the Intranet. Fortunately, it is fairly easy to migrate from a small Web server to a larger, high-performance Web server as the system usage increases over time.

Web servers are available for both Windows NT and Unix from Microsoft, Netscape, and a number of other companies. The Web server selection is limited by the server operating system. In the discussion that follows, we describe Web servers for Unix, Windows NT, Windows 95, and Macintosh operating systems.

It is expected that in the next two to five years, stand-alone Web servers will be replaced by servers that are an integral part of the operating system. In addition, these Web servers will handle many tasks that require custom programming today, such as seamless connection to databases, video and audio processing, and document management.

UNIX Web Servers

One of the best and oldest Web servers for Unix-based machines is the National Center for Supercomputing Application's (NCSA) HTTP Web Server. Much of the Internet's growth is primarily due to the popularity of this server, which is free. NCSA is committed to the continued development of its Web server, which provides both common gateway scripting (CGI) capabilities and server side includes (SSI) software. SSI software is used by Web servers to display and/or capture dynamic (changing) information on an HTML/XML page. For example, SSI can be used to display a counter showing the number of visitors to a Web site. The NCSA Web server also allows the creation of virtual servers on the same machine. The virtual servers can have their own unique universal resource identifier (URL). This is useful, for example, for assigning a different IP address to different departments using the same machine.

Netscape Communications Corporation offers the most popular commercial Unix-based Web servers: Enterprise Server and FastTrack Server. Enterprise Server software is designed for building large Intranets. FastTrack is easier to install than Enterprise Server but it offers less functionality. FastTrack is well suited for companies that plan to build small- to medium-sized Intranets. These Web servers are also available for Windows NT.

Windows NT Web Servers

As discussed in the previous section, both Netscape's Enterprise Server and FastTrack Server are available for Windows NT. These servers should be considered when both Windows NT and Unix-based machines are used as servers.

If cross-platform Web server software is not needed, Microsoft's IIS should be considered. This server is used to drive Microsoft's Internet site, and it works well for a large organization. At the time of this writing, Microsoft offers this server free of charge. It also comes bundled with Microsoft's Windows NT Server software. IIS is easy to install and allows new users to be added to the Intranet with minimal effort. IIS comes with an FTP server.

WebSite Professional, from O'Reilly & Associates, Inc., is another popular Windows NT Web server. WebSite Professional is very easy to install. This server has built-in search capabilities, a Web site management tool, and the popular HTML editing tool, HotDog.

Macintosh Web Server

Not many Web servers are available for Macintosh computers. If your organization does not have a Macintosh network already in place, then it advisable to avoid installing a Mac-based Intranet server.

The largest market share of the Macintosh Web server market belongs to the WebStar server from Quarterdeck Corporation. This server is a mature product and is a good choice for a Mac-based Web server. WebStar is very easy to install and maintain.

NetWare Web Server

The NetWare Web Server from Novell is an excellent choice for companies that have a NetWare network already in place. When using a NetWare Web server, IPX to IP translation software must be installed on each client machine. When this is done, an IP address does not need to be assigned to each client machine.

2.1.4.5   Desktop Clients Running TCP/IP

TCP/IP must be installed on each client machine running on the Internet.[2] To use an Internet-based application (such as a Web browser) on a Windows-based machine, a TCP/IP stack must be present. Windows 95, Windows NT, and IBM's OS/2 Warp operating systems include the TCP/IP protocol suite. Most Unix-based systems use TCP/IP as their main network communication protocol.

If a company has Windows 3.1 clients, they should consider upgrading the clients to Windows 95 or Windows NT. Many of the advanced Internet and Intranet applications are only available for Unix, Windows 95, and Windows NT operating systems. If the Windows 3.1 clients can not be upgraded, then TCP/IP software must be installed for each client. One of the more popular TCP/IP software applications for Windows 3.1 is Trumpet Winsock. Trumpet Winsock can be downloaded from the Internet for free.

2.1.4.6   Web Browsers

The last component needed to make a functional Intranet is a Web browser. There are two major choices for a browser: Microsoft's Internet Explorer or Netscape's Navigator.

It is expected that Netscape will retain its dominance in the Unix market, since Microsoft has chosen not to support a Unix platform. Therefore, if the Intranet must support Unix, Macintosh, and Windows clients, and the company requires standardization on a single browser, Netscape's Navigator product is the only viable option.

Browsers, as we know them today, will probably not exist in a few years. For example, it is expected that eventually Microsoft will integrate the browser's functionality into its business application software (such as Word, Excel, etc.) and operating system.

2.1.4.7   Intranet Component Summary

In this section, the basic components of an Intranet were examined. We recapitulate below some of the key concepts covered in this section:

- Intranets are based on a client–server network computing model. By definition, the client side of a network requests information or services and the server side responds to a client's requests.
- The physical components of an Intranet include network interface cards, cables, and computers.
- Suites of protocols, such as TCP/IP and IPX/SPX, manage data communication for various network technologies, network operating systems, and client operating systems.
- IPX to IP translation programs provide NetWare users with the ability to build an Intranet without running the TCP/IP suite of protocols on their network.
- Windows-based Intranets are easier and less expensive to deploy than Unix-based Intranets.
- Netscape and Microsoft provide both low- and high-end server software products designed to meet the needs of large, medium, and small organizations.
- Netscape's Navigator and Microsoft's Internet Explorer provide advanced browser features for Intranet applications.

2.1.5   Intranet Implementation

2.1.5.1   Information Organization

After the physical components of the Intranet are in place, the next step is to design the information content of the Intranet and/or Internet Web pages. This task involves identifying the major categories and topics of information that will be made available on the Intranet. Information can be organized by department, function, project, content, or any other useful categorization scheme. It is advisable to use

---

[2]If the network does not support TCP/IP, a gateway application that translates TCP/IP for the network operating system protocol must be used.

cross-functional design teams to help define the appropriate informational categories that should be included on the corporate Web site. The following types of information are commonly found on corporate Intranet homepages:

- What's new
- Corporate information (history and contacts)
- Help desk and technical support
- Software and tools library
- Business resources
- Sales and marketing information
- Product information
- Human resources related information (benefits information, etc.)
- Internal job postings
- Customer feedback
- Telephone and e-mail directory
- Quality and system maintenance records
- Plant and equipment records
- Finance and accounting information
- Keyword search/index capability

2.1.5.2 Content Structure

After the main topics of information to be displayed on the corporate Web page(s) have been identified, the flow and manner of presentation on the Intranet must be developed. Four primary flow models are used to structure the flow of presentation at an Intranet Web site: linear, hierarchical, non-linear (or Web), and combination information structures.

A linear information structure is similar in layout to a book in that information is linked sequentially, page by page. When a linear layout is used, the Web pages are organized in a "slide show" format. This layout is good for presenting pages that should be read in a specific sequence or order. Since linear layouts are very structured, they limit the reader's ability to explore and browse the Web page contents in a non-sequential (or non-linear) manner.

When a hierarchical layout is used to structure the information, all the Web pages branch off from the home page or main index. This layout is used when the material in the Web pages does not need to be read in any particular order. A hierarchical information structure creates linear paths that only allow up and down movements within the document structure.

A non-linear, or Web, structure links information based on related content. It has no apparent structure. Non-linear structures allow the reader to wander through information spontaneously by providing links that allow forward, backward, up and down, diagonal, and side-to-side movement within a document. A non-linear structure can be confusing, and readers may get lost within the content, so this structure should be chosen with care. The World Wide Web uses a non-linear structure. The advantage of a non-linear structure is that it encourages the reader to browse freely.

The combination Web page layout, as the name implies, combines elements of the linear, Web, and hierarchical layouts. Regardless of the type of flow sequence employed, each Web page typically has links that allow the user to move back and forth between pages and back to the home page.

Over the lifetime of the Intranet, it is likely that the layout and organization of information on the corporate Web pages will change many times. It is often helpful to use flow charting tools to help manage and document the updated information flows. Visio for Windows by Visio Corp. and ABC Flowcharter by Micrografx are two excellent tools for developing flowcharts. In addition, some of the Web authoring tools offer flowcharting and organizational tools to help design and update the information structure on the Web pages.

2.1.5.3   Interface Design

After defining the Intranet's structure, the next step is to define the functionality and user interface. The Intranet design should be consistent with the organization's corporate image. For example, items such as corporate images, logos, trademarks, icons, and related design themes add a familiar look and feel to the content. Where possible, they should be included in the Web page design. It is also advisable to work with the marketing department when designing the web page layouts to ensure that a consistent theme is maintained in all the company communications that will be viewed by the outside world.

A technique called storyboarding is frequently used to design the web page layout. Storyboards are used by film producers, story writers, and comic strip artists to organize the content and sequence of their work. A storyboard depicts the content, images, and links between pages of the Intranet in the form of a rough outline.

Software — such as Microsoft PowerPoint or a similar presentation program — can be used to develop a storyboard and sample Web pages. It is a good idea to test the interface design to ensure that the icons, buttons, and navigational tools are logical and intuitive. An Intranet without intuitive navigational tools is like a road without signs. Just as it would be difficult for drivers to find their way from one city to another without the aid of signs, street names, and directional information, Intranet users will find it difficult to retrieve information without easy to follow categories, buttons, and links. It is often helpful to employ graphic designers and marketing–communications staff to create effective graphics and images for the web site.

2.1.5.4   Free Clip Art and Images

Many icons and navigational signs are available as clip art that comes with word processing, page layout, and presentation software programs. In addition, many Web sites offer images and clip art, which can be downloaded for free. However, the licensing agreements for downloading free images may have restrictions and requirements that should be observed.

2.1.5.5   Intranet Functionality

The required functionality of an Intranet dictates many of the design and user interface features. One of the goals in designing the Intranet should be to improve existing systems and infrastructures. After examining the current information structure, it may become clear which aspects of the structure work well and which ones need improvement.

Workflow processes, document management, and work collaboration are areas that the organization should strive to improve through the use of an Intranet. A workflow analysis should consider ways in which the Intranet can automate various organizational tasks and processes. For example, if a company has a geographically dispersed project team, the Intranet might be used to post and update project information as various tasks are completed. Other team members could then visit the Intranet page at any time to check the project status.

The following checklist is helpful when developing a list of the functions that need to be supported by the Intranet:

Functionality Checklist
- The user interface must be intuitive and tested
- The Intranet's design should support continuous updates
- The Intranet may need to be integrated with database management systems to allow users to access information (such as customer and product data)
- The Intranet should support existing (legacy) applications, as needed
- The Intranet should have built-in directories, such as corporate telephone numbers and e-mail addresses
- The Intranet should incorporate groupware applications
- Support (or future expansion) for on-line conferencing should be considered
- The Intranet should provide division-specific and corporate-wide bulletin boards for electronic postings

- The Intranet should be designed with a document sharing and management process in mind
- The Intranet should foster teamwork and collaboration by enhancing channels of information distribution
- Search engines — which simplify a user's ability to locate and access information — should be made available
- The Intranet should support e-mail
- The Intranet should support (for future expansion) multimedia applications that use text, images, audio, and video
- Automated real-time Web-page generation should be encouraged
- The Intranet should be designed so it can interface, at least potentially, with factory equipment, other manufacturing devices, or other critical legacy systems
- The Intranet should support the automation of organization workflows

2.1.5.6 Content Management

Many organizations struggle with the tasks of information creation, management, and dissemination. They are time consuming and difficult to control. The Intranet alone cannot solve information management problems unless specific Intranet solutions are implemented that directly address the need for document management. The following list identifies content-management tasks that should be considered in the Intranet plan:

- Users must have the ability to easily add or update content on a regular basis
- Users must have the ability to protect their content from changes by other users
- A content-approval process should be defined and in place. This process should encompass ways to manage and control document revisions, especially changes to shared documents

As policies and procedures relating to content management are formulated, it is important to designate responsibilities to specific individuals to ensure that they are put into place and followed. An Intranet style guide should be developed that provides page layout, design elements, and HTML/XML code guidelines. The style guide will help the organization to maintain a consistent look and feel throughout the Intranet's Web pages. The style guide should contain information on where to obtain standard icons, buttons, and graphics, as well as guidelines on page dimensions and how to link to other pages. As part of the style guide, it is helpful to create Web page templates. These templates consist of HTML/XML files, and are used to provide a starting point for anyone interested in developing Web pages or content for the Intranet. Although it is very easy to create a working Web page and to publish it for mass viewing, the real challenge is in producing a well-conceived Web page.

2.1.5.7 Training and Support

After the Intranet is up and running, efforts should be focused on how to maintain the information content and on employee training. Part of the document-management strategy should encompass the selection of content stakeholders. Content stakeholders are individuals in different departments or work groups who are responsible for the creation and maintenance of specific content. Stakeholders can be department managers, team leaders, or content authors and publishers.

Some organizations create a position called a Webmaster. This position is responsible for maintaining and supporting the content published on the Intranet. A good Webmaster should have the following skills:

- Basic Internet skills, including an understanding of e-mail, FTP, and Telnet
- A thorough understanding of HTML/XML document creation
- Experience with CGI programming
- Programming experience with languages such as Perl, C/C++, and Java
- Experience with content creation and the conversion of text and images
- Knowledge of client–server processing
- Experience with server setup and maintenance

- Knowledge of your organization's structure and inner workings
- Organizational and training skills

It is possible that the organization may choose to decentralize the maintenance of the information content. In this case, individuals from various departments might be selected to maintain the content relating to their respective department. These individuals should be trained to handle a variety of maintenance issues. A decentralized approach depends on having more than one individual with the necessary skills available to maintain the web pages. A decentralized support structure gives authors and content owners direct control and responsibility for publishing and maintaining information. This can help prevent bottlenecks in making information available in a timely fashion.

Training for stakeholders, Webmasters, and Intranet users is an important part of an Intranet strategy. Intranet customers and content stakeholders should be trained to understand the Intranet and how it will improve the organization and the way the company does business. They should also be given training on how to create, utilize, and maintain content on the Web page(s). Companies that invest in the education and training of their employees will have a better chance of creating and maintaining a successful Intranet.

2.1.6   Intranet Deployment

Since Intranets are easy to set up, many companies do not realize what the true resource requirements are to maintain the Intranet with up-to-date information. The goal of this section is to provide a realistic perspective on how organizations are most likely to achieve long-lasting benefits from the Intranet.

Some companies invest much more than their competitors in information technology, such as an Intranet, but still fail to effectively compete in the marketplace. Computers alone do not, and cannot, create successful companies. A good start is to empower all employees to contribute to the Intranet. As is true for any collaborative effort, every member is responsible for the overall success of the team.

2.1.6.1   Technological Considerations

The major technological challenges facing the organization after the initial implementation of an Intranet include:

- Converting existing paper documents into electronic documents that employees can access electronically via the Intranet.
- Connecting existing databases to the Intranet so they are accessible by a wide range of computing platforms (such as Windows- and Mac-based systems).
- Coordinating the use of multiple servers used across departmental lines.
- Continuously enhancing the Intranet's features and capabilities to keep employees motivated to use the Intranet.
- Installing security features within the Intranet to prevent unauthorized access to confidential or sensitive information.

Intranet technology, and information technology in general, is changing so fast that keeping up with the latest software and hardware solutions requires a substantial ongoing organizational commitment.

Conversion of Paper Documents Into Electronic Form

The first issue facing companies after the initial Intranet release is how to convert large numbers of existing paper documents into electronic format ready for distribution on an Intranet. There are many tools, such as HTML Transit, that can be used to convert documents from most electronic formats to HTML or XML format. Microsoft's Internet Assistant for Microsoft Word can also be used to easily convert existing Word documents into HTML or XML documents. After paper documents have been converted to HTML or XML and placed on the Intranet, the next challenge is to keep the documents up to date.

TABLE 2.1.1 Intranet Document Tracking Information

Data For Tracking Intranet Documents
Name of document
Document description
Page owner
Type of document (i.e., official, unofficial, personal)
Confidentiality status (i.e., confidential, non-confidential, etc.)
Original publish date
Date document last modified
Frequency of update (i.e., daily, weekly, monthly, etc.)

Obsolete information can frustrate Intranet users and may encourage them to revert to old ways of information gathering (i.e., calling people, walking to various offices, and writing memos). One way to minimize this problem is to create a database containing the document title, date of last change, and frequency of update in a database. Other useful information that can be used to track the status and nature of documents on the Intranet is shown in Table 2.1.1. A program can then be written to search the Intranet for documents that have not been updated recently. The program can then issue e-mail to the document owner to request an update.

Interface to Legacy Database(s)

Connecting databases to the Intranet is not an easy task, and may require additional staff or reassignment of current programming staff. Legacy database vendors are currently working on various Intranet solutions to facilitate the implementation of this requirement.

Companies may need to connect the Intranet to legacy databases in order to access:

- Financial reports (regarding project costs, product costs, the overall financial health of the enterprise, etc.)
- Document-management systems
- Human resources information (e.g., so employees can review details on health care and benefits)

Use of Multiple Servers

As the Intranet becomes more complex, multiple servers will be needed. This is especially true for companies that have a large number of divisions and business units using the Intranet. For example, a product-development group may need to provide team members the ability to search project-specific databases, submit forms to various databases, and to use a private on-line discussion group. The Webmaster may find it impossible to support these service needs in a timely manner. When this happens, companies frequently relegate the task of server maintenance to each respective department.

Over the next few years, installing and using a Web server will become as easy as installing and using word processor software. Web servers will probably become part of the Windows NT server operating system. When each department is responsible for maintaining their own Web server, it is particularly important to choose server software that is easy to install and maintain. A Pentium-class machine running Windows NT server software and Microsoft's IIS is a good choice for small departments. Another way to provide departments with their own domain name and disk space is to use a virtual domain name. Companies use virtual servers to reduce hardware costs. In the case of the Web, an HTTP-based server runs on a server computer. For example, a company may need two types of Web servers, one that allows easy access and one that requires usernames and passwords. In the past, the company would have purchased two different computers to run the Web server software. Today, however, the company can run both servers on the same system — as virtual servers.

Standardizing Hardware and Software

To avoid supporting multiple hardware and software components, it is important to standardize the server software, hardware, HTML and XML editing tools, and browser software. This will help to minimize the potential for unexpected network errors and incompatibilities.

2.1.6.2 Maintaining the Information Content on the Intranet

One of the major challenges organizations must face is how to transition from paper-based systems to computer-based systems, while keeping information up to date.

Automating HTML/XML Authoring

After establishing a policy for the distribution of Intranet documents, it is advisable to develop a set of guidelines that clearly specifies who is responsible for keeping them current. Inaccurate information greatly reduces the effectiveness of the Intranet. If employees lose confidence in the accuracy of the on-line information, they will revert to calling people to find information. Unfortunately, many people tend to ignore the need to update information, irrespective of its form (i.e., electronic or print).

In some cases, the Intranet will contain information that employees must update daily, weekly, or monthly. Spreadsheets can be used to capture highly time-sensitive data. Macros can be written (typically, by a staff programmer) that automatically convert the spreadsheet data into HTML or XML format.

Managing Document Links

In a traditional document-management system, documents often reference one another. In most cases, authors list the applicable references at the top of each new document. Intranets, unfortunately, create a situation where organizations cannot easily control the accuracy of links in documents.

HTML or XML document developers can use links freely and, in many cases, without checking the accuracy of those links. Even if employees test the initial accuracy of their document links, it is difficult to maintain and check the accuracy of those links after the document is released. If you have ever encountered a "broken" link when surfing the Web, you know that it can be frustrating. People depend on links within a Web document to find information. Today, however, there are a few mechanisms available to assure the accuracy of document links. Employees must understand that other people may link to their pages, and that they should not freely move the location of their documents. Employees must view their needs in the total organizational context.

2.1.6.3 Centralized vs. Distributed Control

The implementation of an Intranet is a major change for any organization. Although change is not easy, people are more inclined to modify their behavior when leaders have a clear sense of direction, involve employees in developing that direction, and are able to demonstrate how the Intranet will positively affect the employees' well being. Managers should work with their employees to show that Intranets can free them from the routine aspects of their job. This, in turn, will allow employees to spend more time learning and developing new ideas for the corporation.

Some of the benefits that can be obtained using a distributed model of Intranet control are:

- Employees can tap into the knowledge of everyone in the organization, making everyone a part of a solution.
- The power of any one Webmaster to dictate the Intranet's form and function is limited.
- It empowers departments to create their own information databases and to work with outside customers and vendors.

2.1.7 Intranet Security Issues

By their very nature, Intranets encourage a free flow of information. This means that it is also very easy for information to flow directly from the Intranet to the desktops of those who might seek to gain access to information they should not have. To guard against this situation, adequate security measures should be in place when the Intranet is deployed. In the discussion that follows, we review various security techniques to protect an Intranet from unauthorized external and internal use.

2.1.7.1 Firewalls

The Internet was designed to be resistant to network attacks in the form of equipment breakdowns, broken cabling, and power outages. Unfortunately, the Internet today needs additional technology to prevent attacks against user privacy and company security. Luckily, a variety of hardware and software solutions exist to help protect an Intranet. The term *firewall* is a basic component of network security. A firewall is a collection of hardware and software that interconnects two or more networks and, at the same time, provides a central location for managing security. It is essentially a computer specifically fortified to withstand various network attacks. Network designers place firewalls on a network as a first line of network defense. It becomes a "choke point" for all communications that lead in and out of an Intranet. By centralizing access through one computer (known as a *firewall-bastion host*), it is easier to manage the network security and to configure appropriate software on one machine. The bastion host is also sometimes referred to as a *server*.

The firewall is a system that controls access between two networks. Normally, installing a firewall between an Intranet and the Internet is a way to prevent the rest of the world from accessing a private Intranet. Many companies provide their employees with access to the Internet long before they give them access to an Intranet. Thus, by the time the Intranet is deployed, the company has typically already installed a connection through a firewall. Besides protecting an Intranet from Internet users, the company may also need to protect or isolate various departments within the Intranet from one another, particularly when sensitive information is being accessed via the Intranet. A firewall can protect the organization from both internal and external security threats.

Most firewalls support some level of encryption, which means data can be sent from the Intranet, through the firewall, encrypted, and sent to the Internet. Likewise, encrypted data can come in from the Internet, and the firewall can decrypt the data before it reaches the Intranet. By using encryption, geographically dispersed Intranets can be connected through the Internet without worrying about someone intercepting and reading the data. Also, a company's mobile employees can use encryption when they dial into your system (perhaps via the Internet) to access private Intranet files.

In addition to firewalls, a router can be used to filter out data packets based on specific selection criteria. Thus, the router can allow certain packets into the network while rejecting others.

One way to prevent outsiders from gaining access to an Intranet is to physically isolate it from the Internet. The simplest way to isolate an Intranet is to not physically connect it to the Internet. Another method is to connect two sets of cables, one for the Intranet and the other for the Internet.

Even without a connection to the Internet, an organization is susceptible to unauthorized access. To reduce the opportunity for intrusions, a policy should be implemented that requires frequent password changes and keeping that information confidential. For example, disgruntled employees, including those who have been recently laid off, can be a serious security threat. Such employees might want to leak anything from source code to company strategies to the outside. In addition, casual business conversations, overheard in a restaurant or other public place, may lead to a compromise in security. Unfortunately, a firewall cannot solve all these specific security risks.

It should be noted that a firewall can not keep viruses out of a network. Viruses are a growing and very serious security threat. Prevention of viruses from entering an Intranet from the Internet by users who upload files is necessary. To protect the network, everyone should run anti-virus software on a regular basis.

The need for a firewall implies a connection to the outside world. By assessing the types of communications expected to cross between an Intranet and the Internet, one can formulate a specific firewall design. Some of the questions that should be asked when designing a firewall strategy include:

- Will Internet-based users be allowed to upload or download files to or from the company server?
- Are there particular users (such as competitors) that should be denied all access?
- Will the company publish a Web page?
- Will the site provide telnet support to Internet users?
- Should the company's Intranet users have unrestricted Web access?
- Are statistics needed on who is trying to access the system through the firewall?
- Will a dedicated staff be implemented to monitor firewall security?
- What is the worst-case scenario if an attacker were to break into the Intranet? What can be done to limit the scope and impact of this type of scenario?
- Do users need to connect to geographically dispersed Intranets?

There are three main types of firewalls: network level, application level, and circuit level. Each type of firewall provides a somewhat different method of protecting the Intranet. Firewall selection should be based on the organization's security needs.

Network, Application, and Circuit-level Firewalls

Network-level Firewall

A network-level firewall is typically a router or special computer that examines packet addresses, and then decides whether to pass the packet through or to block it from entering the Intranet. The packets contain the sender and recipient IP address, and other packet information. The network-level router recognizes and performs specific actions for various predefined requests. Normally, the router (firewall) will examine the following information when deciding whether to allow a packet on the network:

- Source address from which the data is coming
- Destination address to which the data is going
- Session protocol such as TCP, UDP, or ICMP
- Source and destination application port for the desired service
- Whether the packet is the start of a connection request

If properly installed and configured, a network-level firewall will be fast and transparent to users.

Application-level Firewall

An application-level firewall is normally a host computer running software known as a proxy server. A proxy server is an application that controls the traffic between two networks. When using an application-level firewall, the Intranet and the Internet are not physically connected. Thus, the traffic that flows on one network never mixes with the traffic of the other because the two network cables are not connected. The proxy server transfers copies of packets from one network to the other. This type of firewall effectively masks the origin of the initiating connection and protects the Intranet from Internet users.

Because proxy servers understand network protocols, they can be configured to control the services performed on the network. For example, a proxy server might allow ftp file downloads, while disallowing ftp file uploads. When implementing an application-level proxy server, users must use client programs that support proxy operations.

Application-level firewalls also provide the ability to audit the type and amount of traffic to and from a particular site. Because application-level firewalls make a distinct physical separation between an Intranet and the Internet, they are a good choice for networks with high-security requirements. However, due to the software needed to analyze the packets and to make decisions about access control, application-level firewalls tend to reduce the network performance.

Circuit-level Firewalls

A circuit-level firewall is similar to an application-level firewall in that it, too, is a proxy server. The difference is that a circuit-level firewall does not require special proxy–client applications. As discussed in the previous section, application-level firewalls require special proxy software for each service, such as ftp, telnet, and HTTP.

In contrast, a circuit-level firewall creates a circuit between a client and server without needing to know anything about the service required. The advantage of a circuit-level firewall is that it provides service for a wide variety of protocols, whereas an application-level firewall requires an application-level proxy for each and every service. For example, if a circuit-level firewall is used for HTTP, ftp, or telnet, the applications do not need to be changed. You simply run existing software. Another benefit of circuit-level firewalls is that they work with only a single proxy server, making it easier to manage, log, and control a single server than multiple servers.

2.1.7.1.1 *Firewall Architectures*

Combining the use of both a router and a proxy server into the firewall can maximize the Intranet's security. The three most popular firewall architectures are the dual-homed host firewall, the screened host firewall, and the screened subnet firewall. The screened-host and screened-subnet firewalls use a combination of routers and proxy servers.

Dual-homed Host Firewalls

A dual-homed host firewall is a simple, yet very secure configuration in which one host computer is dedicated as the dividing line between the Intranet and the Internet. The host computer uses two separate network cards to connect to each network. When using a dual-home host firewall, the computer routing capabilities should be disabled, so the two networks do not accidentally become connected. One of the drawbacks of this configuration is that it is easy to inadvertently enable internal routing.

Dual-homed host firewalls use either an application-level or a circuit-level proxy. Proxy software controls the packet flow from one network to another. Because the host computer is dual-homed (i.e., it is connected to both networks), the host firewall can examine packets on both networks. It then uses proxy software to control the traffic between the networks.

Screened-host Firewalls

Many network designers consider screened-host firewalls more secure than a dual-homed host firewall. This approach involves adding a router and placing the host computer away from the Internet. This is a very effective and easy-to-maintain firewall. A router connects the Internet to your Intranet and, at the same time, filters packets allowed on the network. The router can be configured so that it sees only one host computer on the Intranet network. Users on the network who want to connect to the Internet must do so through this host computer. Thus, internal users appear to have direct access to the Internet, but the host computer restricts access by external users.

Screened-subnet Firewalls

A screened-subnet firewall architecture further isolates the Intranet from the Internet by incorporating an intermediate perimeter network. In a screened-subnet firewall, a host computer is placed on a perimeter network which users can access through two separate routers. One router controls Intranet traffic and the second controls the Internet traffic. A screened-subnet firewall provides a formidable defense against attack. The firewall isolates the host computer on a separate network, thereby reducing the impact of an attack to the host computer. This minimizes the scope and chance of a network attack.

2.1.7.2 CGI Scripting

Web Sites that provide two-way communications use CGI (common gateway scripting). For example, if you fill in a form and click your mouse on the form's Submit button, your browser requests the server computer to run a special program, typically a CGI script, to process the form's content. The CGI script runs on the server computer, which processes the form. The server then returns the output to the browser for display.

From a security perspective, the danger of CGI scripts is that they give users the power to make a server perform a task. Normally, the CGI process works well, providing an easy way for users to access information. Unfortunately, it is also possible to use CGI scripts in ways they were never intended. In some cases, attackers can shut down a server by sending potentially damaging data through the use of CGI scripts. From a security perspective, it is important to make sure that users cannot use CGI scripts to execute potentially damaging commands on a server.

2.1.7.3 Encryption

Encryption prevents others from reading your documents by "jumbling" the contents of your file in such a way that it becomes unintelligible to anyone who views it. You must have a special key to decrypt the file so its contents can be read. A key is a special number, much like the combination of a padlock, which the encryption hardware or software uses to encrypt and decrypt files. Just as padlock numbers have a certain number of digits, so do encryption keys. When people talk about 40-bit or 128-bit keys, they are simply referring to the number of binary digits in the encryption key. The more bits in the key, the more secure the encryption and less likely an attacker can guess your key and unlock the file. However, attackers have already found ways to crack 40-bit keys.

Several forms of encryption can be used to secure the network, including: link encryption, document encryption, secure-sockets layer (SSL), and secure HTTP (S-HTTP). The following sections describe these encryption methods in more detail.

2.1.7.3.1 Public-key Encryption

Public-key encryption uses two separate keys: a public key and a private key. A user gives his/her public key to other users so anyone may send them encrypted files. The user activates his/her private key to decrypt the files (which were encrypted with a public key).

A public key only allows people to encrypt files, not to decrypt them. The private user key (designed to work in conjunction with a particular public key) is the only key that can decrypt the file. Therefore, the only person that can decrypt a message is the person holding the private key.

2.1.7.3.2 Digital Signatures

A digital signature is used to validate the identity of the file sender. A digital signature prevents clever programmers from forging e-mail messages. For example, a programmer who is familiar with e-mail protocols can build and send an e-mail using anyone's e-mail address, such as BillGates@microsoft.com.

When using public-key encryption, a sender encrypts a document using a public key, and the recipient decodes the document using a private key. With a digital signature, the reverse occurs. The sender uses a private key to encrypt a signature, and the recipient decodes the signature using a public key. Because the sender is the only person who can encrypt his or her signature, only the sender can authenticate messages. To obtain a personal digital signature, you must register a private key with a certificate authority (CA), which can attest that you are on record as the only person with that key.

2.1.7.3.3 Link Encryption

Link encryption is used to encrypt transmissions between two distant sites. It requires that both sites agree on the encryption keys that will be used. It is commonly used by parties that need to communicate with each other frequently. Link encryption requires a dedicated line and special encryption software. It is an expensive way to encrypt data. As an alternative to this, many routers have convenient built-in encryption options. The most common protocols used for link encryption are PAP (Password Authentication) and CHAP (Challenge Handshake Authentication Protocol). Authentication occurs at the data link layer and is transparent to end-users.

2.1.7.3.4 Document Encryption

Document encryption is a process by which a sender encrypts documents that the recipient(s) must later decrypt. Document encryption places the burden of security directly on those involved in the communication. The major weakness of document encryption is that it adds a step to the process by which a sender and receiver exchange and receive documents. Because of this extra step, many users prefer to save time by skipping the encryption. The primary advantage of document encryption is that anyone with an e-mail account can use document encryption. Many document encryption systems are available free or for little cost on the Internet.

2.1.7.3.5 Pretty Good Privacy (PGP)

Pretty good privacy (PGP) is a free (for personal use) e-mail security program developed in 1991 to support public-key encryption, digital signatures, and data compression. PGP is based on a 128-bit key. Before sending an e-mail message, PGP is used to encrypt the document. The recipient also uses PGP to decrypt the document. PGP also offers a document compression option. Besides making a document smaller, compression enhances the file security because compressed files are more difficult to decode without the appropriate key. According to the PGP documentation, it would take 300 billion years for someone to use brute force methods to decode a PGP-encrypted, compressed message.

2.1.7.3.6 Secure Socket Layer (SSL)

The Secure Socket Layer (SSL) was developed by Netscape Communications to encrypt TCP/IP communications between two host computers. SSL can be used to encrypt any TCP/IP protocol, such as HTTP, telnet, and ftp. SSL works at the system level. Therefore, any user can take advantage of SSL because the SSL software automatically encrypts messages before they are put onto the network. At the recipient's end, SSL software automatically converts messages into a readable document.

SSL is based on public-key encryption and works in two steps. First, the two computers wishing to communicate must obtain a special session key (the key is valid only for the duration of the current communication session). One computer encrypts the session key and transmits the key to the other computer. Second, after both sides know the session key, the transmitting computer uses the session key to encrypt messages. After the document transfer is complete, the recipient uses the same session key to decrypt the document.

2.1.7.3.7 Secure HTTP (S-HTTP)

Secure HTTP is a protocol developed by the CommerceNet coalition. It operates at the level of the HTTP protocol. S-HTTP is less widely supported than Netscape's Secure Socket Layer. Because S-HTTP works only with HTTP, it does not address security concerns for other popular protocols, such as ftp and telnet.

S-HTTP works similarly to SSL in that it requires both the sender and receiver to negotiate and use a secure key. Both SSL and S-HTTP require special server and browser software to perform the encryption.

2.1.7.4 Intranet Security Threats

This section examines additional network threats that should be considered when implementing Intranet security policies.

2.1.7.4.1 Source-routed Traffic

As discussed earlier, packet address information is contained in the packet header. When source routing is used, an explicit routing path for the communication can be chosen. For example, a sender could map a route that sends packets from one specific computer to another, through a specific set of network nodes. The road map information contained in the packet header is called *source routing*, and it is used mainly to debug network problems. It is also used in some specialized applications. Unfortunately, clever programmers can also use source routing to gain (unauthorized) access into a network. If a source-routed packet is modified so that it appears to be from a computer within your network, a router will obediently perform the packet routing instructions, permitting the packet to enter the network, *unless special precautions are taken*. One way to combat such attacks is simply to direct your firewall to block all source-routed packets. Most commercial routers provide an option to ignore source-routed packets.

2.1.7.4.2 Protecting Against ICMP Redirects (Spoofing)

ICMP stands for Internet Control Message Protocol. ICMP defines the rules routers use to exchange routing information. After a router sends a packet to another router, it waits to verify that the packet actually arrived at the specified router. Occasionally, a router may become overloaded or may malfunction. In such cases, the sending router might receive an ICMP-redirect message that indicates which new path the sending router should use for transmission.

It is fairly easy for knowledgeable "hackers" to forge ICMP-redirect messages to reroute communication traffic to some other destination. The term *spoofing* is used to describe the process of tricking a router into rerouting messages in this way. To prevent this type of unauthorized access, it may be necessary to implement a firewall that will screen ICMP traffic.

2.1.8 Summary

Intranets are being used to improve the overall productivity of the organization. Important Intranet concepts covered in this chapter are summarized below:

- TCP/IP was created because of the need for reliable networks that could span the globe. Because of its reliability and ease of implementation, TCP/IP has become the standard language (or protocol) of the Internet. TCP/IP defines how programs exchange information over the Internet.
- An Intranet is based on Internet technology. It consists of two types of computers: a client and a server. A client asks for and uses information that the server stores and manages.
- Telnet, ftp, and gopher are widely used network programs that help users connect to specific computers and to transfer and exchange files.
- The World Wide Web (or Web) is a collection of interlinked documents that users can access and view as "pages" using a special software program called a browser. The two most popular browser programs are Netscape Navigator and Microsoft Internet Explorer.

- HTML (hypertext markup language) and XML (extended markup language) are used to describe the layout and contents of pages on the Web.
- Java is a new computer programming language that allows users to execute special programs (called applets) while accessing and viewing a Web page.
- A network computer is a low-cost, specialized computer designed to work in conjunction with the Internet and Java application programs.

To be effective, the Intranet must deliver quality information content. To ensure this, management must be in a proactive role in assigning staff who will keep the corporate information reservoirs on the Intranet current and relevant. The following is a checklist of some of the ways to encourage the development of a high-quality Intranet:

- Give users access to document management systems and various corporate databases.
- Distribute the responsibility of maintaining the Intranet to increase the number of staff involved in developing and enhancing Intranet content.
- Create a corporate culture based on information sharing.
- Place employee training at the center of an Intranet deployment strategy.
- Design and implement appropriate security measures as soon as possible.
- Use firewalls to control access to the network.
- Use anti-virus software.
- Implement a security plan that controls the access that employees and outsiders have to the network.
- Design and implement CGI scripts with security in mind.
- Encourage users to encrypt files before sending confidential data across the Internet/Intranet.

2.2 Internet Security

John Braun

The Internet offers the ability for anyone, from an individual computer owner to a major multinational corporation, to make information and computing resources available to the world. Alas, since the Internet was initially designed with ease of communications, rather than security, in mind, care must be taken to ensure that sensitive information is not made available to the world as well. This section will discuss several aspects of protection information that one makes available via a network or the Internet.

2.2.1 Physical Security

Although not as trendy or exciting as some of the exotic attacks that can be made against a network on the protocol level, physical security is nonetheless important. You can have the best security software in the world installed on your network, but it may do you little good if an attacker can waltz up to a key piece of network or computing equipment and disable it!

Key pieces of network hardware, such as routers, firewalls, and servers, should be stored in a secure room with some sort of access control such as a traditional or electronic lock, card reader, or other means which can limit access to authorized individuals. Access to especially sensitive devices should be further restricted by placing them in a locked cabinet. Also, access to buildings that contain these rooms should also be controlled with security guards or other access control so that visitors can't wander about.

Exposed network cables, especially those connected to routers, hubs, and other devices which carry traffic for several other users, should also be physically protected. If an attacker has access to these cables, they could physically cut them and insert equipment that could monitor and even generate network traffic. For maximum protection, network cables should be placed inside of pressurized pipes with sensors placed on various locations along the piping. Network monitoring tools should also be used so that a break in a cable can be identified quickly.

2.2.2   Modems

Modems present two security threats. First, modems offer a channel for data to leave your premises, circumventing security and auditing measures that may be in place for the rest of the network. A review of services that are accessed by modem should be made, and, if possible, this access should be rerouted over a secure internal network. Second, modems offer a potential method for unauthorized individuals to access your network from the outside. Since there may be a need for users who are on the road or working from home to access a network remotely, additional security measures need to be taken for these connections. One measure is a dial-back modem, which will call back the user at a predetermined number before allowing access to the network. Another is a system where each user is provided with an electronic card that displays a random number every few minutes. A similar device that performs the same calculation to produce this number is located on the network one wishes to access. Without the card, and the ability to produce this number, the remote user is denied access.

2.2.3   Data Security

There are many aspects to data security. One is to prevent files from being viewed by someone other than the owner. On a multi-user system, this concern can be addressed by proper system administration. Users should not be allowed access to directories or files that do not belong to them. On a single-user system, the ability to share files over the network should be closely monitored, so that users don't inadvertently allow access of their hard drive contents to anyone who cares to look. Another aspect of data security is to protect the contents of file or network data via encryption. This can be especially important for data travelling over a network, since the data can be broadcast to other terminals or network devices which are not the intended recipients of the data. By using encryption, data that falls into the wrong hands will be unusable unless an encryption key or password is also known. Although security at the TCP/IP level is still a ways off, several third-party products provide the ability to protect and encrypt data.

2.2.4   Passwords

Since passwords usually comprise an initial layer of defense against an attack, they should be chosen and implemented with care. A written policy and/or enforcement by the operating system can help. Passwords should not be dictionary words, should be as long as possible, contain a series of letters, numbers, and other characters, and be changed on a regular basis. When deciding on a policy, care should be taken to balance security needs vs. ease of use. If a policy is too tedious to follow, users may end up writing their passwords down somewhere near their terminal, eliminating any sort of benefit the password policy could have offered.

2.2.5   Workstation Security

Unattended workstations could be a great danger to the entire system, and a security system could be completely wasted if an unauthorized person could access someone else's logged-in workstation. For that reason, users need to be aware of this danger and be properly trained how to secure an unattended workstation, either by logging off or by using a screen saver or screen lock which activates after a short amount of inactivity.

2.2.6   TCP/IP Security

Since the current version of TCP/IP was designed to provide a robust, standard method of moving data on a network, rather than security, one should be aware of several attacks which could compromise network security or availability.

2.2.6.1   IP Spoofing

Spoofing is the act of altering the contents of a TCP or IP packet header in order to trick the remote system into thinking the packet is valid. One trick is to change the source IP address of a packet to an address that is valid on a network behind a firewall or router. Older equipment that would have otherwise blocked the packet will allow it to go through since it appears to be coming from a friendly network. There are attacks where a connection can be hijacked or terminated by combining IP address spoofing with the spoofing of the SEQ and ACK fields in a TCP header. The SEQ and ACK fields help synchronize traffic between two hosts. If these fields are modified by attackers, the attackers can take over connections, while legitimate hosts lose the connection since their packets now appear to be out of order. Additional fields could be activated so that the connection is terminated prematurely.

2.2.6.2 Denial of Service (DoS)

Many DoS attacks take advantage of nuances in the method used to establish a TCP/IP connection. Since connections may take a while to establish, portions of the TCP/IP establishment process include timeouts so that slow equipment or busy networks will not cause a connection attempt to fail. However, a program which intentionally completes only a portion of this negotiation will result in a host waiting for a connection to complete, when it never will. While the host is waiting for the connection attempt to time out, system resources are being used. If enough of these bogus attempts are made, the host will run out of resources, and future connection attempts will be refused.

Another major type of attack involves the sending of single packets, whose contents have been modified to some unexpected or invalid data. This can result in the remote system crashing with a nasty Blue Screen of Death, system bomb, or core dump. One attack sends an ICMP (a special type of IP packet, with no TCP) echo request, also known as a ping, whose data payload is very large. Since a ping packet normally has no data associated with it, some implementations that don't expect this data will grind to a halt when receiving this type of packet. Another attack interferes with the data offset field in the TCP header, so that the remote host is tricked into trying to read packet data where none exists, also causing a crash.

2.3 Virtual Private Networking Solutions

Endre Sara

Today's large corporate networks are geographically distributed and clients or employees need access to corporate information from different locations. The cost of a long-distance dialup session is very high, and it is also not efficient to deploy point-to-point connection between each possible location.

Virtual private network (VPN) is a concept of securely transferring sensitive corporate information between various geographically dispersed sites over a public network, such as the Internet. The market numbers for these services tell a success story and a win/win situation for both providers and users. But, while the market numbers are good, numerous concerns remain. They are:

- There are not enough integration services to help users deploy VPNs
- The products are not yet interoperable
- Security standards are not yet unique
- There are different protocol standards
- Many users are not yet fully comfortable using Internet technologies in the mainline business

Typical users of VPNs are driving the implementation of VPN services. The most important benefits and points to consider are summarized in Table 2.3.1. Weights are not included in this consideration.

This document describes and compares the available standards and products for VPN solutions.

The VPN is a network that uses a private address space which operates over another network infrastructure. It means that the VPN will use the same physical cabling, switches, bridges, and routers, but it uses a different address space. This is accomplished by encapsulating the VPN traffic (which doesn't have to be IP) into secure protocols. The emerging standards concentrate around Layer 2 and Layer 3 protocols.

TABLE 2.3.1 VPN Benefits and Concerns

VPN Applications	Benefits	Points to Consider
Dial access for remote users	Outsource modems reduce dial-in costs	Client software
	Eliminate access lines	Are appropriate tunneling protocols supported in client software?
Connecting branch offices	Reduces number of dedicated lines	Encryption performance issues
	Lets IT managers consolidate central-site WAN equipment	Does VPN access-control system integrate with existing user access privileges?
Extranet	Gives trading partners and customers access to intranet	Does system scale well?
	Makes collaborating with contractors and consultants much easier	Are there tools to handle the administrative burden of adding new users?
New business	Can create just-in-time networks for short-term projects	Interoperability of different VPN equipment
	Can give worldwide sites access much sooner than waiting for leased lines	Management of mixed equipment environment is not easy

2.3.1 Layer 2 Protocols

Layer 2 protocols enable the transfer of data from a remote client to the private network of an enterprise by creating a virtual private network most often across a TCP/IP-based data network. Layer 2 protocols support on-demand, multi-protocol virtual private networking over public networks, such as the Internet. Internet access is provided by an internet service provider (ISP), who wishes to offer services other than traditional registered IP address-based service to dial-up users of the network.

This architecture is transparent to the end systems. In case of connecting two distant local area networks (LANs) through a VPN, the users will notice no difference while their traffic is being encapsulated in IP packets and transmitted to the remote VPN access server, which puts them back to the remote LAN. If a remote user wants to connect to the private network of the enterprise through a VPN connection, his/her computer has to support the implemented VPN protocol to be able to encapsulate the traffic. Although this encapsulation provides some security against intercepting the actual data, additional encryption should be implemented to provide secure communication.

2.3.1.1 Point-to-Point Tunneling Protocol (PPTP)

The PPTP networking technology is defined as an extension to the remote access Point-to-Point Protocol (RFC1171). PPTP is a network protocol that encapsulates PPP packets into IP datagrams for transmission over the Internet or other public TCP/IP-based networks.

After the client has made the initial PPP connection to the ISP, a second dial-up networking call is made over the existing PPP connection. Data sent using this second connection is in the form of IP datagrams that contain PPP packets, referred to as *encapsulated PPP packets*.

The second call creates the VPN connection to a PPTP server on the private enterprise LAN, referred to as a *tunnel*. This is shown in Figure 2.3.1.

The secure communication using the PPTP protocol typically involves three processes, each of which requires successful completion of the previous process:

PPP Connection and Communication — The PPTP client uses PPP to connect to an ISP by using a standard phone line or ISDN line. This connection uses the PPP protocol to establish the connection and encrypt data packets.

PPTP Control Connection — Using the connection to the Internet established by the PPP protocol, the PPTP protocol creates a control connection from the PPTP client to the PPTP server over the Internet. This connection uses TCP to establish the connection and is called a PPTP tunnel.

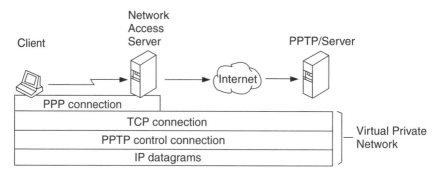

FIGURE 2.3.1   The PPTP tunnel.

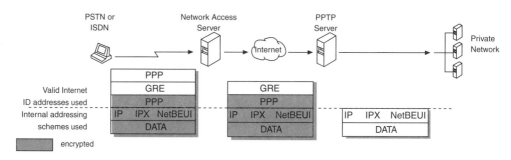

FIGURE 2.3.2   Connecting a dial-up networking PPTP client to the private network.

PPTP Data Tunneling — Finally, the PPTP protocol creates IP datagrams containing encrypted PPP packets, which are sent through the PPTP tunnel to the PPTP server. The PPTP server disassembles the IP datagrams and decrypts the PPP packets, and then routes the decrypted packets to the private network.

Note that the encapsulated PPP packet can contain multi-protocol data such as TCP/IP, IPX, or NetBEUI protocols. Because the PPTP server is configured to communicate across the private network by using private network protocols, it is able to read multi-protocol packets.

Figure 2.3.2 illustrates the multi-protocol support built into PPTP. A packet sent from the PPTP client to the PPTP server passes through the PPTP tunnel to a destination computer on the private network.

PPTP encapsulates the encrypted and compressed PPP packets into IP datagrams for transmission over the Internet. These IP datagrams are routed over the Internet until they reach the PPTP server that is connected to the Internet and the private network. The PPTP server disassembles the IP datagram into a PPP packet and then decrypts the PPP packet using the network protocol of the private network. As mentioned earlier, the network protocols on the private network that are supported by PPTP are IPX, NetBEUI, or TCP/IP.

An ISP's network access server may require initial dial-in authentication. If this authentication is required, it is strictly to log on to the ISP network; it is not related to the PPTP server authentication.

There are different options to provide data encryption between the PPTP client and the server. Microsoft uses the RAS "shared secret" encryption process. It is referred as a shared secret, because both ends of the connection share the encryption key. Under Microsoft's implementation of RAS, the shared secret is the user password. Other encryption methods base the encryption on some key available in public; this method is known as *public key encryption*. Microsoft's PPTP uses the PPP encryption and PPP compression schemes called Microsoft Point-to-Point Encryption (MPPE). The Compression Control Protocol (CCP) used by PPP is used to negotiate encryption. The encryption key is derived from the hashed password stored at both the client and the server. The RSA RC4 standard is used to create the 40-bit session key based on the client password. This key is used to encrypt all data that is passed over the Internet, keeping the connection private and secure.

PPTP is aimed primarily at Internet-based remote access. The main advantages are multi-protocol support and simplicity, because it functions on the Layer 2 level. This can be a preferred solution in a multi-protocol environment, but data security is a concern. The proposed standard does not provide a data encryption solution, although Microsoft has a vendor-specific solution as discussed earlier. The other difference is, compared to Layer 3 solutions, PPTP only provides a single point-to-point connection. It means that there can be no simultaneous Internet access while using a VPN connection. With multi-point tunneling, such as a Layer 3 solution discussed later, a user could have an Internet session at the same time as several VPN connections. This is also an inherent consequence from the PPTP architecture being a client–server model-based solution, while Layer 3 solutions are based on a more general host-to-host model.

2.3.1.2 Layer 2 Forwarding (L2F)

L2F achieves private network access through a public system by building a secure "tunnel" across the public infrastructure that connects directly to a user's home gateway. Multiple corporate networks can use a single local telephone number terminated on a service provider's dialup switch or access server. The access server establishes identity, sets up a private tunnel to the user's home gateway router, and tunnels clients to that gateway. The gateway is responsible for authentication of the remote user, thereby ensuring client control of access security and addressing.

A key component of the virtual dialup service is tunneling, a vehicle for encapsulating packets inside a protocol that is understood at the entry and exit points of a given network. These entry and exit points are defined as a tunnel interfaces. The tunnel interface itself is similar to a hardware interface, but is configured in software.

Figure 2.3.3 shows the format in which a packet would traverse the network within a tunnel.

Tunneling involves the following three types of protocols:

- The passenger protocol is the protocol being encapsulated; in a dialup scenario, this protocol could be PPP, SLIP, or text dialog
- The encapsulating protocol is used to create, maintain, and tear down the tunnel. Cisco supports several encapsulating protocols including the L2F protocol, which is used for virtual dialup services
- The carrier protocol is used to carry the encapsulated protocol; IP will be the first carrier protocol used by the L2F protocol, because of IP's robust routing capabilities, ubiquitous support across different medias, and deployment within the Internet

No dependency exists between the L2F protocol and IP. In subsequent releases of the L2F functionality, Frame Relay, X.25 VCs, and asynchronous transfer mode (ATM) switched virtual circuits (SVCs) could be used as a direct Layer 2 carrier protocol for the tunnel.

Cisco's L2F implementation provides several management features. End system transparency ensures that neither the remote end system nor its corporate hosts should require any special software to use this service. Authentication is provided by dialup PPP, Challenge Handshake Authentication Protocol (CHAP), or Password Authentication Protocol (PAP), including Terminal Access Controller Access Control System Plus (TACACS+) and Remote Authentication Dial-In User Service (RADIUS) solutions, as well as support for smart cards and one-time passwords; the authentication will be manageable by the user independent of the ISP. Addressing will be as manageable as dedicated dialup solutions; the address will be assigned by the remote user's respective corporation, and not the ISP. Authorization will be

IP/UDP	L2F	PPP (Data)
Carrier Protocol	Encapsulator Protocol	Passenger Protocol

FIGURE 2.3.3 Tunneling packet format.

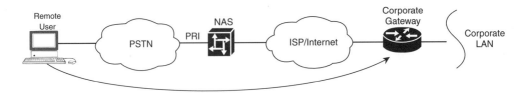

FIGURE 2.3.4   Virtual dialup topology.

managed by the corporation's remote users, as it would be in a direct dialup solution. Accounting will be performed both by the ISP (for billing purposes) and by the user (for charge back and auditing).

Figure 2.3.4 illustrates the topology of a virtual private connection using Cisco's L2F.

In a traditional dialup scenario, the ISP using the NAS in conjunction with a security server follows an authentication process by challenging the remote user for both the username and password. If the remote user passes this phase, the authorization phase can begin.

For the virtual dialup service, the ISP pursues authentication to the extent required to discover the users' apparent identity (and by implication, their desired corporate gateway). No password interaction is performed at this point. As soon as the corporate gateway is determined, a connection is initiated with the authentication information gathered by the ISP. The corporate gateway completes the authentication by either accepting or rejecting the connection. (For example, the connection is rejected in a PAP request in which the username or password are found to be incorrect.) Once the connection is accepted, the corporate gateway can pursue another phase of authentication at the PPP layer. These additional authentication activities are outside the scope of the specification, but might include proprietary PPP extensions, or textual challenges carried within a TCP/IP Telnet session.

For each L2F tunnel established, L2F tunnel security generates a unique random key to resist spoofing attacks. Within the L2F tunnel, each multiplexed session maintains a sequence number to prevent the duplication of packets.

Cisco provides the flexibility of allowing users to implement compression at the client end. In addition, encryption on the tunnel can be done using IPSEC.

There are similar advantages of L2F as of PPTP, because the two solutions are similar, and being merged to a common standard called L2TP. L2F also has a nice advantage of connecting multi-protocol networks, because of its Layer 2 functionality. But again there is no support in the proposed standard for VPN data encryption. Cisco refers to the IPSec standard as a possible encryption method for IP traffic carried with L2F. For non-IP traffic L2F lacks the solution for security. In comparison with Layer 3 solutions, L2F provides only a single point-to-point connection, which makes parallel Internet access impossible while being connected to the VPN.

2.3.1.3   L2TP

The IETF draft for PPTP titled as "Point-to-Point Tunneling Protocol," draft-ietf-pptp-00.txt was submitted to the Internet Engineering Task Force (IETF) in June 1996 by the companies of the PPTP Forum, which includes Microsoft Corporation, Ascend Communications, 3Com/Primary Access, ECI Telematics and US Robotics. Cisco's proposal for L2F was submitted to the IETF for approval as a proposed standard. Northern Telecom, Inc., and Shiva Corporation have announced their support for L2F. At the June 1996 IETF meeting in Montreal, the IETF PPP Extensions working group agreed to combine Cisco's proposal with PPTP proposed by Microsoft Corporation. The emerging proposed standard, Layer 2 Tunneling Protocol (L2TP) is currently drafted by Cisco Systems, Microsoft, Ascend, 3Com, and US Robotics. The latest L2TP draft (09) was submitted in January 1998.

A typical connection scenario would start with the remote user initiating a PPP connection to an ISP via either the PSTN or ISDN. The ISP's access point (LAC) accepts the connection and the PPP link is established. The ISP may now undertake a partial authentication of the end system/user. Only the username field would be interpreted to determine whether the user requires a Virtual dial-up service. It is expected — but not required — that usernames will be structured (e.g., username@company.com). Alternatively, the ISP may maintain a database mapping users to services. In the case of Virtual dial-up, the mapping will name a specific endpoint, the L2TP Network Server (LNS). If a virtual dial-up service is not required, standard access to the Internet may be provided.

If no tunnel connection currently exists to the desired LNS, one is initiated. L2TP is designed to be largely insulated from the details of the media over which the tunnel is established; L2TP requires only that the tunnel media provide packet-oriented point-to-point connectivity. Obvious examples of such media are UDP, Frame Relay PVCs, or X.25 VCs. Once the tunnel exists, an unused slot within the tunnel, a "Call ID," is allocated, and a connect indication is sent to notify the LNS of this new dial-up session. The LNS either accepts the connection, or rejects it. The initial connect notification may include the authentication information required to allow the LNS to authenticate the user and decide to accept or decline the connection. In the case of CHAP, the set-up packet includes the challenge, username, and raw response. For PAP or text dialog, it includes username and clear text password. The LNS may choose to use this information to complete its authentication, avoiding an additional cycle of authentication.

If the LNS accepts the connection, it creates a "virtual interface" for PPP in a manner analogous to what it would use for a direct-dialed connection. With this virtual interface in place, link layer frames may now pass over this tunnel in both directions. Frames from the remote user are received at the POP, stripped of CRC, link framing, and transparency bytes, encapsulated in L2TP, and forwarded over the appropriate tunnel. The LNS accepts these frames, strips L2TP, and processes them as normal incoming frames for the appropriate interface and protocol. The virtual interface behaves very much like a hardware interface, with the exception that the hardware in this case is physically located at the ISP POP. The other direction behaves analogously, with the LNS encapsulating the packet in L2TP, and the LAC stripping L2TP before transmitting it out via the physical interface to the remote user.

At this point, the connectivity is a point-to-point PPP session whose endpoints are the remote user's networking application on one end and the termination of this connectivity into the LNS's PPP support on the other. Because the remote user has become simply another dial-up client of the LNS, client connectivity can now be managed using traditional mechanisms with respect to further authorization, protocol access, and packet filtering.

Accounting can be performed at both the L2TP Access Concentrator (LAC) as well as the LNS. This document illustrates some accounting techniques that are possible using L2TP, but the policies surrounding such accounting are outside the scope of this specification.

For the virtual dial-up service, the ISP pursues authentication only to the extent required to discover the user's apparent identity (and by implication, their desired LNS). This may involve no more than detecting DNIS information when a call arrives, or may involve full LCP negotiation and initiation of PPP authentication. As soon as the apparent identity is determined, a call request to the LNS is initiated with any authentication information gathered by the ISP. The LNS completes the authentication by either accepting the call, or rejecting it. The LNS may need to protect against attempts by third parties to establish tunnels to the LNS. Tunnel establishment can include authentication to protect against such attacks.

L2TP, like other Layer 2 protocols, does not provide any further security for data encryption, but rather refers to Layer 3 encryption techniques for IP traffic, such as IPSec.

Although L2TP seems to be the result of different Layer 2 Tunneling initiatives, it still does not have the widest acceptance in the industry. Its predecessor PPTP is popular, because of the large number of Windows NT users, but IPSec, a general initiative to add security to the IP protocol, has the strongest support by manufacturers and suppliers. In a multi-protocol environment there will still be a need for Layer 2 Tunneling with the addition of encryption for IP traffic (using IPSec). But in an IP-only environment Layer 3 solutions are more effective. Unless augmented with IPSec these Layer 2 solutions cannot support extranets, because extranets require keys and key management.

In comparison with other Layer 2 protocols, L2TP has better features, such as an addition to PPTP and L2F, and the support for ATM or SONET as an underlying transmission medium, which is only planned for the other two protocols.

2.3.2   Layer 3 Tunneling Protocols

In the previously discussed architecture the PPP connection begins at the remote client and terminates at the corporate network's L2TP server, going through the ISP access point and the corporate gateway.

Layer 3 tunneling proposes a different scenario initiating the PPP connection from the remote client, but terminating it at the ISP. This requires the re-encapsulation of the PPP frame and transmitting the Layer 3 information only to the corporate access router. This access router does not need to support any additional standard, but it acts only as a simple router. The difference from traditional dial-up services is that the ISP's IP Gateway will provide the IP address to the client. It can roam with this address as long as the IP Gateway does the reframing of the IP PPP packet and sends it to the corporate gateway as a standard IP packet.

The advantage of the latter scheme is that there is no need to support L2TP at either the remote client end or at the corporate gateway end. The remote client only needs a standard IP stack, and the corporate gateway acts as a simple IP router. In this case if the remote node is a router connecting a sub-network to the corporate network, the packets can be routed just like any other traffic trough the ISP network. In either case an additional functionality is needed to provide security on the IP Layer (network layer). Although L2TP can encrypt the PPP packets on Layer 2, it does not claim to be secure against denial of service attacks or man-in-the-middle attacks (someone modifying the PPP frames in the tunnel).

2.3.2.1   IPSec

IPSec is a protocol suite defined by the IETF working group on IP security to secure communication at the Layer 3 (network layer) between communicating peers. The goal of the IPSec protocol suite is to provide secure tunneled transport of IP data only. Essentially, it takes private IP packets, performs data security functions such as encryption, authentication, and integrity, then wraps these secured packets in other IP packets for transport across the Net. Key management functions also will be a part of the IPSec protocol suite. The IETF has issued five requests for comments — RFC 1825 through 1829. An interesting note is that if IPv6 succeeds in replacing IPv4, IPSec will be the automatic Internet VPN standard since it is integrated into the IPv6 specifications.

Like the Layer 2 VPN protocols, IPSec works as a LAN-to-LAN and dialup-to-LAN solution. It is designed to support multiple encryption protocols, a feature that allows users to choose a desired amount of data privacy. Obviously, IPSec will only be of value to companies that want to tunnel IP exclusively since it doesn't support other data protocols.

There are several different scenarios where IPSec can be used. In case two hosts A and B want to communicate with each other through a firewall, the host A can tunnel packets to the firewall, the firewall can decrypt/authenticate the packets, and send them to B based on its rules. In a different setup there can be a secure tunnel between host A and host B, where the firewall is authorized to act as a key management proxy, and has the capability to decrypt the packets and apply its packet filtering policy. A third setup is a combination of the previous two, where the inner payload is secured from host A to B, and the outer payload is secured and tunneled through the firewall. The advantage of this scheme is that the firewall is able to authenticate packets and decide whether to allow the packet without applying its filtering rules. This is typical of what happens today, where an employee gets into the network via dialup PPP.

There is a different scheme, where packets have to be secured while travelling the Internet. In this case IPSec will secure packets between two or more border routers of a topologically distributed organization. In this case since security associations are set up between the border routers, any traffic should go through these routers. All packets between the two routers must contain valid IPSec, otherwise they will be dropped.

A growing number of VPN, security, and major network companies either support or plan to support IPSec. It is also strongly supported by a user group consisting of manufacturers and suppliers. Although it deals with IP-only traffic, it is the most often recommended or chosen solution to ensure privacy in VPN communication. It can be implemented as a single Layer 3 solution, but it can be implemented over a Layer 2 solution to provide data encryption for IP traffic. It is capable of maintaining multiple tunnels, including simultaneous VPN and public access connection inherently from its general host-to-host model. It can also support extranets with its built-in key management functionality, which is missing in other Layer 2 solutions.

2.3.2.2 Mobile IP

Mobile IP is intended to enable nodes to move from one IP subnet to another. It is just as suitable for mobility across homogeneous media as it is for mobility across heterogeneous media. That is, Mobile IP facilitates node movement from one Ethernet segment to another as well as it accommodates node movement from an Ethernet segment to a wireless LAN, as long as the mobile node's IP address remains the same after such a movement.

Mobile IP introduces the following new functional entities:

Mobile node — A host or router that changes its point of attachment from one network or subnetwork to another. A mobile node may change its location without changing its IP address; it may continue to communicate with other Internet nodes at any location using its (constant) IP address, assuming link-layer connectivity to a point of attachment is available.

Home agent — A router on a mobile node's home network which tunnels datagrams for delivery to the mobile node when it is away from home, and maintains current location information for the mobile node.

Foreign agent — A router on a mobile node's visited network which provides routing services to the mobile node while registered. The foreign agent detunnels and delivers datagrams to the mobile node that were tunneled by the mobile node's home agent. For datagrams sent by a mobile node, the foreign agent may serve as a default router for registered mobile nodes.

A mobile node is given a long-term IP address on a home network. This home address is administered in the same way as a permanent IP address is provided to a stationary host. When away from its home network, a care-of address is associated with the mobile node and reflects the mobile node's current point of attachment. In this case Mobile IP uses protocol tunneling to hide a mobile node's home address from intervening routers between its home network and its current location. The tunnel terminates at the mobile node's care-of address. The care-of address must be an address to which datagrams can be delivered via conventional IP routing. At the care-of address, the original datagram is removed from the tunnel and delivered to the mobile node.

The Mobile IP standard contains a very strong authentication between the mobile node and the home agent to authenticate themselves. The default algorithm is keyed MD5, with a key size of 128 bits. This will result in the mobile node's traffic being tunneled to its care-of address. But the standard does not provide privacy protection; it rather refers to other IP encryption standards, such as IPSec.

The Mobile IP standard as with other Layer 3 standards has the advantage of scalability, security, and reliability, but they are more complex to develop and, inherent from their Layer 3 functionality, they only support a specific protocol, which is IP in this case.

2.3.3 Frame Relay

Traditionally, VPNs were provided in the form of broadband packet switched services, such as Frame Relay, X.25, or ATM. Now, with growth of the Internet as a viable service infrastructure, it is possible to run VPNs over an alternative protocol. As it can be seen from the previously discussed standards, these latter solutions tend to be more popular. The reason is fairly simple: while traditional VPNs are very useful for fixed LAN-to-LAN connectivity, they do not easily accommodate individual users whose only access to the outside world is in the form of their PC, a modem, and the public switched telephone network. VPNs that run over IP are easily accessed by these users.

With any choice of the above-mentioned protocols, virtual circuit connections can be defined between remote locations, and the LAN traffic can be bridged over these circuits. This usually provides an emulated Layer 2 network for the users, which can be used to transmit multiprotocol traffic. Although proper authentication and encryption also has to be taken care of in these solutions, these networks are not as sensitive to security threats, as they are usually a private ATM, Frame Relay backbone of the provider, but in any case less public than the Internet in the previous solutions.

The advantages of this solution are the built-in quality of service (QoS) guarranties that are part of the virtual circuit definitions. Where the bandwidth availability could be argued in the past for an Internet-based VPN solution compared to these solutions, nowadays the ISPs can also provide high-speed Internet connections, especially when they utilize only their backbone network to provide VPN services. The disadvantage of the Frame Relay, ATM, or X.25-based VPN services is that these means of access should be available at each location where the VPN service has to be used. It usually means special equipment, wiring, and additional management needs. The user mobility is not solved with these solutions.

2.3.4   Layer 2 or Layer 3 Comparison

The goal of Layer 2 tunneling protocols is to transport Layer 3 protocols such as AppleTalk, IP, and IPX across the Internet. To achieve this, the architects of PPTP and L2F leveraged the existing Layer 2 PPP standard, which is designed to transport different Layer 3 protocols across serial links. In these schemes, Layer 3 packets are encased in PPP frames, which are then encased in IP packets for transport across the Internet.

From a security standpoint, Layer 2 tunneling protocols are insufficient to be secure VPN solutions on their own. None of these protocols provide the data encryption, authentication, or integrity functions that are critical to maintaining VPN privacy. The L2TP specification disclaims any data security functions and refers IP data security to IPSec, but no serious security provisions or references are made for the other Layer 2 protocols. In addition, none of these protocols provide a mechanism for key management, which limits their scalability.

PPTP and L2F are vendor-specific, proprietary protocols, so interoperability is limited to products from supporting vendors. In contrast, L2TP is a multivendor effort, so interoperability is not as much of a problem. It is important to note that when utilizing tunneling protocols besides IP, users will have to rely on vendor-specific data security features. On the upside, PPTP, L2F, and L2TP can transport multiple protocols. They also function both in LAN-to-LAN and dial-up-to-LAN tunneling modes, allowing them to cover the applications most desired for VPN.

In case of Layer 3 Tunneling there is no need for a globally unique address space, which is a requirement with L2TP for remote client address assignments. This global address space doesn't have to be registered, since it is seen from the corporate network but not visible from the public network (Internet).

Another difference is that the tunneling causes an additional overhead as opposed to Layer 3 Tunneling, which only sends regular IP packets after the ISP's IP Gateway to the corporate network. (The tunneling takes place only between the remote client and the IP Gateway.) This might make Layer 3 solutions more scalable, but with the loss of the additional features, such as the freedom of network protocols that can be used over the PPP layer.

References

Bay Networks: *http://www.baynetworks.com/Solutions/vpn/*
Cisco: *http://www.cisco.com/warp/public/779/servpro/solutions/vpn/*
Microsoft: *http://www.microsoft.com/ntserver/nts/commserv/exec/feature/VPNFeatures.asp*
Shiva: *http://www.shiva.com/remote/vpn.html*
3Com: *http://www.3com.com/enterprise/vpn/*

2.4   Effective Website Design

Karen M. Freundlich

The goal of an effective website should be to achieve the desired results using the best available technology for the job. Many webmasters fall prey to the seduction of fitting the goals of their site to the sophisticated tools now freely available. A measurable objective should be narrowly defined before the site is even designed. As in any good novel, the plot must be carefully laid out before the writing and editing take place. This segment will review four steps to consider when creating a website to provide a business solution.

In today's world of "what you see is what you get" website creation tools it is easy to construct sites similarly to what we did as children, creating anything out of the available blocks or Tinkertoys in the bag. Instead, really effective websites, especially ones focused on financially rewarding e-commerce, must avoid the temptation to "*build*." Measurable goals must be defined, the production and maintenance efforts determined, a system for measuring effectiveness put in place, and the intuitive layout tested and determined before the "building" tools are contracted. This methodical process will more readily guarantee the site's success.

2.4.1 Goals Defined

The goals of the business behind the site must be strictly defined. The website should be regarded as a tool to run the business; a store, not just a storefront. E-commerce software is often accompanied by e-commerce consulting for just this reason. It is simply not enough to have a clean, attractive, well-filled store; the products must be suited to the market and carefully distributed according to supply and demand. It is the same in the virtual world of the internet. There is a destructive misconception that everything on the internet occurs in shortened time, for example, that a "web year" is only four months. Like any computer, the web is run by humans, and just like the chain is as strong as the weakest link, the speed of the web is only at fast as the humans that run it. Admittedly, the web can reduce communication times among the parties, but the thinking and planning process to design the systems takes the same time on the part of the human. Trying to rush this is guaranteed to cause errors, especially in business decisions. The web cannot make informed, intelligent business decisions more quickly. You must engage the best minds to provide the best solutions. Only when the goals are clearly defined to achieve the business solutions should one proceed to the smaller details of implementing them.

2.4.2 Production and Maintenance Efforts

Once the goals are defined, the next step is to determine what levels of production and maintenance are needed to meet the goals. Since, presumably, the website is blazing a trail, it is prudent to use the least complex solutions to meet your goals, then provide a term for evaluating the results and to better define the needs, and finally upgrade according to your growth. It is useful to review and critique the approaches used by similar or parallel industries. The web makes competitor research very accessible. The site should be produced with maintenance objectives easily implemented. For instance, an administrator page can be included which allows password-protected users to perform a variety of maintenance tasks on-line. The inclusion of the maintenance goal in the site design will assure that a comprehensive and complete job is achieved. Quality control should be an important objective, best planned for in advance.

2.4.3 A System for Measuring Effectiveness

In order to determine if the site is meeting its goals, there must be inherent systems in place to measure the results. It takes much less effort to build your site with systems to track measurable objectives than it is to try to figure out how and what to measure once the site and systems are designed. The types of measures will vary based on the goals, but one way to approach the problem is to design the reports that will communicate the results. Once the report's answers to questions are put in writing, it is much easier to determine what pieces of information will be needed. The pieces will usually reside in two places. First, they will consist of direct responses from the users, such as orders and return mail. This type of information is designed during the production of the site and should be carefully constructed to provide measurement. Second, there will be data stored on the server's log report of all activity on the site. Know in advance what these data are because they can provide important demographic and navigational information that can be strategically used to provide measurement of the site's objectives. If the site's hyperlinks are carefully designed, much information about the user's thought process while navigating the site can be gleaned from the log report's data regarding the order in which the pages were requested from the server. A much more efficient evaluation of the site's effectiveness can be achieved when measurable objectives are included as part of the site design.

2.4.4   Intuitive Layout

Finally, no web browser likes to be "lost in the funhouse." Chances are, there was enough work expended just to find your site via a search engine, word of mouth, or reputation. Once there, the user wants to find solutions with speed and comfort. If they came to browse, the information must be presented and found quickly. Everyone knows how easy it is to get lost, frustrated, and eventually one leaves without much thought of returning. Intuitive, concise, and neatly laid out site design will make the user feel comfortable there. Remember, unlike traditional stores, it is much easier to "walk out" of a website. And even though good help is hard to find in those traditional stores, web users don't expect that they will even need help. The design should be uncluttered, the directions and navigation should be intuitive, and there should always be a link that will allow the user to return to the home page. Remember that the original goal behind the hyperlink was to allow non-linear access to information. All of your website's offerings should not be shown on one page. A carefully designed linking system will allow the user to access information in an intuitive way, without having to muddle through the clutter. Overuse of graphics and animation should be avoided. Remember to use the best available tools to meet the objective without burdening the user or the computer resources with extraneous efforts.

2.5   Web-enabled Data Warehousing

Dermot Murray

2.5.1   Introduction

Since the early 1980s, business analysts have identified the inherent value in analyzing the huge amounts of data generated in production online transaction processing (OLTP) systems. Hidden deep inside such data repositories lies key information that can make a product or service more marketable, that can make a customer more profitable, and that can make processes more efficient. This process of analyzing production data in order to unearth that critical business intelligence is known as decision support systems (DSS). The problem has always been getting at that information in such a way that it adds value to the business as a whole. William H. Inmon promoted the concept of data warehousing[3] in the 1980s as a means of separating operational systems from DSS in order to extract the data necessary for business intelligence without impacting mission critical processing systems. Since then, there has been a huge growth in the market for data warehousing and decision support tools, with all of the major database vendors such as Oracle, IBM, and Sybase developing products to satisfy the demand.

This paper examines the next logical step in the evolution of data warehousing technology; i.e., Web-enabling the applications that provide access to this key business data. With the growth in the use of intranets, extranets, and the Internet in general, such Web-enabled data warehousing products have revolutionized the way business analysts generate the reports and charts they need in order to analyze the trends and patterns in the operational data. We will explore the concepts behind Web-enabled data warehousing and look at the technology that makes it all happen.

---

[3]*The Essential Client/Server Survival Guide* — Orfali, Harkey, and Edwards, (John Wiley, 1996).

2.5.2   Data Warehousing Overview

2.5.2.1   Concepts

The concept behind data warehousing first emerged in the client/server platform environment of the 1980s. Although Inmon first started writing about data warehousing in 1981, it wasn't until the early 1990s that industry giants such as IBM, Oracle, and Sybase started taking the technology seriously. Today, the worldwide market for data warehousing products is estimated at over $30 Billion.[4] What drives the popularity of data warehousing is the need for executives and business analysts to gain competitive advantage from analyzing trends and patterns hidden within the mountains of data that companies generate in their day-to-day transactions. Data warehousing is the process of extracting data from various OLTP systems into a centralized format that can be analyzed using DSS and executive information systems (EIS) tools, which provide more business-specific and powerful queries for higher level managers and executives. Collectively, these tools are known as online analytical processing (OLAP) or multidimensional analysis (MDA). The data warehouse itself can either be a single or distributed specialized database management system (DBMS) that contains replicated data from different sources within the organization. This data is usually extracted from internal data production sources such as OLTP and enterprise resource planning (ERP) systems but increasingly from external sources such as Dow Jones, Reuters, and even the Internet itself. The data is typically cleansed and transformed to a format that can be analyzed by DSS tools. Information about the content and format of the data is stored as "metadata" in information directories that can be accessed by both business analysts and database administrators (DBA) alike. Figure 2.5.1 demonstrates the various steps involved in data warehousing.

2.5.2.2   Advantages

The main advantage of a data warehouse is that it provides executives with up-to-date data that can be queried for reporting and analysis in order to assist them in making strategic decisions about the operations of their organizations. The SQL queries generated by OLAP tools are typically very sophisticated and can contain multiple criteria such as sales by region, state, and customer. The results of these queries are displayed as reports or charts that can then be presented in a concise form to executives who typically don't want all of the detail that is contained in the original production data. Due to their multiple-criteria nature, these queries are usually long-lived and may even result in lost or stray processes that could jeopardize the data integrity of a real-time production environment such as OLTP or ERP. Therefore, data warehouses are usually separate DBMSs that store replicated and transformed copies of the production data and are not required to provide the same fast response times that are necessary for mission-critical OLTP systems. However, the new genre of operational data stores[5] provides faster response to DSS/EIS queries by using near real-time transactional data for the most current query and analysis. In addition, DSS/EIS tools provide drill down capabilities that allow executives to get more detailed information on a particular trend or subject matter by generating more detailed queries on that subject. Data warehouses also support data mining tools that provide analysts with the ability to discover unexpected patterns in data by using fuzzy logic searches.

2.5.2.3   Data Marts

A very popular form of data warehousing is the data mart, which is typically a subset of the data warehouse that contains data of specific interest to a particular department, such as marketing, sales, or human resources. These data marts are less expensive to build and maintain than enterprise-wide data warehouses and offer immediate business value to the departments that they service. However, the very independence that data marts provide tends to act as an obstacle to true enterprise integration in information reporting and analysis. While the debate between centralized enterprise-wide data warehousing and decentralized data marts has consumed industry advocates for the last ten years, a compromise concept of dependent

---

[4]*Database Solutions White Paper* — Palo Alto Management Group, Inc., July 1998.
[5]*Data Warehousing Management/ Productivity Tools* — Datamation White Paper, 1997.

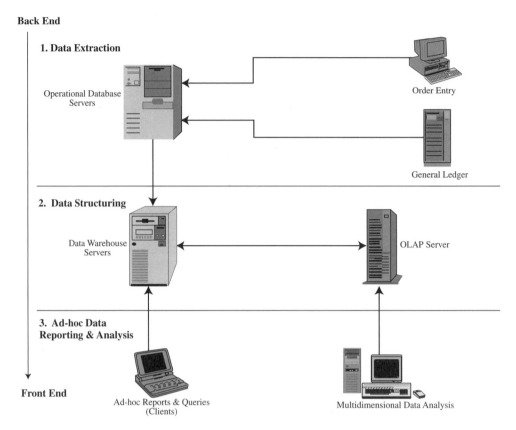

FIGURE 2.5.1   Data warehouse/OLAP environment.

data marts has potentially bridged the gap. This hybrid solution provides for the extraction of data from the centralized data warehouse to the departmental data marts so that each department is working off the same enterprise data, albeit formatted to their individual needs.[6]

2.5.2.4   Future Growth

In spite of the technical and organizational challenges that data warehouse implementations pose, the ultimate benefits that the technology offers has led to a huge growth in its adoption. For instance, 90% of the Global 2000 companies had either implemented or planned to implement data warehouses by June 1998.[7] The future of data warehousing also seems pretty secure with market estimates expecting to grow exponentially to over $100 billion by 2002 (Figure 2.5.2).[8] One of the biggest factors in the expected growth of data warehouse implementation is the ease of access to timely reports and queries that Web-based data warehouse tools provide.

2.5.3   Web-enabled Data Warehousing

2.5.3.1   Benefits

Perhaps the most perplexing aspects of implementing a data warehouse strategy have been the escalating costs of installing and maintaining the GUI-based clients that are used to generate the queries. According

---

[6]The Middle Ground — *CIO Magazine,* January 1999.

[7]Data Warehousing Is Worth The Investment — *InternetWeek,* June 1998.

[8]Database Solutions White Paper — Palo Alto Management Group, Inc., July 1998.

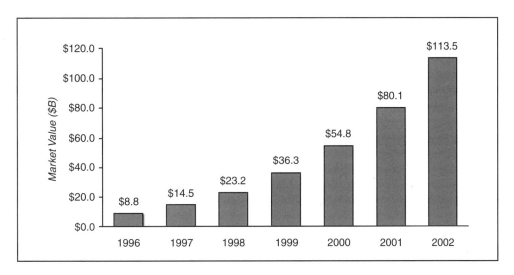

FIGURE 2.5.2   Worldwide data warehousing solutions market (Source: Palo Alto Management Group).

to a Meta Group, Inc. study in 1997, the average cost of implementing a data warehouse project was $1.9 million,[9] which accounted for up to 19% of the total IT budgets of those companies surveyed. One solution to minimizing the costs associated with data warehouse implementations is Web-enabled data warehousing, a concept that has grown in popularity since it first appeared in the mid 1990s. These tools have given IT departments a more cost-effective option of allowing access to the enterprise-wide data warehouses for a larger group of users. By allowing users to generate queries and reports through their Web browsers, IT departments can roll out data warehouse installations in a much shorter period of time by simply allowing the users to have the appropriate level of access to the DSS application server. With the advent of virtual private networking (VPN) and subsequent extranet technology, various levels of data warehouse access can even be provided to external users such as suppliers and customers. This is particularly critical in today's supply chain environment where companies have to work closely with their strategic partners, i.e., suppliers and customers, and therefore have to provide a certain level of access to their enterprise data warehouses for reporting and analysis. This Web-enabled approach has had the effect of reducing the costs associated with hardware and software by using the "thin" client approach, as opposed to the "fat" client requirements of the previous client/server model. For instance, the Aberdeen Group was able to cut costs from $1000 per seat for client/server DSS tools to $50 for Web-based DSS access, primarily by consolidating most of the software and hardware costs at the server.[10] This server-centric model also has the effect of reducing software licensing costs by switching from a per-seat basis to a more cost-effective server-based licensing model. The move away from dependence on fat desktop clients to the thin browser-based client model may also allow the new breed of Internet appliances, such as mobile phones and network computers, to be able to access information from data warehouses over the Internet. For instance, the advent of "push" technology could allow a sales executive to be paged if sales figures for a certain region fall below a predefined threshold.

2.5.3.2   Making it Happen

Web-enabled data warehousing involves a combination of HTML, XML, HTTP, and mobile component-based technology, typically Java or ActiveX. In this model, a user can generate a SQL query using a HTML/XML form that embeds the control information for the search criteria into a HTML/XML query and sends it to the remote Web server that in turn passes the request to a Web Gateway server. The Gateway server converts the HTML/XML query into an OLAP-specific request and passes it to the OLAP

---

[9]The Middle Ground — *CIO Magazine,* January 1999.

[10]Warehousing And The 'Net — Marriage Made In Heaven — *Insurance & Technology,* July 1997.

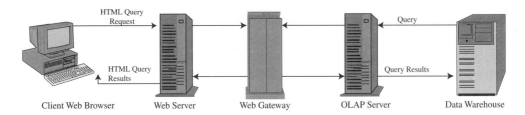

FIGURE 2.5.3   Web-based client/server architecture.

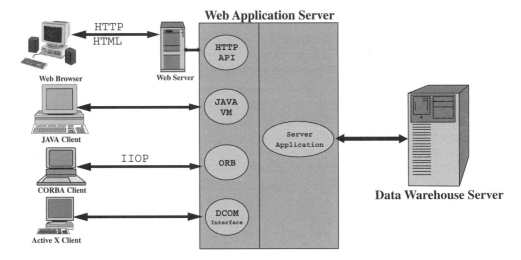

FIGURE 2.5.4   Object-based architecture (Source: Adapted from *DB2 Magazine*).

server, which executes the query directly against the data warehouse. Typically the Web gateway and OLAP Server are bundled into the vendor-supplied DSS Server application package. When that application retrieves the resulting data, it sends the data embedded in an HTML/XML file back to the Web server which forwards it back to the Web client.[11] Some DSS servers also store predefined DSS objects such as reports and queries that can be executed by the user to generate more standardized reports. This type of interaction is similar to three-tier client/server architecture and in fact most industry analysts would argue that Web-enabling traditional client/server applications are just the next step in the evolution of client/server — the intergalactic model (see Figure 2.5.3).

The interaction between the DSS server application and data warehouse is provided using distributed-object-computing protocols, such as CORBA, IIOP, or DCOM, which provide the client/server state management unlike HTTP, which is a stateless protocol.[12] In some cases, the DSS server downloads client-based applications, such as Java applets or ActiveX controls, that actually run within the Web browser's memory space and interact directly with the application server. The purpose of this is to provide an improved graphical user interface to the end-user in situations where plain old HTML is not sufficient for multi-dimensional data visualization, such as graphs and charts. These applets also provide more robust object-based communications between client and application server by using CORBA Internet Inter-ORB Protocol ("IIOP") and Java remote method invocation ("RMI"), with HTTP only being used for downloading the applets to the client. Figure 2.5.4 demonstrates the different levels of interaction

---

[11]*A Methodology For Client/Server And Web Application Development* — Roger Fournier (Yourdon Press, 1998).
[12]Building Web Information Systems — *Byte Magazine*, July 1998.

between the client, application server, and data warehouse server using both the Web and object-based protocols such as CORBA and DCOM.

The advent of Web-enabled data warehousing has spawned the growth of information delivery, whereby specific business information is "pushed" to information consumers at predefined intervals. In effect, consumers subscribe to information by searching the metadata stored in the information directories using Java/ActiveX agents or Web-based search engines in order to locate the information that is of interest to them.[13] The information delivery itself can either be schedule-driven, meaning that the appropriate decision support objects are executed at user-defined intervals and the resulting information delivered to a Web server, or event-driven when particular business events occur. These user-defined events, also known as triggers, could also indicate that new data has arrived from the source DBMS. "Push" technology is also being incorporated into the channel method of delivering information, whereby a business analyst can subscribe to a channel, which in turn delivers the most up-to-date information on a particular subject.[14]

2.5.3.3 Obstacles and Limitations

Although Web-enabled data warehousing has revolutionized the distribution of enterprise-wide business information, it has not been the panacea for all data warehouse access problems. There are limitations to what "power users" can do with Web-enabled DSS tools and in those cases, the traditional desktop approach is still being used. Such functionality as drill-down analysis and multidimensional queries are still not as effective with the Web-based tools and so users who need this functionality still require the more mature client/server tools to perform these analyses. As Web-enabled tools become more sophisticated, this problem will be rectified, but in the meantime, IS managers must deal with the reality that both Web-enabled and desktop data warehouse access must coexist for the foreseeable future.

Ironically, the ActiveX and Java components that are designed to make user interaction with the data warehouse environment possible can cause incompatibility problems if the end-user's Web browser cannot support them. This might be the case if the browser is an older version or if the particular downloadable component only works with either Netscape or Internet Explorer. Of course, the solution to this problem is to standardize Web browsers and plug-in components, a daunting task for organizations with potentially thousands of users, but nevertheless manageable.

Security has always been a problem with enabling external users to access corporate information. As mentioned earlier, the growth in supply-chain management has dictated that companies work in tandem with their suppliers and customers in order to develop their products and services. This trend has forced companies to open up their corporate information systems, including data warehouses, to these external users. Of course, this has led to fears that such mission-critical and confidential information could be compromised in transit between users across a public network such as the Internet. Another potential security hole is the remote access required by mobile users, such as sales agents, who need to pull up reports using dial-up connections over the Internet. Improvements in corporate firewall and VPN technology, coupled with better encryption algorithms and Public Key Infrastructure, have eased these fears. However, organizations implementing Web-enabled data warehousing must constantly be vigilant not only with potential hackers from the outside but also with internal users who should only have the level of access to reports and data that they require. One potential solution to this problem that has been promoted since Web-enabled data warehousing began is to distribute the data among data marts, thereby limiting the users' access to only the data stored in those data marts.[15]

Like every new technology, Web-enabled data warehousing has its downside as illustrated above. However, with vendors addressing these issues all the time, the usage of this technology has had a phenomenal growth rate and while traditional client/server warehouse access is not going away any time soon, Web-enabled DSS tools are making strong inroads into the marketplace.

---

[13]Building Web Information Systems — *Byte Magazine,* July 1998.
[14]Warehouses Webicum: Evolution Of A Species — *DB2 Magazine,* April 1998.
[15]Just Browsing Thanks — *CIO Magazine,* October 1996.

2.5.4   Vendors

This burgeoning sector of the industry has produced a number of leading products from both established and startup vendors. While most of the major database vendors have been very proactive in providing Web-based access to their data warehouse products, the market for packaged applications has come mostly from startup vendors like MicroStrategy. As there are too many vendors to mention in this market segment, I will discuss a sample of the products being offered by these different categories of vendors.

2.5.4.1   Major Database Vendors

Oracle Corp., arguably the biggest name in the relational database market, has its presence thanks to the Oracle Express suite, which includes the Express Server and Express Web Agent modules. This product was actually acquired from Information Resources in 1995 and accounted for 21% of the OLAP market in 1998.[16] The Oracle Express Server is the actual back-end OLAP engine that performs the end-user queries. This is the component that provides the interaction between the Express OLAP Server and the Web server. This module takes advantage of the "Network Computing Architecture" model, which is Oracle's blueprint for a three-tier thin client environment, and the Express Stored Procedure Language cartridge to provide the communication between the Web server and the Express Server. The Express Agent Developer's Toolkit, which comes with the Agent, supports the development of both HTML pages and components such as Java and ActiveX to produce customizable reports and analyses for the end user.

For their part, IBM provides Web access to the DB2 OLAP Server through the use of its Net.Data development platform. This product supports standard SQL statements as well as C++ and Java enablers that allow developers to write macros that automate SQL queries. On the server side, it supports FastCGI, which is a high-performance Web server interface that provides better performance for Net.Data applications. In addition, IBM integrated its OLAP Server with Hyperion's Essbase Web Gateway, which is a Web application server component similar to Oracle's Web Agent.

As for the other major database vendors, it appears that they have been slow to fully integrate their OLAP products with Web access, including Microsoft. For their part Informix has tried to buy its way into the Web access market by acquiring Red Brick Systems, who in turn had teamed up with Web application specialists Caribou Lake[17] to deliver Web-based OLAP products. Therefore, this is still a market segment where the niche DSS vendors have been able to take the lead, at least for now.

2.5.4.2   Startup DSS Vendors

MicroStrategy Inc., the market leader in Web-based OLAP products, develops a suite of DSS applications that interacts with DBMS platforms such as Oracle, Informix, and DB2 among others. This suite includes the DSS Web Server and DSS Broadcaster products that use the Internet and the World Wide Web as the communications medium. DSS Web Server, at version 5.5 as of this writing, is a Web-based interface that works with the company's DSS Server OLAP engine to allow users to generate reports and analyses over the Web. Because the company uses ActiveX and Java components in DSS Web Server 5.5, they have been able to provide features such as drill down analysis and report pivoting, that were only available in their Windows-based DSS Agent product. Its DSS Broadcaster product is an application that allows for customized information delivery to be pushed to Web-based clients such as browsers and Internet appliances. The product won the "IT Manager's Choice" Product of the Year in the data warehouses and data marts category organized by *Datamation* magazine.[18] In addition, the company announced 1999 first quarter revenue of $35 million, an increase of 80% over the corresponding quarter in 1998, gaining 55 new clients, including First USA Bank, France Telecom, and Kmart.[19]

InfoSpace, Inc. is the developer of SpaceOLAP, a fully Java-compliant solution that provides client/server-like interfaces for reporting and analysis. It includes a Java Application Server that resides on the Web server

---

[16]End-User Query and Reporting Tools — *ComputerWire PLC.,* March 1998.

[17]PR Newswire Association, Inc, December 1998.

[18]*Datamation* — February 1999.

[19]MicroStrategy's Revenue, Profit Soars — *InformationWeek,* April 1999.

and supports data extracts from Oracle Express, DB2 OLAP, and Hyperion Essbase OLAP servers. Administration is provided through the SpaceOLAP Administrator module, while the SpaceOLAP Designer lets users and developers create customized reports and graphs based on the results of their queries. Reports, or "Presentations" as they are referred to by the company, can be displayed in HTML or Java applet supporting format, supporting pivot and drill down capabilities as well as multidimensional analyses.[20]

Information Builders services the Web-enabled data warehousing market through its WebFOCUS product line. WebFOCUS provides reporting and publishing via the Web from both legacy databases and data in ERP applications. The company takes advantage of Java applets that provides for customized reporting and a Java Developers Workbench for designing customized reports from a Web browser. Among the other startup companies that have made a name for themselves in the Web-enabled data warehousing market are Cognos Corporation, the makers of Impromptu and PowerPlay, and Information Advantage, Inc., developers of the DecisionSuite OLAP product line.

2.5.4.3 Established DSS Vendors

The more established data warehousing companies such as Prism Solutions and Brio Technology have also developed Web-based interfaces into their OLAP products. The Brio Enterprise Server suite provides two modules that allow the user to extract information from data warehouses via the Web — the OnDemand Server and the Broadcast Server, competitor products to MicroStrategy's DSS Web Server and DSS Broadcaster, respectively. Prism Solutions added Web-enabled functionality to its Warehouse Directory product in 1997 when it introduced the Web Access module, which allows users to view and query the Directory from their Web browsers. Having been acquired by Ardent Software, Inc. in April 1999, this module has since been integrated as a standard feature of the Warehouse Directory package and indeed, Ardent's Vice President Peter Fiore sees Extensible Markup Language (XML) as the future conduit for sharing access to corporate data warehouses.[21]

The following table summarizes the list of vendors and products mentioned above.

Vendor	Product(s)	Technologies Supported
Oracle Corp.	Express Server (Web Agent)	HTML, Java, ActiveX
IBM	Net.Data, DB2 OLAP (Hyperion Essbase)	SQL, C++, Java, CGI
MicroStrategy	DSS Server OLAP, DSS Web Server, DSS Broadcaster	HTML, Java, ActiveX
InfoSpace	SpaceOLAP	Java, HTML
Information Builders	WebFOCUS	Java
Brio Technology	Enterprise Server (OnDemand Server, Broadcast Server)	CGI, Java Runtime Environment (JRE)
Prism Solutions	Warehouse Directory (Web Access module)	Java, XML

2.5.5 Future Trends

The trend to add more functionality to Web-based DSS tools will continue to grow as more companies realize the benefits of implementing Web-based access to their corporate data warehouses. However, don't expect to see the demise of pure client/server-based DSS tools anytime soon as analysts predict that a coexistence of both forms of access will prevail for a time.[22] As mentioned earlier, the main reason is that there are still features and functionality available with mature client/server-based tools and not with Web access tools. DSS vendors are working hard to add that functionality to their Web-based offerings so that in the future, even power users will be able to get the reports and analyses they need via the Web. Overall,

---

[20]SpaceOLAP — *DBMS,* August 1997.

[21]Ardent's Peter Fiore Is Passionate About Data Warehousing — *InfoWorld,* May 1999.

[22]Self Storage; Lower Data Warehouse-Management Costs With Web Access — *Communications News,* August 1998.

the emphasis will shift away from desktop access and more toward the centralized "intergalactic" Web-based model as data warehouse implementations become larger and more users require access.

2.5.5.1   Web Farming

One area of interaction between data warehouses and the Web that is set to grow steadily in the next few years is the concept of Web Farming. This is defined as "systematic business intelligence by farming the information resources of the Web, so as to enhance the contents of a data warehousing system."[23] Essentially, it is the process of using the Web not as a means of distributing data warehouse information to the end user but as a means of obtaining the raw data that goes into the warehouse itself. Such external data sources include commercial databases, e.g., the IBM database of patents, and public databases such as The Electronic Data Gathering, Analysis, and Retrieval (EDGAR) operated by the U.S. Security and Exchange Commission to track publicly traded companies.[24]

Similar to extracting production data from OLTP systems into data warehouses, Web data must be refined to be suitable for use by the warehouse applications. This process, known as acquisition can vary depending on the sources of the Web content and is crucial in ensuring the usability and reliability of the data content as it becomes part of the decision support structure. While this sector is still in its infancy, it won't be long before the major vendors turn their expertise to developing products that will make this form of data collection more reliable and efficient.

2.5.6   Conclusion

The value of using the World Wide Web to disseminate decision support information is evident from the growth in the use of Web-based DSS and OLAP tools. By providing access to corporate data warehouses over the Web, companies are empowering a larger user base with the necessary tools to analyze the business data. This in turn has opened up the decision-making process to more people within the organization. The lower costs of rolling out and maintaining Web-based data warehouse access has made this medium very attractive to companies that wish to gain the most value from their data warehouses. In addition, the trend to open up information systems to the customers and suppliers who are part of the supply chain has made easier access to data warehouse information more critical to these external users.

While this technology has grown in leaps and bounds and looks set to take over as the predominant form of data warehouse access, IS departments will still have to grapple with coexistence between Web-based and client/server-based access. But with the improving feature set of the various products being offered by the many vendors in this arena, Web-enabled data warehouse access will become the *de facto* access medium for this powerful business information.

References

Carrickhoff, Rich — SpaceOLAP (*DBMS*, August 1997).

ComputerWire PLC — End-User Query and Reporting Tools (March 1998).

Datamation White Paper — Data Warehousing Management/Productivity Tools. (Datamation, 1997).

Davis, Beth — MicroStrategy's Revenue, Profit Soars (*InformationWeek,* April 1999).

Fournier, Roger — *A Methodology for Client/Server and Web Application Development* (Yourdon Press, 1998).

Hackathorn, Richard — *Web Farming For The Data Warehouse* (The Morgan Kaufmann Series in Data Management, 1998).

Hackathorn, Richard — Rouging the Web for Your Data Warehouse (*DBMS*, August 1998).

Koch, Christopher — The Middle Ground (*CIO Magazine*, January 1999).

Orfali, Robert. Harkey, Dan. Edwards, Jeri — *The Essential Client/Server Survival Guide* (John Wiley & Sons, Inc., 1996).

---

[23]*Web Farming For The Data Warehouse* — Richard D. Hackathorn (Morgan Kaufmann Publishers, 1998).

[24]Routing The Web For Your Data Warehouse — *DBMS*, August 1998.

Palo Alto Management Group, Inc., — Database Solutions White Paper (July 1998).

PR Newswire Association, Inc. — Red Brick and Caribou Lake Software Team Up (December 1998).

Reinauer, Rob — Self Storage; Lower Data Warehouse-Management Costs with Web Access (*Communications News*, August 1998).

Row, Heath — Just Browsing Thanks (*CIO Magazine*, October 1996).

Schroeck, Mike — Data Warehousing is Worth the Investment (*InternetWeek*, June 1998).

Schwartz, Susana — Warehousing And The 'Net — Marriage Made In Heaven (*Insurance & Technology*, July 1997).

Scott, Jim — Warehousing Over the Web (Association for Computing Machinery, *Communications of the ACM*, September 1998).

Vizard, Michael — Ardent's Peter Fiore Is Passionate About Data Warehousing (*InfoWorld*, May 1999).

White, Colin — Building Web Information Systems (*Byte Magazine*, July 1998).

White, Colin — Warehouses Webicum: Evolution of a Species (*DB2 Magazine*, April 1998).

2.6   E-commerce Technologies: A Strategic Overview

Mihir Parikh

> The changes sweeping through electronic communications will transform the world's economies, politics, and societies — but they will first transform companies.
>
> —Frances Cairncross, in *The Death of Distance*, 1997, p. 119.

E-commerce is a result of these changes. In simple terms, e-commerce is defined as a business conducted over the Internet with the use of computer and communications technologies. There are three major types of e-commerce: business to business (B2B) e-commerce, business to consumer (B2C) e-commerce, and electronic markets. *B2B e-commerce* deals with one business providing products and services to another business typically as a part of the supply chain or as an enabler of business processes. Examples of B2B e-commerce include an auto-part manufacturer supplying an auto company, or a bank providing credit card payments and other financial services to a retailer. For many years, electronic data interchange (EDI) handled B2B transactions in many companies. Now, Web-based open system applications are replacing proprietary EDI systems. *B2C e-commerce* deals with a business providing products and services to consumers at the end of the supply chain. Examples include a book retailer selling books to a reader (Amazon.com) and a broker providing financial trade executions to an individual investor (E*Trade). *Electronic markets* provide marketspaces on the Internet, as opposed to marketplaces in the physical world, where buyers and sellers can meet and exchange products and services. Electronic markets are of two types: consumer markets and business hubs. Ebay, priceline.com, and accompany.com are typical examples of electronic markets for consumers. Chemdex, Metalsite.com, and Ultraprise.com are typical examples of electronic markets for businesses.

Recently, International Data Corporation (IDC) estimated that yearly worldwide e-commerce would increase to more than $1 trillion by year 2003.[25] IDC estimated that by then non-U.S. countries would account for half of worldwide e-commerce. Are the current businesses ready for it? Well, not really. A recent survey conducted by the Cutter Consortium found that 65% of companies did not have an overall e-commerce strategy and nearly 25% lacked even a basic business and implementation plan for e-commerce.[26] A major reason for this is that a few companies have aligned their business strategies with information technology (IT) strategies. A large percentage of companies still do not involve IT in high-level strategic planning. IT in these companies is relegated as a support function. The companies have failed to recognize the changing role of IT from business support to business enabler. The strategic implications of IT and the use of IT as a strategic weapon are not well understood.

---

[25]International Data Corporation Report, NET062899PR.htm, June 28, 1999.

[26]*InternetWeek*, September 6, 1999. Page 29.

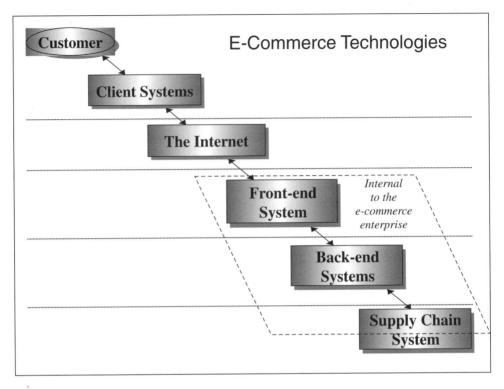

FIGURE 2.6.1 E-commerce technologies

Information technologies are key enablers of e-commerce. A combination of many information technologies ranging from a Web browser to logistics management systems makes e-commerce possible. Identifying and implementing the right technologies to execute business processes is critical to succeed in the e-commerce space.

2.6.1 E-commerce Technologies

Figure 2.6.1 shows different categories of e-commerce technologies. All of these technologies are required and play significant roles in managing and conducting business over the Internet.

2.6.1.1 Client Systems

Client systems reside on customers' computers. Customers utilize these technologies to participate in e-commerce activities.

Web browser and cookies: The most common of all client systems is a Web browser. Web browsers provide an interface through which a user can view information on the Internet. In the last five years, Web browsers have evolved from a small software using simple hypertext markup language (HTML) to a very sophisticated software that uses new technologies such as Java, ActiveX, VRML, XML, and different types of plug-ins and Web applications in addition to a much advanced version of HTML. Most web browsers utilize cookies whereby they transmit basic information about a user to the Web server on the other end for identification purposes. Such identification can be used to personalize services for the user. Multiple levels of cookies are used. Low-level cookies simply provide user name and password, while high-level cookies may include information about credit card, mailing address, previous purchase patterns, and browsing habits.

Communications software and hardware: Communications software and hardware help customers' computers to connect to the Internet via one of the multiple modes including modem, cable modem, satellite links, and local area networks.

Plug-ins: Plug-ins are independent software applications utilized to show special data files within a Web browser. Plug-ins enable a Web browser to show a multimedia presentation or to play a streamed audio piece. Shockwave Flash, RealPlayer, and Adobe Acrobat Reader are plug-ins.

Software agents: Agents are other types of independent software that can assist users in carrying out some specific activities. Such activities include filtering information, searching the Web to find the right information, and comparison shopping.

Biometric identification: Biometric identification is a technology that uses a measurable physical characteristic to recognize the identity, or verify the claimed identity, of an enrollee.[27] It utilizes physical characteristics such as fingerprints, facial design, iris patterns, retina patterns, hand geometry, signature verification, and voice recognition instead of keys, passwords, and plastic cards. Several advantages, including reduction in fraud and never losing the identification, have prompted several large companies such as IBM and Motorola to invest heavily in biometric identification. Biometric identification is one of the fastest-growing areas on client systems and security.

Push technologies: These publish and subscribe technologies enable delivery of possibly useful information without the recipient asking for it. Push technologies utilize extensive user profiles containing preferences of each user to match and deliver information over the Internet. Push evolved as an alternative to the Web (pull technology) from PointCast's personalized broadcasting technology in 1996. In the last three years, it has gone through the full length of the hype curve. Recently, it has returned with reasonable expectations and clear understanding of what the technologies can and cannot do. Several companies, including Marimba and BackWeb, provide push technology solutions.

2.6.1.2 The Internet

Many advanced telecommunications technologies have been utilized to create and operate the Internet. Most prominent are optical fiber, routers, digital switches, Synchronous Optical Network (SONET) rings, asynchronous transfer mode (ATM), frame relay, Transmission Control Protocol (TCP), and Internet Protocol (IP). These hardware and software technologies provide the backbone lines and the transmission rules to carry out information exchange over the Internet. As the Internet is growing, new technologies are emerging to increase speed and volume of data transmission. The continuing convergence of textual data with audio and video data over the Internet is prompting new technologies to support quality of service (QoS) which recognizes the differences between data types and assigns appropriate priority for transmission.

2.6.1.3 Front-end Systems

Front-end systems are the ones with which the customers interact. They provide a face to an e-commerce business.

Web pages: Web pages are the data files containing HTML-coded information. In the early days of e-commerce, Web pages were static and generated by human Web programmers. Now, in most e-commerce sites, Web pages are dynamic and are generated by Web page management systems. These systems work with other front-end systems and back-end systems to develop HTML-coded content that the customers receive.

Traffic management: Due to a special sales event or some top news, sometimes an unexpected number of visitors go to a Web site all at once. Such an unexpected load puts pressure on the Web server and dramatically reduces its speed and increases download time. If the load extends for a longer period of time, it crashes the system. To avoid such a mishap, traffic management tools are used. These reduce massive congestion by spreading traffic load on multiple servers and increase overall network efficiency.

Search engines: Several types of search engines are utilized by e-commerce companies. Some search engines provide capabilities to search a specific product based on description, features, or other information. Some search engines provide capabilities to search and locate Web content based on key words or phrases. Search engines can be used to search a specific Web site, a regional part of the Web, or the

---

[27]Digital Imaging: Connecticut Biometric Imaging Project. http://www.dss.state.ct.us/digital.htm

whole Web. Often directory engines are used along with search engines to automatically categorize Web content for future searches.

Site servers: Site servers are the most comprehensive tools for e-commerce. They provide support ranging from creating an electronic storefront to controlling and managing it. Site servers include support for site building, standard code sharing, code library development and management, dynamic Web page generation, product promotion, product cataloging, order taking and processing, securing transactions, and managing payment systems. Site servers help provide product information, dynamic pricing information, marketing and promotion, shopping cart services, tax calculations, shipping and handling calculations, and automated post-sales follow-up. In addition, they capture market demographic information and coordinate with back-end systems. Several off-the-shelf products are available to support small business shopping services including IBM's Net.Commerce and Microsoft's Site Server. However, a major e-commerce site requires custom application development to support the above-discussed activities.

Shopping engines: These types of software enable customers to find and compare different products or services on features and prices. Some shopping engines also include product reviews from independent agencies such as *Consumers Digest*. Some shopping engines also provide reviews from and discussion with current users of the products to help others make more informed purchase decisions. It is a very useful tool in e-retailing. Amazon.com's Junglee division and Inktomi are the two leading companies that provide shopping engines.

Customer relationship management (CRM): Through a marketing agreement, E-commerce merchants pay between $0.90 to $2.67 per visitor referred by a portal.[28] As the search cost for every new customer is staggering, increasing customer satisfaction and loyalty is crucial to maintain the current customer base and for e-commerce success. CRM systems provide the critical customer support and services. They provide a database of frequently asked questions (FAQ), searchable knowledge base, multiple way (e-mail, Internet telephony, video conferencing, etc.) to assist shoppers in real time, follow-up support, order tracking, return processing, after-sales services, and warranty processing. They also help maintain profiles of buyers that contain their shopping behaviors and preferences. The currently available CRM-related technologies include enterprise portal, mobile computing, net telephony, desktop video conferencing, speech recognition, call center systems, and data warehousing.

Personalization: As the number of e-commerce businesses increases, a key differentiator would be how well an e-commerce business customizes its storefront for each customer. Knowing the customer and his or her preferences will improve customer service and retention. This personalization or mass customization requires creating user profiles from purchase patterns and browsing patterns, and applying business rules and inference on the information collected in the user profiles to create new knowledge about the users. Key technologies used for personalization are databases, dynamic Web pages, business rules, inference engines, cookies, and push technologies.[29] In some cases, data warehousing and mining technologies can also be used for personalization.

Security: More than half of B2C e-commerce transactions are paid with credit cards. Protecting transmission of credit card and other private, confidential information over the Internet and securing the information on merchants' Web servers are two of the most pressing issues in e-commerce. Encryption of data through secure sockets layer (SSL) and private and public keys are the most commonly used technologies to secure Internet transmission of confidential data.

2.6.1.4 Back-end Systems

While the front-end systems manage the interface with customers, the back-end systems carry out operations and manage e-commerce organizations.

Enterprise Resource Planning (ERP): ERP systems are commercial software packages utilized to integrate information flowing through different functions of an organization, such as financial and accounting, human resources, sales and customer service, supply chain, etc. ERP systems coordinated

---

[28]International Data Corporation Report, May 4, 1999. NET050499PR.htm
[29]The Web gets personal. *Byte,* special section on E-business Technology, June 1998.

with front-end systems can capture orders, provide order confirmation, accept payment, check credit cards for approval, process coupons and other promotions, handle billing and invoicing, control inventory and procurement, integrate with payment systems, and coordinate with fulfillment systems for order execution. Before ERP systems, these processes were handled by various, independent information systems indigenous to different functional divisions in the organization. Information systems from one functional division were often not compatible with the information systems in other functional divisions. This brought inefficiencies in business processes and overall higher costs. ERP systems promise seamless integration and easy information flow among different functional divisions. However, successful implementation of an ERP system has been one of the major critical issues in the utilizing ERP systems.

Databases/Data Warehouses/Data Mining: Databases and data warehouses are at the center of running a business in this information economy, especially an e-commerce business. They provide repositories for information collected through business processes. This information is the lifeblood of organizations and its optimum use is very important. Several emerging database technologies such as multidimensional databases provide holistic perspective and better understanding of the information. When used in conjunction with data mining technologies, e-commerce businesses can find and exploit hidden relationships and buying behaviors to increase market share and sales.

2.6.1.5 Supply Chain Systems

These systems work with business alliance partners and other enablers. They provide smooth transfer of information with partners to carry out outsourced business processes. Some of these systems are external to e-commerce organizations and implemented by the partners.

Supply chain management: Most emerging e-commerce companies are not vertically integrated. They depend on many partners and intermediaries on both sides of the supply chain (a value addition sequence through which raw material flows to become a finished product). Supply chain management systems help businesses coordinate their processes with suppliers, manufacturers, raw material providers, shippers, distributors, and associated retailers. In e-commerce businesses, greater efficiencies are achieved by moving information rather than actual products along the supply chain. Actual products are generally delivered directly to the consumer by the product manufacturers without any supply chain intermediaries handling or storing the products. Supply chain management systems help control product life cycle, forecast demand, arrange advanced scheduling, plan manufacturing and distribution, and enable order promising and processing. Several companies provide supply chain software. The leaders are i2 Technologies, Manugistics, and Numetrix. Several ERP companies are also moving into this area by extending ERP capabilities.

Payment systems: A majority of payments, almost 89%, over the Internet is conducted through credit cards and checks. Payment systems help e-commerce businesses coordinate with banks and credit card companies to approve credit card purchases and clear checks. Some e-commerce companies also utilize these systems to work with soft cash providers such as CyberCash and electronic money.

Fulfillment/Logistics management: These systems coordinate with logistics partners such as FedEx, UPS, and independent warehouse operators. These systems work with back-end and front-end systems to help determine shipping and handling charges, delivery terms, delivery schedule, order tracking, freight management, custom and excise duty clearance, and other fulfillment issues.

2.6.2 Strategic Challenges

In the first chapter of the e-commerce storybook, the technology largely drove business models. Now the business models are driving technology.

—Peter G. Keen, *Computerworld,* September 13, 1999

2.6.2.1 High Cost of Small Errors

In e-commerce, there is little room for errors. In most cases, you do not get a second chance. Integrating right strategies with right technologies and continuously improving competitive position are important for survival and success. Any error in technology or strategy implementation can lead to a serious loss in market position and the ability to carry out the business in future. Once, eBay's stock price dipped more than 50%

largely due to recurring, unexpected shutdowns of its Web site. Most of these shutdowns were not more than a few hours long. However, persistence and a strong commitment by the company's top management to improve technological infrastructure lead to a rebound in the stock price. E-commerce businesses pay a very high cost for small errors. As an e-commerce business expands, the probability of making such errors increases. While most managers and leaders assume that being there first is the key to success, a study has found that many pioneers fail and most current leaders are not pioneers.[30] The study found that five factors (vision, persistence, commitment, innovation, and asset leverage) are critical to success. These factors often lead to making fewer errors and quickly correcting the errors when they are made.

2.6.2.2 Building Relationships

Technology is a double-edged knife. While it provides unprecedented advantages to you, it also provides the same benefits to your competitors and future competitors. It enables your existing competitors to quickly react to your moves. It also reduces the barriers to entry for new competitors. Assuming that if you build it, they will come and stay is one of the fallacies of e-commerce. To pre-empt this, you have to build a strong customer base and retain it. Building a strong customer base requires building relationships and providing useful services and content to your customers. Often it is measured in terms of "stickiness" of the Web site, that is, how long a visitor stays on the Web site. Media Metrix reports eBay (125.5 average minutes per user), E*Trade (66.5), Microsoft sites (66.0), and Yahoo sites (64.6) are the top four stickiest Web sites.[31] It is not a coincidence that the companies with the stickiest sites are the most successful in e-commerce.

2.6.2.3 Speed

The speed of doing business has increased tremendously with the Internet. Speed is required in growth, in decision making, in adapting to the changing conditions, and in supporting and servicing customers. An e-commerce company has to continuously innovate and improve its business processes and Web-based storefront. It is always vulnerable to the quick imitation of its processes and its innovative shopping features by its competitors. Often not only the Web site designs but also business models are copied overnight by competitors. This also requires building flexibility in the front-end and back-end systems to adapt to the continuously changing conditions and stay ahead of competitors. Therefore, industry experts recommend an open and adaptive architecture for enterprise information systems.[32]

Operationally, when a customer visits an electronic storefront, performance of the Web site becomes an important issue. First of all, the customer wants quick downloading of the Web pages and immediate response to any search queries. If the Web site is slow, the customer will very likely move on to another competing store. This invariably happens when an unexpected number of customers come to the Web site at the same time. Several technologies can help improve speed of the Web site. Use of traffic management tools can balance the load on a Web server and increase the speed of interaction. Scalability of the hardware and software provide quick integration of additional resources. The use of better search engines and shopping engines can also increase the speed of searching the requested product from millions of product profiles stored in the databases. Good shopping engines can also help identify related, complementary products for cross selling and up selling.

2.6.2.4 Security

With the growth of e-commerce, more and more business processes and databases are put on the Internet. This is required to improve customer service and increase organizational efficiency. However, this also makes the processes and databases vulnerable to malicious forces including business spies, computer hackers, and disgruntled former employees. Having complete control over who gets to see what and who

---

[30]First to Market, First to Fail? Real Causes of Enduring Market Leadership. By Gerard Tellis and Peter Golder. *Sloan Management Review*, 32(2), 1996, 65-75.

[31]Snapshot: Sticky Sites. *Computerworld*, September 13, 1999. Page 42.

[32]Designing a growing back end. *InfoWorld*, August 23, 1999. Pages 34-35.

gets to change what is extremely important. Security management software, virus protection software, and intrusion detection software can help increase security of the Web site.

2.6.2.5 Technology Evolution

E-commerce technologies are in a constant state of flux. E-commerce businesses have to continuously evaluate emerging technologies and adapt them quickly to stay competitive. As the technologies are constantly evolving, few standards exist. The ability to choose a right technology with a strong future becomes a critical skill in managing e-commerce.

2.6.3 Emerging Trends for the Future

> What is my ROI (Return on Investment) on e-commerce? Are you crazy? This is Columbus in the New World. What was his ROI?
>
> —Andy Grove, Intel Chairman

New technologies are emerging every day. Some of these technologies may become a killer app for e-commerce. While it is almost impossible to predict them, using standard measures (such as ROI) to evaluate may fail, too. However, several underlying trends may help to identify and evaluate the right technologies.

The drop in the cost of computing continues. Storage, processing, and distribution cost of information is decreasing to a level where the cost is less than the value of information. This has lead to the development of new enterprises that provide free products and services in return for information and loyalty. On the hardware side, we have FreePC, PeoplePC, eMachine+Compuserve alliance, etc. On the applications side, we have HotMail, when.com, Yahoo, etc. On the Internet access side, we have NetZero, Freeserve, etc. On the Web site hosting side, we have GeoCity, Tripod, etc. More and more of these types of e-commerce businesses will continue to rise, making it difficult for the current businesses to compete.

While the power of computer processors is going up, their sizes are decreasing. In addition, computer processors are now used in many devices and products (such as autos, refrigerators, washing machines, dishwashers, etc.). As computing technologies continue to expand to household products and appliances, communications technologies will soon follow. These products and appliances, when connected to the Internet, will create new e-commerce opportunities. For example, a refrigerator in the future may be able to automatically buy groceries for you. The jar of milk is put in the refrigerator in one location where there is a sensor that notices how much milk is left. As soon as milk reaches the reorder level (determined based on your consumption pattern), the refrigerator will automatically connect with a grocery store on the Web (maybe NetGrocer, PeaPod, or WebVan) and place an order for milk.

New software applications are emerging every day. Each one makes e-commerce business processes more efficient and effective. This will enable more and more people to go online for their shopping, entertainment, business and home management needs. New e-commerce models will rise to support these changes in customer behavior.

2.7 Internet Protocols

John Braun

The purpose of this section is to describe concepts that are essential to understanding how Internet protocols work. Basic addressing at the device or client level will be described, followed by protocols that are used to exchange information among these devices. Although the term device may seem vague at first, the types of devices that communicate via the Internet have matured from simple text-based information, to the use of audio, video, animation, and other forms of communication. Once the basics of communication are laid out, some of the specific protocols and applications that utilize these basics will be discussed. Security aspects of these protocols and applications will be covered. Useful search tools that can help locate information on the Internet will be covered, and a discussion of some of the major industry players will conclude this section.

2.7.1 Addressing for Internet

All devices on a network that supports Internet Protocol (IP) have a unique numeric address, 32 bits in length.

The most common way of representing a device's IP address is by using a "dotted quad" — four decimal numbers ranging from 0 to 255, separated by periods. This results in a theoretical range of 0.0.0.0 to 255.255.255.255.

There are currently five classes of IP addresses:

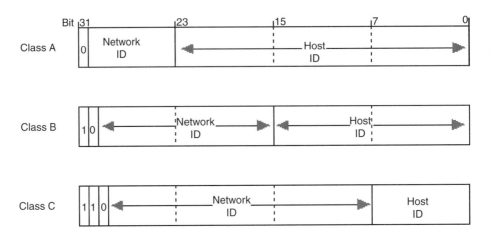

A Class D address has the following format:

    1110 MMMMMMMMMMMMMMMMMMMMMMMMMMMM(28)

A class E address has the following format:

    1111 XXXXXXXXXXXXXXXXXXXXXXXXXXXX(28)

N = Network portion of address
L = Local portion of address
M = Multicast address
X = Undefined

This results in the following valid ranges:

Class	Begin	End
A	0.0.0.0	127.255.255.255
B	128.0.0.0	191.255.255.255
C	192.0.0.0	223.255.255.255
D	224.0.0.0	239.255.255.255
E	240.0.0.0	255.255.255.255

Class D and Class E addresses, which have the first (high order) bits of the address set to 111, are classified as part of an undefined "extended addressing mode."

There are a few addresses that have special purposes and should not be assigned to a device.

If the network ID portion of an address is set to all zeros, it means "this network." For example, address 0.0.0.42 would mean host 42 on this network.

If the host ID portion of an address is set to all ones, it means "all hosts." For example, the address 199.27.24.255 would mean all hosts on the 199.27.24 network.

The class A address with a network ID of 127 is defined as a loopback address, where any packet sent by the host will be returned to the host without being sent across the network. This can be used for testing purposes, or as a sanity check to determine if one's TCP/IP implementation is working properly. The typical value used for this purpose is 127.0.0.1.

References

Postel, J., "Internet Protocol," RFC 791, USC/Information Sciences Institute, September 1981.
Reynolds, J. and J. Postel, "Assigned Numbers," RFC 943, USC/Information Sciences Institute, April 1985.
*** Latest is RFC 990

2.7.1.1 DNS

The domain name system (DNS) is a global network of computers that can translate a numerical IP address to a human-readable name, and also translate a name to the corresponding IP address. This makes navigation of a TCP/IP network much easier. For example, www.crcpress.com is much easier to remember than 199.29.24.3, the IP address which corresponds to this address.

Before DNS, each computer on a network would have to maintain a large file (typically called hosts) with all known IP address and name pairs. This is obviously impossible to do now with the sheer number of hosts on the Internet, but can still be useful for small, private TCP/IP networks which are not directly connected to the Internet.

When configuring a client to access DNS services, multiple DNS servers should be specified, if available. Modern TCP/IP implementations are smart enough to try another DNS server if the initial one is unavailable.

A DNS client will submit a DNS request to a server, and receive one of three types of reply. The client will be told that the lookup was successful and be given the name, that the server couldn't perform the lookup but knows another server that may, or that the lookup failed.

References

RFC 1035 — Domain names — implementation and specification. P.V. Mockapetris. Nov-01-1987

2.7.2 Communication Protocols in Internet

There are many protocols used on the Internet. Most are classified at either the lower layers (Layer 3, Network and Layer 4, Transport) of the OSI model, or at the application level (Layer 9).

IP
RFC 791
The IP in TCP/IP refers to the Internet Protocol used at the network layer. The basic purpose of this protocol is to try to deliver packets. It does not offer such services as acknowledgment, retransmission, error correction, flow control or guarantee of order of delivery. This is the job of higher-level protocols. The advantage of IP is that it offers a common framework for devices to communicate.

An IP header looks like this:

Version	IHL	Type of Service	Total Lengh
Identification		Flags	Fragment Offset
Time to Live		Protocol	Header Checksums
Source Address			
Destination Address			
TCP Header, Then your data ********************************			

UDP

RFC 768

User Datagram Protocol (UDP) is a connectionless protocol that provides a means to send data with a low protocol overhead. Only the source port, destination port, length, and checksum are added to the raw data. However, it does not guarantee delivery or protection from duplicates. For applications where performance is critical, a small loss of data is not critical, and link is known to be reliable, UDP may be used. Streaming audio or video are examples of applications where high performance is more important than a possible, but potientially correctable, loss of data.

TCP

RFC 793

Transmission Control Protocol (TCP) offers a more robust, connection-oriented method for reliably sending data. Flow control and multiplexing are supported. Unlike UDP, TCP supports the concept of a continuous stream of data between two hosts. Unlike IP, TCP will make multiple attempts to deliver data if the initial attempt fails. If the integrity of data is critical, TCP should be used.

A TCP header looks like this:

Source Port		Destination Port	
Sequence Number			
Acknowledgment Number			
Data Offset	Reserved	Windows	
Checksum		Urgent Point	
Your Data *** next 100 octets ************************			

2.7.3 Information Transfer in Internet

A socket is a virtual communications channel that is established between two hosts. It can use either UDP or TCP for the transport protocol. Each socket has a unique descriptor, and multiple sockets can reside on the same port. This allows multiple clients to take advantage of a service on a single port without the server getting confused. In order to prevent congestion on a single port, many application-level protocols have a control connection on a known port, and then negotiate another port for subsequent data transfer.

2.7.4 Types of Internet Access

There are two basic types of Internet access. One is a full-time direct connection; the other is via a telephone line with a modem. The disadvantage of a direct connection is mostly cost in both dollars and network maintenance. The advantage is speed, where T1 (1.5 Mb/sec) or T3 (45 Mb/sec) rates are common measures. The disadvantage of a modem connection is reliability (line noise) availability (busy signals) and speed. Certain schemes such as v.90 can achieve up to 56k bps download speeds. The advantage of a modem connection is the cost of both equipment and service, with 56k bps modems available for under $300 U.S., and unlimited service for around $20 U.S. a month.

Integrated services digital network (ISDN) can provide 128 kb/sec transfer rates. It never seemed to catch on due to the difficulty in configuring the equipment and inconsistent pricing plans across the country. Many providers charge a flat rate, with some adding charges for each unit of time or unit of data sent, whereas a local telephone line typically allows unlimited usage.

Asymmetric digital subscriber line (ADSL) is a relative newcomer that takes advantage of existing twisted-pair wiring, and can reach speeds of 6 Mb/sec for downloading, and 640 kb/sec for uploading. It is being deployed in major cities, but it remains to be seen how widespread the service will become.

Cable modems are slowly becoming available, and offer up to 10 Mb/sec transfer rates. Typically, the upstream connection consists of high-speed fiber, with the final connection to the cable modem being made with coax.

There are hybrid solutions, such as a satellite (with speeds of 400 kb/sec and higher) that uses a phone line for uploading data, and a satellite for data download. This can be a good solution for scenarios such as surfing the web, where the amount of data sent to request information is much less than bandwidth-intensive data types, such as graphics and sound, which comprise the data.

There are two major protocols used for establishing a TCP/IP connection over phone lines. Point to Point Protocol (PPP) is the more modern method, and PPP software is included with nearly every major operating system. Serial Link Internet Protocol (SLIP) is an older standard that is being phased out in favor of PPP.

2.7.5 Internet E-mail

SMTP — STD10
POP — STD53
IMAP4 — RFC2060

There are a few standards for sending and receiving Internet e-mail. The most popular protocol for receiving e-mail is Post Office Protocol (POP) which defaults to TCP port 110. A newer protocol, Internet Message Access Protocol (IMAP4), used for receiving e-mail, resides on TCP port 143.

The most common protocol used for sending e-mail is Simple Mail Transfer Protocol (SMTP) which defaults to TCP port 25. Note that POP can also be used for sending e-mail, but this feature is an optional extension of the POP protocol, and is not supported by many e-mail clients and servers.

SMTP

The SMTP service allows the sending of e-mail by providing relevant information, in a specific order, to a SMTP server. The conversation between the client (sender) and the server consists of human-readable text commands that are assigned four-letter codes, and three-digit code numbers as responses. A typical exchange may look like this:

```
C: HELO megacorp.com                       (sender identification)
S: 250 smtp.conhugeco.com                  (server identification)
C: MAIL FROM:<johnbraun@megacorp.com>      (who is this from?)
S: 250 OK
C: RCPT TO:<dilbert@anotherdomain.net>     (who is the recipient?)
S: 250 OK
C: DATA                                    (ready for data?)
S: 354 Go ahead…                           (sure)
C: Hey Dilbert get to work!
C: .                                       (end of input)
S: 250 OK
C: QUIT                                     (sender all done)
S: 221 Bye…                                 (server says seeya)
```

This system works well enough, with servers cooperating when mail needs to be forwarded to a domain outside of their own. Messages can be delivered in minutes, if all the servers in a message's path respond in a timely manner.

Although standard SMTP does not have any provision for confirmation of receipt (there are proposed standards, but they are not widely implemented yet) problems encountered along the path of the message are usually reported to the sender. If a server finds that another server is unavailable, it will usually try to send the message several times before giving up. If a server crashes while processing mail, the message may get lost forever. For critical documents, it may be wise to ask the receiver to confirm receipt.

One problem with current SMTP implementations is that they don't confirm the validity of the sender's address. This has led to massive abuse by junk e-mail senders, who really don't want you to respond via e-mail, anyway. An attempt to reply will result in getting a message saying that the (bogus) return address doesn't exist, or that it has already been shut down.

There are proposed methods of authenticating an entity wishing to use an SMTP server, in hopes of reducing spam and other evils. There are also solutions, mainly using public key encryption and digital signatures, to confirm the identity of the sender. These features are not available at the protocol level yet.

POP

The POP protocol is used to retrieve e-mail for a specific user. Session establishment can be done with a simple username and password scheme, or can optionally use more sophisticated means like APOP. APOP never sends the user's password over the connection, instead applying the MD5 algorithm to the password and other time-sensitive data. Not all POP servers support APOP, and will return an error message if this method of login is attempted but not supported.

After a message is retrieved, it is usually deleted. Some POP servers allow the user to keep their e-mail on the server after it has been retrieved, but this is not guaranteed. The IMAP4 protocol is better suited for keeping e-mail on a server.

Some POP servers offer the ability to send e-mail as well as receive it. The advantage of using POP for sending e-mail is that it requires users to identify themselves, thereby reducing the generation of spam and other unauthorized e-mail. It also eliminates the need to maintain a separate SMTP server. The disadvantage of this method is that not all e-mail clients and servers support sending via POP.

IMAP4

IMAP4 is a newer protocol for retrieving e-mail. It offers a richer set of features than POP, including keeping e-mail on a remote server, searching e-mail before retrieval, and a greater number of authentication schemes.

The option of keeping e-mail on a server helps reduce resource requirements on the client, but increases the need for more disk space, as well as regular backups to ensure the historical data is preserved. This scheme can benefit devices such as portable computers, network computers, and portable digital assistants, which may have limited storage capacity. It can also offer convenience to mobile users, since their mail can be stored on the server, rather than being spread among multiple clients.

2.7.6   Telnet in Internet

Telnet — STD8

Telnet is a service that resides on TCP port 23 and offers terminal services to remote users. The client and server can negotiate features of the connection, which can range from a simple ASCII exchange, to enhanced services such as cursor control and styled text.

A telnet client can be used to interact with other TCP services that exchange text or binary data, provided that the telnet client allows one to specify the port one wants to connect to. For example, a telnet client can connect to an SMTP server on port 25 and send e-mail. This can be handy for debugging or investigative purposes.

2.7.7   File Transfer in Internet

FTP

Request for Comments: 959
RFC 990 — Port #

File Transfer Protocol (FTP) provides a reliable, standard way to transfer both text and binary files between hosts. There are two types of connections when using FTP, a control connection and a data connection. A control connection is used for the client and server to exchange commands and status information. The default port for a control connection is 21. The default client data port is 21, and the default server data port is 20.

In most cases, a data port with a value outside the range of commonly known services is selected for security reasons. The client can request a specific data port via the PORT command, or can ask the server to select one with the PASV command.

Although FTP makes every attempt to deliver a file, circumstances beyond the user's control (modem disconnect, network failure) may cause the transfer to be interrupted. Most, but not all, FTP implementations support a restart mode where file transfer can begin at a point other than the beginning of the file, allowing data transfer to begin at the point of interruption and complete.

2.7.8 News and Usenet

NNTP — RFC977

The most common way of distributing news on the Internet is by using the Network News Transfer Protocol (NNTP). The system, originally created to exchange messages in a university environment, has evolved to a worldwide messaging system with thousands of distinct topics of interest, called newsgroups.

Each newsgroup name consists of several elements, with the general classification at the beginning of the name, and the specific subject matter at the end. For example, the group comp.sys.mac.hardware.video is a computer-related group about video cards for Mac systems.

The most popular hierarchies are sometimes referred to as the "big seven" and include comp (computing), misc (miscellaneous), news (newsgroup-related), rec (recreation), sci (science), soc (social), and talk (discussion). There are many other hierarchies for alternative and localized content.

Clients to allow one to read and post news are available as both stand-alone applications, or integrated with other Internet clients such as a web browser. Sophisticated clients can allow one to organize messages about a similar topic into threads, or filter messages based on certain criteria. Although most messages pertain to the topic of the newsgroup, there are those who "spam" the newsgroups with ads or other material unrelated to the group.

Care should be taken when configuring your news client where it asks for your e-mail address. Your address is normally added to any messages that you post, with the intent of making it easier for others to make a personal reply. Alas, there are those who scan the newsgroups for e-mail addresses, which are then used for the purpose of sending junk e-mail. A good strategy is to add some text to your address that a human would know to delete before making a personal reply, such as john@ihatespam.mega-corp.com.

A dedicated news host is required before clients can post and retrieve messages. Typically, a news host will server a large group of clients, and will exchange articles with another upstream host. The connection between hosts is referred to as a feed, and is meant for propagating articles and control messages, and not meant for direct client connections. If all hosts along a certain path agree to exchange the same group, a message posted in a particular group on one host will eventually propogate to all other news hosts. Hosts can also choose to restrict the groups they carry, which can help conserve disk space and bandwidth.

Care must be taken in deciding which groups a host should carry, and how long the articles should remain on the server before being deleted or expired. Inaccurate estimates can result in a dreaded disk full error, which causes many hosts to reject articles until someone clears some disk space.

2.7.9 Mailing Lists in Internet

A mailing list is a method of allowing Internet users to communicate about specific topics via e-mail. To join a mailing list, one sends an e-mail to a special address, and indicates a desire to join the list, as well as what e-mail address should receive further information from the list. The subscriber will then receive information on how to submit messages to the list, and how to perform administrative functions, such as being removed from the list. If the list is unmoderated, any message submitted to the list will be sent to all other subscribers. If the list is moderated, an administrator will decide which submissions should be distributed to other members. Messages may be sent one-by-one, or grouped and sent in a digest. Choosing to receive messages in digest form is a good idea for busy lists, lest your mailbox gets filled with hundreds of messages.

http://www.lsoft.com/listserv-hist.stm
http://www.wrt.tcu.edu/wrt/wri/files/barnes.htm

listserv distribute protocol information — rfc1429.txt

2.7.10   Information Search in Internet

2.7.10.1   Spiders

There are several tools, referred to as "spiders" which basically "crawl" through a web site, acquiring some or all of the information on each page. They then allow users to search through this information, in hopes of finding a resource relating to the topic of interest.

Many spiders have evolved into portals, which not only allow you to find information drawn from the Internet, but also provide links to other commonly used services, from maps to news headlines to package tracking.

Some of the more popular spiders are:

Alta Vista	http://www.altavista.digital.com
Excite	http://www.excite.com
Infoseek	http://www.infoseek.com
Lycos	http://www.lycos.com
WebCrawler	http://www.webcrawler.com

2.7.10.2   Category Indexes

Another method of finding what you are looking for is to use a tool where sites are scanned by actual humans, and then placed into one or more catagories. This method of locating information is a good complement to using a spider, since a spider may return too much information to be useful. The most popular of these services is Yahoo at http://www.yahoo.com.

2.7.11   Netscape and Microsoft

Netscape (NASDAQ: NSCP) and Microsoft (NASDAQ: MSFT) are two of the major players in the commercial Internet client and server markets. Netscape, established in 1994, was the first company to offer a commercial Internet browser, followed by server products. Microsoft was late to the party, but now also offers a full suite of Internet client and server products. Whereas Microsoft server products run on Windows, and their client programs span Windows, Mac, and UNIX, Netscape offers server products on both the Windows and UNIX platforms, and client programs for Windows, Mac, and UNIX.

Netscape has evolved from being a browser-only company, to one that now depends on server software and access to their NetCenter site for revenue. This is no doubt due to Microsoft's giving away their Internet Explorer browser. Netscape eventually made the source code to their browser available, in an attempt to gain back some browser market share by opening up their product.

Although the tight integration of Microsoft's browser and server products with the Windows operating system can offer increased functionality, it also tends to lock one into Windows. This can be a problem when attempting to use Microsoft products with more standards-based solutions like those from Netscape. Proprietary technologies like ActiveX controls and VBScript don't always work well with Microsoft products.

In the fast-moving world of the Internet, it is hard to predict which, if either, company will win. The strength of Netscape is that they provide solutions for a wider variety of systems, especially in the UNIX realm, and that they don't lock you into a single environment. The strength of Microsoft is that their products are on almost all desktop systems. For a solution where both client and server products are guaranteed to come from Microsoft, the increased functionality of being closely linked to Windows can be worth it.

To put things in perspective, the most popular web server application as of this writing is the freeware Apache web server, with almost 50% market share.

Intranet References

The following web sites provide comprehensive sources of information and links to other sites for reference material relating to the Internet:

http://www.cisco.com
http://www.intel.com
http://www.microsoft.com
http://www.sun.com
http://www.wcom.com

Glossary of Intranet Acronyms and Terms

ARPAnet	Earliest packet switched network; the progenitor to today's Internet.
ASCII	Plain text file, containing only regular keyboard characters.
BCP	Bridging Control Protocol, used to configure bridge protocol parameters on both ends of a point-to-point link.
C/C++	High-level programming language allowing program control of system hardware, and designed for code portability on all types of computers. C++ is a programming language that extends the Object Oriented capabilities of C.
CA	A Certificate Authority: a third-part which can attest that you are on record as the only person with the key associated with a personal digital signature.
CCITT	International Telegraph and Telephone Consultative Committee.
CHAP	Challenge Handshake Authentication Protocol: a commonly used protocol for link encryption. Authentication occurs at the data link layer and is transparent to end users.
CGI	Common Gateway Scripting: used to support two-way browser communication.
Digital Signature	Used to validate the identity of the file sender's e-mail.
DNS	Domain Name Server protocol, part of TCP/IP.
ECP	Encryption Control Protocol, part of the PPP suite.
Ethernet	LAN protocol as specified in the IEEE 802.3 standard.
Firewall	Collection of hardware and software that interconnects two or more networks and, at the same time, provides a central location for managing security.
FTP	The File Transfer Protocol: a standard protocol for transferring and copying files from one computer to another.
Kbps	1,024 bits per second.
Home Page	The first page that users see when they access a particular Web site.
HTML	HyperText Markup Language: used to describe the layout and contents of pages on the Web. The Hypertext Markup Language (HTML) is the language of the World Wide Web.
ICMP	Internet Control Message Protocol: defines the rules routers use to exchange routing information.

ICMPv6	Internet Control Message Protocol Version 6.
IGMP	The Internet Group Management Protocol is used by IP hosts to report host group clusters to neighboring multicast routers.
IGRP	Interior Gateway Routing Protocol, part of TCP/IP.
IPCP	IP Control Protocol, used to configure IP parameters on both ends of the PPP link.
IPX	Internetwork Packet Exchange: Novell's implementation of the Xerox Internet Datagram Protocol (IDP), used to define a set of rules for coordinating network communication between network components.
ISDN	Integrated Services Digital Network: provides digital communications circuit that allows transmission of voice, data, video, and graphics at very high speeds (from 56 Kbps to 128 Kbps), over standard communication lines.
ISP	Internet Service Providers act as middlemen, renting time to other users who want to access the Internet.
JAVA	A computer programming language that allows users to execute special programs (called applets) while accessing and viewing a Web page. Java is designed for creating animated Web sites.
LAN	Local Area Network.
LCP	Link Control Protocol: configures and tests the data link connection, and is part of the PPPsuite.
MARS	Multicast Address Resolution Server.
Mbps	1,000,000 bytes per second.
MPEG	Motion Picture Experts Group defining standards for handling video and audio compression.
OS/2	A high-performance, multi-tasking workstation operating system.
OSPF	Open Shortest Path First, link-state routing protocol used for IP routing.
PAP	Password Authentication Protocol: used for transparent session authentication occurring at the data link layer.
PGP	Pretty Good Privacy is a free (for personal use) e-mail security program developed in 1991 to support public-key encryption, digital signatures, and data compression. PGP is based on a 128-bit key.
PPP	Point-to-Point Protocol is one of the major protocols used to connect to the Internet. PPP is newer and faster than SLIP.
PKE	Public Key Encryption allows a sender to encrypt a document using a public key, which the recipient decodes using a private key.
POP3	Post Office Protocol version 3, allowing dynamic workstation access to mail drops on a server host.
PPTP	Point-to-Point Tunneling Protocol.
RIP	Routing Information Protocol: maintains network exchange and topology information.
RFC	Request for Comments: discussion notes, recommendations, and specifications for the Internet.
Router	This hardware device can be used to filter out data packets-based specific selection criteria. Thus, the router can allow certain packets into the network while rejecting others.
S-HTTP	Secure HTTP: a protocol developed by the CommerceNet coalition that operates at the level of the HTTP protocol.

SLIP	Serial Line Internet Protocol: one of the major protocols used to connect to the Internet. It predates PPP.
SMTP	Simple Mail Transfer Protocol: specifies the format and delivery handling of electronic messages.
SPX	The Sequenced Packet Exchange protocol defines a set of rules for coordinating network communication between network components.
SSI	SSI software is used by Web servers to display and/or capture dynamic (changing) information on an HTML page.
SSL	The Secure Socket Layer (SSL) was developed by Netscape Communications to encrypt TCP/IP communications between two host computers.
Subnet	Partition on an IP network based on class, involving use and movement of a subnet mask.
Surfing	Term used to describe accessing (through a Web browser) a chain of documents through a series of links on the Web.
T1	A leased line which can support transmission speed to 1.54 Mbps.
TCP/IP	The Transmission Control Protocol/Internet Protocol: specifies the rules for the exchange of information within the Internet or an Intranet, allowing packets from many different types of networks to be sent over the same network.
TCM	Total quality management involves creating systems and workflows that promote superior products and services.
Telnet	This service allows connection to a remote Internet host so that programs can be executed from a remote computer.
UDP	User Datagram Protocol provides message service for TCP/IP.
UNIX	UNIX is an operating system well suited to the Internet's open system model.
URL	Millions of documents that are distinguished by a unique name called a URL (Uniform Resource Locator), or more simply, a Web address. The URL is used by Web browsers use to access Internet information
UUCP	Unix to Unix Copy: allows files to be copied from the Unix system to another.
XML	Extended Markup Language: designed to provide a self-descriptive, platform-independent mechanism for exchanging management information between applications.
Web Browser	Allows you to traverse and view documents on the World Wide Web.
Windows NT	Provides a high-performance, multi-tasking workstation operating system.
WWW	The World Wide Web — or Web — is a collection of seamlessly interlinked documents that reside on Internet servers. The Web is so named because it links documents to form a web of information across computers worldwide.

3

Network Management and Administration

Joe Ghetie
Bellcore

Kornel Terplan
Industry Consultant and Professor

Endre Szebenyi
Industry Consultant

Takeo Hamada
Fujitsu Laboratories America

Hiroshi Kamata
OKI Electric

Stephanie Hogg
Telsta Research

Carel Marsman
CMG

Introduction

It is not enough to develop, implement, and rollout new technologies by telecommunications service providers. These technologies should be properly administered and managed. Over a five-year period, management and administration would take up to 85% of operating expenses; acquiring the technology, just 15%. It is a very important metric for cost-justifying investments into management and administration.

This segment addresses administration and management issues. Management concepts outline the basics of managers and managed entities. Concepts include several management models, such as central and decentral, concentrated and distributed, and the use of hierarchical schemes supported by umbrella managers. Also, open management is addressed, including the open systems conceptual model, associated systems concepts, and requirements for open management systems. Distribution of management processes and functions will play a key role in future management solutions. Managed entities must be connected with element managers and management platforms using in-band or out-of-band communication schemes. This contribution gives examples for both alternatives.

Administration and management are usually an afterthought when considering the deployment of innovative technologies. This contribution tries to bring the technology deployment with the selection and implementation of management solutions into synchronization. Each technology that is considered innovative, such as frame relay, FDDI/CDDI, Switched MultiMegabit Data Service (SMDS), ATM, Sonet/SDH, Cable, mobile and xDSL, is investigated for how far management and administration solutions are available and implementable. In particular, the availability and structure of MIBs (management information base) are analyzed. In most cases, MIBs support most of fault, configuration, performance, security, and accounting management functions. MIBs in combination with SNMP managers do useful work for history-type of data visualization, analysis, and reporting. State-of-the-art technology needs additional management tools and applications that help with real-time decision support.

Management and administration depend to a large extent on management standards. There are two principal groups: standards for enterprise-level administration and management, and standards for specific telecommunications environments. The management standards contribution focuses on enterprise-level standards, such as SNMP, RMON, and DMI, first. Components of telecommunications standards are also discussed in some depth, e.g., CMIP, Corba, and DCOM. This contribution prepares the readers for telecommunication network management (TMN) and Telecommunications Information Networking Architecture (TINA), and for better understanding management framework products and management applications.

TMN is a very simple model for streamlining management and administration. It uses four layers in addition to the network element's layer at the bottom. Management processes, fucntions, and tools may be categorized in accordance with these layers. This TMN contribution goes into depth and discusses various TMN models (information, fucntional, and physical), TMN elements (operations systems function, workstation functions, mediation functions, Q adapter function, network element function), TMN internal and external interfaces (Q3, Qx, X, F, and M), and the most appropriate use of Data Communication Network (DCN).

TINA goes one step behind TMN and offers four dimensions of considerations: life cycle management, computational infrastructure, partitioning by layers and domains, and functional representing fault, configuration, accounting, performance, and security management. TINA and TMN can work together, but they are not identical. TINA puts more weight on service fulfillment and service assurance. Also, resources are described in much more depth. TINA can be tailored to the needs of particular service providers.

The TeleManagement Forum offers guidance for deploying and re-engineering telecommunications business processes. This contribution uses the basic business model of breaking down support processes into two dimensions: life cycle of services, such as fulfillment, service assurance, and billing processes, then hierarchy of services, such as customer care processes, service development and operations processes, and networks and systems management processes. This contribution handles all 16 principal support processes, individually. Also, their links to each other, to the customers, and to the physical networks are addressed.

Management frameworks are the heart of support systems for telecommunications providers. They consist of an application platform and of management applications. This contribution outlines the principal attributes, such as architecture, application programming interfaces, protocol support, hardware and software platforms, graphical user interface, application programming interfaces, management functions supported, security modules, modeling capabilities, and internal systems services. For both telecommunications and enterprise environments, framework products are listed, and a few of them, such as OpenView from Hewlett-Packard, TNG from Computer Associates, FrontLine Manager from ManageCom, NetExpert from Objective Systems Integrators, and TeMIP from Compaq/Digital are analyzed in some depth. Over the next couple of years, frameworks are expected to embed the best of suite management applications with the result of full functionality to implement operations, business, and marketing support systems.

It is expected that telecommunications service providers and their customers will connect their management systems and applications. The name of this concept is customer network management (CNM). This contribution outlines the joint work, principal management processes and functions, and also legal issues. If successful, information exchange between providers and customers can be accelerated, and duplicated functions can be eliminated.

Service management is in the center of the next contribution. Service quality may be improved when certain management functions are outsourced to third parties. This contribution details the drivers for outsourcing and critical success factors of outsourcing alliances. Reporting on principal services metrics is key in all relationships. This contribution deals with practical examples for performance indicators and service level reports. Service management means more than element management. Service management is targeting more consolidated metrics in the TMN architecture.

Web technology is going to change the way management and administration systems work. Using Java applets and components of Web-Base Enterprise Management (Wbem) standards, management and administration can be unified and simplified. This contribution handles Web basics (URL, Web server, Web browser, HTML, XML, and HTTP), evolving standards (Java and Wbem), and many application examples from framework vendors and management application vendors. This technology is expected to penetrate and change the way present operations, business, and marketing support systems work.

Support systems of telecommunications providers represent a very complex but increasingly significant segment of the communications industry. This contribution starts with the market drivers for support systems, such as network complexity, customer in focus, more standards, very high growth rates, deregulation, and convergence, followed by startegical benefits of such support systems. This contribution also identifies the suppliers of support systems, such as consulting companies, computer manufacturers, equipment manufacturers, software companies, and outsourcers. The remaining part of this contribution focuses on positioning products in terms of supporting markets (voice, data, Internet, cable, and wireless), supporting management areas (customer care and billing, provisioning and order processing, and network operations management), and compliance to TMN layers, such as business, service, network, and element management layers.

Intranets are penetrating both the telecommunications providers and enterprise infrastructures. This contribution targets the management of these kinds of networks. In detail, it identifies sensitive components that may cause congestion or bottlenecks. Special emphasis is on log file analysis, wire monitoring, look-through measurements, traffic shapers to conserve bandwidth, and on administering Web server farms. For each subject area, product examples are also included. In most cases, Web content is expected to drive decisions about resource facilities and equipment reservation/allocation.

3.1   Management Concepts

Joe Ghetie

The last decade of the past millennium witnessed one of the most dramatic advancements of communications technologies and services in human history. Communication, as a way of conveying and exchanging management information, had found in the Internet one of the best examples of the explosive growth with a tremendous impact on the current and future abilities of humans to share information.

The dream of universal access to information, the dream of a giant village, the dream of fast, reliable, content-rich information exchange are today closer to reality than anybody has anticipated. Data communications, video communications, and both wired and wireless communications media have increased our ability to control through communications large, global enterprises and businesses.

Network and systems management are specialized systems targeting, monitoring, and controlling the vast array of network and computing systems resources used in communications, manufacturing, commerce, finance, banking, and education, as well as in research and development.

Management systems were born out of necessity to prevent, diagnose, configure, and solve problems raised by the size, complexity, and heterogeneity of multivendor, multiprotocol, and multitechnology environments that characterize the underlying network and computing systems.

Although management systems are value-added components to communications technologies, they are as vital as the transmission, switching, and operations systems in order to supervise and maintain the normal information exchange.

3.1.1   Management and Data Communications

Management systems aimed at monitoring and controlling communications systems represent conceptual design and associated infrastructure that, essentially, resemble particular implementation of open systems.

3.1.1.1   Communications General Model

A simplified view of any point-to-point communication assumes an information source (sending party) and the information destination (receiving party). The communication takes place over a transmission

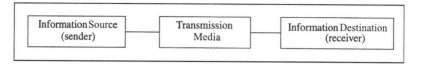

FIGURE 3.1.1   Communication network conceptual model.

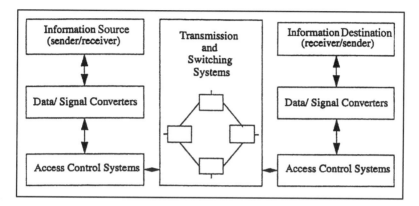

FIGURE 3.1.2   Communication network conceptual model.

media which can be a pair of copper wires, coaxial cable, fiber optic, or a wireless media such as radio, microwave, satellite, or infrared rays (Figure 3.1.1).

The information source can be a telephone set, a computer, a TV pattern, a facsimile, or an instrumentation process. The information can be another telephone set or a computer, a TV set, a fax machine, or a control panel.

In order to be transmitted, the native information sources, voice, computer or instrumentation data, graphics, or video images require successive data/signal conversions, according to adopted communications media and transmission technologies, and a rigorous security control of the access to shared network resources. Therefore, new components such as data/signal converters and access control systems should be added to the communication model. This communication can be asymmetric, i.e., taking place only in one direction, or it can be symmetric, i.e., taking place bidirectionally (Figure 3.1.2).

As a further consideration, the box representing the transmission media becomes more than a single conduit; a mixture of transmission/transport components and switching components in the form of circuits, links, nodes, routers, and switches participates in the design of a shared network environment.

3.1.1.2   Network Management General Model

The task of management, as derived from the general model of communications, is very clear: to be able to supervise, monitor, and control all the components that participate in the process of communications from the source to destination. That might include various computer hosts and terminals as sources/destinations of information, the devices performing data/signal conversions (protocol converters, emulators, concentrators, multiplexers), devices required to control the access to the network (security access, authorization, encoding, encryption), and all the components used in transmission, switching, and routing (Figure 3.1.3).

The task is not only clear but quite challenging when the list of actual devices is spelled out. Many dozens of different technologies implemented in hundreds of different components, developed, designed, and manufactured by thousands of vendors, are all potential subjects of management systems, especially when it comes to the point of providing end-to-end, enterprise-wide management services from monitoring, diagnostics, control, and reporting.

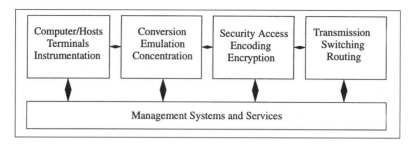

FIGURE 3.1.3   Network management conceptual model.

PCs, workstations, minicomputers, servers, mainframe computers, terminals, test equipment, phones, PBXs, TV sets, set-top boxes, cameras, modems, multiplexers, protocol converters, CSU/DSUs, statistical multiplexers, packet assembler dissassemblers, ISDN adapters, NIC cards, codecs, data encoders, data compression devices, gateways, front-end processors, line trunks, repeaters, regenerators, matrix switches, DCS/DACs, bridges, routers, and switches just begin a list of devices which should be or might be managed.

The management picture is complete only if we consider, in addition to the management of network resources, the management of computing systems resources such as thousands of different businesses, users, systems applications, databases, and complex, specialized, large operations systems.

All the information collected and exchanged in conjunction with management operations is translated in management data which is manipulated using techniques similar to those employed by data communications networks. However, substantial differences exist between data communication exchange and management information exchange to claim a specialized technical field, specialized communication protocols, information models, and specialized skills to design and operate management systems and interpret fault, performance, configuration, or security management information.

The following subsections will explain what is peculiar in management systems, the major requirements appended to management systems, the management paradigms adopted in management, and the historic and technical evolution of management systems.

3.1.2   Management Requirements

The diversity of managed resources, as found in traditionally distinct fields of communication such as voice, data, and video communications, generate different views on what should be the management functions and management requirements associated with management systems.

3.1.2.1   High-level Management Functions

Regardless of the diverse management views, three high-level management functions top the list: **monitoring**, **controlling**, and **reporting**. Monitoring represents the continuous collection of management information about the status of management resources, delivered in the form of events and alarm notifications when the threshold attached to managed resource parameters is exceeded. Controlling is the targeted attempt of the manager or management application to change the status or configuration of selected managed resources. Reporting consists of delivering and displaying the management information in an accessible form for reading, viewing, searching, and ultimately interpreting the reported information.

In practice, several other functions are associated with management systems and management applications according to particular business needs such as provisioning, service activation, capacity planning, network/systems administration, inventory management, backup and recovery management, and management operations automation. Many of these complex functions include or are built on basic monitoring, controlling, and reporting.

3.1.2.2   High-level Users Management Requirements

Based on the users' perspective on management, we can derive a set of high-level requirements associated with management, as listed below:

a. Ability to monitor and control end-to-end network and computing systems components.
b. Remote access and configuration of managed resources.
c. Ease of installation, operation, and maintenance of the management systems and their applications.
d. Secure management operations, user access, and secure transfer of management information.
e. Ability to report meaningful management-related information.
f. Real-time management and automation of routine management operations.
g. Flexibility regarding systems expansion and ability to accommodate various technologies.
h. Ability to backup and restore management information.

3.1.2.3   Driving Forces behind Management Technologies

Although the term of "network management" gained a clear acceptance only in the mid 1980s with the advancement of IBM management tools (later incorporated into the IBM NetView family of management products), network management was equally driven by the development of telecommunications, data communications, and computing systems networking. For telecommunications and data communications, the management technologies were concentrate on management of transmission and switching equipment (hardware devices, connections, circuits) along with conversion and access control devices. In the case of computing systems, the management technologies were concentrate on managing large computing system resources (hardware, interfaces, memory, data storage devices, etc.) and applications/databases.

With the convergence of telecommunications and computing systems, which embraces various technologies commonly known as computer telephony integration (voice over Internet is one of the most recent developments), the common point of these major fields becomes the network which connects these systems and the management of large data communications networks. This will be the dominant factor of the networks of the future.

3.1.2.4   Justifying Network Management Investment

It is well known that management systems are perceived as overhead cost. However, the cost of not being able to prevent major network and systems problems or to quickly find and restore a system to normal functionality is even higher and can be crippling for many businesses relying on information exchange.

The following reasoning can be used in justifying the investment in network management. Some points can be quantified and used as a basis for a front-end analysis when selecting management systems.

a. Reducing downtime of critical components of networks and computing systems.
b. Controlling the corporate networks as strategic investment assets.
c. Controlling the performance, growth, and complexity of user application.
d. Improving services in customer support and security of data transfer.
e. Controlling the cost of information technology deployment and operations.

3.1.3   Management Paradigms

Before analyzing the capabilities or the openness expected from management systems, we have to understand the fundamental paradigms used in management and the views associated with these paradigms.

3.1.3.1   Management Basic Model

Conceptually, the management systems are based on a simple model. In this model, management is the interaction/cooperation between two entities: the **managing entity** and the **managed entity**. The management entity represents a management system, a management platform, and/or a management application. The managed entity represents the managed resources. Looking at this simple model, it is

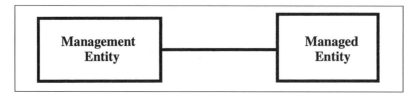

FIGURE 3.1.4   Management basic model.

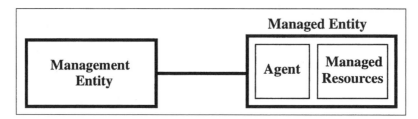

FIGURE 3.1.5   Manager-agent model.

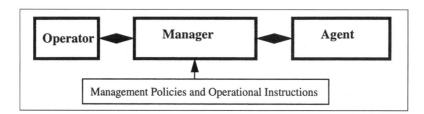

FIGURE 3.1.6   Real manager-agent model.

important to note its similarity to the basic communication model presented at the beginning of this chapter (Figure 3.1.4).

In order to communicate with the managed resources, which do not have any native mechanism to pass management information, there is a need to create an intermediary component, the agent. The agent is also called management agent or managed agent. The manager is the management entity, while the agent hides the interaction between the manager and the actual managed resources (Figure 3.1.5).

The manager–agent model is very common, used in describing the interaction between the management entity and the managed entity at a high level. This is the reason that all the paradigms natively created for management purposes closely follow the manager–agent model. In reality, the manager–agent model is more complex (Figure 3.1.6).

The complexity becomes more evident when we consider the interactions between the manager or the management applications and the human operators. Other components, less visible but also very important because they shape the nature of interactions between managers and agents, are the **management policies** and the **operational instructions** given to the manager and implicitly to the operator.

There are other paradigms such as client–server and applications–object server that can be used for management information exchange. Natively, these paradigms have been conceived for building distributed applications or distributed object environments. Nevertheless, these general paradigms can be applied for management and there are products that use variations of these paradigms for management purposes.

3.1.3.2   Management Views and Associated Models

Management assumes, as a primary function, the communication between the managing entity and managed entity. The management communication is based on the request–reply paradigm. The manager

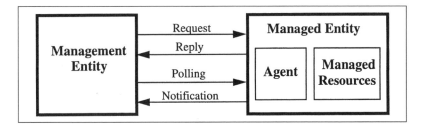

FIGURE 3.1.7 Manager-agent communication model.

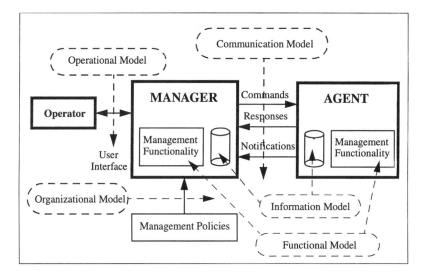

FIGURE 3.1.8 Manager-agent relationship models.

will request from the agent specific management information and the managed entity, through the agent, will reply with a message containing the information requested. If the request–reply communication is used continuously in order to reach each agent and the corresponding managed objects, the mechanism is called polling and it is primarily used in the management of Internet environments based on the Simple Network Management Protocol (SNMP) (Figure 3.1.7).

The request–reply mechanism is considered a synchronous communication mechanism, i.e., the manager expects an answer from the agent in a limited time frame before taking any action. If the reply is not received, a request for retransmission should be initiated by the manager.

There is an additional mechanism for communication between the manager and agents, called notification. The notification is an asynchronous mechanism initiated by the agent that communicates to the manager important changes in the status of managed resources which require either manager attention or intervention.

When building management systems, there are many aspects that should be taken into consideration. In addition to the communication model, several other models are used in conjunction with the manager–agents relationship, as follows: architectural model, organizational model, functional model, and informational model (Figure 3.1.8).

The architectural model deals with the design and structure of the components participating in the management process, i.e., the manager or managers and the agents supplying management information according to the network topology. The manager can be designed as a management platform that consists of a management framework and a suite of management applications providing the actual management functionality such as configuration, fault, and performance management. More details will be provided in the sections dealing with management platforms.

The operational model deals with the operator's interface to the management system and specifies the nature and the type of interactions available to the user such as controlling managed objects, displaying and searching for specific events, dialog with the systems, and alerting the operator in case of critical alarms. Most of the operational specifications are included in the product's technical specifications such as user guide, administrative guide, etc.

The functional model refers to the structure of management functions performed by the management system through management applications. The functional model is considered a layered model where basic management functions such as configuration, fault, performance, security, and accounting management are the foundation of the functional model. Several other management functions such as trouble ticket administration, help desk, provisioning/service activation, and capacity planning consist of a combination of the basic management functions. At the pinnacle of the functional model, there are applications performing complex functions such as alarms/events correlation, expert systems, and management automation.

The organizational model is tightly linked to the overall management policies and operational procedures. This model specifies management domains, partition of management realm among the management operators, access of the user to the management systems, customer-based network management, interchangeability of the roles between managers and agents, and the overall cooperation between the manager and other managers or management applications.

The information model, although mentioned at the end of this list, is critical in handling all the management aspects. Given the variety of managed resources, in order to support their management in a common way, there is a need of an abstraction of managed resources in the form of a common information model, known by both manager and agents. The management information model establishes the basis for defining, naming, and registering the managed resources. Managed objects are considered abstractions of physical and logical managed resources. Therefore, the term of managed objects implies the use of an information model. Access to the managed resources is allowed only through the use of managed objects. The conceptual repository of management information is called management information base (MIB). When we refer to a particular MIB, that means a collection of managed object definitions that describe a particular management domain or environment. The definition of managed objects is standardized and on this basis a manager implementing a particular protocol and information model can communicate with distributed agents which implement the same MIB.

3.1.3.3 Management Domains

Historically, as we mentioned earlier, the notion of network management was launched by IBM. The IBM NetView products were in fact a combination between mainframe systems management and network management. Since then, the concept of management has evolved. At the beginning, the management products have reflected the division, typical to most of the businesses, between network and computing systems management. With the advent of management platforms, the difference between network and systems management is blurring since the nature of the application and not the platform framework will determine the use of management systems.

Currently, it is commonly accepted that two major management domains can be considered when discussing the nature of managed resources: managing **physical resources** and **logical resources**. Physical resources are considered all the hardware components of the telecommunications and data communications networks that participate in the process of exchanging information. This management domain is known as **network management**. The management of computing systems' physical resources such as processors, memory, input/output interfaces, and storage devices, are considered part of systems management.

The management of logical resources is built around **applications management** and **databases management**, both associated with computing systems. Service management, user management, management of distributed transaction services, and data flow management are also considered system management of logical resources (Figure 3.1.9).

There is a separate domain which deals with the management of specific logical resources, i.e., the protocols used in standards-based communications. Layered protocols, layered service primitives, and

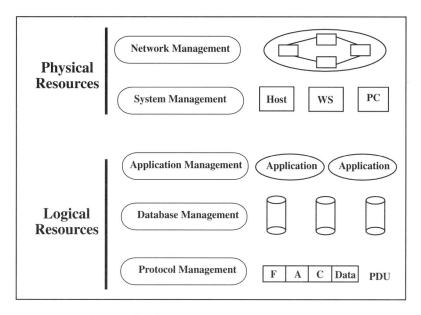

FIGURE 3.1.9 Management domains classification.

embedded management services are examples of protocol management. This type of management is applied to interfaces of particular technologies such as ATM, SONET, and WDM in the form of embedded channels or embedded **layer management entities** (LMEs). This type of management is conceptualized in the OSI Basic Reference Model, the foundation of standardized layered architecture and management.

3.1.4 Open Management Systems

In order to evaluate the management systems, there is a need for a reference model. This reference model is the open system and its corresponding model, the open management system.

3.1.4.1 Open Systems Conceptual Model

The open systems conceptual model assumes a design of systems modeled by the presence of four entities and by the relationship between these entities: **application platform, applications**, **application programming interface** (APIs), and **platform external interface** (PEI). This model can be applied to any computing system as part of the overall design and implementation. What makes any computing system (which runs software programs or applications) an open system is the separation of applications from the applications platform through APIs (Figure 3.1.10).

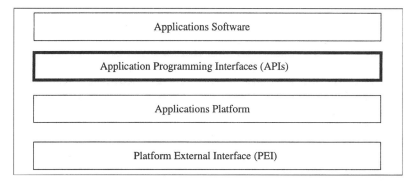

FIGURE 3.1.10 Open systems conceptual model.

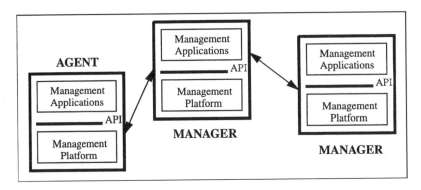

FIGURE 3.1.11   Open management system.

3.1.4.2   Open Management Systems Concept

The open systems conceptual model can also be applied to management systems, i.e., to managers and managed agents. In this case, the applications will be specialized management applications providing fault, configuration, performance, security, and accounting management. The management platform is a management framework which consists of, in addition to the computing platform, specific management services such as event management services, communication services, graphical user interface services, or database services (Figure 3.1.11).

As mentioned earlier, key components to open systems are the APIs. In this case, the APIs are specific management APIs that allow the development of management applications by using specific management platform services. Last but not least, the management platforms are not isolated; they communicate with managed agents (as a minimum) or with other management systems which may be modeled as open management systems. The platform external interface, in this case, will be an open standardized interface with well defined management operations, services, and protocols.

3.1.4.3   Requirements for Open Management Systems

Four high-level requirements characterize open systems and open management systems: **operability**, **interoperability**, **portability**, and **scalabilty** (Figure 3.1.12).

Operability represents the ability of management systems to provide easy installation, operations, and maintenance, as well as adequate reliability and performance. Interoperability represents the ability of management platforms to transparently exchange management information with managed agents or peers' management systems. Portability expresses the ability of management platforms and/or management systems applications to be ported to a different environment (computing platform) with minimum changes or no changes. Scalabilty refers to the ability of management systems to be expanded in coverage, user domain, and management functions without the need to change the initial design.

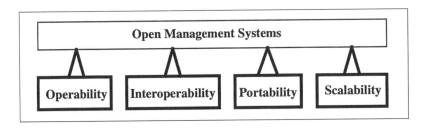

FIGURE 3.1.12   Open management system major requirements.

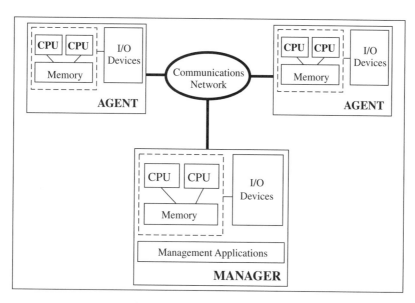

FIGURE 3.1.13 Management of distributed systems.

3.1.5 Distributed Management Systems

Most of the computing systems, telecommunications, and data communications networks are distributed, i.e., interconnected by a communication network designed to transfer information and messages related to specific business needs. Management means managing various network and systems resources which in most instances are physically separated. Therefore, by its very nature, the management is distributed.

A system is considered autonomous (it can be simple processor or multiple processor based) if the processes that constitute the system share the same memory. In contrast, distributed systems consist of interconnected autonomous systems with no shared memory. Since any networked computing environment is inherently a distributed system, the management of these systems is also inherently distributed.

3.1.5.1 Distributed Network and Computing Systems

The true nature of management, as distributed or centralized, is determined not by the physical distribution of its components (managers and agents) but by the centralization and processing of management information (Figure 3.1.13).

If the system is designed to collect all the management information from all the agents (which constitute the management domain) in one point, we deal with a centralized type of management. If the collection of management information takes place in several interconnected processes and the information may be held in distributed databases, we deal with distributed management systems.

3.1.5.2 Distributed Management Systems Architectures

In a truly distributed management system, multiple management users or operators as management clients access the management server through a local or a wide area network. The actual manager runs the management applications and it is the holder of a MIB for a particular management domain. Each manager is responsible for the agents that are part of his/her domain.

The ability to exchange management information between servers (managers), keep in synchronization the shared MIB information, take over the management domain of a failed manager, and of the operators to interact with multiple managers, creates a truly distributed management system architecture (Figure 3.1.14).

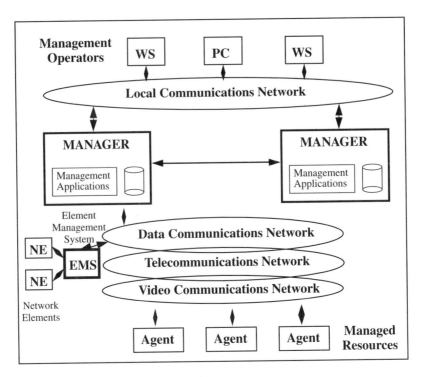

FIGURE 3.1.14 Distributed management systems architecture.

We have to emphasize that in all these examples we assume that the manager has a high degree of remotely accessing and configuring agents, with each agent acting as a management agent for a collection of managed objects and management processes.

3.1.5.3 In-band and Out-of-band Management Systems

All of the diagrams were presented in order to introduce the management concepts, and the properties of the management field included representation of interconnecting networks that carry management information. It is important to emphasize that these networks have rarely been designed as management-only infrastructures. Most of the management systems use for management information exchange the very network that carries the business-related data, voice, or video information. For the purpose of management, specialized protocols, operations, and application entities have been created and used. However, the management information is carried on the same physical infrastructure and on the same communication stack as the business information. In this case, we deal with the **in-band** type of management. This is a very cost-effective solution. However, there are some issues. By sharing the same service channels, the management information may take a significant chunk of the available bandwidth, and this may affect the overall performance of data exchange. That puts restrictions on how much and how often information is collected.

This is the reason that some management systems are built using **out-of-band channels**. The out-of-band management solutions may include unused bandwidth from a current channel allocation. A good example is the use of the low-band portion (50Hz–200Hz) of the voice grade channels as a dedicated data channel for management purposes. This solution is used for the management of the modems that share the same infrastructure with the voice communications. Other solutions consist of reserving a bit from the normal bit stream (for example, T1 multiplexer) to create a dedicated data channel for management purposes or assigning fields in each of the transmitted frames or cells for management purposes as it happens in the SONET and WDM technologies.

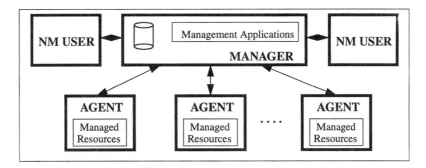

FIGURE 3.1.15 The "single manager" topological framework.

3.1.6 Management Systems Topological Frameworks

At a very high level, the architectural model of management systems is understood as the relationship between the main components of management systems, the managers and agents. The accepted term for the architectural layout of the management network is called topology and in most cases follows the business network infrastructure.

The topological view is the basis for the representation of the network and systems components using graphical user interfaces. An elaborate collection of graphical icons representing logical and physical resources, the links between these icons and the colors associated with the status of managed resources allow the management operators or the users to have a view of the management components along with their status.

Three major topological frameworks are considered when designing management systems: **single manager**, **manager of managers**, and **network of managers**.

3.1.6.1 The Single Manager

The single manager topology framework uses one management system which concentrates the collection and processing of management information from various managed resources such as routers, bridges, multiplexers, matrix switches, etc. Thus, the manager is the only point of exercising control over the network.

The system playing the role of the manager is usually a monolithic application that performs management operations and stores the management information received from all the managed resources. The single manager topology is fully centralized. Historically, most of the management systems started as host-based, centralized systems. Today, they still represent the most common topological framework (Figure 3.1.15).

Regarding the single manager framework, we emphasize its weaknesses as follows: concentration of network management functions and applications in one point; limitation of the number of resources to be managed (lack of scalability); and the high vulnerability of these systems when the manager fails. This topological framework is used for the management of small- to medium-size networks and systems.

3.1.6.2 The Manager of Managers

The manager of managers (MOM) topology is a logically centralized framework with distributed control capabilities. The MOM acts as a single integration point for several distributed element management systems (EMSs) (Figure 3.1.16).

The actual management of managed resources/devices is provided by the EMSs that monitor and control a particular management domain, which may consist of a group of network components and associated applications. Usually, EMSs are designed to manage a family of similar products built around a particular technology. In other instances the management domain is determined by geographical, administrative, or jurisdictional considerations.

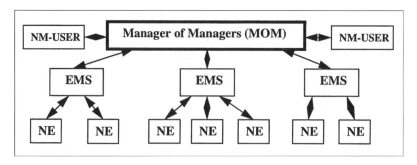

FIGURE 3.1.16   The "manager of managers" topological framework.

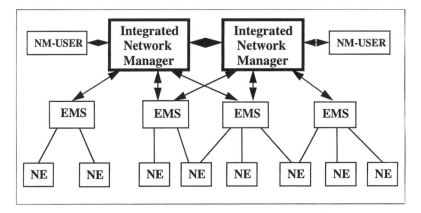

FIGURE 3.1.17   The "network of managers" topological framework.

This topology is used for medium and large networks. Only vital, critical information such as alarms, security alerts, and capacity planning-related information is elevated to the level of MOM, which acts as a management integrator.

3.1.6.3   The Network of Managers

The network of managers topological framework provides fully distributed management based on cooperative management between integrated network managers (INMs).

In this topological framework, management information can be exchanged between peer managers. Each INM is responsible for the management of its own grand domain. Cooperative links between INMs allow management information exchange. More than that, each INM can take over the management functions of an adjacent manager. Within each domain, the INM acts as the focal point of distributed management provided by several EMSs (Figure 3.1.17).

3.1.6.4   The Management Platforms

Management platforms do not represent a new topological framework; thus, they can be used in any of the topologies described in this section. The management platforms are designed as open management systems to allow the development and operation of portable distributed management applications. By employing, as part of the platform framework, advanced management services such as directory, security, and time services, in addition to basic communication, event management, graphical user interface, and database services, the management platforms can manage large, heterogeneous, multivendor, multitechnology, and multiprotocol environments.

Several management platform components such as graphical user interfaces, management databases, and management applications can be distributed among several computing platforms. Multiple management

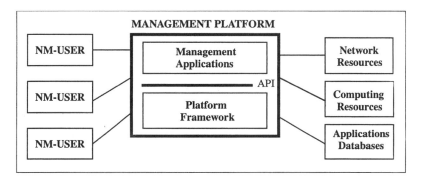

FIGURE 3.1.18 Management platform framework.

platforms can communicate to each other in order to manage large administrative domains (Figure 3.1.18).

3.1.7 Management Systems Evolution

Although the management systems have been established as specialized systems to manage large and complex network and computing systems since the mid 1980s, distinct events and distinct phases can be identified in the technical and chronological evolution of management systems.

3.1.7.1 Management Systems Technical Evolution

The first phase in the management systems development is exemplified by **passive monitoring systems** targeted solely toward network components management and providing test, instrumentation, and protocol analysis results. This was characteristic for the management systems designed in the late 1970s and early 1980s.

The next major phase was the build-up of **element management systems** (EMSs) which provide monitoring and controlling capabilities of individual systems. Acting as stand-alone systems, the EMSs target the management families of network elements, equipment, and hosts. Generally, the EMSs contain a single management application bundled with the computing platform, forming the actual run-time operational management environment. These types of management systems were typical in the 1980s and they covered the management of modems, multiplexers, T1 multiplexers, matrix switches, etc. In most cases, the management was limited to one type of equipment provided by a single vendor (Figure 3.1.19).

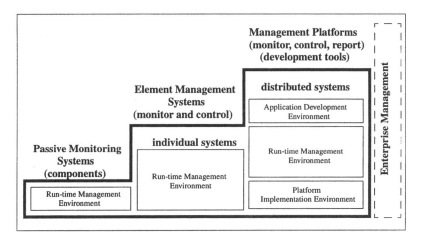

FIGURE 3.1.19 Management systems evolution.

Management platforms are the third major generation in the history of management systems development that go beyond the run-time operational environment which characterized the earlier stages. By adding an application development environment with tools and APIs which allow multiple and portable applications to run on top of management platform framework, the management platforms have embraced many of the concepts of distributed open management systems.

The run-time environment is represented by common management services provided by the platform and reflects the overall operability aspects of a management platform. The development environment includes the run-time environment and provides portability for management applications and integration of these management applications with the platform services. In order to develop management applications that use platform management services, the development environment should include the run-time environment. In addition, the complexity of management platforms requires an implementation environment for testing and conformance to standards or vendor specifications. The implementation environment provides means to assess the management platforms' interoperability. More details about management platforms will be provided in the following two sections.

3.1.7.2 Management Systems Chronological Events

In order to fully understand the evolution of management systems, it is very important to provide a more detailed, chronological order of all the events that ultimately shaped the field of management. It is well known that, traditionally, management systems trail the development of new network or computing technologies. In many instances, management solutions are afterthoughts, ad-hoc, and patched solutions. That creates later problems because the management is modeled, designed, and processed in systems outside of the technologies that are managed. The technical design and the economies of a later-added management are always inferior to the one delivered as part of the native technology. This is a major shortcoming of overall network and computing systems management and it will very likely persist in a wild competitive environment where the designers and manufacturers of new technologies continue to rush their products to market even though the management of that technology is missing or is not mature enough. This situation explains the existing differences in achieving standardization for network and computing systems management.

The following list of events, in chronological order, attempts to capture the evolution of management systems.

IBM mainframe-based data extraction and performance analysis tools, 1980–1985
IBM NetView, and integrated systems and network management, 1986
AT&T Unified Network Management Architecture (UNMA), 1987
DEC, Enterprise Management Architecture (EMA), 1987
Element Management Systems (EMSs) from Timeplex, Paradyne, Codex, GDC, 1987-1989
OSI Management International Standardization (starting in 1989)
Network management platform concept advanced by DEC, HP, IBM, OSF, 1989
Adoption of graphical user interface for network/systems topological display, 1990
Internet SNMP-based recommended standards adoption, 1990
OSF Distributed Management Environment (DME) proposal and subsequent RFPs, 1990
SUN SunNet Manager, Unix-based workstation management solution, 1990
OSF DME technology candidates selection, 1991
ITU-T, ANSI T1M1, ETSI contribution toward TMN interface standardization, 1992
Commercial management platforms, HP OpenView, Digital DECmcc, Tivoli TME, 1992
IBM SystemView blueprint and SystemsView for AIX platform, 1993
Home-grown management platforms from NetLabs, Cabletron, OSI, 1994
OSF DME delivered as the network management option (NMO), a major failure, 1994
HP OperationsCenter and HP AdministrationCenter systems management platforms, 1994
NM Forum, SPIRIT management platform requirements, version 1, 1995
DSET provides stand-alone management applications and agent development tools, 1995

Tivoli Systems TME acquired by IBM, integration plan with SystemsView for AIX, 1996
AT&T OneVision management platform-based integration solutions is proposed, 1996
HP OpenView DM 4.x, advanced distributed management versions, 1996-1997
Computer Associates Unicenter TNG management platforms, 1997
SUN Solstice Enterprise Manager platform and family of products, 1997

By the end of 1997 most of the surviving management platforms had become mature products although they were still far from the ideals of fully open, distributed management platforms. Evidently, the biggest failure was the standardization of management platform framework, as a collection of interchangeable management services. Few management platforms are designed and built according to the advanced features of object-oriented information modeling.

With all of these shortcomings the management platforms are the best hope to build enterprise-wide management systems where integration of management solutions and scalability are two major issues.

3.1.7.3 Management Platforms for Enterprise-wide Management

In the previous sections we indicated the complexity and difficulties confronting the management field. We also indicated the shortcomings in the historical development of management systems. Management platforms do not solve these shortcomings overnight but they bring a flexible approach to the management of multivendor, multitechnology, multiprotocol networks and systems environments. This is why a close look at the design of management platforms is necessary.

Management platforms, either as autonomous or interworking management systems, should provide several basic management functions. First of all, a management platform has to communicate with the external world through platform external interfaces, i.e., it has to provide communications services. Next, management information is exchanged in the form of management events that have to be stored, processed, and named. Therefore, there is a need for event management services. Furthermore, the events, as related to the management of network or computing systems resources, should be displayed (after the necessary processing) by using graphical user interfaces. Management information and all the components associated with management are organized using managed object models. Therefore, there is a need for a service that provides manipulation of managed objects. Ultimately, the management information about network configurations and managed resource status and parameters has to be stored in databases. Such database service may allow near real-time presentation of the status of the managed systems and components. Additional management operation services are needed to support all of the other platform services. In addition to these management services, there is a need for other distributed management services such as time service (synchronization), directory service (naming), and security services (Figure 3.1.20).

User interface services provide support for presentation of management information and support for interactions between users/human operators and distributed management applications used to manage network and computing systems. These services support both graphical user interfaces (GUIs) and asynchronous command line interfaces, by providing network and computing systems layout display based on visual icons, windows environment manipulation, on-line information, and general support for common applications development. The user interface is the most visible point of integration between various platform components and management applications.

Event management services provide common services to other platform management services and to the management applications running on top of management platforms. The events can be generated by network/computing systems, components state changes, systems errors, applications, and by users/operators. Common event operations include event collection, event logging, event filtering, and event administration.

Management communications services, either object-based or message-based, provide support for communications interfaces, management protocols, and communications stacks used to carry management information. Primarily, this support targets standardized management protocols such as Simple Network Management Protocol (SNMP), Common Management Information Protocol (CMIP) and Services (CMIS), and Remote Procedure Calls (RPCs).

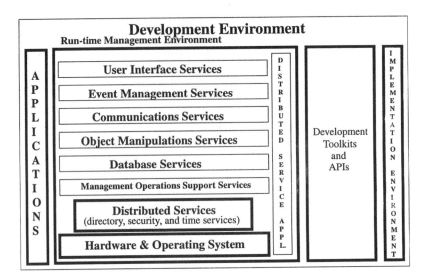

FIGURE 3.1.20 Management platform architectural components.

Object manipulation services provide support for information exchange between objects as abstractions of physical and logical resources ranging from network devices and computing systems resources to applications and management services. Primarily, this support targets object interfaces as defined by the OMG Common Object Request Broker Architecture (CORBA) and OMG Interface Definition Language (IDL). Object manipulation includes operations on the MIB, object support services providing location transparency for objects exchanging requests and responses, persistent storage for MIBs, and support for object-oriented applications development.

Database services provide support for management data storage and retrieval along with its integration with various platform services and management applications. Management information can include dynamic instance information related to configuration events, fault events, or performance events, to historical and archive information for security audit trail. The database services include database management systems (DBMSs), standardized database access and retrieval mechanisms such as structured query language (SQL), database concurrence mechanisms, and data backup mechanisms.

Management operations support provides common services to the management platform core and to the management applications running on top of the platforms. It includes management of the background processes associated with the platform hardware and operating system, and the handling of management applications.

Regarding management applications running on top of the management platforms, they can be classified in two major groups: core management applications and resource-specific applications. Core applications include functional management applications to manage specific management functional areas (SMFAs) such as configuration, fault, performance, security, and accounting management. Core applications also include compound applications which provide cross-area functionalities, as required by user needs and business considerations (e.g., trouble ticketing). Resource-specific applications, as the name indicates, are built specifically to manage particular network devices or computing system components.

The management applications development environment mirrors the run-time environment and includes specific development tools for user interfaces, event management, communications, object manipulation, and database operations. The APIs for various platform management services are critical components in the development environment. The development environment includes the run-time environment since the newly developed management applications are tested against the run-time environment.

The management implementation environment mainly refers to the overall acceptance testing of management platforms as they are used in the management of real network and computing systems. Implementation testing is different from the testing done on various platform hardware and software components during their development and production. The implementation environment covers effective systems testing based on test criteria, test procedures, and test tools used to operate, maintain, and troubleshoot complex distributed systems, which are tailored to management platforms.

As we mentioned earlier, the management platform consists of management services and the actual management applications that run on top of management services. Both management services and management applications make use of the computing hardware and operating system. Although the design of the hardware and operating system is outside the scope of management platform design, it is important to understand the differences between various computing systems and to understand the alternatives in selecting hardware and operating systems. In a truly open system environment, the platform services and applications are supposed to be hardware and operating system independent.

3.2 Management of Emerged and Emerging Technologies

Kornel Terplan

3.2.1 Introduction

Telecommunications services offered by domestic and international providers are based on a mixture of emerged and emerging technologies. Emerged technologies include private leased lines on T/E-basis, ISDN, traditional voice networks, message switching, packet switching, and SS7-based signalling. Emerging technologies are the following: frame relay, FDDI, ATM, SDH/Sonet, SMDS, wireless, cable, and xDSL.

Due to rapid progress made in technologies and infrastructures, the number of choices to offer certain telecommunications services is continuously increasing. In layered communications structures, technology usually occupies the lower layers. It is true for both TMN and OSI layers. If IP-based services are the goal, this service can be deployed in various ways. The supporting lower-layer infrastructure is going to be reassessed. Besides the traditional IP-ATM-Sonet/SDH combination, other options are also under consideration. This alternative is conservative, less risky from the engineering point of view, and achievable now. It may, however, lead to low efficiency and to high costs. Enhanced frame relay may substitute ATM everywhere, offering lower costs at good quality of service (QoS). But, in certain areas, there are no QoS standards available. ATM transport may eliminate the Sonet/SDH-layer offering ATM ring functions similar to distributed ATM switches. There are just a few vendors who consider this option. ATM/IP hybrids on the basis of Sonet/SDH would reduce the number of routers, with the result of lower management expenses. This technology is in the test phase, still unproven. IP over Sonet/SDH eliminates the ATM layer completely. If megarouters are at cost parity with ATM switches, this alternative would be the low-cost IP delivery solution. This technology is unproven; operating costs of megarouters are difficult to predict. Optical IP would be the lowest cost delivery of IP services. In this case, IP is directly connected to the optical subnetwork of Layer 1, neither using ATM nor Sonet/SDH. It is the least proven technology; there are serious concerns about fault management with this alternative.

This contribution investigates the manageability of both emerged and emerging technologies. Each technology discussed will be introduced briefly without details. Emphasis is on the availability of management information bases (MIBs), management protocols used, and management products usually deployed.

3.2.2 Foundation Concepts for Networking Technologies

The majority of emerged and emerging technologies has a few basic foundation principles. These will be addressed in this segment. The basics for this segment can be found in more details in (BLAC94) and (TERP98).

3.2.2.1 Connection-oriented and Connectionless Communications

Communication systems that employ the concepts of circuits and virtual circuits are said to be connection-oriented. Such systems maintain information about the users, such as their addresses and their ongoing QOS needs. Often, these types of systems use state tables that contain rules governing the manner in which the user interacts with the network. While these state tables clarify the procedures between the user and the communication network, they do add overhead to the communication process.

In contrast, communication systems that do not employ circuits and virtual circuits are said to be connectionless systems. They are also known as datagram networks and are widely used throughout the industry. The principal difference between connection-oriented and connectionless operation is that connectionless protocols do not establish a virtual circuit for the end user communication process. Instead, traffic is presented to the service provider in a somewhat *ad hoc* fashion. Handshaking arrangements, mutual confirmations are minimal and perhaps nonexistent. The network service points and the network switches maintain no ongoing knowledge about the traffic between the two end users. State tables as seen with connection-oriented solutions are not maintained. Therefore, datagram services provide no *a priori* knowledge of user traffic and they provide no ongoing current knowledge of the user traffic — but they introduce less overhead.

3.2.2.2   Physical and Virtual Circuits

End users operating terminals, computers, and client equipment communicate with each other through a communication channel called the physical circuit. These physical circuits are also known by other names, such as channels, links, lines, and trunks. Physical circuits can be configured wherein two users communicate directly with each other through one circuit, and no one uses this circuit except these two users. They can operate this circuit in half-duplex or full-duplex. This circuit is dedicated to the users. This concept is still widely used in simple networks without serious bandwidth limitations.

In more complex systems, such as networks, circuits are shared with more than one user-pair. Within a network, the physical circuits are terminated at intermediate points at machines that provide relay services on another circuit. These machines are known by such names as switches, routers, bridges, gateways, etc. They are responsible for relaying the traffic between the communicating users. Since many communication channels have the capacity to support more than one user session, the network device, such as the switch, router, or multiplexer is responsible for sending and receiving multiple user traffic to/from a circuit.

In an ideal arrangement, a user is not aware that the physical circuits are being shared by other users. Indeed, the circuit provider attempts to make this sharing operating transparent to all users. Moreover, in this ideal situation, the user thinks that the circuit directly connects only the two communicating parties. However, it is likely that the physical circuit is being shared by other users.

The term "virtual circuit" is used to describe a shared circuit wherein the sharing is not known to the circuit users. The term was derived from computer architectures in which an end user perceives that a computer has more memory than actually exists. This other, additional virtual memory is actually stored on an external storage device.

There are three types of virtual circuits:

- **Permanent virtual circuits (PVC)** — A virtual circuit may be provisioned to the user on a continuous basis. In this case, the user has the service of the network any time. A PVC is established by creating entries in tables in the network nodes' databases. These entries contain a unique identifier of the user payload which is known by various names, such as a logical channel number (LCN), virtual channel identifier (VCI), or virtual path identifier (VPI).
- Network features such as throughput, delay, security, and performance indicators are also provisioned before the user starts with operations. If different types of services are desired, and if different destination endpoints must be reached, then the user must submit a different PVC identifier with the appropriate user payload to the network. This PVC is provisioned to the different endpoint, and perhaps with different services.
- **Switched virtual circuits (SVC)** — A switched virtual circuit is not preprovisioned. When a user wishes to obtain network services to communicate with another user, it must submit a connection

request packet to the network. It must provide the address of the receiver, and it must also contain the virtual circuit number that is to be used during the session. SVCs entail some delay during the setup phase, but they are flexible in allowing the user to select dynamically the receiving party and the negotiation of networking parameters on a call-by-call basis.

- **Semi-permanent virtual circuits (SPVC)** — With this approach, a user is preprovisioned, as in a regular PVC. Like a PVC, the network node contains information about the communicating parties and the type of services desired. But these types of virtual circuits do not guarantee that the users will obtain their level of requested service. In case of congested networks, users could be denied the service.

In a more likely scenario, the continuation of a service is denied because the user has violated some rules of the communications. Examples are higher bandwidth demand and higher data rates than agreed with the supplier.

3.2.2.3 Switching Technologies

Voice, video, and data signals are relayed in a network from one user to another through switches. This section provides an overview on prevalent switching technologies.

Circuit switching provides a direct connection between two networking components. Thus, the communicating partners can utilize the facility as they see it — within bandwidth and tariff limitations. Many telephone networks use circuit switching systems. Circuit switching provides clear channels; error checking, session establishment, frame flow control, frame formatting, selection of codes, and protocols are the responsibility of the users. Today, the traffic between communicating parties is usually stored in fast queues in the switch and switched on to an appropriate output line with time division multiplexing (TDM) techniques. This technique is known as circuit emulation switching (CES). In summary:

- Direct connection end-to-end
- No intermediate storage unless CES used
- Few value-added functions
- Modern systems use TDM to emulate circuit switching

Message switching was the dominating switching technology during the last two decades. The technology is still widely used in certain applications, such as electronic mail, but it is not employed in a backbone network. The switch is usually a specialized computer. It is responsible for accepting traffic from attached terminals and computers. It examines the address in the header of the message and switches the traffic to the receiving station. Due to the low number of switching computers, this technology suffers under backup problems, performance bottlenecks, and lost messages due to congestion. In summary:

- Use of store-end-forward technology
- Disk serves as buffers
- Extensive value-added functions
- Star topology due to expense of switches

Packet switching relays small pieces of user information to the destination nodes. Packet switching has become the prevalent switching technology of data communications networks. It is used in such diverse systems as private branch exchanges (PBXs), LANs, and even with multiplexers. Each packet only occupies a transmission line for the duration of the transmission; the lines are usually fully shared with other applications. This is an ideal technology for bursty traffic. Modern packet switching systems are designed to carry continuous, high-volume traffic as well as asynchronous, low-volume traffic, and each user is given an adequate bandwidth to meet service level expectations.

The concept of packet and cell switching is similar; each attempts to process the traffic in memory as quickly as possible. But cell switching is using much smaller PDUs relative to packet switching. The PDU size is fixed with cell switching. The PDU size may vary with packet switching. In summary:

- Hold-and-forward technology
- RAM serves as buffers
- Extensive value-added-functions for packet, but not many for cells

Switching will remain one of the dominating technologies in the telecommunications industry.

3.2.2.4 Routing Technologies

There are two techniques to route traffic within and between networks: source routing and non-source routing. The majority of emerging technologies use non-source routing.

Source routing derives its name from the fact that the transmitting device — the source — dictates the route of the protocol data unit (PDU) through a network or networks. The source places the addresses of the "hops" in the PDU. The hops are actually routers representing the internetworking units. Such an approach means that the internetworking units need not perform address maintenance, but they simply use an address in the PDU to determine the destination of the PDU.

In contrast, non-source routing requires that the interconnecting devices make decisions about the route. They don't rely on the PDU to contain information about the route. Non-source routing is usually associated with bridges and is quite prevalent in LANs. Most of the emerging new technologies implement this approach with the use of a VCI. This label is used by the network nodes to determine where to route the traffic.

The manner in which a network stores its routing information varies. Typically, routing information is stored in a software table, called a directory. This table contains a list of destination nodes. These destination nodes are identifiers with some type of network address. Along with the network address (or some type of label, such as a virtual circuit identifier) there is an entry describing how the router is to relay the traffic. In most implementations, this entry simply lists the next node that is to receive the traffic in order to relay it to its destination.

Small networks typically provide a full routing table at each routing node. For large networks, full directories require too many entries, and are expensive to maintain. In addition, the exchange of routing table information can impact the available bandwidth for user payload. These networks are usually subdivided into areas, called domains. Directories of routing information are kept separately in domains.

Broadcast networks contain no routing directories. Their approach is to send the traffic to all destinations.

Network routing control is usually categorized as centralized or distributed. Centralized uses a network control center to determine the routing of the packets. The packet switches are limited in their functions. Central control is vulnerable; a backup is absolutely necessary, which increases the operating expenses. Distributed requires more intelligent switches, but they provide a more resilient solution. Each router makes its own routing decisions without regard to a centralized control center. Distributed routing is also more complex, but its advantages over the centralized approach have made it the preferred routing method in most communications networks.

3.2.2.5 Multiplexing Technologies

Most of the emerged and emerging technologies use some form of multiplexing. Multiplexers accept low-speed voice or data signals from terminals, telephones, PCs, and user applications and combine them into one higher-speed stream for transmission efficiency. A receiving multiplexer demultiplexes and converts the combined stream into the original lower-speed signals. There are various multiplexing techniques:

Frequency Division Multiplexing (FDM) — This approach divides the transmission frequency range into channels. The channels are lower frequency bands; each is capable of carrying communication traffic, such as voice, data, or video. FDM is widely used in telephone systems, radio systems, and cable television applications. It is also used in microwave and satellite carrier systems. FDM decreases the total bandwidth available to each user, but even the narrower bandwidth is usually sufficient for the users' applications. Isolating the bands from each other costs some bandwidth, but the simultaneous use outweighs this disadvantage.

Time Division Multiplexing (TDM) — This approach provides the full bandwidth to the user or application, but divides the channel into time slots. Each user or application is given a slot and the slots are rotated among the attached devices. The TDM multiplexer cyclically scans the input signals from the entry points. TDMs are working digitally. The slots are preassigned to users and applications. In case of no traffic at the entry points, the slots remain empty. This approach works well for constant bit rate applications, but leads to waste capacity for variable bit rate applications.

Statistical Time Division Multiplexing (STDM) — This approach allocates the time slots to each port on a STDM. Consequently, idle termninal time does not waste the capacity of the bandwidth. It is not unusual for two to five times as much traffic to be accomodated on lines using STDMs in comparison to a TDM solution. This approach can accomodate bursty traffic very well, but does not perform too well with continuous, nonbursty traffic.

Wave Division Multiplexing (WDM) — WDM is the optical equivalent of FDM. Lasers operating at different frequencies are used in the same fiber, thereby deriving multiple communications channels from one physical path. There is a more advanced form of this technology with even better efficiency, called Dense Wave Division Multiplexing (DWDM).

3.2.2.6 Addressing and Identification Schemes

In order for user traffic to be sent to the proper destination, it must be associated with an identifier of the destination. Usually, there are two techniques in use:

An explicit address has a location associated with it. It may not refer to a specific geographical location but rather a name of a network or a device attached to a network. For example, the Internet Protocol (IP) address has a structure that permits the identification of a network, a subnetwork attached to the network, and a host device attached to the subnetwork. The ITU-T X.121 address has a structure which identifies the country, a network within that country, and a device within the network. Other entries are used with these addresses to identify protocols and applications running on the networks. Explicit addresses are used by switches, routers, and bridges as an entry into routing tables. These routing tables contain information about how to route the traffic to the destination nodes.

Another identifying scheme is known by the term of label, although other terms may be more widely used. Those terms are logical channel number (LCN) or virtual circuit identifier (VCI). A label contains no information about network identifiers or physical locations. It is simply a value that is assigned which identifies each data unit of a user's traffic.

Almost all connectionless systems use explicit addresses, and the destination and source addresses must be provided with every PDU in order for it to be routed to the proper destination.

3.2.2.7 Control and Congestion Management

It is very important in communication networks to control the traffic at the ingress and egress points of the network. The operation by which user traffic is controlled by the network is called flow control. Flow control should assure that the traffic does not saturate the network or exceed the network's capacity. Thus, flow control is used to manage congestion.

There are three flow control alternatives with emerged and emerging technologies:

- Explicit flow control — This technique limits how much user traffic can enter the network. If the network issues an explicit flow control message to the user, the user has no choice but to stop sending traffic or to reduce traffic. Traffic can be sent again after the network has notified the user about the release of the limitations.

- Implicit flow control — This technique does not absolutely restict the flow. Rather, it recommends that the user reduce or stop traffic it is sending to the network if network capacity situations require a limitation. Typically, the implicit flow control message is a warning to the user that the user is violating its service level agreement with the internal or external supplier regarding network congestion. In any case, if the user continues to send traffic, it risks having traffic discarded by the network.

- No flow control — Flow control may also be established by not controlling the flow at all. Generally, an absence of flow control means that the network can discard any traffic that is creating problems. While this approach certainly provides superior congestion management from the standpoint of the network, it may not meet the performance expectations of the users.

3.2.3 Management Solutions for Emerged Technologies

The present status for emerged technologies, such as private leased lines, voice networks, SS7-based signalling techniques, message and packet switching, can be summarized as follows:

- Proprietary solutions dominate: this means that the management protocols selected are controlled by the supplier of the equipment or facilities vendors.
- Re-engineered by SNMP: many of the equipment vendors include SNMP agents into their devices to meet the requirements of customers. The SNMP agents provide information for performance management and reporting, but they usually do not change the real-time processing of status data within the devices.
- The management structures are very heterogeneous: in most cases, these structures are hierarchical including a manager of managers. This manager is using a proprietary architecture. Most of the interfaces to element managers and managed devices are proprietary.
- TMN has a very low penetration: suppliers have recognized the need for a generic standard, but they are not willing to invest heavily into supporting it. Some of the providers go as supporting the Q3-interface.
- Operating Support Systems are heavy: the legacy-type OSSs support the emerged technologies on behalf of the suppliers well, but they are flexible enough to address future needs. They lack in separating operations functionality from operations data, in using flexible software, and in separating network management from service management.

3.2.4 Emerging Technologies

This segment gives an overview on telecommunication technologies that are either in use but still considered new technology, or are considered for near-future implementation. These technologies include frame relay, FDDI, SMDS, ATM, Sonet/SDH, and mobile and wireless communications. The capabilities of these technologies are presented using the same format, including technology description flowed by evaluating management capabilities.

3.2.4.1 Frame Relay

The purpose of a frame relay network is to provide an end user with a high-speed virtual private network (VPN) capable of supporting applications with large bit-rate transmission requirements. It gives a user T1/E1 access rates at a lesser cost than can be obtained by leasing comparable T1/E1 lines. It is actually a virtually meshed network.

The design of frame relay networks is based on the fact that data transmission systems today are experiencing far fewer errors and problems than they did decades ago. During that period, protocols were developed and implemented to cope with error-prone transmission circuits. However, with the increased use of optical fibers, protocols that expend resources dealing with errors become less important. Frame relay takes advantage of this improved reliability by eliminating many of the now unnecessary error checking and correction, editing, and retransmission features that have been part of many data networks for almost two decades.

Frame relay has been working for many years. It represents a scaled-down version of LAPD. The flexibility of assigning bandwidth on demand is somewhat new. Frame relay is one of the alternatives of fast packet switching.

MIB availability

The frame relay objects are organized into three object groups:

- Data link connection management interface group
- Circuit group
- Error group

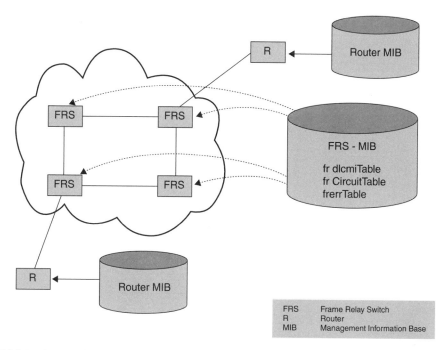

FIGURE 3.2.1 Communications paths between manager, agents, subagents, and managed objects.

These groups are stored in tables in the MIB and can be accessed by the SNMP manager. Figure 3.2.1 illustrates where SNMP operates, and lists the names of these three groups.

The frDlcmiTable contains 10 objects. Their purpose is to identify each physical port at the unified network interface (UNI), its IP address, the size of the DLCI header that is used on this interface, timers for invoking status and status inquiry messages, the maximum number of DLCIs supported at the interface, whether or not the interface uses multicasting, and miscellaneous operations.

The frCircuitTable contains 14 objects. Their purpose is to identify each PVC, its associated DLCI, if the DLCI is active, the number of BECNs and FECNs received since the PVC was created, statistics on the number of frames and octets sent and received since the DLCI was created, the DLCI's Bc and Be, and miscellaneous operations.

The third table is the frErrTable containing 4 objects. Their purpose is to store information on the types of errors that have occurred at the DLCI (unknown or illegal), and the time the error was detected. One object contains the header of the frame that created the error.

SNMP-based and proprietary solutions compete for management. Basically, each physical and logical component can be managed by periodically polling the PDUs in the MIB. Any powerful management platform can accomodate frame relay management, but the polling overhead over the wide area should be carefully controlled.

3.2.4.2 Fiber Distributed Data Interface (FDDI)

FDDI was developed to support high-capacity LANs. To obtain this goal, the original FDDI specifications stipulated the use of optical fiber as the transport media, although it is now available on twisted pair cable (CDDI). FDDI has been deployed in many corporations to serve as a high-speed backbone network for other LANs, such as Ethernet and Token Ring.

Basically, the standard operates with 100 Mbps rate. Dual rings are provided for the LAN, so the full speed is actually 200 Mbps, althrough the second ring is used typically as a backup to the primary ring. In practice. most installations have not been able to utilize the full bandwidth of FDDI. The standard defines multimode optical fiber, although single mode optical fiber can be used as well.

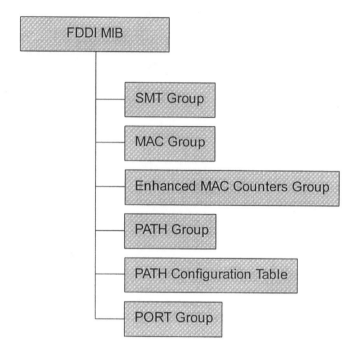

FIGURE 3.2.2   FDDI management information base.

FDDI was designed to make transmission faster on an optical fiber transport. Due to the high-capacity 100 Mbps technology, FDDI has a tenfold increase over the widely used Ethernet, and a substantial increase over Token Ring. FDDI was also to extend the distance of LAN interconnectivity. It permits the network topology to extend up to 200 km (14 miles).

FDDI II, which is able to incorporate voice, is not receiving enough industry interest.

FDDI is actually not a new technology, but internetworked FDDIs offer new alternatives in metropolitan areas competing with other technologies, such as frame relay and SMDS.

MIB Availability

The FDDI MIB has the following five groups (Figure 3.2.2):

- SMT Group — Contains a sequence of entries, one for each SMT implementation. The entries describe the station I.D., the operation I.D., the highest and lowest version I.D., the number of MACs in this station or concentrator, and other configuration and status information.

- MAC Group — A list of MAC tables, one for each MAC implementation across all SMTs. Each table describes the SMT index and the MIB-II If Index associated with this MAC, and status and configuration information for the given MAC.

- Enhanced MAC Counters Group — The MAC Counters table contains a sequence of MAC Counters entries. Each entry stores information about the number of tokens received, number of times TVX expired, number of received frames that have not been copied to the receive buffer, number of TRT expirations, number of times the ring entered operational status, and threshold information.

- PATH Group — Contains a sequence of PATH entries. Each entry starts with an index variable uniquely identifying the primary, secondary, and local PATH object instances. The entries also store information on different min and max time values for MACs in the given PATH.

- PATH Configuration Table — A table of path configuration entries. This table contains a configuration entry for all the resources that may be in this path.
- PORT Group — Contains a list of PORT entries. Each entry describes the SMT index associated with this port, a unique index for each port within the given SMT, the PORT type, and then neighbor type, connection policies, the list of permitted PATHs, the MAC number associated with the PORT (if any), the PMD entity class associated with this port and other capabilities and configuration information.

FDDI has its own management capabilities, defined in SMT, but they have never really taken off. Instead, suppliers are concentrating on SNMP capabilities. Using the MIB of FDDI agents, any SNMP manager can be used to manage FDDI.

3.2.4.3 Switched Multi-megabit Data Service

SMDS is a high-speed connectionless packet switching service which extends LAN-like performance beyond a subscriber's location. Its purpose is to ease the geographic limitations that exist with low-speed wide area networks. SMDS is designed to connect LANs, MANs, and also WANs.

The major goals of SMDS are to provide high-speed interfaces into customer systems, and at the same time allow customers to take advantage of their current equipment and hardware. Therefore, the SMDS operations are not performed by the end user machine; they are performed by customer promises equipment (CPE), such as a router.

SMDS is positioned as a service. If SMDS is considered unto itself, it is a new technology; it offers no new method for designing or building networks. This statement is emphasized because SMDS uses the technology of DQDB (dual queue dual bus), and then offers a variety of value-added services, such as bandwidth on demand, to the SMDS customer.

SMDS is targeted for large customers and sophisticated applications that need a lot of bandwidth, but not permanently. Generally, SMDS is targeted for data applications that transfer a lot of information in a bursty manner.

However, applications that use SMDS can be interactive. For example, two applications can exchange information interacting through SMDS, such as an X-ray, a document, etc. The restriction of SMDS is based on the fact that SMDS is not designed for real-time, full-motion video applications. Notwithstanding, it does support an interactive dialog between users, and allows them to exchange large amounts of information in a very short time. For example, it takes only one to two seconds for a high-quality color graphic image to be sent over a SMDS network. For many applications, this speed is certainly adequate.

MIB Availability

Figure 3.2.3 shows the location of the MIB in relation to the SMDS network. The SIP layers are also known in this figure to aid in reading the following material. The MIB is organized around managed objects which are contained in major groups. The groups, in turn, are defined in tables. As shown in the figure, at the bottom, the major entries in the MIB are the SMDS address, which is the conventional 60-bit address preceded by the 4 bits to signify individual group addresses. Thereafter, the groups are listed with their object identifier name. These names will be used in this segment to further describe the entries.

The sipL3Table contains the layer 3 (L3-PDU) parameters used to manage each SIP port. It contains entries such as port numbers, statistics on received traffic, information on errors such as unrecognized addresses, as well as various errors that have occurred at this interface.

The sipL2Table, as its name implies, contains information about layer 2 (L2-PDU) parameters and the state variables associated with information on the amount of the number of level 2 PDUs processed, error information such as violation of length of PDU, sequence number errors, MID errors, etc.

The sipDS1PLCP Table contains information on DS1 parameters and state variables for each port. The entries in the table contain error information such as DS1 severely erred framing seconds (SEFS), alarms, unavailable seconds encountered by the PLPC, etc.

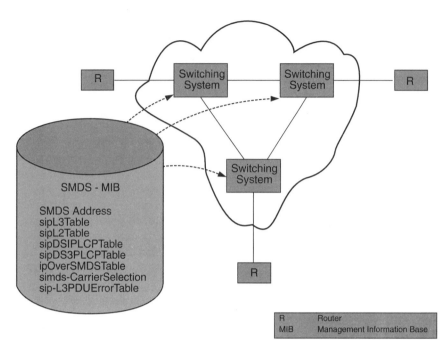

FIGURE 3.2.3   Structure of SNMP-based management services.

The sipDS3PLPC Table contains information about DS3 interfaces and state variables for each SIP port. Like its counterpart in DS1, this table contains information on severity-erred framing seconds, alarms, unavailable seconds, etc.

The ipOverSMDS Table contains information relating to operations of IP running on top of SMDS. It contains information such as the IP address, the SMDS address relevant to the IP station, addresses for ARP resolution, etc.

The smdsCarrierSelection group contains information on the inter-exchange carrier selection for the transport of traffic between LATAs.

Finally, the sipL3PDU Error Table contains information about errors encountered in processing the layer 3 PDU. Entries such as destination error, source error, invalid BAsize, invalid header extension, invalid PAD error, BEtag mismatch, etc. form the basis for this group.

As mentioned earlier, SNMP is used to monitor the MIBs and report on alarm conditions that have occurred based on the definitions in the MIBs.

SMDS management is expected to be ambitious. Switching systems are intelligent devices with the need of real-time decision making. In such situations, SNMP is not necessarily the right choice. However, SNMP may be used to transmit MIB entries to the manager for performance reporting.

3.2.4.4   Asynchronous Transfer Mode

The purpose of ATM is to provide a high-seed, low-delay, multiplexing and switching network to support any type of user traffic, such as voice, data, or video applications. ATM is one of four fast relay services. ATM segments and multiplexes user traffic into small, fixed-length units called cells. The cell is 53 octets, with 5 octets reserved for the cell header. Each cell is identified with virtual circuit identifiers that are contained in the cell header. An ATM network uses these identifiers to relay the traffic through high-speed switches from the sending CPE to the receiving CPE.

ATM provides limited error detection operations. It provides no retransmission services, and few operations are performed on the small header. The intention of this approach — small cells and with

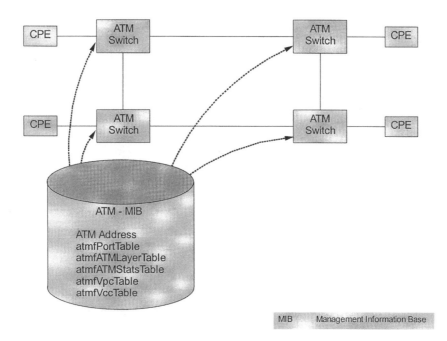

FIGURE 3.2.4 The ATM management information base.

minimal services performed — is to implement a network that is fast enough to support multimegabit transfer rates.

ATM is a new technology. ATM is supposed to be the foundation of providing the convergence, multiplexing, and switching operations. ATM resides on top of the physical layer.

MIB Availability
The ATM Forum has published a MIB as part of its Interim Local Management Interface Specification (ILMI) (Figure 3.2.4). The ATM MIB is registered under the enterprise node of the standard SMI in accordance with the Internet. MIB objects are therefore prefixed with 1.3.6.1.4.1.353.

Each physical link (port) at the UNI has a MIB entry that is defined in the atmfPortTable. This table contains a unique value for each port, an address, the type of port (DS3, SONET, etc.) media type (coaxial cable, fiber, etc.), status of port, (in service, out of service, etc.) and other miscellaneous objects.

The atmfAtmLayerTable contains information about the UNIs physical interface. The table contains: the port id, maximum number of VCCs, VPCs supported and configured on this UNI, active VCI/VPI bits on the UNI, and a description of public or private for the UNI.

The atmVpcTable and atmVccTable contain similar entries for the VPCs and VCCs, respectively, on the UNI. These tables contain the port id, VPI or VCI values for each connection, operational status (up, down, etc.), traffic shaping and policing descriptors (to describe the type of traffic management applicable to the traffic), and any applicable QoS that is applicable to the VPI or VCI.

The ATM Forum has defined two aspects of UNI network management:

- ATM layer management at the M plane, and
- Interim Local Management Interface (ILMI) specification.

M-Plane Management — Most of the functions for ATM M-plane management are performed with the SONET F1, F2, and F3 information flows. ATM is concerned with F3 and F4 information flows.

ILMI — Because the ITU-T and the ANSI have focused on C-plane and U-plane procedures, the ATM Forum has published a interim spefication called ILMI. The major aspects of ILMI are the use of SNMP and a MIB. The ILMI stipulates the following procedures. First, each ATM device supports the ILMI, and one UNI ILMI MIB instance for each UNI. The ILMI communication protocol stack can be SNMP/UDP/IP/AAL over a well-known VPI/VCI value. SNMP is employed to monitor ATM traffic and the UNI VCC/VPC connections based on the ATM MIB with the SNMP get, Get-Next, Set, and Trap operations.

3.2.4.5   Sonet and SDH

Sonet/SDH is an optical-based carrier (transport) network utilizing synchronous operations between the network components. The term Sonet is used in North America, and SDH is used in Europa and Japan. Attributes of this technology are:

- A transport technology that provides high availability with self-healing topologies
- A multivendor that allows multivendor connections without conversions between the vendors' systems
- A network that uses synchronous operations with powerful multiplexing and demultiplexing capabilities
- A system that provides extensive OAM&P services to the network user and administrator

Sonet/SDH provides a number of attractive features when compared with current technology. First, it is an integrated network standard on which all types of traffic can be transported. Second, the Sonet/SDH standard is based on the optical fiber technology which provides superior performance in comparison to microwave and cable systems. Third, because Sonet/SDH is a worldwide standard, it is now possible for different vendors to interface their equipment without conversion.

Fourth, Sonet/SDH efficiently combines, consolidates, and segregates traffic from different locations through one facility. This concept, known as grooming, eliminates back hauling and other inefficient techniques currently being used in carrier networks. Back hauling is a technique in which user payload is carried past a switch that has a line to the user and sent to another endpoint. Then, the traffic to the other user is dropped, and the first users' payload is sent back to the switch and relayed back to the first user. In present configurations, grooming is eliminated, but expensive configurations, such as back-to-back multiplexers that are connected with cables, panels, or electronic cross-connect equipment are required.

Fifth, Sonet/SDH eliminates back-to-back multiplexing overhead by using new techniques in the grooming process. These techniques are implemented in a new type of equipment, called an add-drop multiplexer (ADM).

Sixth, the synchronous aspect of Sonet/SDH makes for more stable network operations. These types of networks experience fewer errors than the older asynchronous networks, and provide much better techniques for multiplexing and grooming payloads.

Seventh, Sonet/SDH has notably improved OAM&P features relative to current technology. Approximately 5% of the bandwidth is devoted to management and maintenance.

Eighth, Sonet/SDH employs digital transmission schemes. Thus, the traffic is relatively immune to noise and other impairments on the communications channel, and the system can use efficient TDM operations.

Sonet have been around a couple of years. The technology is not completely new, but its implementation is new.

MIB Availability

The Sonet/SDH MIB consists of eight groups. Each of the following groups have two tables: the Current Table and the Interval Table (Figure 3.2.5).

- The Sonet/SDH XXX Current Table contains various statistics that are being collected for the current 15-minute interval. The Sonet/SDH XXX Interval Table contains various statistics being collected by each system over a maximum of the previous 24 hours of operation. The past 24 hours may be broken into 96 completed 15-minute intervals. A system is required to store at least 4 completed 15-minute intervals. The default value is 32 intervals.

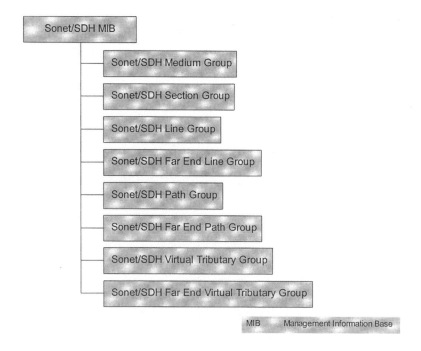

FIGURE 3.2.5 The Sonet/SDH management information base.

- The Sonet/SDH Medium Group: Sonet/SDH interfaces for some applications may be electrical interfaces and not optical interfaces. This group handles the configuration information for both optical Sonet/SDH interfaces and electrical Sonet/SDH interfaces, such as signal type, line coding, line type, and the like.
- The Sonet/SDH Section Group: This group consists of two tables:

    The Sonet/SDH Section Current Table and

    The Sonet/SDH Section Interval Table

    These tables contain information on interface status, counters on errored seconds, severely errored seconds, severely errored framing seconds, and coding violations.
- The Sonet/SDH Line Group: This group consists of two tables:

    The Sonet/SDH Line Current Table and

    The Sonet/SDH Line Interval Table

    These tables contain information on line status, counters on errored seconds, severely errored seconds, severely errored framing seconds, and unavailable seconds.
- The Sonet/SDH Far End Line Group: This group may only be implemented by Sonet/SDH (LTEs) systems that provide for far end block error (FEBE) information at the Sonet/SDH Line Layer. This group consists of two tables:

    The Sonet/SDH Far End Line Table and

    The Sonet/SDH Far End Line Interval Table
- The Sonet/SDH Path Group: This group consists of two tables:

    The Sonet/SDH Path Current Table and

    The Sonet/SDH Path Interval Table

    These tables contain information on interface status, counters on errored seconds, severely errored seconds, severely errored framing seconds, and coding violations.

- The Sonet/SDH Far End Path Group: This group consists of two tables:
    The Sonet/SDH Far End Path Current Table and
    The Sonet/SDH Far End Path Interval Table
- The Sonet/SDH Virtual Tributary Group: This group consists of two tables:
    The Sonet/SDH VT Current Table and
    The Sonet/SDH VT Interval Table

For SDH signals, virtual tributaries are called Vcs instead of Vts.

VT1.5   VC11
VT2   VC12
VT3   none
VT6   VC3

These tables contain information on virtual tributaries width and status, counters on errored seconds, severely errored seconds, severely errored framing seconds, and unavailable seconds.

- The Sonet/SDH Far End VT Group: This group consists of two tables:
    The Sonet/SDH Far End VT Current Table and
    The Sonet/SDH Far End VT Interval Table

The operation, administration, and maintenance (OAM) functions are associated with the hierarchical, layered design of Sonet/SDH. Figure 3.2.6 shows the five levels of the corresponding OAM operations, which are labeled F1, F2, F3, F4, and F5. F1, F2, and F3 functions reside at the physical layer; F4 and F5 functions reside at the ATM layer.

The Fn tags depict where the OAM information flows between two points, as shown in Figure 3.2.7.

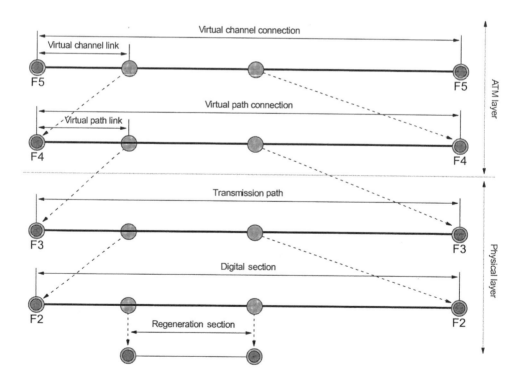

FIGURE 3.2.6   Relationships in ATM layers.

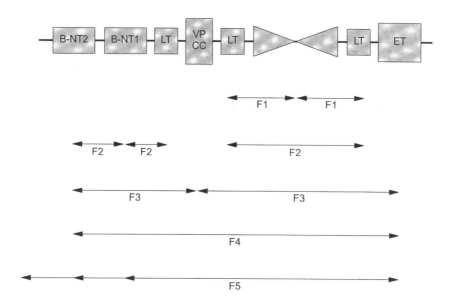

FIGURE 3.2.7   Information flows in B-ISDN.

The five OAM flows are the following:

F5: OAM information flows between network elements performing VC functions. From the perspective of a BISDN configuration, F5 OAM operations are conducted between B-NT2/B-NT1 endpoints. F5 deals with degraded VC performance, such as late arriving cells, lost cells, cell insertion problems, etc.

F4: OAM information flows between network elements performing VP functions. From the perspective of a BISDN configuration, F4 OAM flows between B-NT2 and ET. F4 OAM reports on an unavailable path or a virtual path (VP) that cannot be guaranteed.

F3: OAM information flows between elements that perform the assembling and disassembling of payload, header, and control (HEC) operations, and cell delineation. From the perspective of a BISDN configuration, F3 OAM flows between B-NT2 and VP cross connect and ET.

F2: OAM information flows between elements that terminate section endpoints. It detects and reports on loss of frame synchronization and degraded error performance. From the perspective of a BISDN configuration, F2 OAM flows between B-NT2, B-NT1, and LT, as well as from LT to LT.

F1: OAM information flows between regenerator sections. It detects and reports on loss of frame and degraded error performance. From the perspective of a BISDN, F1 OAM flows between LT and regenerators.

3.2.4.6   Mobile and Wireless Communication

The purpose of a mobile communications system is the provision for telecommunications services between mobile stations and fixed land locations, or between two mobile stations. There are two forms of mobile communications: cellular and cordless.

The best approach is to examine their major attributes and compare them to each other. First, a cellular system usually has a completely defined network which includes protocols for setting up and clearing calls and tracking mobile units through a wide geographical area. So, in a sense, it defines a UNI and a NNI. With cordless personal communications, the focus is on access methods in comparison to a closely located transceiver — usually within a building. That is to say, it defines a rather geographically limited UNI.

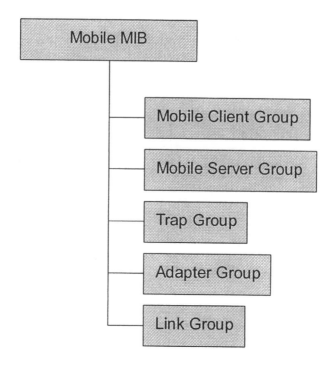

FIGURE 3.2.8 Mobile management information base.

Cellular systems operate with higher power than do the cordless personal communications systems. Therefore, cellular systems can communicate within large cells with a radius in the km range. In contrast, cordless cellular communication cells are quite small, usually in the order of 100 meters.

Cellular will continue to be a preferred medium for the consumer and business market. PCS will emerge as a driver for wireless. Telecommunications providers will use PCS to help reduce cellular churn, joining the customers closer to the vendor by providing end-to-end service. Wireless data connectivity, driven by lighter and smaller equipment capable of being carried by humans as well as in vehicles, will drive wireless connectivity requirements deeper and further into the network infrastructure.

The technology is not new, but implementation took some time. Cordless systems are undergoing rapid technological changes. Different protocols and standards are used.

MIB Availability

The premise of the Mobile MIB is the following:

- Network managers require more control over remote network workstations.
- To manage these remote machines, the network manager requires data on remote network access.
- Some of this data can be supplied via SNMP from the remote workstation.
- Other important information is available only on the local network at the point where the connection was created.
- This requires, therefore, two important SNMP agents with associated Mobile MIB components, one at the remote workstation and the other at the local network connection point.

The following groups of information have been created (Figure 3.2.8):

- The Mobile Client Group contains information to be relayed from the remote workstation to a network management system on the attached network. This group stores information on mobile client name, description, location, phone number, power management configuration, connection hardware and software, client type (CPU, RAM, disk, video), system software, date, time, network adapter used, and configuration and statistics.

- The Mobile Server Group contains information to be relayed from the local network server to a network management system on the same network. This group stores information about the network server where remote connections can be originated, such as server's name, remote network hardware, slot and port number, server uptime, connection speed, service type, and traffic statistics.

- The Trap Group describes SNMP trap types for remote workstations, such as mobile computer docked, undocked, suspended, resumed, pcmcia inserted, and a trap table. This is a table of alerts, which can be sent to the specified ipaddress using the specified protocol by setting the value of the mobileTrapTrigger object to the index of an entry in this table.

- Adapter Group contains information about network adapters, including hardware information and type of connection.

- Link Group provides data about mobile network links, such as link status and link performance.

Management is more difficult than with wirelines due to the fact that more information is processed in real-time or near real-time. In most cases, proprietary solutions dominate. The implementation of TMN and other standards is extremely slow.

3.2.4.7 Digital Subscriber Line Technologies

The enabling technology is digital subscriber line (xDSL), a scheme that allows mixing data, voice, and video over phone lines. There are, however, different types of DSL to choose from, each suited for different applications. All DSL technologies run on existing copper phone lines and use special and sophisticated modulation to increase transmission rates (ABER97).

Asymmetric digital subscriber line (ADSL) is the most publicized of the DSL schemes and is commonly used as an ideal transport for linking branch offices and telecommuters in need of high-speed intranet and Internet access. The word asymmetric refers to the fact that it allows more bandwidth downstream (to the consumer) than upstream (from the consumer). Downstream, ADSL supports speeds of 1.5 to 8 Mbit/s, depending on the line quality, distance, and wire gauge. Upstream rates range between 16 and 640 Kbit/s, again depending on line quality, distance, and wire gauge. For up to 18,000 feet, ADSL can move data at T1 using standard 24-gauge wire. At distances of 12,000 feet or less, the maximum speed is 8 Mbit/s.

ADSL delivers a couple of other principal benefits. First, ADSL equipment being installed at carriers' central offices offloads overburdened voice switches by moving data traffic off the public switched telephone network and onto data networks — a critical problem resulting from Internet use. Second, the power for ADSL is sent by the carrier over the copper wire with the result that the line works even when their local power fails. This is an advantage over ISDN, which requires a local power supply and thus a separate phone line for comparable service guarantees. The third benefit is again over ISDN; ADSL furnishes three information channels — two for data and one for voice. Thus, data performance is not impacted by voice calls. Rollout plans are very aggressive with this service. Its widespread availability is expected for the end of this decade.

Rate-adaptive digital subscriber line (RADSL) has the same transmission limits as ADSL. But as its name suggests, it adjusts transmission speed according to the length and quality of the local line. Connection speed is established when the line synchs up or is set by a signal from the central office. RADSL devices poll the line before transmitting: standards bodies are deciding if products will constrain the line speed. RADSL applications are the same as with ADSL and include Internet, intranets, video-on-demand, database access, remote LAN access, and lifeline phone services.

High-bit-rate digital subscriber line (HDSL) technology is symmetric, meaning that it furnishes the same amount of bandwidth both upstream and downstream. The most mature of the xDSL approaches, HDSL has already been implemented in the telco feeder plant — the lines that extend from central office to remote nodes — and also in campus environments. Because of its speed — T1 over two twisted pairs of wiring, and E1 over three — telcos commonly deploy HDSL as an alternative to T1/E1 with repeaters.

At 15,000 feet, HDSL operating distance is shorter than ADSLs, but carriers can install signal repeaters to extend its useful range (typically by 3000 to 4000 feet). Its reliance on two or three wire-pairs makes it ideal for connecting PBXs, interexchange carrier POPs (point of presence), Internet servers, and campus networks. In addition, carriers are starting to offer HDSL to carry digital traffic in the local loop, between two telco central offices and customer premises. HDSL's symmetry makes this an attractive option for high-bandwidth services like multimedia, but availability is still very limited.

Single-line digital subscriber line (SDSL) is essentially the same as HDSL with two exceptions: It uses a single wire pair and has a maximum operating range of 10,000 feet. Since it is symmetric and needs only one twisted pair, SDSL is suitable for applications like video teleconferencing or collaborative computing with identical downstream and upstream speeds. Standards for SDSL are still under development.

Very high-bit rate digital subscriber line (VDSL) is the fastest DSL technology. It delivers downstream rates of 13 to 52 Mbit/s and upstream rates of 1.5 to 2.3 Mbit/s over a single wire pair. But the maximum operating distance is only 1000 to 4000 feet. In addition to supporting the same applications as ADSL, VDSL, with its additional bandwidth, could potentially enable carriers to deliver high-defition television (HDTV). VDSL is still in the defition stage.

A number of critical issues must be resolved before DSL technologies achive widespread commercial deployment. Standards are still under development. Modulation techniques, such as carrierless amplitude phase (CAP) and discrete multitone (DMT) have been separated by standards bodies. Some other problems include interoperability, security, eliminating interference with ham radio signals, and lowering power systems requirements from the present 8 to 12 watts down to 2 to 3 watts. A nontechnical but important factor will be how well carriers can translate the successes they have realized in their xDSL technology trials into market trials and then to commercial deployments.

MIB Availability

MIB definitions are in the works with standard committees and with vendors. The present defition is not yet complete, but as a guideline is very useful for further steps.

The MIB is presently structured into eight groups:

- xdlsDevIfStats group provides statistics specific to the xDSL link. Statistics are collected on a per port basis and on specific intervals. Hence, such a table is indexed by the xdslDevifStatsIfIndex and xdslDevIfStatsInterval. Also, statistics are grouped into remote and central statistics. "Remote" means that the statistics are collected by the device at the customer premises. "Central" means that the statistics are collected by the device located at the central office. The objects which are not divided into these two groups are related to both ends of the xDSL link.

- xdslDevIfConfig group provides configuration information specific to a xDSL device, or system. The table is indexed by an object which corresponds to ifIndex. These ifIndex entries themselves denote and identify specifc xdsl interfaces on the board or module. Also, the configuration parameters are grouped into two broad categories, up and down. Up reflects the upstream direction (from customer premises to the central office). Down reflects the downstream direction (from the central office to the customer premises).

- xdslRemoteSys group's description is identical with the mib-2 system MIB.

- xdslRemoteDTEStatus group provides status information about the DTE port of the DSL RTUs.

- xdslDevMvlIfConfig group provides configuration information specific to a xDSL (MVL) device or system. The table is indexed by an object which corresponds to ifIndex. These ifIndex entries, themselves, denote and identify specific xdsl(Mvl) interfaces on the board or module.

- xdslDevNAPCustomerAccount group provides customer accounting information on each DSL port. Network access providers can accurately bill their end station DSL customers by the amount of usage. The table is indexed by ifIndex and xdslDevNAPCustomerAccountInterval. The ifIndex identifies specific xdsl interface on the device and xdslDevNAPCustomerAccountInterval specifies the accounting information for the current day or for the previous day. Customer data excludes all traffic used for management purposes.

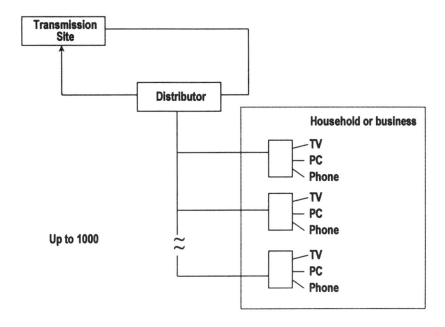

FIGURE 3.2.9 Use of cable technology in subscriber environments.

- xdslLinkUpDownInformation group contains the reason why the DSL Link went down. This information is obtained when the DSL Link comes up.
- xdslRemoteInjection group identifies the processes at the remote site, indicating injection types and various traps.

Management is more difficult than with other technologies due to the fact that more information is processed in real-time or near real-time. In most cases, proprietary solutions dominate. The implementation of TMN and other standards is extremely slow.

3.2.4.8 Cable Technnology

Cable service providers can enter the competition for voice and data services. Depending on the country, there are millions of households and businesses with cable television connections. In the majority of cases, cable television is a distribution channel supporting only one-way communication. But, using cable modems, channels could be provided for two-way-communications, allowing consumers to send back data or use the cable for phone conversations.

Figure 3.2.9 shows the structure of such an arrangement. The structure is simple requiring just a few additional networking devices. The components are:

- The equipment at the transmission site generates television signals and houses switches that route phone calls and monitor the network.
- Parallel fiber optic lines spread out over the area. If a line is out, traffic moves almost instantly to another route.
- Electronic nodes convert signals for transmission via coax cables. One node can serve approximately 500 to 1000 homes.
- Coaxial cable runs into a box on the side of the home. Electronics in the box split signals, sending phone traffic over regular phone lines, and television signals over cable lines to the television set.
- Phones would work on the current wiring. Television could be interactive, allowing signals to flow back up the cable line. Personal computers also could be connected through cable-modem, which transmits information up to 1000 times faster than phone lines.

Turning cable television systems into local phone networks will be not easy or cheap. Plans are ambitious to compete with local service providers. A basic cable system starts out as copper coax that carries signals in a line from the head-end — where TV signals are generated — to each consumer. Signals go one way; in the case of cable cuts or other damages, all the consumers from that point are out of service.

A system designed to handle phone calls and interactive TV looks much different. The main trunks of the network are set up in interconnecting rings of fiber-optic lines, which can carry thousands of times more information than a copper line. If a line is cut, traffic moves to another ring in a microsecond and practically no one is losing service. Because phone calls are two-way, a phone-cable system must be two-way and contain the sophisticated switches that send calls to the right places.

About 40% of today's cable TV systems are still copper. The other 60% have some fiber in their trunk lines, but most of them need upgrades. Just a few are set up in rings and have switches.

Power could cause another problem. Telephone lines have just enough electricity running through them to power the phone and the network, independent of the power grid. If a storm knocks out the main electrical grid, most of the time the phone still works. Cable lines don't carry power. If electricity goes out, so does cable. So would a phone hooked to an unpowered cable network. Cable lines, however, are capable of carrying power. But adding power from a node over the coaxial cable could add noise. There is a debate today in the industry about whether to get power from the cable-phone network, or use another solution, like attaching back-up batteries to the side of the households.

Once the pieces of the network come together, cable companies will face other issues. One is number portability — allowing consumers to keep their phone numbers even if they change phone service from the local phone company to the cable company. Right now, if somebody changes carriers, the number must be changed. State or federal regulation is expected to challenge that. Cable-phone systems also will have to prove to a sceptical public that they can be as reliable as current phone systems, which almost never break down. Cable systems also have to overcome a reputation for going out too frequently.

From a technology standpoint, the competition in the high-speed data services market will initially be between a dedicated architecture and a shared architecture (Figure 3.2.2). Both ISDN or other dedicated solutions, like xDSLs and cable-LAN services require infrastructure upgrades, and that means significant investments on the part of service providers. While most performance comparisons focus on peak bandwidth, other aspects of network usage, such as customer density and average session time, can also affect cost and quality of service.

The type of upgrade that is needed to deliver high-speed data over cable networks should be fairly obvious: cable television transmission is typically one way, but a data and phone network must permit two-way traffic. Limited forms of data service are possible by using the telephone line as a return route, but this is not a long-term solution. Some cable networks already have migrated to a hybrid fiber-coax infrastructure, but many operators are still immersed in the upgrading process. Once the networks are tailored for two-way traffic, broadband LAN technology will be incorporated so that digital data can be transmitted over a separate channel. Most of the vendors that supply technology to the cable operators will use an Ethernet-like approach, where the consumer's computer will have to be fitted with a network interface card and a cable modem for accessing the cable LAN. Access speed will depend upon the peak rate of the cable modem and the volume of traffic on the cable LAN. Subscribers will likely experience connection speeds that vary according to such factors as usage.

Cable LANs will operate full duplex with two channels, each channel sending data in a different direction. For customers using Internet or intranet applications, one of these channels would connect to an Internet or intranet POP router, which would then forward data packets from all users on the neighborhood LAN to and from all other systems on the Internet or on intranets. The other channel would receive data from the Internet or intranet service. Cable LANs are unlike ISDN-based services in that they furnish full-time connections, rather than switched or dial-up links. The obvious advantage is that it eliminates the need for dedicated transmission and switching resources. Users can access the cable LAN when needed, and only the cable network makes use of the Internet or Intranet point of presence.

A shared cable LAN requires only a single connection to the Internet or intranet provider. With ISDN, the lines of many individually served subscribers have to be multiplexed and concentrated, and the number of subscribers online at any given time is limited by the number of connections between the provider and the telecommunication network.

In terms of geographical coverage, cable-LANs can be quite large. At the beginning, they are intented to be implemented for residential broadband services. The future target is, however, corporate internetworking.

Currently, there are no standards governing the transmission of data over cable LANs, but that has not slowed down equipment suppliers. They bring proprietary products to the market as rapidly as possible.

The cable modem, which is used to connect the consumer's PC with a coax drop linked to the two-way, broadband cable LAN, is specifically designed for high-speed data communications. When the modem is in operation, cable television programming is still available. But the return path that the cable modem uses is limited and must be shared among all digital services, including interactive television, video on demand, telephony, videoconferencing, and data services. One problem with the cable modem is the immaturity of existing devices. The majority of manufacturers are targeting peak bandwidth between 128 Kbps and 30 Mbps.

Competing with cable technology are, first of all, ISDN and emerging technologies, such as ADSL, RADSL, HDSL, SDSL, and VDSL. They all use existing wires to corporations and to residential customers. Because ISDN delivers both voice and data, it makes use of circuit-switching and packet-switching technology. For high-speed data services, the local telephone plant must be upgraded so that two-way digital transmission — over the existing copper pairs that ISDN also relies upon — is possible. In addition, ISDN-capable digitial switches must be installed at the telecommunications central office. ISDN may offer faster data transmission than analog modems, but the dial-up access model is still the same. From the data transmission perspective, this is an inefficient approach because circuit switching requires the service provider to dedicate resources to a customer at all times. In other words, it is not consistent with the bursty nature of data services. The result is wasted bandwidth.

Implementing an ISDN network could end up being more costly than deploying cable LANs. In telecommunications networks, e.g., an assessment of the incremental costs of providing ISDN Internet or intranet access includes the cost of initializing a subsciber's ISDN circuit at the central office; the cost of multiplexers, concentrators, and terminal servers; and possibly the cost of T1 lines from the telco to the Internet or intranet provider.

Cable-LAN access might prove to be less expensive for data and online services. It is assumed that cable television providers are facing less financial risks than telecommunications providers when deploying data services. Lower costs translate into consumer savings. The cost per bit of peak bandwidth (ROGE95) in providing Internet or Intranet access is significantly lower for hybrid fiber-coax networks than it is for ISDN — about 60 cents for a 4 Mbps residential service as opposed to S 16 for a 128 Kbps service. Shared fiber-coax networks also compete well against dedicated ISDNs on average bandwidth and peak bandwidth. A 4 Mbps residential cable service, for example, can provide the same average bandwidth and about 32 times the peak bandwidth of a 128 Kbps ISDN service for 40% less money.

While deployment costs for both technologies continue to decline, ISDN deployment still runs several hundred dollars more per subscriber than cable-LAN deployment. This difference, in addition to the higher performance of cable LANs, will be important factors as Internet, intranet, and online service access become a commodity product. Yet, the cable-LAN approach and the cable services industry have their shortcomings, too. Cable-LAN modems and access products are still proprietary, so once an operator has selected a vendor, it is likely to remain locked in. Also, shared networks function properly only when subscribers' usage habits are well known. The more subscribers deviate from an assumed usage profile, the more performance is likely to deteriorate. Big changes in usage could impact the cost of providing acceptable service levels over neighborhood cable-LANs.

Finally, the cable industry itself is greatly fragmented in many countries. This inhibits coordinated efforts which are vital to the rapid deployment of services. Although telecommunications providers are in competition against each other, they will work better as a group if they act quickly in deploying the resources needed to make ISDN universally available.

In order to avoid these bottlenecks and other fault- and performance-related problems, cable service providers need a powerful management solution. It requires a complete rethinking of presently used solutions. The management solution should address high volumes of events, alarms, and messages, failsafe operations, traffic control and management, quality assurance of services, and collection of accounting data.

MIB Availability

MIB definitions are in the works with standard committees and with vendors. The present definition is not yet complete, but as a guideline is very useful for further steps.

The MIB is structured at present into six groups:

- The docsDevBase group extends the MIB-II "system" group with objects needed for cable device system management. It includes the device role, date and time, serial number, and reset conditions.
- The docsDevNmAccessGroup provides a minimum level of SNMP access security to the device by network management stations. The access is also constrained by the community strings and any vendor-specific security. The management station should have read-write permission for the cable modems.
- The docsDevSoftware group provides information for network downloadable software upgrades. It includes file identification, administration status, operational status, and current version.
- The docsDevServer group provides information about the progress of the interaction between the CM and CMTS and various provisioning servers. It includes boot state, DHCP, server time, configuration file, and TFTP configuration parameters.
- The docsDevEvent group provides control and logging for event reporting. This group offers very detailed information on control parameters, syslog details, throttle conditions, severity codes, and priorities. It also offers entries for vendor-specific data.
- The docsDevFilter group configures filters at link layer and IP layer for bridges data traffic. This group consists of a link-layer filter table; docsDevFilterLLCTable, which is used to manage the processing and forwarding of non-IP traffic; an IP packet classifier table, docsDevFilterIpTable, which is used to map classes of packets tp specific policy actions; and a policy table, docsDevFilterPolicyTable, which maps zero or more policy action tables. At this time, the MIB specifies only one policy action table, docsDevFilterTosTable, which allows the manipulation of the type of services bits in an IP packet based on matching certain criteria. The working group may add additional policy types and action tables in the future, for example, to allow QoS to modem service identifier assignment based on destination.

Management is more difficult than with other technologies due to lack of management capabilities in managed devices. In most cases, proprietary solutions dominate. The implementation of TMN and other standards is extremely slow.

3.2.4.9 Management Solutions

The present status for emerging technologies can be summarized as follows:

- Proprietary solutions dominate: it means that the management protocols selected are controlled by the supplier of the equipment or facilities vendors. Working Groups do not exist, with the exception of the ATM Forum.
- Re-engineered by SNMP: many of the equipment vendors include SNMP agents into their devices to meet the requirements of customers. The SNMP agents provide information for performance management and reporting, but they usually do not change the real-time processing of status data within the devices. Most equipment is extremely intelligent. SNMP is simply too slow for certain decisions.

- The management structures are very heterogeneous: in most cases, domains are managed. Domain boundaries may be defined by the geography or by the family of managed objects. Most of the interfaces to element managers and managed devices are still proprietary. But, at least, the interfaces are going to be defined clearly by TMN.

- TMN has a very low penetration: suppliers have recognized the need for a generic standard, but they are not willing to invest heavily into supporting it. Some of the providers support the Q3-interface. There are no implementation examples for the higher layers of TMN. The network element and network management layers are well understood.

- Definition of MIBs: Each technology requires the definition of public and private MIBs. It is merely a question of time until each of the emerging technologies will be using MIB as the basis of management.

- Operating Support Systems are in the transitioning process: the legacy-type OSSs will give place to the new OSSs that are based on separating operations data from operating, the implementation of very flexible software, and separating network management from service management.

3.2.5 Summary

TMN will change the positions of managing emerged and emerging technologies. Customers will show little interest for the management solutions at lower layers. The primary target is to offer management capabilities at the service and business management layers. It needs significant extensions to existing management products. This segment of the handbook showed the status of management capabilities at element and network management layers only. In order to offer metrics for higher layers, MIB extensions and eventually new tools are required. If TMN layering combined with CNM, customers and providers may share the same management tools.

References

ABER97   Aber, R.: xDSL Supercharges Copper, *Data Communications*, March 1997, p. 99-105.
BLAC94   Black, U.: *Emerging Communications Technologies, Prentice-Hall Series in Advanced Communication Technologies*, Englewood Cliffs, NJ, 1994.
ROGE95   Rogers, C.: Telcos versus Cable TV: The Global View, *Data Communications*, September 1995, p. 75-80.
TERP98   Terplan, K.: *Telecom Operations Management Solutions*, CRC Press, Boca Raton, 1998.

3.3 Commercial Network and Systems Management Standards

Kornel Terplan

Industry standards help suppliers, vendors, and customers to communicate with each other more efficiently, but finding the common denominator of all interests is not easy. Service creation, provisioning, assurance, delivery, and service management cannot wait for fully completed standards. This chapter summarizes the most important management standards for multimedia communications. Besides TMN, CMIP, SNMP, RMON, OMA, and DMTF, Web-based standards are also introduced that may help to unify and simplify management frameworks and applications. The TeleManagement Forum may help a lot to improve the interoperability between multiple suppliers by providing practical guides, specifications, solution sets, and conformance documents.

3.3.1 Manager–Agent Relationship

Protocols always represent an agreement between the communicating parties. In the area of network management, management–agent relationships have been frequently implemented. Figure 3.3.1 shows

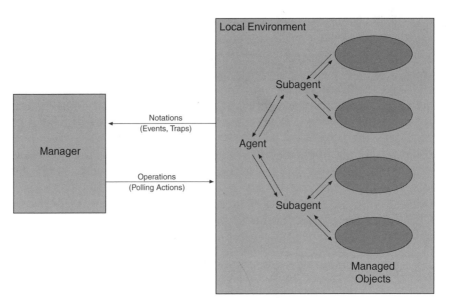

FIGURE 3.3.1   Communications paths between manager, agents, subagents, and managed objects.

this relationship. The agent side may be hierarchical by implementing subagents into specific devices. Depending on the nature of the management protocol, either the manager or the agents starts the dialog. There are always exceptions for high-priority networking events.

In case of CMIP, eventing techniques are used. Assuming that the agents are intelligent enough to capture, interpret, filter, and process events, they will notify the manager about alarming conditions. Of course, the manager is allowed to interrupt, and send inquiry-types of messages to the agents.

In case of SNMP, the manager polls the agents periodically. The agents respond by sending an abundance of information on device status and performance. Usually, the agents wait for the poll unless unusual events occur in the network. For such events, special traps can be defined and implemented.

3.3.2   Common Management Information Protocol (CMIP)

TMN protocols include OSI protocols such as CMIP and FTMP, ISDN, and Signalling System No. 7 protocols. They are organized into protocol suites or profiles for specific TMN interfaces. Functions and protocols support TMN services which include:

- Traffic management
- Customer management
- Switching management
- Management of transport networks
- Management of intelligent networks
- Tariffing and changing

The primary protocol is Common Management Information Protocol (CMIP). The estimated overhead scares both vendors and users away with the exception of the telecommunication industry, where separate channels can be used for management. CMIP is event driven, assuming processing capabilities at the agent level. Once fault and performance thresholds are exceeded, the manager is alarmed by the agent. This is similar to SNMP traps.

Open Systems Interconnected network management follows an object-oriented model; physical and logical real resources are managed through abstractions known as managed objects (MO). Management

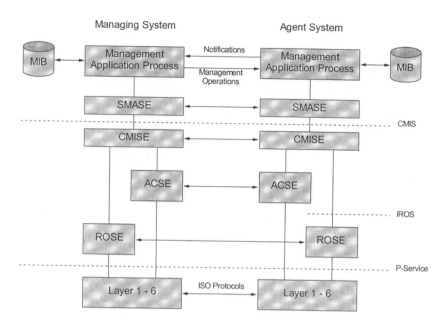

FIGURE 3.3.2 Application layer protocols and services.

systems also need managed objects that do not represent anything but exist for the needs of the management system itself. Most are handled by applications in agent roles and are accessed by applications in manager roles in order to implement management policies. The global collection of management information is consolidated in MIBs; each agent handles a part of it in its Management Information Tree. Information in the manager–agent interaction is conveyed through the management service/protocol common management information services (CMIS) and CMIP. In the context, agent, managed system, and managed node are synonymous; manager, managing application, and management stations are synonymous as well. CMIP is part of the OSI management framework. Its elements are:

- CMIS/CMIP defines the services provided to management applications and the protocol providing theinformation exchange capability
- Systems management functions specifying all the functions to be supported by management
- Management information model that defines the MIB
- Layer management that defines management information, service, and functions to specific layers

Based on the seven-layer model, network-management applications are implemented in layer 7. Layers 1 through 6 contribute to network management by offering the standard services to carry network management-related information. Figure 3.3.2 shows the structure for the application layer. In particular, four system management application entities (SMAE) are very useful:

For generic use: Association control service element (ACSE) and
                  Remote operations service element (ROSE)
For specific use: Common management information service element (CMISE)
                  System management application service element (SMASE)

Communication services in the OSI management model are provided by the CMIS. The service is realized through CMIP over a full or lightweight OSI stack. In the OSI world, two mappings are defined:

- a service over a full OSI stack (CMIP) and
- a connectionless one over Logical Link Control 1 (LLC1)(CMOL)

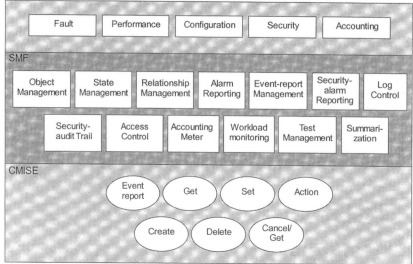

FIGURE 3.3.3 Overview of OSI management structure.

In the Internet world, there is a third mapping to provide the service over TCP/UDP using a lightweight presentation protocol (CMOT). CMOT and CMIP applications are portable on each other's stack with the same API, but will not work over CMOL.

Figure 3.3.3 shows the complete structure of CMISEs and SMFs. The overall goal is to support typical network management functions, such as fault, configuration, performance, security, and accounting management. SMFs or a group of SMFs support specific management functions. For the communication between entities, CMISE is used. Figure 3.3.4 summarizes these service elements between SMASE and CMISE. Each of the service elements are defined in Table 3.3.1.

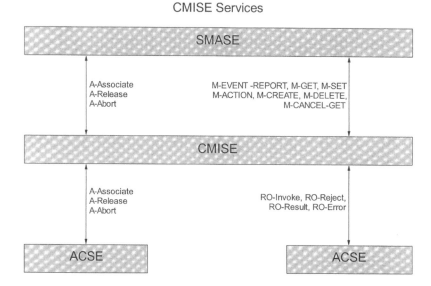

FIGURE 3.3.4 Information exchange using service elements.

TABLE 3.3.1   Definition of Service Elements

Service	Type	Definition
		Management Notification Service
M-event-report	Confirmed/unconfirmed	Reports on an event to a peer CMISE service user
		Management Operations Service
M-get	Confirmed	Requests retrieval of information from peer CMISE user
M-set	Confirmed/unconfirmed	Requests modification of information from peer CMISE user
M-action	Confirmed/unconfirmed	Requests that peer CMISE user perform some action
M-create	Confirmed	Requests that peer CMISE user create an instance of a Managed Object
M-delete	Confirmed	Requests that peer CMISE user delete an instance of a Managed Object
M-cancel-get	Confirmed	Requests that peer CMISE user cancel outstanding invocation of M-get service

Further details can be found in (TERP96) — Effective Mgmt of LANs, McGraw-Hill. The strengths of CMIP include:

- General and exensible object-oriented approach
- Support from the telecommunication industry
- Support for manager-to-manager communications
- Support for a framework for automation

Weaknesses of CMIP are:

- It is complex and multilayered
- High overhead is the price of many confirmations
- Few CMIP-based management systems are shipping
- Few CMIP-based agents are in use

CMIP assumes the use of the OSI stack for exchanging CMIP protocol data units. In layer 7, there are also other applications which may be combined with CMIP.

3.3.3   Simple Network Management Protocol (SNMPv1, SNMPv2, and SNMPv3)

In the SNMP environment, the manager can obtain information (Figure 3.3.5) from the agent by polling managed objects periodically. Agents can transmit unsolicited event messages, called traps, to the manager. The management data exchanged between managers and agents is called the management information base (MIB). The data definitions outlined in SMI must be understood by both managers and agents.

The manager is a software program residing within the management station. The manager has the ability to query agents using various SNMP commands. The management station is also in charge to interpret MIB data, construct views of the systems and networks, compress data, and maintain data in relational or object-oriented databases. A traditional SNMP manager is shown in Figure 3.3.6 (STAL99).

This figure shows typical functions that are not only valid for managers, but also for managed agents. Figure 3.3.7 shows the typical functional blocks of SNMP agents.

In a traditional manager, the SNMP engine contains a dispatcher, a message processing subsystem, and a security subsystem. The dispatcher is a simple traffic manager. For outgoing PDUs, the dispatcher accepts PDUs from applications and performs the following fucntions. For each PDU, the dispatcher determines the type of message processing required — it may be different for SNMP versions 1, 2, and 3 — and passes the PDU on to the appropriate message processing module in the message processing subsystem. Subsequently, the message processing subsystem returns a message containing that PDU and

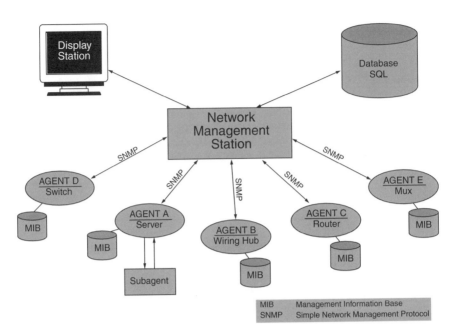

FIGURE 3.3.5   Structure of SNMP-based management services.

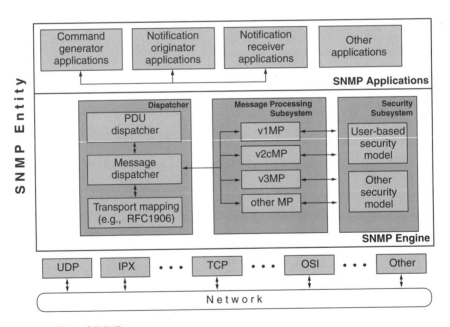

FIGURE 3.3.6   Traditional SNMP manager.

including the appropriate message headers. The dispatcher then maps this message onto a transport layer for transmission. For incoming messages, the dispatcher accepts messages from the transport layer and performs the following functions. The dispatcher routes each message to the appropriate message processing module. Subsequently, the message processing subsystem returns the PDU contained in the message. The dispatcher then passes this PDU to the appropriate application.

The message processing subsystem accepts outgoing PDUs from the dispatcher and prepares these for transmission by wrapping them in the appropriate message header and returning them to the dispatcher.

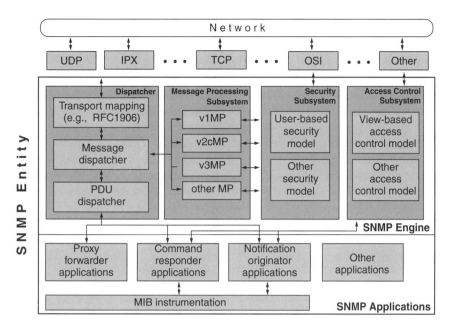

FIGURE 3.3.7   Traditional SNMP agent.

It also accepts incoming messages from the dispatcher, processes each message header, and returns the enclosed PDU to the dispatcher. An implementation of the message processing subsystem may support a single message format corresponding to a single version of SNMP, or may contain a number of modules, each supporting a different version of SNMP.

The security subsystem performs authentication and encryption functions. Each outgoing message is passed to the security subsystem from the message processing subsystem. Depending on the services required, the security subsystem may encrypt the enclosed PDU, and it may generate an authentication code and insert it into the message header. The processed message is then returned to the message processing subsystem. Similarly, each incoming message is passed to the security subsystem from the message processing subsystem. If required, the security subsystem checks the authentication code and performs decryption. An implementation of the security subsystem may support one or more distinct security models.

The SNMP engine for a traditional agent has all the components found in the SNMP engine for a traditional manager, plus an Access Control Subsystem. This subsystem provides services to control access to MIBs for the reading and setting of management objects. These services are performed on the basis of the contents of PDUs. An implementation of the security subsystem may support one or more distinct access control models. The security-related functions are organized into two separate subsystems: security and access control. This is an excellent example of modular design, because the two subsystems perform quite distict functions and therefore it makes sense to allow standardization of these two areas to proceed independently. The security subsystem is concerned with privacy and authentication, and operates on SNMP messages. The access control subsystem is concerned with authorized access to management information, and operates on SNMP PDUs.

For both SNMP managers and agents, there are a number of components that constitute the management functionalities. These components are listed in Table 3.3.2 (STAL99).

In terms of MIBs, there are continuing changes. In addition to standard MIBs, such as MIB I and II (Table 3.3.3), the IETF has defined a number of adjunct MIBs covering hosts, routers, bridges, hubs, repeaters, FDDI networks, AppleTalk networks, frame relay networks, switches, ATM nodes, mobile components, and applications.

TABLE 3.3.2   Components of SNMP Entity

Dispatcher — Allows for concurrent support of multiple versions of SNMP messages in the SNMP engine. It is responsible for (1) accepting PDUs from applications for transmission over the network and delivering incoming PDUs to applications; (2) passing outgoing PDUs to the message processing subsystem to prepare as messages, and passing incoming messages to the message processing subsystem to extract the incoming PDUs; and (3) sending and receiving SNMP messages over the network.

Message Processing Subsystem — Responsible for preparing messages for sending, and for extracting data from received messages.

Security Subsystem — Provides security services such as the authentication and privacy of messages. This subsystem potentially contains multiple security models.

Access Control Subsystem — Provides a set of authorization services that an application can use for checking access rights. Access control can be invoked for retrieval or modification request operations and for notification generation operations.

Command Generator — Initiates SNMP Get, GetNext, GetBulk, and/or Set request PDUs and processes the response to a request that it has generated.

Command Responder — Receives SNMP Get, GetNext, GetBulk, and/or set request PDUs destined for the local system as indicated by the fact that the contextEngineID in the received request is equal to that of the local engine through which the request was received. The command responder application will perform the appropriate protocol operation, using access control, and will generate a response message to be sent to the request's originator.

Notification Originator — Monitors a system for particular events or conditions, and generates Trap and/or Inform messages based on these events or conditions. A notification originator must have a mechanism for determining where to send messages, and which SNMP version and security parameters to use when sending messages.

Notification Receiver — Listens for notification messages, and generates response messages when a message containing an Inform PDU is received.

Proxy Forwarder — Forwards SNMP messages. Implementation of a proxy forwarder application is optional.

TABLE 3.3.3   MIB II Structure

11 Categories of Management (2) Subtree	Information in the Category
System (1)	Network device operating system
Interfaces (2)	Network interface specific
Address translation (3)	Address mappings
Ip (4)	Internet protocol specific
Icmp (5)	Internet control message protocol specific
Tcp (6)	Transmission protocol specific
Udp (7)	User datagram protocol specific
Egp (8)	Exterior gateway protocol specific
Cmot (9)	Common management information services on TCP specific
Transmission (10)	Transmission protocol specific
Snmp (11)	Snmp specific

In terms of SNMP, the following trends are expected. SNMP agent-level support will be provided by an even greater number of vendors. SNMP manager-level support will be provided by only a few leading vendors in the form of several widely-accepted platforms. Management platforms provide basic services, leaving customization to vendors and users.

Wider use of intelligent agents is also expected. Intelligent agents are capable of responding to a manager's request for information and performing certain manager-like functions, including testing for thresholds, filtering, and processing management data. Intelligent agents enable localized polling and filtering on servers, workstations, and hubs, for example. Thus, these agents reduce polling overhead and management data traffic, forwarding only the most critical alerts and processed data to the SNMP manager.

RMON MIB will help to bridge the gap between the limited services provided by management platforms and the rich sets of data and statistics provided by traffic monitors and analyzers.

The strengths of SNMP include:

- Agents are widely implemented
- Simple to implement
- Agent-level overhead is minimal
- Polling approach is good for LAN-based managed object
- Robust and extensible
- Offers the best direct manager-agent interface

SNMP met a critical need; it was available and implementable at the right time. The weaknesses of SNMP include:

- Too simple, does not scale well
- No object-oriented data view
- Unique semantics make integration with other approaches difficult
- High communication overhead due to polling, in particular for WAN-based management objects
- Many implementation-specific (private MIB) extensions
- No standard control definition
- Small agent (one agent per device) may be inappropriate for systems management

SNMP is being continuosly improved and extended. SNMPv2 addresses many of the shortcomings of version 1. SNMPv2 can support either a highly centralized management strategy or a distributed one. In the latter case, some systems operate both in the role of manager and agent. In its agent role, such a system will accept commands from a superior manager; these commands may deal with access of information stored locally at the intermediate manager or may require the intermediate manager to provide summary information about subagents. The principal enhancements to SNMPv1 provided by version 2 fall into the following categories (STAL99):

- Structure of management information is being expanded in several ways. The macro used to define object types has been expanded to include several new data types and to enhance the documentation associated with an object. A noticeable change is a new convention has been provided for creating and deleting conceptual rows in a table. The origin of this capability is from remote monitoring (RMON).

- Transport mappings help to use different protocol stacks to transport the SNMP information, including user datagram protocol, OSI connectionless-mode protocol, Novell internetwork (IPX) protocol and Appletalk.

- Protocol operations with the most noticeable change of including two new PDUs. The GetBulkRequest PDU enables the manager to efficiently retrieve large blocks of data. In particular, it is powerful in retrieving multiple rows in a table. The InformRequest PDU enables one manager to send trap-type information to another.

- MIB extentions contain basic traffic information about the operation of the SNMPv2 protocol; this is identical to SNMP MIB II. The SNMPv2 MIB also contains other information related to the configuration of SNMPv2 manager to agent.

- Manager-to-manager capability is specified in a special MIB, called M2M. It provides functionality similar to the RMON MIB. In this case, the M2M MIB may be used to allow an intermediate manager to function as a remote monitor of network media traffic. Also, reporting is supported. Two major groups, Alarm and Event, are supported.

- SNMPv2 security does include a wrapper containing authentication and privacy information as a header to PDUs.

The SNMPv2 framework is derived from the SNMP framework. It is intended that the evolution from SNMP to SNMP2 be seamless. The easiest way to accomplish this is to upgrade the manager to support SNMPv2 in a way that allows the coexistence of SNMPv2 managers, SNMPv2 agents, and SNMP agents. In order to map commands mutually into the target protocol, proxy agents are used (STAL99).The actual implementation of the proxy agent depends on the vendor; it could be implemented into the agent or into the manager.

The key new feature of SNMPv3 is better security. The design goals for v3 can be summarized as follows (STAL99):

- Use existing work, for which there is implementation experience and some consensus as to its value. As a result, the SNMP architecture and SNMPv3 security features rely heavily on SNMPv2u and SNMPv2*.

- Address the need for secure Set request messages over real-world networks, rectifying the most important deficiency of SNMPv1 and SNMPv2.

- Design a modular architecture that will (1) allow implementation over a wide range of operational environments, some of which need minimal, inexpensive functionality and some of which may support additional features for managing large networks; (2) make it possible to move portions of the architecture forward in the standards track even if consensus has not been reached on all pieces; and (3) accomodate alternative security models.

- Keep SNMP as simple as possible, despite the many necessary and useful extensions.

Based on these design goals, developers have made the following design decisions (STAL99):

- Architecture: An architecture should be defined which identifies the conceptual boundaries between the documents. Subsystems should be defined which describe the abstract services provided by specific portions of a SNMP framework. Abstract service interfaces, as described by service primitives, define the abstract boundaries between documents, and the abstract services that are provided by the conceptual subsystems of a SNMP framework.

- Self-contained documents: Elements of procedure plus the MIB objects that are needed for processing a specific portion of a SNMP framework should be defined in the same document, and as much as possible, should not be referenced in other documents. This allows pieces to be designed and documented as independent and self-contained parts, which are consistent with the general SNMP MIB module approach. As portions of SNMP change over time, the documents describing other portions of SNMP are not directly impacted. This modularity allows, for example, security models, authentication and privacy mechanisms, and message formats to be upgraded and supplemented as the need arises. The self-contained documents can move along the standards track on different time lines.

- Remote configuration: The security and access control subsystems add a whole new set of SNMP configuration parameters. The security subsystem also requires frequent changes of secrets at the various SNMP entities. To make this deployable in a large operational environment, these SNMP parameters must be able to be remotely configured.

- Controlled complexity: It is recognized that manufacturers of simple managed devices want to keep the resources used by SNMP to a minimum. At the same time, there is a need for more complex configurations which can spend more resources for SNMP and thus provide more functionality. The design tries to keep the competing requirements of these two environments in balance and allows the more complex environment to logically extend the simple environment.

- Threats: The security models in the security subsystem should protect against the principal threats, such as modification of information, masquerade, message stream modification, and disclosure. They do not need to protect denial of service and traffic analysis.

SNMPv3 is secure against the following threats:

- Modification of information: Any entity could alter an in-transit message generated by an authorized entity in a way that effects unauthorized management operations, including the setting of object values. The essence of this threat is that an unauthorized entity could change any management parameter, including those related to configuration, operations, and accounting.
- Masquerade: Management operations that are not authorized for some entity may be attempted by that entity by assuming the identity of an authorized entity.
- Message stream modification: SNMP is designed to operate over a connectionless transport protocol. There is a threat that SNMP messages could be reordered, delayed, duplicated, or replayed to effect unauthorized management operations. For example, a message to reboot a device could be copied and replayed later.
- Disclosure: An entity could observe exchanges between a manager and an agent and thereby learn the values of managed objects and learn of notifiable events. For example, the observation of a set command that changes passwords would enable an attacker to learn the new passwords.

But SNMPv3 is not intended to secure against the following two threats:

- Denial of service: An attacker may prevent exchanges between a manager and an agent.
- Traffic analysis: An attacker may observe the general pattern of traffic between managers and agents.

SNMPv3 is still in its initial implementation phase. If fully implemented, the power of SNMP will be significantly improved.

3.3.4   Remote Monitoring (RMON1 and RMON2)

The RMON MIB will help to bridge the gap between the limited services provided by management platforms and the rich sets of data and statistics provided by traffic monitors and analyzers. RMON defines the next generation of network monitoring with more comprehensive network fault diagnosis, planning, and performance tuning features than any current monitoring solution. The design goals for RMON are (STAL99):

- Off-line operation: In order to reduce overhead over communication links, it may be necessary to limit or halt polling of a monitor by the manager. In general, the monitor should collect fault, performance, and configuration information continuously, even if it is not being polled by a manager. The monitor simply continues to accumulate statistics that may be retrieved by the manager at a later time. The monitor may also attempt to notify the manager if an exceptional event occurs.
- Preemptive monitoring: If the monitor has sufficient resources, and the process is not disruptive, the monitor can continuously run diagnostics and log performance. In the event of a failure somewhere in the network, the monitor may be able to notify the manager and provide useful information for diagnosing the failure.
- Problem detection and reporting: Preemptive monitoring involves an active probing of the network and the consumption of network resources to check for error and exception conditions. Alternatively, the monitor can passively — without polling — recognize certain error conditions and other conditions, such as congestions and collisions, on the basis of the traffic that it observes. The monitor can be configured to continuously check for such conditions. When one of these conditions occurs, the monitor can log the condition and notify the manager.
- Value-added data: The network monitor can perform analyses specific to the data collected on its subnetworks, thus offloading the manager of this responsibility. The monitor can, for instance,

TABLE 3.3.4   RMON MIB Groups for Ethernet

Statistics Group	Features a table that tracks about 20 different characteristics of traffic on the Ethernet LAN segment, including total octets and packets, oversized packets, and errors.
History Group	Allows a manager to establish the frequency and duration of traffic observation intervals, called "buckets." The agent can then record the characteristics of traffic according to these bucket intervals.
Alarm Group	Permits the user to establish the criteria and thresholds that will prompt the agent to issue alarms.
Host group	Organizes traffic statistics by each LAN node, based on time intervals set by the manager.
HostTopN Group	Allows the user to set up ordered lists and reports based on the highest statistics generated via the host group.
Matrix Group	Maintains two tables of traffic statistics based on pairs of communicating nodes; one is organized by sending node addresses, the other by receiving node addresses.
Filter Group	Allows a manager to define, by channel, particular characteristics of packets. A filter might instruct the agent, for example, to record packets with a value that indicates they contain DECnet messages.
Packet Capture Group	This group works with the Filter Group and lets the manager specify the memory resources to be used for recording packets that meet the filter criteria.
Event Group	Allows the manager to specify a set of parameters or conditions to be observed by the agent. Whenever these parameters or conditions occur, the agent will record an event into a log.

TABLE 3.3.5   RMON MIB Groups for Token Ring

Statistics Group	This group includes packets, octets, broadcasts, dropped packets, soft errors and packet distribution statistics. Statistics are at two levels: MAC for the protocol level and LLC statistics to measure traffic flow.
History Group	Long-term historical data for segment trend analysis. Histories include both MAC and LLC statistics.
Host Group	Collects information on each host discovered on the segment.
HostTopN Group	Provides sorted statistics that allow reduction of network overhead by looking only at the most active hosts on each segment.
Matrix Group	Reports on traffic errors between any host pair for correlating conversations on the most active nodes.
Ring Station Group	Collects general ring information and specific information for each station. General information includes: ring state (normal, beacon, claim token, purge); active monitor; and number of active stations. Ring Station information includes a variety of error counters, station status, insertion time, and last enter/exit time.
Ring Station Order	Maps station MAC addresses to their order in the ring.
Source Routing Statistics	In source-routing bridges, information is provided on the number frames and octets transmitted to and from the local ring. Other data includes broadcasts per route and frame counter per hop.
Alarm Group	Reports changes in network characteristics based on thresholds for any or all MIBs. This allows RMON to be used as a proactive tool.
Event Group	Logging of events on the basis of thresholds. Events may be used to initiate functions such as data capture or instance counts to isolate specific segments of the network.
Filter Group	Definitions of packet matches for selective information capture. These include logical operations (AND, OR, NOT) so network events can be specified for data capture, alarms, and statistics.
Packet Capture Group	Stores packets that match filtering specifications.

observe which station generates the most traffic or errors in network segments. This type of information is not otherwise accessible to the manager that is not directly attached to the segment.

- Multiple managers: An internetworking configuration may have more than one manager in order to achieve reliability, perform different functions, and provide management capability to different units within an organization. The monitor can be configured to deal with more than one manager concurrently.

Table 3.3.4 summarizes the RMON MIB groups for Ethernet segments. Table 3.3.5 defines the RMON MIB groups for Token Ring segments. At the present time, here are just a few monitors that can measure both types of segments using the same probe.

RMON is very rich on features and there is the very real risk of overloading the monitor, the communication links, and the manager when all the details are recorded, processed, and reported. The

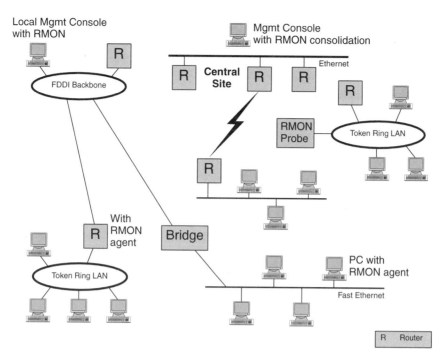

FIGURE 3.3.8   RMON probes in LAN segments.

preferred solution is to do as much of the analysis as possible locally, at the monitor, and send just the aggregated data to the manager. This assumes powerful monitors. In other applications, monitors may be reprogrammed during operations by the managers. This is very useful when diagnosing problems. Even if the manager can define specific RMON requests, it is still necessary to be aware of the trade-offs involved. A complex filter will allow the monitor to capture and report a limited amount of data, thus avoiding overhead on the network. However, complex filters consume processing power at the monitor; if too many filters are implemented, the monitor will become overloaded. This is particularly true if the network segments are busy, which is probably the time when measurement is most valuable.

Figure 3.3.8 shows the RMON probes in the segments. RMON probes can be implemented in three different ways:

- Probe as a standalone monitor
- Probe as a module of hubs, routers, and switches
- Probe as a software module in Unix, NT operatings systems, or in PC workstations

Each of these alternatives shows benefits and disadvantages.

1.   *Probe as a Standalone Monitor*
The benefits of this alternative are:

- Excellent performance
- Support of all functions
- Availability in various options, such as stackable or rack-mountable

The disadvantages are:

- High costs for an average LAN segment
- Multiple probes are required, when segmentation of LANs is deployed by switches without using probes-ports
- Most advanced LAN technologies might not be supported right away

2. *Probe as a Module of Hubs, Routers, and Switches*

The benefits of this alternative are:

- It represents a very convenient solution because the networking components have already been deployed
- The costs are much lower than in case of standalone probes
- The integration of probes into switches is less expensive than deploying an individual probe in each switched segment

The disadvantages are:

- The networking components need an upgrade or customization to support the probes
- Upgrades are not always economical
- The performance might become a problem
- The conformance to standards may not be met
- Problems with the probe may impact the performance of the networking component
- The RMON functionality may be very much limited; not all RMON indicator groups are supported
- Processing programs are very simple in comparison to standalone probes
- Integration with management platforms is usually incomplete
- RMON modules may be provided by different vendors, which may lead to incompatibility problems

3. *Probe as a Software Module in Unix, NT Operatings Systems, or in PC Workstations*

The benefits of this alternative are:

- Much lower costs in comparison to any other alternatives
- Performance is good when run on RISC or Pentium processors
- Scalability and extendability are excellent
- Support of state-of-the technology is easier than with other alternatives
- Combination of supervising Ethernet and Token Ring is possible
- Outband access to probes is possible with proper configuration

The disadvantages are:

- Purchase of adapters and additional workstations may be required
- The user is responsible for the installation

The extension of this alternative might be utilized also for switched LANs. RMON may be installed into the adapter of end-user devices. Usually, just the filter- and packet capture groups are supported at the end-user device level. The other groups are supported by the collector. The overhead is minimal; performance impacts in the switch and in the end-user devices are minimal.

RMON probes are extremely helpful for collecting data on Web site accesses and activities. Independently, how probes are implemented, vendors of probes are expected to work together. Standards are continuously improved offering even more functionality of capturing and processing Web sites-relevant data.

The existing and widely used RMON version is basically a MAC standard. It does not give LAN managers visibility into conversations across the network or connectivity between various network segments. The extended standard is targeting the network layer and higher. It will give visibility across the enterprise. With remote access and distributed workgroups, there is a substantial intersegment traffic. The following functionalities are included:

- Protocol distribution
- Address mapping

- Network layer host table
- Network layer matrix table
- Application layer host table
- Application layer matrix table
- User history collection table
- Protocol configuration.

Protocol distribution and protocol directory table — The issue here was how to provide a mechanism that will support the large number of protocols running on any one network. Current implementations of RMON employ a protocol filter which analyzes only the essential protocols. RMON2, however, will employ a protocol directory system which allow RMON2 applications to define which protocols an agent will employ. The Protocol Directory Table will specify the various protocols a RMON2 probe can interpret.

Address Mapping — This feature matches each network address with a specific port to which the hosts are attached. It also identifies traffic-generating nodes/hosts by MAC, Token Ring, or Ethernet address. It helps identify specific patterns of network traffic. Useful in node discovery and network topology configurations. In addition, the address translation feature adds duplicate IP address detection, resolving a common troublespot with network routers and virtual LANs.

Network-layer Host Table — Tracks packets, errors, and bytes for each host according to a network-layer protocol. It permits decoding of packets based on their network layer address. In essence, permitting network managers to look beyond the router at each of the hosts configured on the network.

Network-layer Matrix Table — Tracks the number of packets sent between a pair of hosts by network layer protocol. The network manager can identify network problems quicker by using the matrix table which shows the protocol-specific traffic between communicating pairs of systems.

Application-layer Host Table — Tracks packets, errors, and bytes by host on an application-specific basis (e.g., Lotus Notes, e-Mail, WEB, etc.) Both the application-layer host table and matrix table trace packet activity of a particular application. This feature can be used by network managers to charge back users on the basis of how much network bandwidth was used by their applications.

Application-layer Matrix Table — Tracks packet activity between pairs of hosts by application (e.g., pairs of hosts exchanging internet information).

Probe Configuration — Currently, vendors offer a variety of proprietary means for configuring and controlling their respective probes. This complicates interoperability. The Probe Configuration Specification, based on the Aspen MIB, defines standard parameters for remotely configuring probes — parameters such as network address, SNMP error trap destinations, modern communications with probes, serial line information, and downloading of data to probes. It provides enhanced interoperability between probes by specifying standard parameters for operations, permitting one vendor's RMON application the ability to remotely configure another vendor's RMON probe.

User History Collection Group — The RMON2 history group polls, filters, and stores statistics based on user-defined variables, creating a log of the data for use as a historical tracking tool. This is in contrast to RMON1, where historical data is gathered on a predefined set of statistics.

After implementation, more and more complete information will be available for performance analysis and capacity planning.

3.3.5 Desktop Management Interface (DMI) from Desktop Management Task Force (DMTF)

Basically, SNMP may be utilized to manage systems, assuming system components accomodate SNMP agents. But there are no MIBs yet that describe principal indicators for management purposes. An important emerging standard for desktop management is the Desktop Management Interface (DMI). The Desktop Management Task Force (DMTF) has defined the DMI to accomplish the following goals:

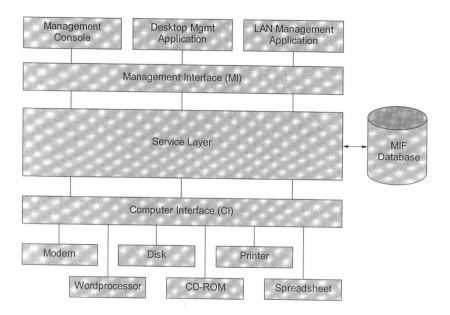

FIGURE 3.3.9 Structure of the desktop management interface.

- Enable and facilitate desktop, local, and network management
- Solve software overlap and storage problems
- Create a standard method for management of hardware and software components using MIFs (management information format)
- Provide a common interface for managing desktop computers and their components
- Provide a simple method to describe, access, and manage desktop components

The scope of management under DMTF includes CPUs, I/Os, mother boards, video cards, network interface cards, faxes, modems, mass storage devices, printers, and applications. Figure 3.3.9 shows the structure of this standard. There is a clear separation between the managed components (CI) and the services offered for management (MI). The commands are similar to SNMP, but they are not identical.

3.3.6 Object Management Architecture (OMA) from OMG

OMA permits the cooperation of distributed objects independently from their locations. OMA has not been designed and structured only for network management. It is a general purpose standard. In particular, it helps to design and implement distributed applications. This architecture has gained interest because the OMA technology is going to be increasingly used for end-user devices.

The organizational model is using a peer-to-peer approach with the result that communicating objects are equivalent in their importance. Figure 3.3.10 shows this architecture.

The communication model is based on CORBA (Common Object Request Broker Architecture). The Object Request Broker is the coordinator between distributed objects. The broker receives messages, inquiries, and results from objects, and routes them to the right destination. If the objects are in a heterogeneous environment, multiple brokers are required. They will talk to each other in the future by a new protocol based on TCP/IP. There is no information model available; no operations are preddefined for objects. But an object does exist containing all the necessary interfaces to the object request broker. For the description, the interface definition language (IDL) is being used. There are no detailed MIBs for objects because OMA is not management specific.

The functional model consists of the Object Services Architecture. It delivers the framework for defining objects, services, and functions. Examples for services are instanciation, naming, storing objects'

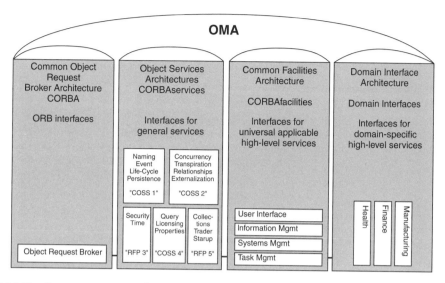

FIGURE 3.3.10 Open management architecture from OMG.

attributes, and the distribution/receipt of events and notification. CORBA services and facilities represent more generic services; they are expected to occur in multiple applications or they are used in specific applications. The driving force beyond designing common facilities for systems management is the X/Open Systems Management Working Group. The Managed Set Service, defined by this group, encourages grouping of objects in accordance of their management needs, with the result of easier administration. In the future, more services are expected to be defined; the next is an Event Management Service that expands the present Object Event Service by a flexible mechanism of event filtering.

DCOM is the heart of Microsoft's ActiveOSS product suite. Basically, DCOM is an integration infrastructure designed to facilitate communication between software components operating on the same host or with DCOM on multiple networked hosts. It was originally developed to create interoperability between components. It is the most widely deployed component object model. Active OSS acts as a centralized management and translation point for an OSS network. Conceptually, applications ride on top of the framework, but communicate through it. DCOM abstracts various application interfaces into objects, basically mapping the functions of the application into a common model that can be stored in a database. The common model allows the various applications to communicate in a uniform manner within the framework or across multple networked frameworks.

3.3.7 Standards for Web-based Systems and Network Management

In July 1996, five major vendors announced an initiative to define *de facto* standards for Web-based Enterprise Management (WBEM). This effort, spearheaded by Microsoft, Compaq, Cisco, BMC, and Intel, was publicly endorsed by over 50 other vendors as well. The initial announcement called for defining the following specifications:

- HyperMedia Management Schema (HMMS): an extensible data description for representing the managed environment that was to be further defined by the Desktop Management Task Force (DMTF).
- HyperMedia Object Manager (HMOM): data model consolidating management data from different sources; a C++ reference implementation and specification, defined by Microsoft and Compaq, to be placed in the public domain.
- HyperMedia Management Protocol (HMMP): a communication protocol embodying HMMS, running over HTTP and with interfaces to SNMP and DMI.

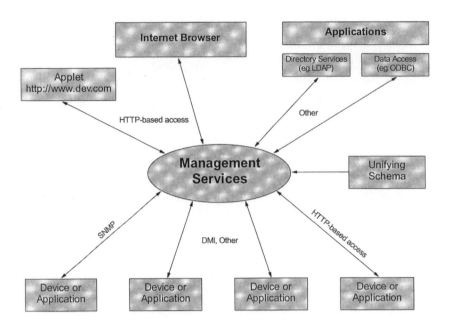

FIGURE 3.3.11 Use of the object manager to coordinate management services.

SunSoft has also announced a programming environment for developing Web-based network and systems management software. This environment, called Solstice Workshop, consists of a Java Management API (JaMAPI), a small footprint database, and a Java programming environment. Solstice Workshop's big drawing card is its extensibility and the popularity of Java's "write once run anywhere" appeal. JaMAPI requires Java, whereas HMMP/HMMS/HMOM specifies HTML/HTTP, although Java is not specifically excluded.

Among these two efforts, the WBEM is certainly the broadest in scope, addressing not only protocol issues, but also data modeling and extensible data description. While JaMAPI includes object class definitions, it does not go as far as data description.

The complete solution is envisioned in Figure 3.3.11 including both major directions of webification.

The initial euphoria over WBEM and JaMAPI is starting to wear off, and it is time for doing the hard work of pounding out specifications and, more importantly, building products. Customer demand will push Web-based management to its limits, but disillusion is sure to set in if a lack of progress becomes obvious on the standards' front. But there are several emerging products that have been developed with an eye for supporting current and future standards, that bring to market a practical approach to take advantage of Web-based management.

The DMTF has developed a common information model (CIM) to take advantage of object-based management tools and provide a common way to describe and share enterprise-wide management information. Using HMMS as an input, the new model can be populated by DMI 2.0 and other management data suppliers, including SNMP and CMIP, and implemented in multiple object-based execution models such as JaMAPI, CORBA, and HMM. CIM will enable applications from different developers on different platforms to describe and share management data, so users have interoperable management tools that span applications, systems, and networks, including the Internet.

CIM is a conceptual information model for describing management that is not bound to a particular implementation. This allows for the interchange of management information between management systems and applications. This can be either "agent to manager" and "manager to manager" communications, which provide for distributed system management.

In a fully CIM-compliant world, it should be possible to build applications such as Service Level Agreement Tracking applications using management data from a variety of sources and different management systems such as TME, OpenView, ManageWise, SMS, etc. The management data would be collected, stored and analyzed using a common format (CIM) while allowing property extensions providing "value add."

There are two parts to CIM: the CIM specification and the CIM schema.

The CIM specification describes the language, naming, meta schema, and mapping techniques to other management models such as SNMP MIBs, and DMTF MIFs, etc. The meta schema is a formal definition of the model. It defines the terms used to express the model and their usage and semantics. The elements of the meta schema are classes, properties, and methods. The meta schema also supports indications and associations as types of classes, and references as types of properties.

The CIM schema provides the actual model descriptions. The CIM schema supplies a set of classes with properties and associations that provide a well-understood conceptual framework within which it is possible to organize the available information about the managed environment.

The CIM schema itself is structured into three distinct layers:

- Core schema — an information model that captures notions applicable to all areas of management
- Common schema — an information model that captures notions common to particular management areas but independent of a particular technology or implementation. The common areas are systems, applications, databases, networks, and devices. The information model is specific enough to provide a basis for the development of management applications. This schema provides a set of base classes for extension into the area of technology specific schemas. There are currently four common schemas:

  1. Systems
  2. Applications
  3. Networks (LAN)
  4. Devices

- Extension schemas — Represent technology-specific extensions of the common schema. These schemas are specific to environments, such as operating systems (for example, UNIX or Microsoft Windows).

More schemas are planned for definition in the areas of directory enabled networks (DEN), service level agreements, and distributed application transaction measurement (DATM). Others will follow.

The formal definition of the CIM schema is expressed in a managed object file (MOF) which is an ASCII file that can be used as input into a MOF editor or compiler for use in an application.

The unified modeling language (UML) is used to portray the structure of the schemas. Techniques to develop UML (VISIO) files from MOF files are being developed.

This is the first time in this industry that a common method of describing management information has been agreed upon and followed through with implementation. Other efforts have failed because of the lack of industry support. Because the model is implementation independent, it does not provide sufficient information for product development. It is the specific product areas — applications, system, network, database, and devices — and their product-specific extensions that produce workable solutions.

Status of current CIM-related projects: The CIM TDC is defining rules and categories for an information model that provides a common way to describe and share enterprise-wide management information. The group will define a meta schema or basic modeling language; a core schema, the base set of classes specific to systems, networks, and applications; and a common schema, a base set of platform-neutral, domain-specific extensions of core schema.

CIM will take advantage of emerging object-based management technologies and ensure that the new model can be populated by DMI and other management data suppliers including SNMP and CMIP. The CIM is being designed to enable implementations in multiple, object-based execution models such as CORBA and common objects model (COM), and object-based management technologies such as JaMAPI (Java Management API).

3.3.8   Lightweight Directory Access Protocol (LDAP)

Directory services are fast becoming the key to the enterprise, allowing applications to locate the resources they need and enabling network managers to authenticate end users. Corporate networkers need to be aware about what LDAP is capable of, where it is headed, and what it was never intended to do.

LDAP was intended to offer a low-cost PC-based front end for accessing X.500 directories. Due to high overhead and acceptance delays of X.500, LDAP has emerged to fill the gap, somehow expanding its role. It rapidly became the solution of choice for all types of directory services applications on IP networks. LDAP applications can be loosely grouped into three categories: those that locate network users and resources, those that manage them, and those that authenticate and secure them. Network managers who want to put the protocol to work need to go into detail, coming to terms with standard components and features. This protocol can save companies time and money. It can help network managers keep pace with the rising demand for directory services. New applications appear almost every day. But there are limits to what a protocol can do for distributed computing. It cannot store all the types of information that network applications need. Knowing the difference between LDAP facts and fiction is the only way to avoid potential pitfalls.

3.3.8.1   Attributes of LDAP

The current specification comprises eight features and functions (HOWE99):

- Information model: Organized according to collections of attributes and values, known as entries, this model defines what kinds of data can be stored and how that data behaves. For example, a directory entry representing a person named Jim Fox might have an attribute called sn (surname) with a value "Fox." The information model, inherited almost unchanged from X.500, is extensible: almost any kind of new information can be added to a directory.

- LDAP schema: Defines the actual data elements that can be stored in a particular server and how they relate to real-world objects. Collections of values and attributes — representing such objects as countries, organizations, people, and groups of people — are defined in the standard, and individual servers can define new schema elements as well.

- Naming model: Specifies how information is organized and referenced. LDAP names are hierarchical; individual names are composed of attributes and values from the corresponding entry. The top entry typically represents a domain name, company, state, or organization. Entries for subdomain, branch offices, or departments come next, often followed by common name entries for individuals. Like the LDAP information model, the naming model derives directly from X.500. Unlike X.500, LDAP does not constrain the format of the namespace; it allows a variety of flexible schemes.

- Security model: Spells out how information is secured against unauthorized access. Extensible authentication allows clients and servers to prove their identity to one another. Confidentiality and integrity also can be implemented, safeguarding the privacy of information and protecting against active attacks such as connection hijacking.

- LDAP functional model: It determines how clients access and update information in a LDAP directory, as well as how data can be manipulated. LDAP offers nine basic functional operations: add, delete, modify, bind, unbind, search, compare, modify distinguished name, and abandon. Add, delete, and modify govern changes to directory entries. Bind and unbind enable and terminate the exchange of authentication information between LDAP clients and server, granting or denying end-users access to specific directories. Search locates specific users or services in the directory tree. Compare allows client applications to test the accuracy of specific values or information using entries in the LDAP directory. Modify distinguished name makes it possible to change the name of an entry. Abandon allows a client application to tell the directory server to drop an operation in progress.

- LDAP protocol: Defines how all the preceding models and functions map onto TCP/IP. The protocol specifies the interaction between clients and servers and determines how LDAP requests and responses are formed. For example, the LDAP protocol stipulates that each request is carried in a common message format and that entries contained in response to a search request are transported in separate messages, thus allowing the streaming of large result sets.
- Application program interface (API): Details how software programs access the directory, supplying a standard set of function calls and definitions. This API is widely used on major development platforms running C, C++, Java, Javascript, and Perl.
- LDAP data interchange format (LDIF): Provides a simple text format for representing entries and changes to those entries. The ability helps synchronize LDAP directories. LDIF and the LDAP API, along with scripting tools like Perl, make it easy to write automated tools that update directories.

LDAP directories and operating systems are melding to create intelligent environments that can locate network resources automatically. Examples include:

- Active Directory and Windows NT (Microsoft)
- HP-Unix and LDAP (Hewlett-Packard)
- Sun Solaris and LDAP (Sun Microsystems)
- Irix and LDAP (Silicon Graphics)
- Digital Unix and LDAP (Compaq)

In this new role as operating system add-on, LDAP furnishes a way to locate printers, file servers, and other network devices and services. LDAP makes these services standard, more accessible, and in many cases, more powerful and flexible. LDAP is also starting to play a critical role in network management, where it can be a great help to network administrators. Without LDAP, managers and administrators have to maintain duplicate user information in many specific and separate directories across the network. With LDAP, it is possible to centralize this information in a single directory accessed by all applications (Figure 3.3.12). Of course, replacing key legacy applications with LDAP-enabled ones takes time, but big changes are already underway.

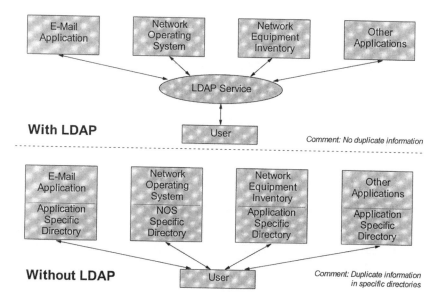

FIGURE 3.3.12   Directory centralization with LDAP.

LDAP also has an important role to play in tighter security, with the directory acting as gatekeeper and deciding who has access to what. In this capacity, LDAP performs two critical jobs. First, it serves as an authentication database. Second, once the identity of a user has been established, it controls access to resources, applications, and services using stored policies and other information. LDAP also permits corporate networkers to use their directories to implement PKI (public key infrastructure) security. From the user's point of view, LDAP provides the directory in which certificates of other users are found, enabling secure communication. From the administrator's point of view, LDAP directories are the way in which certificates can be centrally deployed and managed.

3.3.8.2 Limitations of LDAP

LDAP has three major limitations. First, the protocol cannot and will not make relational databases redundant. It lacks the heavy update, transaction processing, and reporting capabilities of these products. Nor does it offer two-phase committs, true relational structure, or a relational query language like SQL. Using LDAP to implement an airline reservation system would be a serious mistake. Second, it is not reasonable to expect that LDAP serves as a file system. Its information model is based on the simple pairing of attributes and values. Thus, it is not well suited to binary large object data that is managed by typical file systems. It is also not optimized for write performance and is unable to furnish byte-range access to values, both critical features of a file system. Finally, it does not have the locking semantics needed to read- and write-protect files. Third, LDAP is not a stand-in for DNS, which may well be the world's largest distributed database. Although LDAP's abilities are more or less a superset of DNS's — whose biggest job is translating names like home.netscape.com into IP addresses — there is a very good argument for not penetrating tasks of DNS; DNS is working fine. Also, LDAP cannot contend with the connectionless transport that DNS usually runs over. Ultimately, LDAP may have a role in managing and augmenting the information found in DNS. For example, it could link contact information to host information, but it cannot take the place of the DNS database itself.

In summary, LDAP has its place among the succesful network management tools.

3.3.9   Summary

There are multiple standards for network management. All of them have advantages and disadvantages, and, of course, also different application and implementation areas. Telecommunications suppliers and customers will have to live with multiple standards. The question is how these standards can seamlessly interoperate. There are basically three alternatives:

- Management gateway: The interoperability is realized by a special system responsible for translating management information and management protocols. Looking at the practical realization of such a gateway, it is important to target the use of OMA for both OSI- and Internet-based management. Many existing object specifications for management could be taken over by the OMA-based management.
- Platforms with multiple architecture: The interoperabiltity is realized by a multilingual platform, understanding multiple protocols. Protocol conversion is not necessary. Management information can be interpreted and transformed by the platform or by applications. Different architectures are supported simultaneously, but without deep integration.
- Agent with multiple architectures: The interoperability is realized at the agent level. In this case, the management agent understands multiple protocols and languages. It requires some intelligence for the agent. If selected, agent software must be implemented in many, practically in all, networking components. This number is considerably higher than in the case of management platforms.

There is a new group — Joint X/Open TeleManagement Forum Inter-Domain Management Group — that addresses in particular the interoperability between OSI–Management, Internet–Management, and OMG–OMA. This type of work takes a lot of time. In the meantime, practical solutions are absolutely

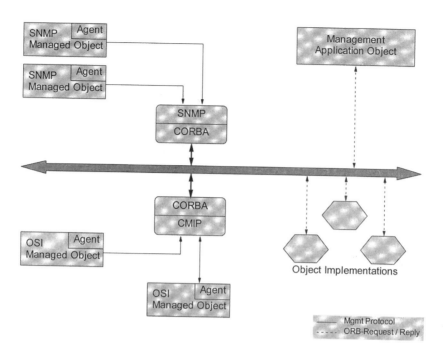

FIGURE 3.3.13 Using an object request broker to connect multiple management protocols.

necessary. In most cases, gateways deliver the quickest solutions. Such a possible solution with management gateways is shown in Figure 3.3.13. CORBA plays in both gateways, related to OSI–CMIP and Internet–SNMP, an important role.

Standardization is absolutely necessary to ensure interoperability of various components of communication systems. This chapter has laid down the basics. Management frameworks and platforms may support some of the standards, but there is no product which supports all of them.

Open database connectivity (ODBC) is an application programming interface (API) allowing a programmer to abstract a program from a database. When writing code to database, the user usually has to add code that talks to a database using a particular language. If the user wants his/her program to talk to an Access, Fox, and Oracle database, code the program with three different database languages. This can cause some problems.

When programming to interact with ODBC, the user only needs to talk the ODBC language (a combination of ODBC API function calls and the SQL language). The ODBC Manager will outline how to contend with the type of database the user is targeting. Regardless of the database being used, all of the calls will be to the ODBC API. All that the users need to do is install an ODBC driver specific to the type of database selected.

Directory enabled networking (DEN) is a specification to save information about network devices, applications and users in central directories. DEN addresses the integration of application and user-level directory information with network control and management, building on open Internet standards, such as LDAPv2 and WBEM/CIM. The CIM initiative is being extended to meet the needs of the DEN intiative. In the future, management applications will have access to authoritative information on the relationships among network elements, services, and individuals, so that preferences, privileges, and profiles can be enforced according to enterprise policies but with personal customization, and so that policies governing network applications can make use of directory-based information.

In summary, information will be consistent in DEN directories and in CIM management systems.

References

HOWE99 Howes, T.: LDAP: Use as Directed, *Data Communications*, February 1999, p. 95-103.

STAL99 Stalling. W.: *SNMP, SNMPv2, SNMPv3 and RMON 1 and 2*, Third Edition, Addison Wesley Longman, New York, 1999.

TERP96 Terplan, K.: *Effective Management of Local Area Networks*, Second Edition, McGraw-Hill, New York, 1996.

3.4   Telecommunications Management Network (TMN)

Endre Szebenyi

3.4.1   Introduction

A telecommunications management network (TMN) is a real, object-oriented, up-to-date, widely applicable network management model, defined by a number of standards and based upon the seven-layer OSI communications model. (Functions and architecture of a TMN are discussed in Paragraph 3.4.3.)

As a consequence of applying the open systems interconnection (OSI) communications protocol architecture, TMN is quite similar to the CMIP-based OSI network management described in Section 3.4.2.2.2. They are, however, not identical: TMN has been developed to be more future-oriented. (More or less, services of the OSI Management can be considered as a subset of TMN services.)

TMN network management standards have been and are conceptualized to satisfy the widest possible range of demands known today, taking streamlining requirements completely into consideration, and with the cooperation of a wide range of standard institutions involved. Institutions participating in standardization of TMN or being connected with these activities include:

- ITU/T
- ETSI
- ISO
- Network Management Forum
- ANSI, etc.

The idea of handling managed objects by network management systems in an object-oriented manner arose in the early 1980s, and later on, the conceptualization of the OSI management also began. Study of the TMN concept was started in 1985 by ITU. ITU Study Group IV aimed at working out a more comprehensive and standardized way of handling intelligent network elements in an object-oriented network management system. As the first formal result, Recommendation M.30 was published in 1988. M.30 summarized the fundamental principles of TMN. Later on, several revisions had been introduced, and M.30 was replaced by Recommendation M.3010 in 1991 [1]. The first recommendation defining the fundamentals was followed by several others discussing every important aspect of TMN, such as network management models, network management functions, standard interfaces and protocols, the way of managing standard network architectures (e.g., SDH), interconnecting networks and network management systems, etc. TMN standards have also been developed and approved by Committee T1 of ANSI.

At the moment, however, the scope of the TMN standards is not complete; their development is still in progress. New standards are planned for intelligent network management and for B-ISDN, etc. We especially have to note that a number of recommendations required to manage ATM networks are also currently being developed.

The series of TMN-related ITU-T Recommendations is illustrated in Table 3.4.1. The most important, relevant ANSI standards include: T1.204, T1.208, T1.210, T1.214, T1.214a, T1.215, T1.224, T1.227, T1.228, and T1.229. The TMN-related ITU-T and ANSI standards generally correspond to or rely on various ISO standards of the series ISO 9595, 9596, 10040, 10164 and 10165.

TABLE 3.4.1 The Series of TMN-related Recommendations

Number of Recommendations	Title of Recommendation
M.3000	Overview of Telecommunications Management Recommendations (supersedes M.30)
M.3010	Principles for a Telecommunication Management Network
M.3020	TMN Interface Specification Methodology
M.3100	Generic Network Information Model
M.3180	Catalogue of TMN Management Information
M.3200	TMN Management Service: Overview
M.3207.1	TMN Management Service: Maintenance Aspects of B-ISDN Management
M.3211.1	TMN Management Service: Fault and Performance Management of the ISDN Access
M.3300	TMN Management Facilities Presented at the F Interface
M.3400	TMN Management Functions
G.773	Protocol Suites for Q-interfaces for Management of Transmission Systems
G.774	Synchronous Digital Hierarchy (SDH) Management Information Model for the Network Element View
G.774.01	Synchronous Digital Hierarchy (SDH) Performance Monitoring for the Network Element View
G.774.02	Synchronous Digital Hierarchy (SDH) Configuration of the Payload Structure for the Network Element View
G.774.03	Synchronous Digital Hierarchy (SDH) Management of Multiplex Section Protection for the Network Element View
G.774.04	Synchronous Digital Hierarchy (SDH) Management of the Subnetwork Connection for the Network Element View
G.774.05	Synchronous Digital Hierarchy (SDH) Management of Connection Supervision Functionality (HCS/LCS) for the Network Element View
G.784	Synchronous Digital Hierarchy (SDH) Management
I.751	Asynchronous Transfer Mode (ATM) Management of the Network Element View
Q.811	Lower Layer Protocol Profiles for the Q3 Interface
Q.812	Upper Layer Protocol Profiles for the Q3 Interface
Q.821	Specifications of Signaling System No. 7. — Q3 Interface Stage 2 and Stage 3, Description for the Q Interface — Alarm Surveillance
Q.822	Specifications of Signaling System No. 7. — Q3 Interface Stage 1, Stage 2, and Stage 3, Description for the Q Interface — Performance Management
Q.824	Specifications of Signaling System No. 7. — Q3 Interface Stage 2 and Stage 3, Description for the Q Interface — Customer Administration — Common Information
Q.824.0	Specifications of Signaling System No. 7. — Q3 Interface Stage 2 and Stage 3, Description for the Q Interface — Customer Administration — Common Information
Q.824.1	Specifications of Signaling System No. 7. — Q3 Interface Stage 2 and Stage 3, Description for the Q Interface — Customer Administration — Common Information
Q.824.2	Specifications of Signaling System No. 7. — Q3 Interface Stage 2 and Stage 3, Description for the Q Interface — Customer Administration — Integrated Services Digital Network (ISDN) Supplementary Services
Q.824.3	Specifications of Signaling System No. 7. — Q3 Interface Stage 2 and Stage 3, Description for the Q Interface — Customer Administration — Integrated Services Digital Network (ISDN), Optional User Facilities
Q.824.4	Specifications of Signaling System No. 7. — Q3 Interface Stage 2 and Stage 3, Description for the Q Interface — Customer Administration — Integrated Services Digital Network (ISDN) — Teleservices

TMN network management systems are conceived to be capable of managing:

- Telephone networks
- LAN and WAN data communication networks
- ISDN networks
- Mobile networks
- Value added and intelligent network services
- Advanced broad-band digital networks, such as:
  - SONET/SDH networks
  - ATM networks
  - B-ISDN networks, etc.

In order to facilitate the operation, practical TMN implementations generally have a graphical user interface (GUI).

The scope of *practical implementations of TMN network management* systems is very limited *as yet.* TMNs are expected to become more widely used in the course of the next decade. In the case of SONET/SDH networks, however, due to the favorable circumstances and special requirements, TMN network management systems are likely to be predominant even in our days. Recent efforts aim at developing network management systems in the scope of which the TMN concept could also be applied to ATM, ISDN, B-ISDN, cellular mobile networks, and other types of communications systems [2–6].

3.4.2 Network Management Key Concepts

3.4.2.1 Evolution

Theory and practice of network management had emerged in the early days of telecommunication, long before sophisticated broadband data communication technologies, such as SONET/SDH, ATM, etc., were developed. Enforced by the growing demand on network quality and the increasing operational costs (more precisely: by the necessity to reduce these costs), management tools in local area (LAN) and wide area (WAN) networks increasingly came into use; moreover, in a relatively simple form, means of network management were applied in the analogue voice communication (telephone) networks as well.

With the increasing volume of computers and other intelligent devices, the interdependent system elements had to be interconnected by different communication tools (data communication lines, LANs, etc.). Large-scale networks with complex topology have evolved in this way. However, if a network does not operate reliably, if possible errors cannot be identified by simple methods, and if operational parameters of the network cannot be kept under control and be modified if necessary, users will not get much benefit from the network established at considerable expense and at great pains.

The development of network management systems was motivated by an intention to eliminate the problems listed above.

Eventually, network management is aimed at allowing authorized personnel to monitor or change important operational parameters of the network by using one or more management terminals. In practice, network management systems are made up of a suitable software, the scope of which basically extends to all the intelligent elements of the network and to some supplementary hardware elements located at the nodes of the network.

The advantage and result of applying a network management system can be summarized as follows:

- Increased safety (decreased time required for error detection, troubleshooting and error correction, increased efficiency of the efforts aimed at error correction, the possibility to reroute network traffic automatically if some parts of the network failed to operate correctly);
- Increased security (controlled and regulated network access pertaining to every user and according to the predefined authentications and authorizations);
- New services provided for the network users (acquiring billing/traffic information, etc.);
- Effective, computerized system monitoring (recording accounts and billing information, tracking and evaluating network load and performance, as well as error statistics, supporting network development strategies, etc.).

In conclusion, the objectives of network management systems are

- Decreasing operational costs
- Improving the quality of services

Achieving these objectives is equally in the interests of the

- Users
- Network operators

- Service providers
- Communications equipment manufacturers

Operating a network management system practically means performing the following essential steps:

a) Acquiring and collecting characteristic data about the elements of the network, about their operation and interrelation in the network (operating conditions, performance, fault conditions and types of the eventual malfunctions, relationship with the neighboring elements, traffic and load parameters, etc.);

b) Storing and evaluating the collected data by the appropriate data processing center of the system;

c) Controlling network operation (modifying functionality of some elements in the network if necessary) as a conclusion and result of the evaluation.

Insofar as a network management system is also required to perform tasks of service and/or business management (see Section 3.4.3.3), additional steps as part of the operation may be necessary, such as:

d) Recording service contracts and managing customer services;

e) Elaborating business plans and technical designs;

f) Modeling and simulating technical and/or financial processes, etc.

Note that a claim to perform operational steps listed under items a) to f) above necessitates the availability of a sophisticated network architecture, the presence of appropriately constructed network elements, and a suitable network management system as well. It means efficient network management techniques can be regarded as results of a joint and inseparable development of both LAN/WAN communications technology and the network management methods themselves. Network management is furnished with more and more effective tools by the evolving communications technology.

As actual network solutions concerns, not all of the above steps have been commonly realized up to now.

Chronology of evolution is represented by the sequence of items a) to f) at the same time. In the first stage of development, "network management" was limited to simply visualizing fault conditions (for example, by lighting appropriate error lamps on the equipment) and supervising them directly by the operating personnel. According to the present sense of the terminology, this type of operation cannot be considered as network management at all. The first significant step in the way to achieve efficient network supervision was by gathering fault conditions and other system parameters by computer(s) and storing them automatically. The availability of the stored data provided a convenient opportunity for their programmed evaluation. The increase in network size and volume of data resulted in the improvement of the evaluation algorithms.

The gathered data and their evaluation then allowed the proper network operation to be restored automatically in the case of certain system faults (e.g., by means of stand-by/hot stand-by system elements or through a predetermined reconfiguration of the system). Controlling network operation by a network management system was only applied in sophisticated LANs for a considerable period of time. Traditional WAN networks, however, based upon PDH technology widely used even today, offered a limited number of means for a suitable network control.

As an effect of the new communication technologies (such as SONET/SDH and ATM) the usual contrast between LANs and WANs, data and voice communication tends to vanish or disappear. This is an important moment promoting the evolution of network management methods and probably resulting in a homogeneous network management theory. Real network management products suitable to accomplish service management, technical as well as business planning, and modeling tasks are appearing nowadays.

TMN is an up-to-date management standard allowing construction of network management systems with the most comprehensive set of network management functions exploiting the facilities of the most up-to-date communications technologies.

3.4.2.2 Standard Network Management Systems

In general, different types of networks (LANs, WANs, ISDN, voice and data communication systems as well as public and private networks, etc.) set different requirements for the network management.

As seen before, in the course of the technical development, management systems became suitable for completing an increasing number of tasks. At the beginning, however, they comprised individual, manufacturer-specific hardware and software components, and, as a consequence, network management systems of different manufacturers developed in specific, diverse directions.

In order to avoid this confusing situation, in accordance with the philosophy of recent open systems, interworking of different networks should be ensured: networks and their management systems have to be operated in a multivendor environment. Consequently, in order to meet this requirement, substantial features of the network management systems (interfaces, protocols, system architecture, etc.) have to be defined by international standards. The importance of standardization is still growing.

The wide range of recommendations elaborated and/or accepted by the major international standards institutes (ANSI, CCITT and ITU-T, which became the successor of CCITT in 1993, ISO, IEC, ETSI, etc. and, especially, the Network Management Forum) aim at users' internationally approved independence from equipment manufacturers. Practically, in spite of the efforts to standardize crucial system characteristics (similar to some other fields of hardware/software solutions), there are several management systems competing with each other on the market showing a limited degree of standardization. Systems based on *de facto standards* are also often regarded as standard systems.

The most frequently used network management systems of the 1990s were as follows:

- SNMP-based management
- CMIP-based (OSI) management
- TMN

3.4.2.2.1 SNMP-based Management

The network management standard, Simple Network Management Protocol (SNMP) referring to the applied protocol, has been defined by the Internet Community for managing networks which implement TCP/IP communications protocol, such as Ethernet LANs and Internet segments. SNMP is a de facto standard, and has been available since 1990. Standardization of its second version (SNMPv2) was requested in 1993 and concluded in 1994 [7-11].

Actually, SNMP is a set of protocol standards defining the rules of information exchange between the entities of the network management software; that is, between the manager, the agent, and the management information base (MIB).

The *manager* is a software element running in the network management station and plays the role of a mediator between the human operator and the network management system itself.

The *agent* is the software element installed in the intelligent managed network elements and represents the managed resources of those elements.

MIB is the logical database containing local network management information residing in each of the agents. The manager is in command of a superset of the agents' MIBs. The MIB may include a number of standard objects. Types of standard objects are arranged (in accordance with their standard name syntax) in a hierarchic tree structure applied in TCP/IP management, making up the structure of management information (SMI). Each node of the tree represents a group of managed objects, and each group includes a number of objects. Each object may have values. The actual value of a managed object reflects the present state of the managed resource represented by the object in hand. Object values should comply with the type definition of the given object. The allowed object types are also defined, such as integer, bit string, etc. The involved standard objects are described with a standardized syntax, defined in ITU-T Recommendation X.208, known as ASN.1 (Abstract Syntax Notation One). It means SMI is similar to that of CMIP-based OSI management, but the objects defined in SNMP are different (see Section 3.4.2.2.2).

The manager may send requests to the agents periodically in order to collect actual information about their state (*polling*), or to modify variables in the agents' MIB. Accordingly, types of managers' requests are *get* and *set*. The agent may also send a *trap* to the manager automatically in order to notify it of an event, if necessary.

In spite of the availability of traps, SNMP is basically a polling-based, not event-driven protocol. It is principally devised to provide network element and network level management functionality. As opposed to OSI network management (CMIP) or TMN, the object model of SNMP is less flexible and less efficient. SNMP cannot be regarded as a truly object-oriented network management system. As the differences between SNMPv1 and SNMPv2 are of concern, one of the improvements is that SNMPv2 (contrasted with SNMPv1) can support communication between manager entities.

One of its advantages, however, is, that it can be easily operated and does not require sophisticated hardware resources. As a practical consequence, in spite of all its disadvantages, SNMP network management systems are also significant competitors of OSI network management, and have been more widely used up to now. As a trend in development of communications technology, it has to be mentioned that efforts are made to ensure interoperability of the various network management systems, such as SNMP, OSI/CMIP, and TMN.

SNMP network management systems are primarily applied for managing:

- LANs and corporate networks based upon TCP/IP protocol
- Internet network segments

Notwithstanding that TMN is expected to be the ideal management system of both SDH and ATM networks, existing ATM network products predominantly allow application of SNMP because of the lack of appropriate TMN standards at the moment.

3.4.2.2.2   CMIP-based Management

The CMIP-based OSI management defines a real object-oriented network management system based upon the seven-layer OSI communications protocol architecture. (The OSI Reference Model is defined in the CCITT/ITU-T X.200 series of recommendations, and in ISO Standard 7498 [12].) The standardization of the OSI protocol model began in the late 1980s, and has not been fully finished up to now. The first OSI management standards were defined by ISO; later they were adopted and developed by CCITT/ITU-T (X.700 series of recommendations) and other standards institutes [13, 14].

CMIP-based OSI network management systems can be applied to manage:

- Local area networks (LANs)
- Corporate networks and private wide area networks (WANs)
- National and international networks

The seventh (application) layer *protocol applied by the OSI management is the Common Management Information Protocol (CMIP).* In a CMIP-based OSI management environment, the user's application process (the operation of which is based upon the manager/agent principle) is provided with the Common Management Information Service (CMIS) as an application program interface (API) by the so-called Systems Management Application Entity (SMAE), which is implemented in the seventh (application) layer of the seven-layer ISO/OSI communications model [15, 16].

(Management operation is based upon the manager/agent principle. The functions of manager, agent and MIB are basically similar to those of SNMP, or more characteristically to those of TMN. See Sections 3.4.2.2.1 and 3.4.3.5, respectively.)

The following are major elements of the OSI management application program interface:

- Common management information service element (CMISE)
- Remote operation service element (ROSE)
- Systems management application service element (SMASE)

- Association control service element (ACSE)
- File transfer, access, and management (FTAM)

The *CMISE* is responsible for generating basic standard requests and processing answer messages as defined by the CMIS. CMIS may be used by an application process in a centralized or decentralized management environment to exchange information and commands for the purpose of systems management. (See ITU-T Recommendation X.710.)

The *ROSE* controls and supervises interactions between remote entities of a distributed application, where these interactions can be modeled and supported as remote operations. A remote operation is requested by one entity; the other entity attempts to perform the remote operation and then reports the outcome of the attempt. (See ITU-T Recommendations X.219 and X.229.)

The *SMASE* provides systems management services in support of specific management functions. (See ITU-T Recommendation X.750.)

The *ACSE* is responsible for performing initial negotiation in order to decide if data connection can be established and made available for data communication. According to the definition of Recommendation X.217: "ACSE provides basic facilities for the control of an application association between two application-entities. The ACSE includes two optional functional units. One functional unit supports the exchange of information in support of authentication during association establishment. The second functional unit supports the negotiation of application context during association establishment." (See ITU-T Recommendations X.217 and X.227.)

FTAM organizes and manages file access for application purposes, according to the specifications of ASN.1. (See ITU-T Recommendation X.209.)

In terms of the manager/agent principle, the manager as a software system element may send management operations in the form of CMIS requests to any of the software agents via the CMIP management protocol. The agent forwards these requests to the pertaining managed objects (MOs) which represent the physical and logical resources to be managed, and executes them on the appropriate MOs.

CMIP may also provide event-driven reports for the manager and can identify substantial events that influenced the state of the managed objects.

CMIP/CMIS requests, issued by the manager to the agent in order to initiate an operation, are the following:

- M-Get   (gets an attribute value of one or more managed objects)
- M-Set   (sets/modifies the attribute value of one or more managed objects)
- M-Action   (performs a specific action on one or more managed objects)
- M-Create   (creates a new managed object)
- M-Delete   (deletes one or more managed objects)
- M-Cancel-Get   (cancels a previously requested and currently outstanding M-Get operation)

In addition, the agent may send to the manager:

- M-Event-Report(notifies the manager of an event that occurred in the managed object)

Selection of managed objects that have to be affected by a given CMIP/CMIS operation is facilitated by scoping and filtering. Scoping entails the identification of the managed objects to which a filter is to be applied. Filtering entails a set of tests to each member of the group of the previously scoped managed objects to extract a subset of them.

Object model of the OSI management is based upon CCITT/ITU-T Recommendation X.722, widely known as GDMO (Guidelines for the Definition of Managed Objects). Structure of Management Information (SMI) is described in CCITT/ITU-T Recommendation X.720 [17, 18]. SMI is similar to that of SNMP; however, the involved standard objects and their attributes are different. The standard objects in OSI management are also described by using ASN.1 syntax, defined in ITU-T Recommendation X.208.

CMIP is a real event-driven protocol; the GDMO object model is more comprehensive than that of SNMP. It means that OSI management is more suitable for managing large and complex networks.

Application of CMIP-based OSI network management systems is spreading gradually. Their significance is expected to grow in the future (particularly for telecommunications service providers). However, TMN, as a more comprehensive, OSI-based, standardized management system seems to be a strong competitor of CMIP.

3.4.3 Functions and Architecture of a TMN

The functions and architecture of a TMN can be considered in several dimensions. Each dimension is defined (or will be defined) by existing (or developing) standards.

ITU-T Recommendation M.3010 discusses

- Functions associated with TMN
- Aspects of TMN architectures
- TMN Logical Layered Architecture

Functions associated with a TMN are classified in terms of the OSI Management; accordingly, five management functional areas are defined in Section 3.4.3.1.

As general TMN architecture concerns, three basic aspects are considered in ITU-T M.3010:

- TMN functional architecture
- TMN information architecture
- TMN physical architecture

In addition, M.3010 describes four plus one management layers.

All of these aspects will be discussed here in a didactic sequence, differently from their order in M.3010.

3.4.3.1 Functional Architecture

The TMN functional architecture is based upon a number of TMN function blocks. These represent the appropriate functions required by the TMN to fulfill its general function of network management. These are actually performed by the elements of the physical architecture of TMN (see Section 3.4.3.2). According to Recommendation M.3010, some of the function blocks are partly in and partly out of TMN. These function blocks are the workstation function block, the Q adaptor function block, and the network element function block. It means that these function blocks (besides those functions defined by the Recommendation M.3010) may include a range of functionality (and provide them for the TMN) not covered by TMN recommendations.

Recommendation M.3010 defines five function blocks:

- Operations systems function (OSF) block
- Network element function (NEF) block
- Workstation function (WSF) block
- Mediation function (MF) block
- Q adaptor function (QAF) block

Data communication requirements of these function blocks are satisfied by the

- Data communication function (DCF)

The *OSF block* processes information related to telecommunications management for the purpose of monitoring/coordinating and/or controlling telecommunication functions including management functions.

The *NEF block* communicates with the TMN for the purpose of being monitored and/or controlled. The NEF also includes telecommunications functions that are the subject of management, but not part of the TMN. These functions are represented to the TMN by the NEF.

The *WSF block* provides the means to interpret TMN information for the user, and vice versa. Parts of the WSF may also be located outside of the scope of TMN.

The *MF block* acts on information passing between an OSF and a NEF or QAF to ensure that the information should conform to the expectations of the function blocks attached to the MF.

The *QAF block* is used to connect a non-TMN compatible, NEF-like function block with an OSF, or a non-TMN compatible, OSF-like function block with a NEF. It means that information between a TMN-compatible function block and a non-TMN-compatible function block will be translated by the QAF. A given part of the functions of the QAF is definitely out of the TMN.

Each of the above function blocks is composed of basic functional components, which have been identified as the elementary building blocks of the TMN.

DCF is applied to transfer information between the TMN function blocks.

Pairs of the TMN functional blocks exchanging management information are separated by reference points; that is, reference points are boundaries between two management function blocks.

Three classes of reference points are defined in terms of TMN, namely:

- Reference points q
- Reference points f
- Reference points x

In addition, two further classes of non-TMN reference points are considered:

- Reference points g
- Reference points m

Definition of these reference points will be explained in Section 3.4.3.2, in connection with the physical architecture of TMN.

3.4.3.2 Physical Architecture

TMN physical architecture describes physical building blocks (physical elements) of a TMN network management system and contains exact rules for their relationships. The relating standard is CCITT/ITU-T Recommendation M.3010.

M.3010 defines the physical elements of TMN as:

- Operations system (OS)
- Workstation (WS)
- Network element (NE)
- Data communication network (DCN)
- Mediation device (MD)
- Q adaptor (QA)

Furthermore, M.3010 also defines rules of data interchange between physical elements as

- Standard interfaces

Standard interfaces define the

- Protocol suite
- Messages carried by the protocol

The *OS* performs operations system functions (OSFs, see Section 3.4.3.1). It practically is the heart itself, responsible for managing the network by controlling the operation of their elements. As its realization concerns, the OS is basically made up of one or more data processing centers, performing the task of gathering information from the network elements and processing them according to the functions of the network management system. (Suitably, the operations system meets open systems' requirements.)

Connected to one and the same network, more than one operations system may participate in the management process. Based upon the principle of task division, they may perform a predefined set of tasks and be interconnected through the data communication network.

Workstations perform Workstation Functions (WSFs). They are the physical representations of the necessary man-machine interfaces by means of which the operators can communicate with the TMN. Workstations are preferably computers themselves, equipped with efficient graphic capabilities in order to be able to meet the operators' requirements.

NEs perform Network Element Functions (NEFs). NEs are elementary, manageable telecommunication devices (e.g., switches, multiplexers, cross-connects) situated at the nodes of the network to be managed. They provide for the appropriate functions in network operation, and usually can be identified with a single address by the operations system of the TMN. The pieces of equipment forming network elements are intelligent devices controlled by their own microcomputers and equipped with standard interfaces. Through the standard interface the network element can transfer messages toward the operations station and inform it about the elements' actual state and receive control commands from it. Generally, network elements are devices meeting the requirements of TMN recommendations; however, network elements not conforming to TMN standards can also form part of the telecommunication network.

The *DCN* is a network that performs data communication function (DCF). It transmits messages required to perform management functions between the OS and the NEs. Information is exchanged through the DCN, using standard protocols as determined by standard interfaces. CCITT/ITU-T Recommendation M.3010 defines the data communication network of a TMN network management system in abstract terms. In practice, this network can be realized separately from the managed telecommunication network (e.g., it can be made up of leased lines, make use of a public data network etc.), or even (partly or entirely) by the managed network itself. If the DCN serving for management purposes is realized independently from the network managed, this will result in the advantage that troubles in the managed telecommunication network will not deteriorate functionality of the management system. An obvious disadvantage of this method as compared to the second one is higher investment and maintenance costs and a more complex overall system. A practical way of realization must always be determined with the requirements met in the possible highest degree. Management DCNs are often realized by the managed network (due to the relatively simple structure and the mostly small size) in the case of LANs, and it is almost the exclusive solution for networks built up of SDH/SONET technology. This latter is justified by the high reliability of the SDH/SONET networks (ensured by the reliable network elements and by redundant network topologies), and furthermore by the availability of communication channels reserved for management purposes in the SDH/SONET frame structure.

Mediation Devices perform mediation functions (MFs). They may adapt the information passing between the OS or the DCN and those TMN-compatible elements located in the network, which still require appropriate operations (storage, adaptation, filtering, etc.) to be performed on the exchanged data. It should be noted that confusion may often be observed concerning the interpretation of MF. Users are namely inclined to refer to the QA as a mediation device.

QAs perform Q adaptor functions (QAFs). They accomplish information exchange between the OS or the DCN applying standard protocols and the eventual non-standard NEs located in the network. For instance, mediation devices may establish interconnection between a standard DCN and the connected non-standard network elements with Q_x interfaces.

Standard interfaces define the ways and rules of information exchange that can be carried out through the reference points of a TMN network. Standard interface definitions include descriptions of the hardware interfaces, descriptions of the communication protocols as rules of data exchange, as well as the information model; that is, the system-level strategy of interworking between elements taking part in communication. Each of the interface definitions involves the definition of the appropriate *protocol families* and protocols.

Reference points are abstract concepts representing theoretical boundaries between the physical elements of a TMN. There are a number of well-defined standard reference points in a TMN network. They are indicated by lower-case letters, whereas the appropriate capitals symbolize the corresponding interfaces. The reference points and interfaces defined in TMN are:

- Reference point "q" may be located between any two of the following function blocks: OSF, QAF, MF, and NEF. It means reference point "q" may be found between either the OS and a NE, or between the OS and a QA, or between the OS and a mediation device, or between any of the above elements and the DCN or between two OSs belonging to the same TMN. The interface "Q" allows data exchange through reference point "q." At present, two types of "Q" interfaces are distinguished; these are interfaces Q_3 and Q_x. The appropriate "Q" protocol suites are the Q_3 and Q_x protocols. "Q_3" protocols are complete implementations of the seven-layer OSI reference model, whereas "Q_x" protocols only comply with the three bottom layers of the OSI model. Q_3 is applied to connect functional elements being fully TMN-compatible, while Q_x can be used in special cases when the Q_3 interface cannot be implemented (due to the presence of any equipment not complying with TMN).
- A reference point "f" is located between function blocks OSF or MF and a WSF. It means that reference point "f" can be found between the DCN (or a mediation device) and a WS connected to it. The corresponding interface "F" allows data exchange through reference point "f" (that is, the interface "F" is applied in cases when the WS is not connected directly to the OS, but through the DCN). The appropriate protocol suite is the "F" protocol.
- A reference point "x" is located between OSFs of two TMNs or between the OSF of a TMN and the OSF-like functionality of another, non-TMN network. It means that a reference point "x" may be found between the DCNs of two TMNs, or it lies between a TMN and another managed system not complying with the TMN standards. A corresponding interface "X" allows data exchange between the two network management systems. The appropriate protocol suite is the "X" protocol.

In addition, Recommendation M.3010 defines two further reference points, lying outside of the area of the standard TMN elements:

- Reference point "g" is located between the human user and the WSF
- Reference point "m" is located between the QAF and an element that does not conform to TMN recommendations

The specifications of standard interfaces, primarily those of the family of interface "Q" and the corresponding protocols, are highly significant regarding the operation of TMN. A number of CCITT/ITU-T recommendations deal with the specification of TMN protocols (M.3020, M.3300, Q.811, Q.812, etc.). It should be noted, however, that the scope of standards specifying standard interfaces, including the "Q" interface, is not complete yet.

An example of a simplified physical architecture for a TMN as demonstrated in the ITU-T Recommendation M.3010 is shown in Figure 3.4.1. A more illustrative representation of the physical architecture of a TMN can be seen in Figure 3.4.2.

3.4.3.3 Logical Layered Architecture of a TMN

According to the TMN terminology, OSFs of the network management are broken down into four hierarchy layers. Each layer of the given hierarchy defines an appropriate group of management operations. The layers are built upon each other; they (and the appropriate operations) are closely interrelated.

TMN Standard M.3010 defines the following four layers of the OSF:

- Network element management layer
- Network management layer
- Service management layer
- Business management layer

OSFs in these layers interact with OSFs in the same or other layers within the same TMN through a reference point "q_3," and with ones of another TMN through reference point "x."

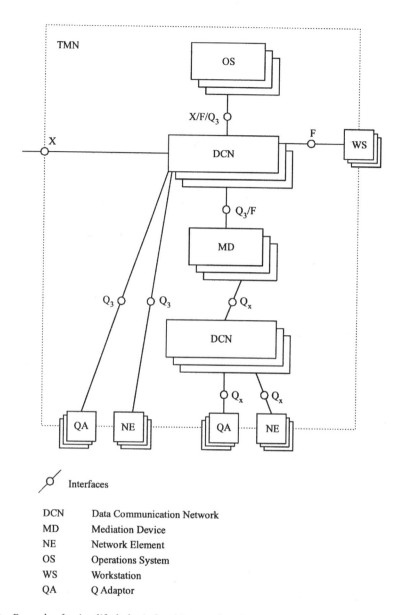

Interfaces

DCN	Data Communication Network
MD	Mediation Device
NE	Network Element
OS	Operations System
WS	Workstation
QA	Q Adaptor

FIGURE 3.4.1 Example of a simplified physical architecture for a TMN, according to ITU-T M.3010.

Furthermore, as network element management is based on the data collected about the respective elements of the network, *network elements* themselves *can be considered as the lowest layer in the hierarchy.* As contrasted with the four OSF layers, layer of the network elements does not involve OSF, but it is concerned with the NEF.

Let us briefly review the functions included in the given layers of the hierarchy, from the bottom to the top.

Network elements are basic components of the managed network, installed as physical devices, specified by standard functions and interfaces, capable of delivering data on their operation, and providing means to be controlled in a specified way by the management system. The concept of NEs is clearly defined by TMN standards (see Section 3.4.3.2).

The network element management layer manages each network element on an individual or group basis. NE management includes gathering data on each of the network elements and controlling them

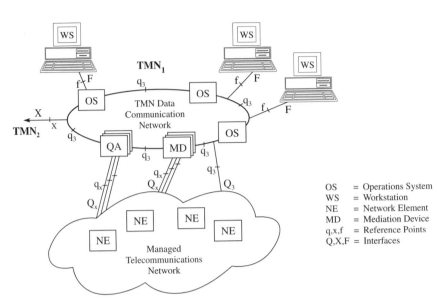

FIGURE 3.4.2 Illustrative representation of the physical architecture of a TMN.

individually. At this layer, decisions on changing the state of any individual NEs must rely on information about the same element, and cannot depend on the state of any other NEs or that of the entire network. Basic fault management as well as performance management operations, such as monitoring and displaying fault conditions or traffic performance of any single network elements, as well as taking elementary actions to eliminate an error (e.g., switching over to an auxiliary channel inside of the same network element, etc.) are performed by the NE management layer. (See also Sections 3.4.1, 3.4.2, and 3.4.3.)

The network management layer has the responsibility for the management of the whole network as supported by the element management layer. This layer transgresses competence of network element management and is responsible for the interconnection and cooperation of all the network elements in the managed system. Tasks of this layer include configuration management (both static and dynamic; see Section 3.4.3), event, fault, and performance management on the network level (all these employing a system approach, e.g., by applying error/performance correlation and evaluation algorithms), as well as security management (monitoring user requests and taking appropriate actions in order to prevent any unauthorized access to the network).

Service management is concerned with and responsible for the contractual aspects of services provided to customers or available to potential new customers. Service management aims at establishing relations between services provided by the network and requirements of the users or customers. Clients and service contracts are recorded, customers and the appropriate service parameters are related, quality of service is traced, clients' complaints are registered and reported, and new orders are accepted and processed, etc. at this management layer.

Business management has responsibility for the total enterprise. It deals with technical and business concerns as an organic complex of the network operators' activity, and has responsibility for the total enterprise. Functions included in this layer are billing and accounting, maintenance management, cost control, controlling spare parts inventory, designing new network elements and/or new network services, technical modeling and optimization, and planning and evaluating profitability of new investments, etc. The business management layer comprises proprietary functionality. In order to prevent access to their functionality, OSFs in the business management layer do not generally have "x" reference points and "X" interfaces. It is included in the TMN architecture to simply facilitate the specifications required of the other management layers. Nevertheless, it may still be necessary that the business management layer interacts with other management or information systems. The task of these interactions should be solved by special software solutions.

Logical layers of the management hierarchy are not defined by the existing standards in full detail. Tasks and processes of the different layers and rules of data exchange between them are not exactly determined at present.

The structure of layers is schematically covered by recommendations M.30 and M.3010. In technical literature, discussing functionality of network management is often confined to simply listing management functions without detailing their hierarchy.

In present practice, standard solutions are mostly restricted to realize network element management and network management layers. Frequently used, standard management functions (see Section 3.4.3.4) essentially correspond to the tasks of these two management levels. As yet, relatively few companies provide actual solutions for service management. Rather, network management products suitable for performing tasks of business management are now mostly under development.

As a consequence, existing tools promising to solve higher level tasks of network management can be regarded as manufacturer-specific, proprietary software products.

The layered architecture of the operations system functions as defined in ITU-T Recommendation M.3010 is shown in Figure 3.4.3.

In technical references, logical layers of the management functions are often illustrated by a pyramid (see Figure 3.4.4), indicating that the greatest amount of elementary data can be found at the bottom level, but the degree of their complexity (as they are processed) increases by going up through the layers.

It is perhaps not useless to note that a remarkable similarity may be established between the pyramid-shaped functional network management model and the structure of a typical business management information hierarchy. In general, this latter may also be divided into several layers between the operative and the top management controlling level. It is also obvious that a close interrelation may be required between business-level network management and the upper controlling level of the relevant business management information system. (Interrelating functions are billing, accounting, cost control, and inventory management, for example.) Establishing on-line interconnection between TMN network management systems and business management information systems may represent one of the most important future developing efforts on the side of software manufacturers of both product families (see Section 3.4.6).

3.4.3.4 Functions Associated with a TMN

As a management functional area of TMN concerns, Recommendation M.3010 lists five management functions and refers to OSI Recommendation X.700 [13]. TMN network management functions are classified by ITU-T Recommendation M.3400 in more detail (see Table 3.4.1). (Remember that streamlined methods of network management are results of a natural evolution, and obviously most items defined as standard TMN functions can also be found in network management systems and standards established before the appearance of TMN. Still, as a new aspect of structuring systems architecture, management functionality is divided into hierarchic layers in terms of TMN.)

Referring to Section 3.4.3.3, we have to point out that existing TMN standards with respect to the layer definitions are rather schematic — especially regarding functions belonging to the higher management layers. Regardless, all data gathered in the scope of the management functions defined by ITU-T M.3400 and listed below may also be related to the functions of the higher management layers, at the least in a condensed and evaluated form. Still, they represent tasks that basically have to be implemented just at the network element and network management layers. As an exception, accounting management may be strictly related to the higher management layers in the architecture.

Standard TMN management functions according to the definitions of Recommendation ITU-T M.3400 are:

- Performance management
- Fault (or maintenance) management
- Configuration management
- Accounting management
- Security management

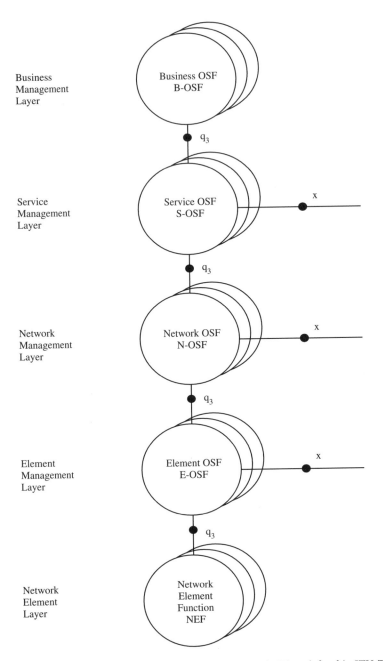

FIGURE 3.4.3   Layered architecture of the operations system functions (OSF) as defined in **ITU-T M.3010**.

Although not considered as an independent management function in itself, and TMN standards do not cover it at the moment, the process of downloading programs from the network management system to the intelligent network elements through the data communication network is closely related to network management functions.

3.4.3.4.1 *Performance Management*
Performance management (sometimes also referred to as traffic and performance management) provides functions to evaluate and report upon the behavior of telecommunication equipment and the effectiveness of the network and/or network elements. It may involve measuring the intensity of data flow along

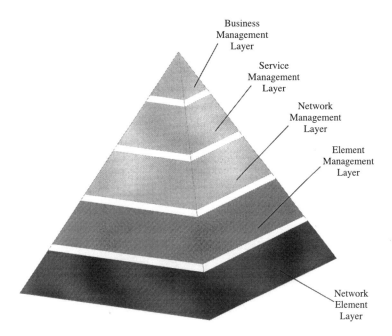

Business Management Layer

Service Management Layer

Network Management Layer

Element Management Layer

Network Element Layer

FIGURE 3.4.4 Pyramidal illustration of TMN Logical Layered Architecture.

the different routes of the network, collecting, evaluating, and displaying data measured in this way, as well as determining efficiency indices and calculating trend analysis. Information gathered and evaluated in this process can be utilized similarly to and in connection with data gained in the scope of the fault management. On the basis of these data, the level of traffic load can be established and it can be determined whether a given network complies with the necessary performance requirements. (If any congestion occurs, overloaded network routes may be relieved by system reconfiguration or by altering the actual routing strategy. Automatic intervention in network operation can be performed by the network management system. If a permanent lack of network capacity has been observed, the decision should be made to start new investments and increase network capacity (see also Section 3.4.2).

3.4.3.4.2 Fault Management

The *fault (or maintenance) management* (sometimes also referred to as event and error management) is a set of functions which enables the detection, isolation, and correction of abnormal operation in the telecommunication network and its environment. Its purpose is to detect and record events that have occurred in different parts of the network, then to establish the cause of these events with the most possible details and accuracy. All these are aimed, first of all, at exploring errors and being able to repair them in the shortest time possible.

Fault management may be confined to simply recording alarm states originating from the separate network elements and generating appropriate error messages in order to inform the operator. A more sophisticated and effective method is to correlate these individual events and evaluate their correlation by means of appropriate algorithms. Correlation algorithms may be needed to identify causes of malfunctions and locate their sources precisely in complex situations. That is, one elementary error may generate a number of error reports — and, oppositely, one error report may refer to several events or alarm states (equipment breakdown, cable rupture, traffic congestion, etc.). If necessitated by the amount or type of failures, automatic loopback tests or other special test routines will be started in order to locate faulty system elements. Analyzing interrelations of the elementary alarm states and evaluating results of the test routines may result in a precise diagnosis even if individual methods have been unsuccessful.

As a result of applying event and error management, a great part of malfunctions can be discovered and eliminated before they may cause any noticeable problems for the users. In some cases, the impact of the emerging failure can be eliminated by an automatic and dynamic reconfiguration of network routes (see also Section 3.4.3). Automatic reconfiguration may generally be effective in meshed networks and (if just a single cable rupture occurred) in a ring network configuration or in case of traffic congestion, etc.

Fault management is also concerned with keeping statistics. Statistics may assist the operator in deciding if a network complies with the appropriate requirements (reliability, performance, etc.) or not.

3.4.3.4.3 Configuration Management

Configuration management provides functions to exercise, control, identify, and collect data from or provide data to Network Elements.

In practice of managing up-to-date, broadband telecommunication networks, *configuration management* generally includes two essential, logically different functions. In particular: static and dynamic configuration management.

Static configuration management involves assigning and unassigning network elements to or from the network logically ("attach" and "detach"), as well as recording, indicating, displaying, and reporting network topology and lists of network equipment along with their system parameters, such as type, topological location, symbolic and physical addresses, etc. In case of small and simple networks (along with recording equipment parameters), static configuration management may also involve *inventory control*; however, in the case of complex networks, this function should be handled separately.

Dynamic configuration management involves establishing the actual routes for the required interconnections via the network. It implies network reconfiguration by establishing a new possible route, if the actual route breaks down, or taking down the route if a request arrived requiring that the connection has to be canceled. In terms of dynamic configuration management, displaying network topology has to reflect the actual routes and connections in the network.

3.4.3.4.4 Accounting Management

Accounting management (sometimes also referred to as billing and accounting management) is a set of functions that enables the use of the network service to be measured and the costs for such use to be determined. Accounting management should provide facilities to collect accounting records and to set billing parameters for the usage of the service.

In the scope of accounting management, time and other characteristics of users' network access is measured and charging data are calculated on the base of several parameters (price lists, subscriber contracts, time of use, utilized services, etc.). Billing and accounting information is collected, classified and recorded. On the base of charging, data bills can be prepared and sent to the customers, income can be calculated and recorded, etc.

3.4.3.4.5 Security Management

Security management involves establishing classes of authentication, checking users' authorization to network access, controlling passwords, and taking other possible measures to prevent the network from any unauthorized access. It may be a special task to protect the management terminals from unauthorized interventions according to the given security requirements. Depending on the special requirements given in accordance with the purpose of a network, functions contained within security management may differ from application to application.

3.4.3.4.6 Downloading

Downloading new software versions, routing tables, cross-connect tables, or other program segments from the management center to the intelligent network elements constitutes an important task of a network management system. Software downloading provides some means to perform remote control of network elements as well. By way of software downloading, for instance, the operation of a network element can be modified, a new software version can be put into use, or the configuration table stored in a network element can simply be modified. One of the practical ways to solve configuration management may be to modify configuration tables by software downloading. While specifications of software downloading in LANs are included in IEEE Standard 803.1, there is no general recommendation for this way of remote control of managed TMN network elements at the moment. (For management of SDH, ITU-T Recommendation G.784 assigns this item for further study.)

3.4.3.5 Information Architecture

The general characteristics of the information model of TMN are discussed by CCITT/ITU-T Recommendations M.3010 and M.3100. Further recommendations deal with the information models of managing SDH and PDH networks (e.g., Recommendation G.774).

Essentially, network management involves the exchange of information between management processes. The TMN network management information model relies, to a great extent, on the network management model OSI/CMIP. The information architecture of TMN is based upon an object-oriented model, applies transaction-oriented information exchanges, and utilizes the so-called agent/manager principle.

The basic concepts used in the definition of the TMN information architecture are similar to those applied in SNMP and OSI/CMIP (see Sections 3.4.2.2.1 and 3.4.2.2.2). They are:

- Managed object (MO)
- Agent
- Manager
- Management information base (MIB)

MOs are abstractions of the *physical or logical* resources to be managed in order to monitor operation of the network and to prevent disorders in network operation. A managed object does not generally correspond to a single network element as defined to be a part of the physical architecture (see Section 3.4.3.2). In most cases, a managed object represents one of the important components of a physical network element (it may be the control unit of a given circuit) or a logical resource (e.g., the status of a basic physical element). Every managed object must have a single unique name; its actual condition is a function of time.

A managed object is defined by

- Its attributes visible at its boundary
- The management operations that may be applied to it
- The behavior exhibited by it in response to management operations
- The possible notifications or messages it may emit during operation

A *manager* is a system element whose task is sending management requests toward the agents for control, coordination, and monitoring purposes, performing operations on the agents by the aids of these requests and receiving messages emitted by the managed objects and forwarded by the agents to the manager. In practice, the manager's part is played by a workstation (that is, the pertaining network management software running in the workstation) which in turn is a part (physical element) of the network management system (see Section 3.4.3.2).

An *agent* is a system element toward which management commands are directed for control, coordination, and monitoring purposes. Agents perform operations on managed objects according to the manager's requests, and forward messages emitted by the managed objects to the manager. In practice, the agent's function is performed by an intelligent network element of the network management system (that is, by the program running in it).

By definition, a many-to-many relation between Managers and Agents may exist. It means that one manager may be involved in the information exchange with several agents, and one agent may exchange information with several managers.

The *management information base* (MIB) is a database containing data pertaining to the managed objects of the system. Similar to the concept applied in the case of SNMP and OSI/CMIP, each of the agents has its own MIB, the manager is in command of a superset of the agents' MIBs (see Section 3.4.2.2.1). The set of rules describing the structure of a management information base is called structure of management information (SMI).

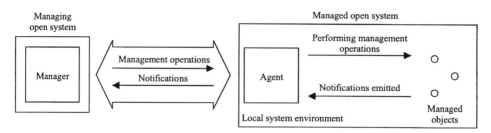

FIGURE 3.4.5   The interaction between manager, agent, and managed objects according to ITU-T M.3010.

The interaction between an agent, a manager, and the managed objects according to ITU-T M.3010 is shown in Figure 3.4.5.

Manager and Agent communicate using standard "Q" protocols built up according to the seven-layer OSI communication model.

The components of a Q protocol are:

- Application interface (command/answer structure)
- Application protocol (the seventh layer of the OSI model)
- Support protocols (layers 4–6 of the OSI model)
- Network protocols (layers 1–3 of the OSI model)

Essential elements of the TMN application interface are highly similar those of the OSI/CMIP management (see Section 3.4.2.2.2). They are:

- Common management information service element (CMISE)
- Remote operation service element (ROSE)
- Systems management application service element (SMASE)
- Association control service element (ACSE)
- File transfer, access and management (FTAM)

The arrangement of Q protocol layers in the seven-layer OSI communication model is shown in Figure 3.4.6.

3.4.3.5.1   The OSI-based Object Model

The object model gives formal description of standard objects applied in the system and defines their relationships. Objects are grouped together to form object classes: members belonging to the same class will have the same characteristics as the given respect of classification concerns. Furthermore, object classes can also be grouped together to form more general object classes. As a consequence, standard objects are arranged in a hierarchical tree structure. Each node of the tree represents a group of managed objects, and each group includes a number of objects.

Basics of TMN object model and information model are described in ITU-T Recommendation M.3010; a catalogue of TMN object classes is given in ITU-T Recommendation M.3180 (see Table 3.4.1). Together with basic object class definitions, M.3010 defines the agent/manager relationship as shared management knowledge (SMK).

In practice, object models, applied in presently used TMN management systems highly rely upon OSI management principles, namely on CCITT/ITU-T Recommendations X.722 (GDMO) and X.208 (ASN.1). See also Section 3.4.2.2

3.4.3.5.2   Distributed Object Models and Service Management

The evolution of communications technologies is reflected by the evolution of TMN and further by the evolution of its object model. Focusing on the second one and considering the hierarchical layers of network management described in Section 3.4.3.3 (representing the most illustrative dimension of this

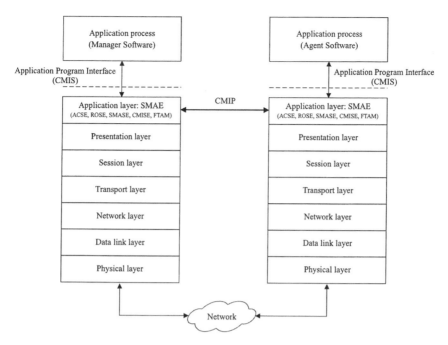

FIGURE 3.4.6 Q protocol layers in the seven-layer OSI model.

evolution), the first steps going up this hierarchy were modeling simple network elements then defining whole network models. The next grade that should now be mastered is service management. One of the main difficulties in this developing phase is that service management sets have different requirements for the object model than management on the element and element management level.

From the point of view of the object model, the main differences between network elements and telecommunication services are that (not intending to be exhaustive): the number of NEs is generally much higher than that of network elements; services are, however, more complicated, more dynamic than NEs, and are of a distributed character while network elements are not.

According to the first two developing steps discussed above, the current OSI-based object model (GDMO) is static, built up using a centralized system approach, and basically preconditions central intelligence. Most of the operations performed on GDMO objects using CMIS/CMIP are relatively simple; these operations generally relate to a group of objects and their extent is determined by the given filter and scope conditions. However, they are very effective in handling a large number of relatively simple objects as required by network element and network level management, of whose requirements they have been designed to meet.

Recent telecommunications technology, taking end-to-end customer service for example, represents a complex, integrated environment which necessitates the availability of service management, and as such, requires interoperability, flexibility, and distributed object processing of the network management system. Service management involves solving the problem of how to integrate and manage different services like telephone, video, Internet, and CATV, and how to manage services passing through the domains of different telecommunications network and/or service providers being geographically dispersed, probably having different service technologies, applying various business and traffic policies, etc.

GDMO is not really suitable for modeling distributed management objects and for realizing the appropriate management functions that are necessary for service management.

To override this problem and to meet the requirements of a distributed system approach, ITU-T defined the Reference Model for Open Distributed Processing (RM-ODP) and Open Distributed Management Architecture (ODMA) in Recommendations X.901-4 and X.703, respectively [19, 20]. RM-ODP lays down the principles and gives basic definition of a distributed system; it does not involve directives for practical implementation. ODMA is aimed at developing OSI management toward a distributed processing environment.

The TeleManagement Forum also tries to standardize an object model for service management. One of the results of this developing effort was OMNIPoint (Open Management Interoperability Point), a standard solution providing object class definition and specifying an integrated management architecture [21, 22]. Recently, the Forum turned its attention to the latest solutions in the field of network management technology, such as integrating CORBA and TMN.

CORBA (Common Object Request Broker Architecture) is a solution for distributed, object-oriented data processing, developed by the Object Management Group [23]. CORBA 1.1 was introduced in 1991, whereas CORBA 2.0 was adopted in 1994. By now, CORBA has become a *de facto* standard in this field.

CORBA defines a general framework for distributed, object-oriented program development. (Network management is one of its possible applications.) CORBA includes an object model; the possible operations on them are furthermore defined and the way these operations may be performed is specified. CORBA also defines a programming language-independent Interface Definition Language (IDL) and the appropriate programming interfaces. The communication mechanism meeting the requirements of the managed objects (as a counterpart of CMIS/CMIP in CORBA) is provided by the Object Request Broker (ORB).

As proven both in theory and by some realized software models, CORBA may effectively be applied for communications service management. It does not mean that traditional OSI object models, such as GDMO (together with CMIS/CMIP) would have to simply be replaced by CORBA at any time. Rather, they should be integrated — GDMO being used for network element- and network-level management — while CORBA is used for the management of services. Principal reasons of this argument are that despite all the advantages of CORBA in service management, GDMO is more effective on the element level, and furthermore, economic reasons require that new developments do not disadvantageously influence utilization of existing implementations.

Integration of CORBA and TMN, however, cannot be solved without any difficulties. The CMIP protocol, the appropriate services defined by CMIS and the GDMO object model, are well different from those protocols, services, and object models represented by CORBA. In order to build facilities of CORBA into TMN, the two different object models, the appropriate management functions and the related operations should be combined. Publications report on special software solutions in this field, and simultaneously, the NMF makes efforts to elaborate standard methods to meet this requirement [24–26].

Nevertheless, recent CORBA applications have to be regarded as proprietary solutions, since neither the way of TMN-CORBA integration, nor service-level TMN objects have been standardized up to now.

One of the most recent results in this respect is Telecommunications Information Networking Architecture (TINA), being developed by the TINA Consortium. TINA specifies a technology-independent information model for telecommunications systems, based upon distributed operations. According to all indications, TINA will not get away from being integrated with TMN [27–29].

3.4.4   Interconnecting Managed Networks

3.4.4.1   Interworking of Different Network Management Systems

As a result of recent technical development, telecommunications networks have become globalized; private and public networks, LANs, and WANs cannot be sharply separated. It means that (in most cases) public network services will actually be realized by a number of network and/or service providers, and by using a number of network elements originating from different vendors and manufacturers. A single end-to-end interconnection, satisfying the users' actual demands, may pass through ISDN, SDH, ATM, mobile network, LAN, WAN, and public and private network paths.

This situation drives both private and public network operators to integrate their network management systems to the highest possible degree. Lately, demand on establishing end-to-end management has also arisen.

As technical aspects of the TMN concerns, increase interworking of different network management systems has to be realized.

According to ITU-T Recommendation M.3010, TMN hierarchies may interact for many reasons, such as:

- To manage the interactions required to provide value-added services
- To manage a number of geographical/functional TMNs as a single TMN
- To provide end-to-end services

As described in Section 3.4.3.2, OSFs located in one and the same TMN may exchange information via the "q" reference point, using a "Q" interface, while different standard TMN systems can be interconnected via the "x" reference points and may exchange information by using standard "X" interfaces. Should we have a network management system not complying with the TMN standards, interworking of a TMN and a non-TMN system may theoretically be established through the "X" interface as well. In the latter case, the non-TMN system must be able to communicate through this "X" interface, handle the appropriate protocol, and process data formats according to the MIB of the TMN. In order to do this, however, sophisticated software solutions should be applied.

In practice, efforts have been made to solve the problem of integrating OSI and SNMP, or TMN and SNMP management. Solutions aimed at this purpose involve the appropriate protocol conversion and the translation of management information [30].

3.4.4.2 Virtual Networks, Customer Network Management

TMN network management systems also enable us to establish virtual networks within the communication routes of the network as a whole.

The users of a virtual network will have an experience similar to using a separate network, being independent of the entire communication system by the aid of which the virtual network is realized. The nodes and the domain of authorized users of a virtual network can be determined in accordance with the user or customer requests, independently of the actual geographical location of the required virtual nodes and that of the user access points (of course, they should be covered by the topology of the entire network) and independently of the relation between the network hierarchy and the locations of the virtual nodes.

A separate management center may be assigned to the virtual network; however, it has to be subordinated to the entire network management system. This means that performing management functions (configuration management, event management, performance management, etc.), with respect to the operation of the whole network, remains the task of the central network management. The management center of the virtual network will have rights to monitor the actual operating parameters of the virtual network, and (by chance) will also have restricted competence to exert active influence on some partial operating parameters as the scope of the virtual network concerns.

Virtual networks will obtain particularly great importance in those cases, when services of public communications networks are used to build up a private network in part or in its entirety. The constituents of the public network, designated to form part of the given private network at the same time, can be regarded as a virtual private network (VPN). In such cases, operators of the private network would like to manage their own communications system as a whole, including those parts being utilized as a VPN.

Standard network management services that can be provided for the customers by Public Data Networks are called Customer Network Management (CNM). The framework of CNM is defined in ITU-T Recommendations X.160–X.162 [31–33]. The given CNM-related ITU-T standards define a set of services and the way of their implementation that may enable the user to manage those parts of its communications network which are actually made available by public service providers. ITU-T Recommendation X.160 defines the architecture for CNM for public data networks. Recommendation X.161 defines services for CNM, whereas management information for CNM service is specified in Recommendation X.162. Furthermore, ITU-T Recommendation M.3010 provides an opportunity for private network operators to realize the concept of CNM in the scope of TMN.

As described in Recommendation M.3010, TMN functionality offered by service providers may be accessed through CNM services, these services being realized as management interactions between users and TMN, or between TMNs. In both cases, reference point CNM (the reference point between the customer and the service provider) defined in Recommendation X.160 has to be realized by TMN reference point x, as defined in M.3010.

Recently, efforts have been made to realize the CNM concept in association with high-speed public data networks [34–36].

3.4.5 Management of SDH/SONET

3.4.5.1 A Brief Survey of SDH and SONET Communications Technologies

Synchronous Digital Hierarchy (SDH) implies the concept of a powerful broadband telecommunications system, defined by a number of international standards, issued by CCITT/ITU-T. The first recommendations, including principles of SDH, were issued by CCITT in 1988 (CCITT G.707, G.708, G.709). SDH is the counterpart of Synchronous Optical Network (SONET), the predecessor of SDH. SONET was originally developed and proposed by Bellcore (U.S.). It is now standardized by ANSI, the base standard of SONET being ANSI T1.105. The concepts and architectures of SDH and SONET are highly similar (SONET may now practically be regarded as a subset of SDH), but there are also differences, first of all concerning standard transmission bit rates defined in the respective signal hierarchies. Recently, close coordination exists between U.S. SONET and international SDH standard bodies.

SDH and SONET are advanced telecommunications technologies — one of their most important features is that they effectively support network management. They have significant advantages over traditional PCM-based, Plesiochronous Digital Hierarchy (PDH) digital data transmission systems.

3.4.5.2 SDH/SONET and TMN

The principles of managing SDH/SONET networks by means of a TMN do not differ from those having been described elsewhere in Section 3. Notwithstanding, regarding SDH/SONET network management, it is important to emphasize the following significant circumstances:

- SDH/SONET are the first communications technologies that have been designed with particular attention to network management aspects;
- In the course of developing TMN standards and among the most important, up-to-date communications technologies, SDH/SONET are the first to have reached an appropriate degree of standardization, allowing a comprehensive TMN network management system capable of meeting the users' actual requirements (of course, all these are prevalent within the unquestionable limits of the present state of TMN standards; see also Sections 3.4.1, 3.4.3.3, and 3.4.3.5.2);
- Application of TMN network management is not only supported by the streamlined systems architecture of SDH/SONET and the existing TMN standards, but also by the fact that in practice, SDH/SONET system devices (add-drop multiplexers, cross-connects etc.) are actually equipped with the appropriate TMN interfaces, hence they are now capable of handling Q protocols and of implementing TMN network management;
- The complexity and high performance of SDH networks obviously require the facilities of TMN network management, and as a consequence, TMN network management is almost the only technology being applied to manage SDH networks in practical implementations [29, 37, 38].

3.4.5.3 Transmitting Management Information Through an SDH/SONET Network

As already mentioned in Section 3.4.3.2, in SDH/SONET networks, management information is usually transferred via the managed network itself; that is, DCN of the management system is actually a part of the network as a whole.

The most important factors of the SDH/SONET technology with respect to TMN network management are:

- The SDH frame structure allows management information to be transferred in data bytes D1, …, D12 of the Section OverHead (SOH in the STM-1/OC-3 frame) alongside the network;
- The characteristic features of SDH/SONET networks (high reliability of the network elements in itself, the availability of rapid and automatic "self-healing" protection mechanisms in redundant

ring topologies and the applicability of error recovery in meshed networks, allowing network communication most likely to be restored in the event of failures) make SDH/SONET networks highly suitable for realizing the DCN of the management system in the managed network itself, without impacting safety of operation (see also Section 3.4.3.4.2).

Getting a nearer view of the SDH/SONET frame, we can find that the Section OverHead (SOH) area of the STM-1/OC-3 SDH/SONET frame contains (and transfers) a number of data bytes, being characteristic of network operation and playing a significant role in network management.

In the SOH, parity control bytes B1 and B2 indicate bit errors of the generator and the multiplexer section, respectively, whereas bytes D1, ..., D12 hold processed management status information. (Each of the bytes in the STM-1/OC-3 SDH/SONET frame represents a communication channel with a bit rate of 64 kbit/sec. It means bytes D1–D12 ensure a total communication capacity of 768 kbit/sec for network management.) The communication channel represented by management bytes D1–D12 can transfer management information from node to node (e.g., from add-drop multiplexer to add-drop multiplexer) along the network itself; furthermore (after the necessary protocol conversions), this information can be "tapped" at the standard Q interface of any node equipment and forwarded to the operations system of the network management center.

The status and alarm messages carried by the SOH can be accessed at the "S" monitoring points of the multiplexers and regenerators. The information made available at points "S" does not appear in an adequately structured form. Therefore, it has to be converted into object-oriented messages by a conversion function included in the given device. The object-oriented messages will then be returned to the STM-N aggregate transfer module and transferred in the SOH from node to node in the network. In order to get local access to these messages, it is necessary to have a standard "Q" interface that performs the necessary protocol conversion and signal interfacing. Output of the "Q" interface may then be connected for example to the local operations system of the network management center. Note, that multiplexers and cross-connects generally do have "Q" interfaces, but regenerators do not.

Figure 3.4.7 illustrates the scheme of the management data flow in the multiplex section of an SDH network.

3.4.5.4 Structure of Managed SDH/SONET Networks

3.4.5.4.1 General Considerations

The physical architecture of a managed SDH network accords with the principles described in Section 3.4.3.2.

It stands to reason, however, that designing a SDH/SONET network management system should be performed, in particular, consideration of the design aspects and architecture of the managed SDH/SONET network itself.

Furthermore (basically as the opposite of the requirement stated above), it is easy to see that, in practice, safety and reliability of the designed SDH/SONET network will essentially be determined by the structure of the pertaining network management system.

From the arguments stated above we can conclude that designing a SDH/SONET communications network and the related network management system should be performed in a close interrelation.

In the case of a simple network architecture, the network management system is also relatively simple. As an example, Figure 3.4.8 shows the scheme of a TMN-managed SDH ring consisting of four network nodes. Operation of the illustrated network management system can briefly be described as follows.

The SDH ring is managed by a TMN. The DCN of the management system is realized by the available (SOH) management channels of the SDH ring, the OS (i.e., the network management center) is represented by a single computer. (This computer also includes the operator's workstation.) The management center is connected to one of the four nodes of the SDH ring through a single "Q" interface.

Taking a more complex network architecture, the structure of the TMN network management system will also be more complex, and its operation will require more complex algorithms. To this topic, the following remarks can be added.

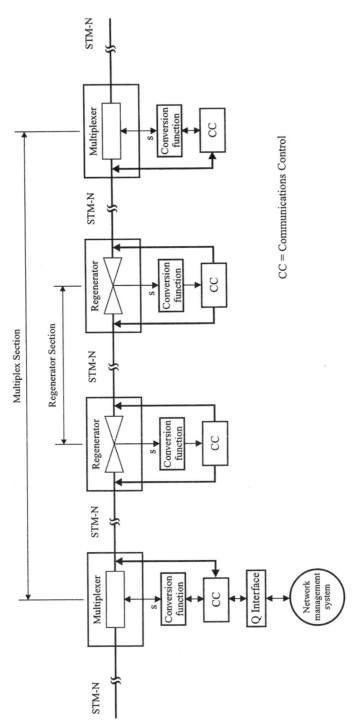

FIGURE 3.4.7   Management data flow in a SDH multiplex section.

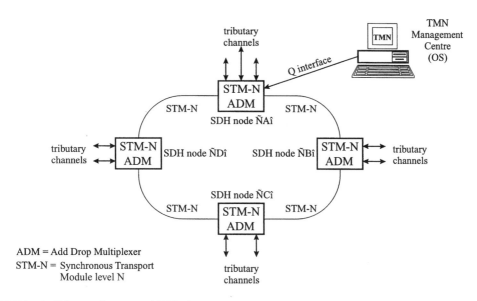

FIGURE 3.4.8 Scheme of a managed SDH ring.

- The "self-healing" protection mechanisms, frequently applied in SDH/SONET rings (aimed at retaining error-free communication in the case of a single failure) are based upon automatic reactions of the network elements and will function without the intervention of the network management system. The task of the TMN is simply to record the occurred events in such cases.
- In meshed networks, it is the TMN's responsibility to select and set up alternative routes in order to eliminate the possible consequences of any network malfunctions. Consequently, management algorithms required in meshed networks are generally more complex than those applied in simple self-healing rings.
- In the course of establishing management systems for hierarchical and/or interconnected networks, the TMN system(s) and the management algorithms must be designed with great care. Very likely, applying more than one OS will be necessary in this case. If so, tasks, access rights, control privileges, and ways of interworking the different OSs (management centers) should be carefully and clearly defined.
- Designing a network management system will be even more difficult if several networks of different topologies, those built up using different network technologies (SDH-PDH) or containing different vendors' equipment should be managed at the same time and in coherence.

3.4.5.4.2 *Hierarchical Networks*
TMN enables us to build up network management systems hierarchically, according to the eventually layered architecture of the network to be managed.

Consequently, a network management system may involve separate network management centers for (as an example) each of the

- Backbone
- Regional
- Sub-regional or local

network layers.

More than one management center may belong to one and the same hierarchy layer (e.g., as a stand-by safety reserve, if necessary); the centers may connect with the DCN at more than one point in order to increase reliability. The number of the operators' workstations may also be more than one. Moreover, some workstations can be installed at remote locations, separated from the network management center. In this case, the workstation has to be connected to the DCN through a standard "F" interface.

The tasks and privileges designated to the management centers of the different hierarchy layers (rights to access data, privileges to control network elements, and to alter network configuration in the different network layers) as well as the rules of how stand-by management centers should take over the functions of network management in case of emergency have to be determined during system design.

Figure 3.4.9 shows the scheme of a hierarchical TMN system managing a layered SDH network.

The illustrated network consists of three layers: the national, regional, and local network layers. Each of them is managed by a separate management center; these in turn are connected to their respective network layers at two network nodes, using two separate communication channels and standard Q_3 interfaces. (Taking a practical example, one of the duplicated interconnections may be realized through the managed network and the second one, in order to increase safety, through an independent communication channel.)

Furthermore, the management centers are interconnected via appropriate communication channels as well, using standard X interfaces. (These interconnections may also be realized through the managed network via its available channels reserved specially for this purpose.)

3.4.5.4.3 *Managing PDH Network Elements*

In practice, network management often has to be extended to some non-SDH system components, such as elements or segments based upon PDH technology, being interfaced with the managed SDH/SONET network. This may be more necessary so that PDH payload may be mapped into the virtual containers of the SDH multiplex structure.

However, it will raise the difficulty that manageability of PDH network elements is fairly limited, both in principle and in practice. This is partly because there is a poor channel capacity available for transferring management information in a PDH network, partly because PDH systems technology does not support dynamic configuration management functions, and partly because PDH network elements cannot handle Q protocols. Management of PDH is not even supported by the existing TMN standards at present. According to Recommendation M.3180, plans exist to cover PDH transmission equipment concerning network-level information model of TMN.

Nevertheless, as up-to-date network management systems are capable of processing element-level network management information in different, non TMN-compatible formats (e.g., in ASCII code), managing PDH network segments with limited functionality is still feasible in practice.

3.4.6 TMN and GIS (Geographic Information System)

The benefits of GIS could be applied to support business processes of telecommunications service providers. The benefits of use are obvious considering that telecommunications services are end-to-end related and include a number of geographical domains, and multiple providers with different networking infrastructures. Managing physical and logical infrastructures requires very accurate documentation. State-of-the-art documentation systems work with digital maps that can be acquired from third-party vendors.

Network management tasks can be defined and supported in various TMN layers. The level of details and the type of information required differ by the layer. The service and business management layers may be using three-dimensional applications that may not be required in the network and element management layers.

The applicability of GIS will be evaluated for three TMN layers. The functional relation of TMN and GIS is shown in Figure 3.4.10.

3.4.6.1 Network Management Layer

The graphical presentation and documentation of physical networks belong to the traditional tasks of CAD/CAM applications. Principal documents include cables (earth and air), cable ducts, vaults and shafts, telco buildings, location of equipment, allocation of ports to wire pairs, and wire pairs to cables,

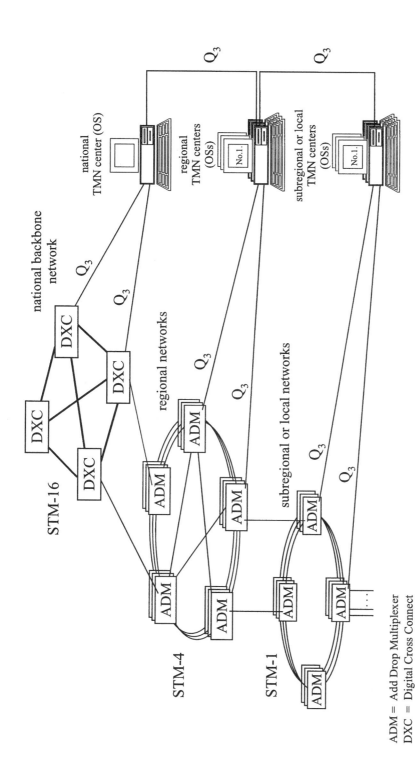

FIGURE 3.4.9 Management of a hierarchical SDH network.

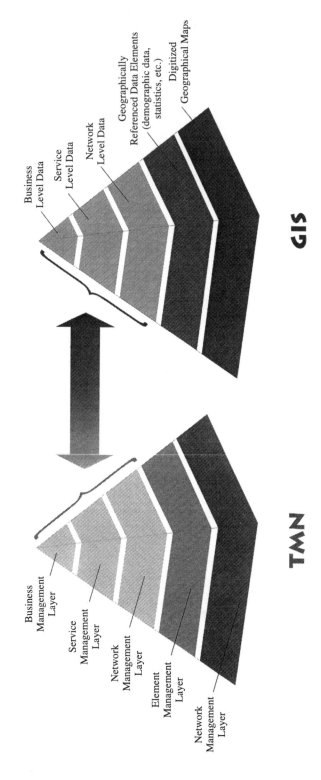

FIGURE 3.4.10 Functional relation of TMN and GIS.

respectively; in addition, other technical attributes of components have to be maintained as well. Details of component attributes are expected to be maintained in the Network Management Layer.

GIS applications are extremely important for wireless communication networks, such as mobile phone systems, URH, and satellites, in analyzing wave diffusion models. Models also include area of coverage and geography of selected areas. Inventory items are part of static configuration management.

Basically, there are two types of configuration management solutions. Dynamic configuration management means on-the-fly configuring of routes on the basis of actual real-time status and performance data of facilities and equipment. It is obvious that GIS applications cannot support this management function by its capabilities of three-dimensional presentation services. Static configuration management includes the logical configuring and changing of routes, equipment configurations, or both; maintaining types and location of equipment; maintaining both physical and symbolic names and addresses; also, other attributes may be maintained and information displayed. In case of relative stability of configuration parameters, GIS applications may be successfully used for displaying logical and physical routes of the network on top of various digital maps.

GIS applications contribute successfully to improve performance, fault, and security management. Performance metrics, load profiles, eventual faults and outages or security violations can be positioned and displayed on top of accurate digital maps. This kind of presentation helps to estimate and quantify impacts on customers and on service areas in real time with the results of expedited problem resolution, load balancing, and security protection.

A GIS is also a great help with preventive and reactive maintenance functions. GIS can pinpoint the location of faults with high accuracy, can highlight the optimal route for workforce to the site, and can search for and assemble the appropriate support documentation.

3.4.6.2 Service Management Layer

Managing services requires accurate data on service configuration, customer level quality metrics and actual metrics agreed upon in Service Level Agreements (SLA). GIS can assist to retrieve information on load profiles, fault summaries, network extensions, and visualization of service areas very fast; also, the correlation of this information with subscribers, contracts, subcontracts, service level agreements, bills, and bill presentment can be done in near real time. The result is that the quality of customer care is improved and customer satisfaction grows.

3.4.6.3 Business Management Layer

Business management requires a number of strategic and tactical decisions on behalf of top management. This is the area where Enterprise Resource Planning (ERP) can be utilized very effectively. It can support decision on complex and inter-related issues, visualization on the basis of GIS may help. Visualization examples may include the view of network segments, the position and size of existing service areas, peering with other service providers income statistics by service areas, statistical background data to support marketing, demographic data by service areas, and expected level of consumption by regions. Most of these data are not available in traditional management and documentation systems. GIS and data warehousing in combination can significantly offer better support to top management of service providers and also of enterprises.

In summary, the meaningful collaboration between TMN, GIS, and ERP (Figure 3.4.11) will ensure a higher quality of information exchange and presentation for service providers.

3.4.7 Trends of Evolution of TMN

The next steps of the evolution of TMN will most likely be driven by the following factors, being characteristic of the global situation of the telecommunications industry early in the 21st century:

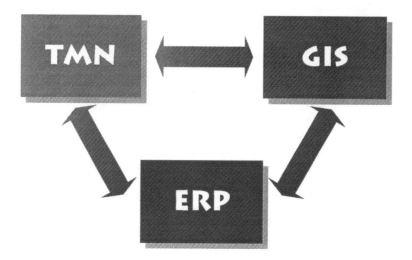

FIGURE 3.4.11 Possible interrelation of TMN, GIS, and ERP systems.

- The growing importance of business aspects of telecommunications network operation, such as rentability of investments, costs, profitability and competitiveness of network services, etc. [39];
- The evolution of the broadband telecommunications technology and the growing prevalence of the most up-to-date network architectures, such as ATM, intelligent networks (IN) [40–42] etc., in both the public and corporate or private network domains;
- The proliferation of the available hardware/software tools in the world of telecommunications;
- The perpetually growing sphere of telecommunications services and their applications, such as mobile services, video and multimedia applications, the increasing area of Internet utilization, such as on-line shopping, etc.;
- The globalization of the world of telecommunications, the integration of LANs and WANs, vanishing of the borders between public and private network domains, voice, data, and video transmission systems, etc.

As concluded from all these general aspects, the following changes probably will occur in the evolution of TMN, concerning its features and functionality:

- Focus of network management will shift from the network element and network-centric view to the service management first, and thereafter to the business management level — it means the relevant standards will be elaborated and the appropriate solutions will gradually appear on the market [39];
- The importance of the end-to-end management and customer management will increase (see Section 3.4.4.2).

Reflecting these functional requirements on the platform of software technology, actual tasks will arise (and have arisen) which have to be completed, such as:

- Defining a standard distributed object model meeting the requirements of service management for TMN (see Section 3.4.3.5.2);
- Defining standards and developing actual solutions for managing up-to-date telecommunications technologies presently not covered by TMN (first of all, as ATM concerns);
- Integrating TMN with the most important foreign platforms in the field of network management such as SNMP (see Section 3.4.4.1);
- Mapping TMN into some more recent telecommunications architectures (such as TINA and IN) theoretically being able to be integrated with TMN (see Section 3.4.3.5.2);

• Integrating TMN with some external information systems being operated outside the scope of telecommunications having however close functional connections with it, such as integrated corporate business management systems (see Section 3.4.3.3) or (representing one of the most recent challenges for TMN) Geographic Information Systems (GIS) [43].

As a result of present and future developments, the field of possible TMN solutions allowed by the whole set of the relating standards will be more and more complex. This situation sets a great task on network operators and/or system designers having to cope with the problem of specifying an appropriate network management system to effectively meet the requirements of their special application. It means design and specification of an actual TMN solution should be performed by selecting the relevant TMN functions from the "endless" set of standard possibilities, not simply declare or require compliance with "standard TMN" [39].

References

[1] CCITT Recommendation M.3010 (1996), *Principles for a Telecommunications Management Network.* International Standard, May, 1996.

[2] Glitho, R. H., Hayes, S., Telecommunications Management Network: Vision vs. Reality, *IEEE Communications Magazine,* 47, March 1995.

[3] Sidor D. J., Managing Telecommunications Networks Using TMN Interface Standards, *IEEE Communications Magazine,* 54, March 1995.

[4] Towle, T., TMN as Applied to the GSM Network, *IEEE Communications Magazine,* 68, March 1995.

[5] Fowler, H. J., TMN-Based Broadband ATM Network Management, *IEEE Communications Magazine,* 74, March 1995.

[6] Mader, W., Ferraris, G., Huber, M. N., Lehr, G., Müllner, W., Rotolo, S., Management of Optical Transport Networks, *NOC '97 (European Conference on Networks and Optical Communication),* Vol. 3., 111, 1997.

[7] Rose, M., McCloghrie, K., *Structure and Identification of Management Information in TCP/IP Based Internets,* RFC1155, May, 1990.

[8] Case, J., Fedor, M., Schoffstall, M., Davin, J., *Simple Network Management Protocol,* Internet RFC1157, May 1990.

[9] Case, J., McCloghrie, K., Rose, M., Waldbusser, S., *Structure of Management Information for version 2 of the Simple Network Management Protocol (SNMPv2),* Internet RFC1442, April, 1993.

[10] Stallings, W., *SNMP, SNMPv2 and CMIP — The Practical Guide to Network Management Standards,* Addison-Wesley, 1993.

[11] Terplan, K., *Effective Management of Local Area Networks,* McGraw-Hill Inc., 1992. Chap. 4.

[12] ITU-T Recommendation X.200 (1994), ISO/IEC 7498-1, *Information Technology — Open Systems Interconnection — Basic Reference Model: The Basic Model.* International Standard, July 1994.

[13] CCITT Recommendation X.700 (1992), *Data Communication Networks — Management Framework for Open Systems Interconnection (OSI) for CCITT Applications.* International Standard, September 1992.

[14] CCITT Recommendation X.701 (1992), ISO/IEC 10040, *Information Technology — Open Systems Interconnection — Systems Management Overview.* International Standard, January 1992.

[15] CCITT Recommendation X.710 (1991), *Common Management Information Service Definition for CCITT Applications.* International Standard, March 1991.

[16] CCITT Recommendation X.711 (1991), *Common Management Information Protocol Specification for CCITT Applications.* International Standard, March 1991.

[17] CCITT Recommendation X.722 (1992), ISO/IEC 10165-4, *Information Technology — Open Systems Interconnection — Systems Management Guidelines for the Definition of Managed Objects.* International Standard, January 1992.

[18] CCITT Recommendation X.720 (1992), ISO/IEC 10165-4, *Information Technology — Open Systems Interconnection — Structure of Management Information: Management Information Model.* International Standard, January 1992.

[19] ITU-T Recommendation X.901-4, *Information Technology — Open Systems Interconnection — Open Distributed Processing — Reference Model.* International Standard.

[20] ITU-T Recommendation X.703, *Information Technology — Open Systems Interconnection — Open Distributed Management Architecture.* International Standard, January 1992.

[21] Network Management Forum, Forum Library, Volume 4: *OMNIPoint 1, Definitions,* Issue 1.0, August 1992.

[22] Network Management Forum, *OMNIPoint Integration Architecture — Delivering a Management System Framework to Enable Service Management Solutions,* Technical Report, Issue 1.0, August 1994.

[23] OMG Group, *The Common Object Request Broker: Architecture and Specification.* Technical Report, December 1993. Revision 1.2.

[24] Manley, A. Thomas, Evolution of TMN Network Object Models for Broadband Management, *IEEE Communications Magazine,* 60, October 1997.

[25] Hashida, Y., Towards Service Operation and Management Technology, *IEICE Transactions on Communications,* E-80-B, 790, June 1997.

[26] Kong, Q., Chen, G., Integrating CORBA and TMN Environments, *NOMS '96 (IEEE Network Operations and Management Symposium),* Vol. 1, 86, 1996, Kyoto.

[27] Strick, L., Wittig, M., Paschke, S., Meinköhn, J., Development of IBC Service Management Services, *NOMS '96 (IEEE Network Operations and Management Symposium),* Vol. 2, 424, 1996, Kyoto.

[28] Larsson, R., Ferrari, L., Use Case Driven Integration of TMN Interface Specification and Application Design, *NOMS '96 (IEEE Network Operations and Management Symposium),* Vol. 2, 434, 1996, Kyoto.

[29] Dr. Ku, B. S., Use of TMN for SONET/SDH Network Management, *NOMS '96 (IEEE Network Operations and Management Symposium),* Vol. 2, 454, 1996, Kyoto.

[30] Motomura, K., Nakamura, N., Aibara, T., Integrated Platform for CMIP-Based and SNMP-Based Management, *IEICE Transactions on Communications,* E-80-B, 861, June 1997.

[31] ITU-T Recommendation X.160 (1994), *Data Networks and Open System Communications Public Data Networks, Maintenance — Architecture for Customer Network Management Service for Public Data Networks.* International Standard, July 1994.

[32] ITU-T Recommendation X.161 (1995), *Data Networks and Open System Communications Public Data Networks, Maintenance — Definition of Customer Network Management Services for Public Data Networks.* International Standard, April 1995.

[33] ITU-T Recommendation X.162 (1995), *Data Networks and Open System Communications Public Data Networks, Maintenance — Definition of Management Information for Customer Network Management Service for Public Data Networks to be used with the CNMc Interface.* International Standard, April 1995.

[34] Aronsheim-Grotsch, J., Berkom D. T., Customer Network Management, CNM, *NOMS '96 (IEEE Network Operations and Management Symposium),* Vol. 2, 339, 1996, Kyoto.

[35] Park, J. T., Lee, J. H., Hong, J. W. K., Customer Network Management System for Managing ATM Virtual Private Networks, *IEICE Transactions on Communications,* E-80-B, 818, June 1997.

[36] Yamamura, T., Tanahashi, T., Hanaki, M., Fujii, N., TMN-based Customer Network Management for ATM Networks, *IEEE Communications Magazine,* 46, October 1997.

[37] Kunieda, T., A Synchronous Digital Hierarchy Network Management System, *IEEE Communications Magazine,* November 1993.

[38] Yamagishi, K., Sasaki, N., Morino, K., An Implementation of a TMN-Based SDH Management System in Japan, *IEEE Communications Magazine,* 80, March 1995.

[39] Willets, K. J., Adams, E. K., TMN 2000: Evolving TMN as Global Telecommunications Prepares for the Next Millennium, *IEICE Transactions on Communications,* E-80-B, 796, June 1997.

[40] Mizuno, O., Shibata, A., Okamoto, T., Niitsu, Y., Models for Service Management Programmability in Advanced Intelligent Network, *IEICE Transactions on Communications,* E-80-B, 915, June 1997.

[41] ITU-T Recommendation I.312 (1992), *Principles of Intelligent Network Architecture.* International Standard, October 1992.

[42] ITU-T Recommendation Q.1200 (1993), *Q-series Intelligent Network Recommendation Structure.* International Standard, March 1993.

[43] Edwards, G., Examining GIS In Relation To The TMN Model, *presented at GIS In Telecoms & Utilities (Conference arranged by IIR Telecoms & Technology)*, September 1997, London.

3.5 TINA

Takeo Hamada, Hiroshi Kamata, and Stephanie Hogg

3.5.1 Introduction[1]

The increasing demand for broadband and sophisticated services, such as universal personal communications and mobile multimedia services, calls for a more flexible and distributed information architecture for advanced telecommunication services and management. Telecommunication Information Network Architecture (TINA) is a software architecture which has been designed by TINA-C since 1995 [1], and a set of architecture and interface specifications has been published.[2] Historically, TINA evolved from two predecessors: IN and TMN. As a service architecture, TINA inherited the concept of network support for telecommunication services from IN. As a management architecture, TINA inherited the concept of distributed, object-oriented representation of network resources from TMN. As its design methodology, TINA adopted ODP principles [2], and also developed its own specification languages for the description of information [3] and computational viewpoints [4,5].

As a result of these architectural heritages, TINA can be understood both from the network service and the service management points of view. For the users of TINA, TINA is a distributed object-oriented system, into which multimedia network services and user applications can be plugged. In IN-like network services, e.g., multimedia version of 800 service, the service components (objects) are provided by the network or the service providers. In collaborative workspace applications, the basic call model and the multi-party session control are provided by the TINA service layer. In either case, TINA services are supported by built-in service management functions in its service architecture [6], and by the service components supporting the service architecture [7].

In this chapter, we focus mainly on the management aspect of TINA, due to the following two reasons; (1) built-in support for management is essential for the integration of call control and management in TINA, and (2) support for telecommunication management needs distinguishes TINA from other enterprise data-communication-oriented network architectures. The TINA management architecture has many facets. It is a set of common goals, principles, and concepts covering the management of services, resources, and parts of the DPE. Due to its broad nature, management is one of the most challenging areas in TINA. TINA imposes the following design goals on the management architecture:

- *Object oriented and distributed:* TINA management architecture takes advantage of a distributed processing environment (DPE), which offers a natural distributed object-oriented environment for both resource and service management. It builds on the object-oriented and distributed approaches of telecommunication management network (TMN) [8] and Open Distributed Management Architecture (ODMA) [9].
- *Service-oriented:* TINA uses a more goal-driven design approach rather than a resource-driven one. In particular, the service management should guarantee quality of service (QoS) as an integrated part of the service.
- *Dynamic and flexible:* The service-oriented nature of TINA requires that resources need to be more dynamically assigned or configured, to support the flexibility required by TINA services.
- *Integrated:* Management is an integrated part of service provision, inherent in the various TINA layers (service, resource, and DPE).

---

[1]This article is a revised and updated version of a paper, An Overview of the TINA Management Architecture, *Journal of Network and Systems Management*, Vol. 5, No. 4, 1997, Plenum Press.

[2]All the TINA-C public documents are available from the TINA-C home page, http://www.tinac.com.

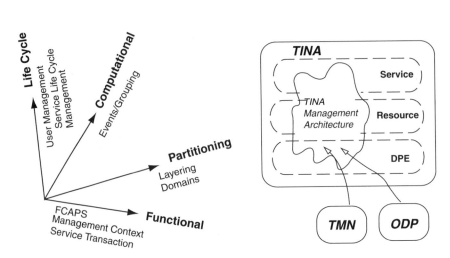

FIGURE 3.5.1   TINA management architecture and conceptual axes. (From Hamada, T., Kamata, H., and Hogg, S., An Overview of the TINA Management Architecture, *Journal of Network and Systems Managements*, 5, 411–435, 1997. Used with permission of Plenum Publishing Corporation.)

The TINA management approach consists of several related disciplines and builds upon other management approaches, such as TMN, ODP, and OmniPoint [10]. The session-oriented, multiparty, multidomain nature of TINA service architecture poses unique problems in its service management. To illustrate the TINA management architecture, this paper is organized around the following four conceptual axes (Figure 3.5.1):

- *Partitioning axis* (Section 3.5.2) represents layering and domain concepts. TINA is partitioned into three layers: service, resource, and DPE. The management architecture is likewise partitioned, into service, resource, and DPE management. It also supports the concept of domain, the management of domains.
- *Functional axis* (Section 3.5.3) represents FCAPS (fault, configuration, accounting, performance, security) functions. To support the FCAPS integrity of a service session, constructs such as management context and service transaction are provided;
- *Computational axis* (Section 3.5.4) represents computational support for management needs. These computational supports are mostly offered by DPE, but we will consider some TINA refinements to event management and grouping concepts; and
- *Life cycle axis* (Section 3.5.5) represents the life cycle issues, including service life cycle (SLC) management and user management, i.e., the life cycle management of consumers.

This chapter intends to give an overview of TINA management architecture, with the main focus on service management and supporting concepts. As a result, some major topics had to be left out. In particular, connection management, which is an important topic in TINA, is not discussed. Pointers to other important topics in TINA are given at the end of this chapter.

3.5.1.1   Relationship between TINA, TMN, and ODP

The relationship between TINA, TMN, and ODP deserves special attention. As we noted earlier, TINA inherited some of its architectural concepts from both TMN and ODP. In fact, there is often more resemblance than difference in TINA to TMN, ODP, and ODMA [9], an ODP-based management architecture. For example, TINA Network Resource Information Model (NRIM) [11] used many network element concepts originated in related TMN documents, such as G.803 [12] and G.805 [13]. There are, however, significant differences between TINA and TMN, as discussed below.

- *Session Concept:* TINA is session-oriented, i.e., many of the constructs and service components are dynamically bounded within the lifetime of service sessions. As a consequence, much of service management activities are also service session-oriented, which is in stark contrast to TMN.

- *Separation between information and computational viewpoints:* By adopting ODP principles, TINA maintains clearer separation between information and computational viewpoints than TMN. In TMN, managed objects (MOs) correspond to an information viewpoint, and a computational viewpoint is represented by manager-agent roles and associated interfaces known as Q, X, and F interfaces. In terms of information-to-computational mappings, therefore, many-to-one mapping is dominant in TMN. In TINA, information objects can be mapped more freely to computational objects. For example, one information object may be mapped to one computational object (one-to-one mapping), or may be mapped to a collection of computational objects (one-to-many mapping). For further discussions on this issue, please see 3.5.4.1, *Computational Aspects of Distributed Management.*
- *Different computational model:* In TMN, GDMO and Common Management Information Protocol (CMIP) in effect defines the computational viewpoint of the TMN system. In TINA, CORBA-based DPE provides its computational viewpoint. Computational objects in TINA are described in CORBA IDL and TINA ODL, an extended version of CORBA IDL, which allows multiple interface objects.
- *Different engineering model:* The above difference in computational model results in different engineering platforms. Usually, TMN systems are CMIP-based, whereas TINA systems are CORBA-based.

Interoperability and migration between TMN and TINA are of serious practical concern. Interoperability and migration issues between TMN and TINA, and between IN and TINA have been extensively studied in the Eurescom P508 project [15,16]. For a recent development on the migration issues, the reader is referred to [14]. For the recent developments of TMN, a recent report [17] gives an excellent survey.

3.5.2 Partitioning Axis

The partitioning axis considers how management may be broken into more manageable problems. TINA supports two types of partitioning: layering and domains. These two concepts are independent. If layering is considered a "horizontal" partitioning, domains may be considered a "vertical" one.

3.5.2.1 Layering

Layering corresponds to the traditional distinction between the resource management and the service management. Resources are network capabilities and equipment. Resource management relates to the management of network resources. We define service management as all the management activities within the service layer. Service management is an inherent part of the service layer.

3.5.2.1.1 Service Management

Service management is becoming increasingly more important as the telecommunication industry evolves toward information and communication services. TINA views service management as a number of management activities, either inherent in its service architecture or provided by a set of management services, allowing the provision of services and their management by users and providers. Management services will use the same service structure as normal services or be based upon DPE and object services. Service management is therefore an intrinsic part of the service layer. TINA service management has the following common requirements:

- *Access management:* User access should provide security, allow customization, and other management capabilities. TINA service architecture [6] entails both access session and subscription management (Section 3.5.5.2.1) to provide required access management.
- *User service management:* Allow users to negotiate management requirements with service providers. A management context approach (Section 3.5.3.1) addresses this issue.
- *Flexibility:* TINA addresses an open market. It is expected that the market and price structure dynamically changes, reflecting today's competitive market. Service management must be flexible to satisfy those needs. This requirement is fulfilled by the management context concept (Section 3.5.3.1).

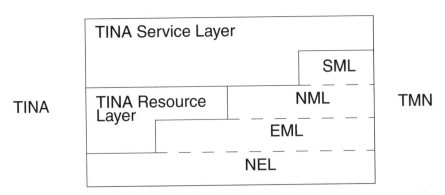

FIGURE 3.5.2 TINA Service Management Principle. (From Hamada, T., Kamata, H., and Hogg, S., An Overview of the TINA Management Architecture, *Journal of Network and Systems Managements*, 5, 411–435, 1997. Used with permission of Plenum Publishing Corporation.)

- *Service provision:* Providers must be able to deploy and withdraw their services.This requirement is answered in the SLC management (Section 3.5.5.1).
- *Controllability:* Providers need to be able to enforce their management policies and to exercise controls over service instances in their domains. For instance, a provider may need to monitor or control the status of service instances to maintain and manage their quality. Management policy (Section 3.5.2.2.3) and life cycle management (Section 3.5.5.1) help satisfy these needs.
- *Multiple business domains:* TINA service management must interoperate across multiple business domains. It must offer a consistent and guaranteed service management quality across these domains. The service management architecture needs to be relatively independent from the business model and the relationships between business entities, as these may vary. The service transaction concept (Section 5.5.3.2) fulfills this requirement.
- *Different QoS levels:* TINA has to offer a guaranteed and more consistent service management quality, such as accountability and fault tolerance, than Internet. Internet technology is very accessible and constantly evolving; on the other hand, its resource usage tends to be wasted and its quality varies. It is the telecommunication provider's ideal to support different QoS requirements of the user in a guaranteed manner as an integrated part of service management (Section 3.5.3.2).

TINA takes a service-oriented approach, starting with service and business-related concerns with the aim of achieving an information and communication service architecture upon a DPE. This provides a natural basis for considering an integrated approach to services and management. In comparison, TMN takes a resource-oriented, bottom-up approach, where service and business requirements are considered last. TMN service management corresponds to a fixed set of functionality on top of the network management layer. Figure 3.5.2 compares the TINA and TMN management models.

3.5.2.1.2 Resource Management
TINA resource management is based on the following principles and requirements:

- *Layers:* TINA borrows layering concepts from TMN, but uses them differently. It does not consider the network element management layer, and combines the network and element management into a single layer that may be used recursively.
- *Domain management capability:* Providers need to be able to enforce their management policies and to exercise management control over resources in their domain (Section 3.5.2.2.3).
- *Policy based:* Unlike service management, resource management tends to be driven by the provider's management policy (Section 3.5.2.2.3) rather than management context (Section 3.5.3.1).

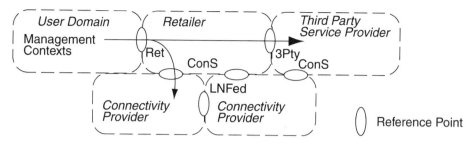

FIGURE 3.5.3 TINA business model and management context. (From Hamada, T., Kamata, H., and Hogg, S., An Overview of the TINA Management Architecture, *Journal of Network and Systems Managements*, 5, 411–435, 1997. Used with permission of Plenum Publishing Corporation.)

- *Multiple business domains:* The interoperability between multiple businesses in the resource layer is considered by management policy and federation (see Sections 3.5.2.2.3 and 3.5.2.2.4). The role of service transaction here needs further consideration.
- *Service layer support:* Provides management services to the service layer which allow information and communication services to set up connections, control their quality, and specify management support.This is part of the service and resource architectures.

3.5.2.2 Domains

TINA assumes that there will be a number of stakeholders acting in various business roles. Stakeholders are real-world commercial entities that provide services or communication resources. Business roles arise from analysis of a business model [6]. The NMF has performed similar analysis [19–21], though it focuses on separate business processes such as billing and trouble ticketing, rather than enterprise level requirements as they are expressed in the TINA business model.

3.5.2.2.1 Administrative Domains

An administrative domain is a portion of a TINA system which is submitted to a single stakeholder's ownership. TINA presumes that each stakeholder will manage and administer the policies and resources of its domain — giving rise to discrete domains that will be required to interoperate. For clarity, the following discussions treat each business role as occupying a separate domain associated with a single stakeholder. In reality, a stakeholder may participate in multiple roles and may administer one or more domains which are based on organizational considerations and need not be coincident with a single business role.

Figure 3.5.3 illustrates the relationship between the TINA business model and management context. Each stakeholder in the diagram constitutes a separate administrative domain, which contact each other through the reference points [18]. The recursive nature of management in TINA is apparent. In this example, the user domain contacts the retailer through the Ret reference point. Similarly, the retailer contacts connectivity providers through the ConS reference point. When a TINA service session is established across multiple administrative domains, management contexts are passed through reference points along with requests for the service session creation. Each context is specific to the service and the domain. For instance, the management context passing through ConS has to contain requirements for resource management.

3.5.2.2.2 Management Domains

A management domain can be modeled by an information object associated with certain management functionality such as accounting, security, or DPE management. Domains are similar to object groups ([22], Section 4.1), in that they both represent a set of objects. Unlike object groups, a domain does not usually provide explicit membership operations, since domain boundaries are usually based upon natural affinities between objects, such as the network resource topology, business stakeholder, or geographical area.

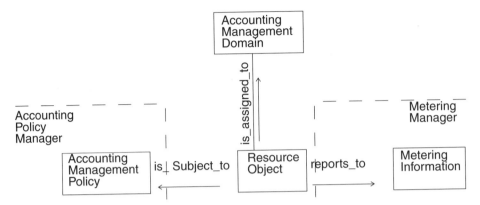

FIGURE 3.5.4   Accounting Management Domain information model. (From Hamada, T., Kamata, H., and Hogg, S., An Overview of the TINA Management Architecture, *Journal of Network and Systems Managements*, 5, 411–435, 1997. Used with permission of Plenum Publishing Corporation.)

The domain concept allows a hierarchical composition of domains. The generic management domain can be specialized with a set of management functions (policies) to become a specific management domain, e.g., an accounting management domain. Domains are by no means disjoint. Not only may an object belong to two or three different management domains, these domains may also relate to each other in different ways. For example, two accounting management domains may belong to two respectively different security management domains, but they both may belong to a single DPE management domain. Each management domain may have its own information model. The information model will usually consist of policy, resource, and manager objects (Figure 3.5.4).

3.5.2.2.3 Management Policy

A management policy is a set of rules governing a particular management function in the domain that is associated with the policy. For example, an accounting management policy may prohibit accounting events from being transported across the domain boundary, due to security or enterprise considerations. This rule effectively forbids on-line billing. A management policy can address many issues, including:

- *Managers:* Management policy determines the number and types of managers deployed within the domain and which resources they manage. Managers responsible for ensuring management activities are in accordance with management policy.

- *Event management policy:* Event management policy dictates acceptable event management options, such as translation, forwarding, and duplication. Policy may limit options, even though they are physically supported, for security or performance reasons. Event management options may limit federation or interoperability between domains.

- *Security features:* Since management actions and events may govern resource access and billing, they are natural targets of attack. As TINA service and resource layers are built on open DPEs, security features are mandatory. For instance, non-repudiation security options may be required for accounting events.

- *Policy applicability rules:* A policy is a set of rules. Rule applicability may vary depending on how the management domain is operated. Each rule is associated with one of three different applicability values: *mandatory (hard)*, *optional (soft)*, or *negotiable*. These applicability values are also used to resolve conflicts when two management domains overlap.

Management policy governs a management domain. By contrast, a service transaction and its associated management context (Section 3.5.3) set management requirements for the delivery of an instance of a service. A management context can take effect only for the duration of the service transaction to which the management context is bound.

3.5.2.2.4 *Federation of Management Domains*

Federation is one use of the management domain concept. With the management policy and its applicability rules, two domain managers can determine whether policy rules can be resolved and the two domains federated. The federation of management domains can be seen as an extension of the third-party call control, giving extra freedom to the management services. There are two major benefits of federation:

- *Sharing of Management Responsibility* means that two federated domains share a part of the management responsibility, such that one of the domains can cover the other's task, adding some fault tolerance.
- *Sharing of Management Resources* means that two federated domains share some management resources which may reduce management cost overheads and allow third party management, for instance a retailer may provide management for small content providers.

3.5.3 Functional Axis

This axis deals with the functional aspect of the management architecture. The key concept in this axis is FCAPS management, which is a well-known acronym of OSI functional classification. The TINA management architecture uses management context and service transactions to support FCAPS management. In this section, we introduce management context and service transactions.

- *Management Context* is a (service) management contract between the user and the provider, which drives (service) management activities in the provider domain.
- *Service Transaction* is a construct to guarantee the integrity of the service session with respect to its FCAPS management. There is some similarity to the database transaction concept, though the purpose of service transactions is to manage the integrity of the service session, not the integrity of data.

Finally, general context configuration management (GCCM) is also explained to show how context can be set-up and negotiated in a TINA environment.

Relationship between management context and service level agreement (SLA) [23] deserves special attention. In essence, management context and SLA services the same purpose, that is, to set up an end-to-end service management guarantee. The difference is in their lifetimes and dynamics. Since TINA is session-oriented, management contexts are dynamically bound to service sessions, thus instances of management contexts and their lifetimes are within those of associated service sessions. Service level agreement, in contrast, is a longer term, more static agreement between operators.

3.5.3.1 Management Context

Each service management context (MgmtCtxt) represents a set of functionality requirements, associated with a particular FCAPS. In general, a typical service transaction is associated with a number of (FCAPS) management contexts. The details of the management context depend on the individual FCAPS management objective. For example, the accounting MgmtCtxt [24] consists of the following information:

- *EventManagementConfiguration* specifies delivery timing and the event channel to be used for accounting event delivery.
- *TariffStructure* is essentially a function. Provider calculates charge/charging rate from accounting event sequence.
- *BillingConfiguration* contains types of billing options such as on-line or shared billing.
- *RecoveryConfiguration* specifies recovery actions to take when the service transaction is aborted for some reason such as network failure.

Figure 3.5.5 illustrates the role of management context in the TINA service architecture. Whenever a service session is joined or started, a management context can be bound to it. The PA could request a preset management context, or the UA could use a default management context, or the user domain (via the PA) can request to negotiate or configure a management context. The resulting context is then bound to a service session, which corresponds to the service transaction explained in the following section.

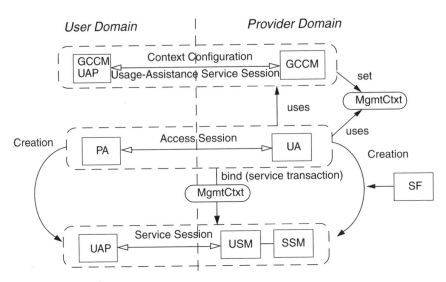

FIGURE 3.5.5   Role of management context in service architecture. (From Hamada, T., Kamata, H., and Hogg, S., An Overview of the TINA Management Architecture, *Journal of Network and Systems Managements*, 5, 411–435, 1997. Used with permission of Plenum Publishing Corporation.)

A service session may be started without a preceding context configuration step. In many cases, a default context specified in the user profile, which is configured within the subscription management, may be sufficient. In all cases, the management context must be compatible with the provider's management policies.

3.5.3.2   Service Transaction

A service transaction is a construct which guarantees that its associated service session satisfies management requirements prescribed by the management contexts. With the service transaction distributing the management contexts, a consistent level of service management throughout multiple administrative domains is achieved.

Figure 3.5.6 illustrates the relationships between service session and service transaction. The service session has two participants, represented by the two respective user session managers (USMs) in the provider domain. Each service transaction is associated with a management context, MgmtCtxt 1 and MgmtCtxt 2, respectively. When service transaction 1 executes, MgmtCtxt 1 is interpreted and translated into resource level operations. The service session manager (SSM) activates the communication session manager (CSM), which activates metering activity for the following communication service session. The USM corresponding to the service transaction passes its notification interface to the metering manager, and the metering manager reports notifications to the USM as required. Requirements for service transaction 2 are met similarly. Note that the two service transactions share the SSM.

The service transaction represents a view of the service session local to the concerned user–provider relationship. The management context in particular represents the contractual agreement between the two parties within the service session. The service transaction is an information object, covering part of the user application (UAP), USM, and SSM. In a sense, it is a virtual framework, which guarantees the completeness of the management requirements.

Execution of a service transaction consists of three phases. Here is an example, using an accounting management context.

1. *Set-up* — The respective management contexts are interpreted by the SSM and the associated USM of the service session and necessary resources such as the accounting log record and event channels are reserved or assigned.
2. *Execution* — The UAP starts running. Following the accounting/billing specified in the accounting management context, accounting events may be logged or reported. If the service guarantees QoS,

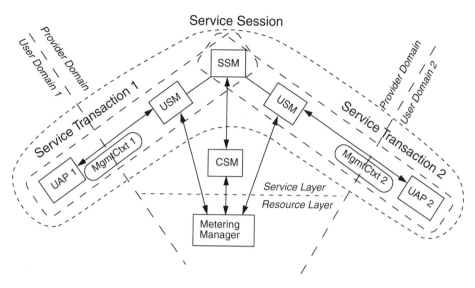

FIGURE 3.5.6 Relationships between service session and service transaction. (From Hamada, T., Kamata, H., and Hogg, S., An Overview of the TINA Management Architecture, *Journal of Network and Systems Managements,* 5, 411–435, 1997. Used with permission of Plenum Publishing Corporation.)

notifications from the performance management may be reported during this execution phase. If the QoS does not to meet the guaranteed quality, it may lead to an early termination of the service transaction.

3. *Wrap-up* — The UAP stops running. Reports from performance management or fault management can be summarized, and are then checked to see if the reported QoS has satisfied the guaranteed level. If it has, the service transaction concludes successfully. If it hasn't, some recovery actions, such as billing compensation, may be considered and actions may be taken by USM and SSM.

3.5.3.3 General Context Configuration Management

The concept of context is widely used in TINA service architecture in various forms. Examples include usage context, accounting management context (AccMgmtCtxt), and security management context (SecMgmtCtxt). The major purpose of these contexts is to associate additional information, represented by a context, with a service instance, so that particular service objectives are achieved. For example, the usage context is used to realize terminal mobility [25]. The management contexts (AccMgmtCtxt, SecMgmtCtxt, etc.) are used to achieve service management goals such as guaranteed accountability and quality of protection.

In any case, despite the difference in their usage and purposes, all these contexts are strikingly similar in their structure and the way they are configured in the service architecture. They need to be agreed between the user and the provider, therefore negotiation is often necessary. We introduce *general context*, which is a generalized class of context in the service architecture. The general context is conceived as a super-class of all the contexts used in the TINA service architecture.

3.5.3.3.1 The General Context Configuration Management Service

General context configuration management (GCCM) is positioned as a usage-assistance service in TINA service architecture. As such, GCCM service should be available across the service level reference points, including the Ret reference point. The objectives of the GCCM service are:

- *Semantics independent*: The GCCM service is semantics independent, which means that GCCM service is available for any type of context, but the semantics of the context needs to be interpreted by other semantic-dependent components used by the GCCM.

- *Negotiation*: The GCCM offers necessary and sufficient negotiation capability for general context management.

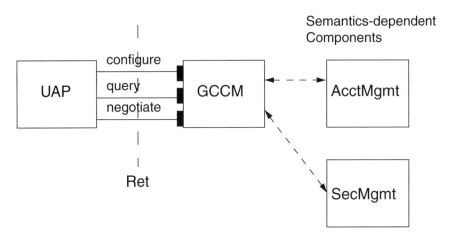

FIGURE 3.5.7   GCCM computational model. (From Hamada, T., Kamata, H., and Hogg, S., An Overview of the TINA Management Architecture, *Journal of Network and Systems Managements,* 5, 411–435, 1997. Used with permission of Plenum Publishing Corporation.)

Figure 3.5.7 illustrates the computational model of GCCM. The GCCM object provides three interfaces:

- *Configure*: This interface is used when the UAP is allowed to set a context without negotiation, or when UAP asks the GCCM to set up a predefined default context. It can be used if a user is invited to a session, or if plug-and-play is preferred and negotiation is not possible.
- *Query*: This interface is used when UAP asks GCCM for the current contents of the contexts, or the life cycle stage (live, dead, when modified, etc.) of the contexts.
- *Negotiate*: This interface is used when UAP and GCCM must negotiate the context.

Although the semantic part of the context is not considered here, three distinctive categories can be recognized, as they are evolutionary steps of the general context concept.

- *Static:* The context consists of a fixed set of attribute–value pairs. The scheme of the context is thus statically fixed. Negotiation has a clear meaning in this case, since the goal of the negotiation is a set of values agreeable to both sides.
- *Dynamic:* The context is still mostly made of the attribute–value pairs, but the scheme is dynamically expandable, i.e., it is allowed to add a new attribute pair on the fly.[1] The negotiation process can be more complex, and more intelligence on both sides (UAP and GCCM) is necessary.
- *Executable:* The context is not only dynamic, but now it can be executable in the provider domain. It may resemble a program more than data, like a shell script or Java program. The context may contain an intelligent mobile agent, which is capable of negotiating for system resources directly with other service components in the provider domain.

Though the contexts conceived so far in TINA service architecture are mostly at the first two levels, it is envisaged that the mobile or intelligent agent type configuration management will be more prevalent in the future [30].

3.5.3.3.2   GCCM Service Usage
The GCCM service is used in a number of places, such as at service instantiation. The following are typical uses of the GCCM service:

- *Prior service subscription:* If a GCCM service is activated prior to the service subscription (or during service subscription modification), the resulting configured context will prevail as the default context, which all the subsequent service instances can use.

---

[1]For example, CORBA Context object (CORBA 2.0, Section 4.5) [9] belongs to this category.

- *At service instantiation:* A GCCM service may be accessed just before a service instantiation, to set a context effective only for the service instance. Once configured, the context can be bound to the associated service session.
- *During a service session:* If a GCCM service is activated during a service session, it may be used to modify the context of the service session dynamically.

The third case may not always be possible, since modification of the context may trigger additional operations, e.g., resource management, to satisfy requirements of the modified context. As a usage assistance service, it can be triggered using normal service requests between domains. The general context can be seen as a contract. To ensure that the contract is agreed upon and valid, it should be possible for the contract to be time-stamped, signed, and made nonreputable. By using security features available from DPE security [26] and service layer security [27, 28], options such as *Time stamp*, *Nonrepudiation*, and *Persistency* should be available to the user of GCCM service.

3.5.4   Computational Axis

The computational axis considers computational aspects of management. The computation viewpoint considers the decomposition of a system into a set of interacting objects that are candidates for distribution.

As TINA is built on DPE, the computation aspect is very important. It provides the basis for the structure of managed and managing systems. The DPE infrastructure provides basic services upon which a more complex management framework can be based.

3.5.4.1   Computational Aspects of Distributed Management

Previous attempts at considering the computation aspects of management have been limited. In TMN, information and computation viewpoints are not clearly distinguished. It effectively uses the Manager and Agent roles as its computational model. ODMA [9] builds on this approach by defining both managers and managed objects as computational objects. It tends to view this as a one-to-one mapping from information to computation objects.

TINA builds from the ODMA approach, but maps from information objects to computational interfaces (rather than objects) and allows more flexible mappings. A managed object may be mapped to one or more computational interfaces and an interface may represent one or more managed objects. Note that a TINA computational object can support both control (or usage) interfaces and management interfaces.

As managers, managed objects, or agents representing managed objects, computational objects can use DPE services, which may be based on generic CORBA common object service (COS) [29]. Inter-working between DPE systems and legacy management systems based on Simple Network Management Protocol (SNMP) and Common Management Information Protocol (CMIP) is important, and NMF [31] and X-Open [32] are working on GDMO to IDL translations. These form the basis of interworking between DPE-based and OSI systems, enabling vertical integration by building TINA service management on top of legacy network management systems. For recent developments on interworking and migration issues see [14].

Computational objects themselves have to be managed, and the computational object management is supported by DPE services (Section 3.5.4.2), service life cycle (Section 3.5.5.1), and object groups. An object group is a "software package," or "subsystem," that contains one or more objects or object groups. It is built, installed, maintained, and managed as a unit. Object groups [22] can be used to structure and control usage and management functions. Unlike an object, however, an object group has internal structure that may be visible and can be distributed.

3.5.4.2   DPE Support Services

Typically a DPE offers a wide range of support services to aid system management. These include various life-cycle management services that support object creation and deletion; security services, which could include encryption, authentication, and authorization; and various basic facilities such as event management and logging.

In general, TINA has similar requirements to the OMG when it comes to considering DPE services and generally has adopted OMG services as the basis of its management services. In some cases, TINA can build on basic DPE services for particular management service requirements. For instance, to locate managed objects in a distributed system, a locator service is required. A locator service could be based on the general CORBA trading service.

As TINA considers telecommunication needs extending from resources to services, including the DPE and DPE services supporting resource and service layers, TINA places particular requirements on its support services. Instead of discussing the many possible services that a TINA system could require, we will focus on event management services and TINA's requirements of them.

3.5.4.2.1 Event Management

Event management is a common issue in both resource and service layers and is common to most functional aspects. What must event management support in a TINA system? Let us consider an accounting management example. Accounting event management is a challenge, since almost all resources are considered accountable, thus generating accounting events.

- *DPE event management service:* we assume that the DPE event management services such as the notification service would guarantee the delivery of accounting events.
- *Interworking issues:* TINA Network Resource Architecture (TINA NRA) accounting event management should be able to interwork with X.734 compliant TMN accounting event management. This requirement is a part of interworking and migration issues in general, but adding this requirement at the event management level would allow flexibility and versatility to migration scenarios.
- *Context-driven event management:* A service management context acts on objects in its scope for the duration of the service transaction with which it is associated. We call this a service management scope, which designates a set of objects that share common service management interests. It is natural that accounting events are collected within this scope, so that correlation and analysis is more efficient as the events should be more relevant semantically.
- *Federation:* The federation in the accounting event management is part of federation issues in general. This level of federation, however, can be done without assuming a complete federation of the accounting management domains, but still allowing the accounting events to cross the domain boundaries.

The underlying event management mechanisms are to be provided by a DPE with event management facilities based on the specifications such as CORBA COS Event Service [33], X/Open Notification Service [34], DPE Notification Service [35], and CORBA Notification Service [36].

3.5.4.2.2 TMN-style Event Management Facilities

TMN provides event management facilities as a set of managed object classes.

- *EventForwardingDiscriminator (EFD)* is defined in X.734/ISO 10164-5. The EFD discriminates on events: that is, the EFD uses given filtering conditions to decide which events are to be forwarded to a particular destination and which are not to be forwarded (and eventually discarded).
- *Log-control function* is defined in X.735/ISO 10164-6 and enables the user of the log-managed object to control its logging activities and to update its logging records.
- *log* is a managed object and is defined in X.721/ISO 10165-2. The log-managed object is a container of logging records with a management interface.

TINA NRA [37] only supports the minimum necessary functionality of its TMN counterparts. There are two reasons for introducing TMN-style event management facilities:

- *Fine-grain interworking/migration:* In general, migration can be characterized as coarse-grained or fine-grained, or as loosely coupled or tightly coupled. In fine-grain interworking, two accounting systems (TINA and TMN) interwork with each other, using these TMN-compliant components.
- *Domain-based management:* TMN-style event management is domain based and independent of service management scope (i.e., it is not service transaction driven).

3.5.4.2.3 Event Management Ladder

How can service level events be correlated with resource level events? This question is of particular importance in TINA, where services can be dynamically configured on demand from the user. Since some of the resource-level accounting activities are triggered by a service transaction and are sustained only for the duration of the service transaction, it is natural that the context for the accounting event management be established for those service-oriented accounting management events. The context is established at the beginning of the service transaction and is resolved at the end of the service transaction.

Figure 3.5.8 illustrates the event management ladder concept. The ladder represents a context for the accounting event management encompassing the accountable objects involved in the service session. Accountable objects are any resources whose use is to be charged. They support some basic management functionality and can generate accounting events. The control messages are passed from the top of the ladder which is a USM within the service layer. Accounting events are passed from one rung of the ladder to the next, starting with the element management layer (EML) in the resource layer at the bottom of the ladder and moving toward the top, so that service-level billing information can be synthesized from the resource-level accounting events.

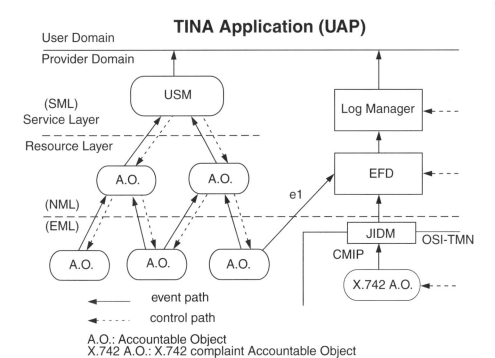

FIGURE 3.5.8 Event Management Ladder (EML). (From Hamada, T., Kamata, H., and Hogg, S., An Overview of the TINA Management Architecture, *Journal of Network and Systems Managements,* 5, 411–435, 1997. Used with permission of Plenum Publishing Corporation.)

The two accounting systems can interwork. For example, an accountable object in the ladder can send an event to the EFD as shown by e1 in the figure. This type of interworking between the two accounting systems is the key to the fine-grain migration from TMN-style accounting management to a TINA-based accounting management system using the event management ladder.

3.5.5 Life Cycle Axis

The life cycle axis deals with service and user life cycles. Service life cycle management considers how providers can deploy services, make them available, and control instance creation and deletion. TINA service life cycle management builds on software life cycle ideas with an emphasis on deploying services in a distributed, object-oriented environment across multiple domains. In a TINA service environment, user starts its life cycle by creating a new account. User life cycle management considers how users are introduced in the TINA service environment, and how they are supported and managed.

3.5.5.1 Service Life Cycle Management

Figure 3.5.9 summarizes the TINA service life cycle model. Every TINA service goes through the states described in the figure. The model focuses on the deployment/withdrawal phases, where three life cycle states corresponding to software states are shown. Further analysis of the service life cycle may reveal more states, as in the Service Instance Life Cycle (SILC). There are finer states within the utilization phase, such that service instances have their own life cycle, separate from the service type.

The following states are identified in the service life cycle:

- *Conceived, not planned:* The service has been conceived but no details about its implementation are known.
- *Not installed:* The service has been planned and does not exist in a TINA environment (although it might have existed in the past).
- *Installed:* The constituent parts of the service exist in a TINA environment but the service cannot be instantiated.
- *Activatible:* The service has the potential for being instantiated.
- *Instantiated:* An instance of the service has been instantiated.

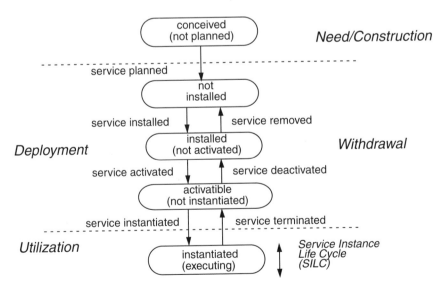

FIGURE 3.5.9 Service Instance Life Cycle (SILC) state transitions. (From Hamada, T., Kamata, H., and Hogg, S., An Overview of the TINA Management Architecture, *Journal of Network and Systems Managements*, 5, 411–435, 1997. Used with permission of Plenum Publishing Corporation.)

Scenario:

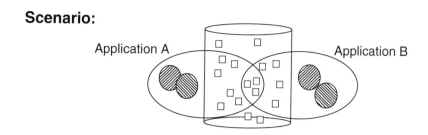

Application A   Application B

Representation:

Service Group A   Service Group B

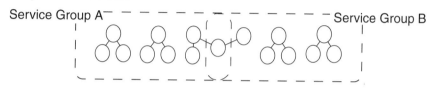

FIGURE 3.5.10   Two Service Groups (SGPs) sharing an Object Group. (From Hamada, T., Kamata, H., and Hogg, S., An Overview of the TINA Management Architecture, *Journal of Network and Systems Managements,* 5, 411–435, 1997. Used with permission of Plenum Publishing Corporation.)

3.5.5.1.1   Service Group

Since deployment and withdrawal of a service implies that a group of objects constituting the service needs to be deployed or withdrawn, it is natural to bundle these as a Service Group (SGP) [38]. Entities may include objects, executable files, data bases, and other resources required for a service to achieve an activatible state. The advantages of this grouping are best illustrated by an example. Consider a database which serves a number of applications, as shown in Figure 3.5.10. There is data in the database used by Application A. A subset of this data is also used by Application B. This situation is shown by the intersection of application data in the figure.

The two applications are each modelled as service groups as shown in the lower part of the figure. Shared data and objects are modelled by the intersection of the two service groups. Should Application A be withdrawn, then it is necessary to remove the data used solely by Application A but to leave those objects and data used by both Application A and Application B. Thus when withdrawing SGP A, the object group in the intersection of the two SGPs must remain.

In the view of the life cycle management, a TINA service is represented by a set of SGPs. Each SGP can be associated with a SGP template, which is the computational representation of service, encompassing the binaries of contained objects and service groups and any configuration information. A SGP template describes how a new service group can be instantiated and installed on a deployment request. A service group or a service group template provides the following properties:

- *Containment:* One or more SGPs may share an entity (e.g., an object group or database). A SGP may contain one or more other SGPs. These complexities are hidden from clients.
- *Scope:* A SGP may span one or more DPEs.
- *Dependencies:* A SGP knows the dependencies between contained entities, such as object groups or other SGPs. A SGP is dependent on the prior installation and activation of contained SGPs and SGPs with which it shares entities.

SGP is similar to the object grouping concept at DPE level. In particular, both support a similar containment hierarchy concept. They are distinguished by two of the properties listed earlier: their scope — a SGP may extend across several DPEs — and the concept of deployment dependency, which is not considered by object groups. The last feature is particularly important in deploying services. For example, a certain object group may need to be installed before others to avoid generating false alarms. Deployment dependency may also be relevant in the withdrawal phase.

3.5.5.1.2 Service Type Management

Service type management (Figure 3.5.11) handles control and management of service types, service type aliases, and service type templates. A service type template gives a generic template of service type, from which individual service instances are created. A service type must be maintained for the active lifetime of a TINA service, and the service type is also used for subscription and service instance creation. Therefore, service type management is an indispensable part of service life cycle management in TINA.

The service type manager is responsible for maintaining a service type during a service's active lifetime, and withdrawing it when the service is turned off. Although some functionality of the service type manager is similar to that of a DPE Trader [39] and CORBA trading service [41], it entails obtaining and distributing unique service type identifiers, registering and distributing service type templates, version control, and sub-domain control, as well as making particular services available.

3.5.5.1.3 Service Instance Management

Service type management and service instance management are both part of the utilization phase of service management. However, instance management covers the dynamic aspects, while service type management covers the static aspects. Service instance management includes creation and deletion of service instances, and monitoring and control of service instances (e.g., stop, resume) during their lifetimes.

The service factory is responsible for creation and deletion of each service instance for a specific service type. A service factory contains a service template which is a service-specific information object that determines how to create (or delete) a service instance. The service instance management operations are applied to the service session through the management interface of their corresponding SSMs.

3.5.5.2 User Life Cycle Management

Since consumer care and support are part of the retailer's business, it is important to consider consumers and their management. TINA has a clear view of consumers in its business model. A consumer role is a business role assumed by a stakeholder, which can be associated with a number of session roles in a service session [6]. Session roles can be associated with access and service usage. Access session roles determine how a person is recognized by the provider, and how a person's access privileges are given from the provider. TINA has a subscription model in which subscribers represent a stakeholder to negotiate contractual relations. An end-user can access TINA services based on subscription information and the user's service profile information created from the subscription management. We separate user life cycle management into two parts: end-user management and subscription management. Figure 3.5.12 shows the relationship between subscriber management, end-user management, and service type management.

3.5.5.2.1 Subscription Management

Subscription management allows a subscriber to establish a contractual relation between a user and a retailer, and it sets up a user service profile and user account information such that the user has access to the subscribed service. It involves the following information objects:

- *Subscription contract* represents the relationship established between consumer and retailer.
- *Subscriber profile* contains all relevant information regarding the subscriber.
- *Subscriber service profile* describes a particular service and its customization for the subscriber.
- *Service assignment group (SAG)* binds a service to one or more end users. SAGs specify the services available to a subscriber's associated users. An end-user may belong to multiple SAGs.

3.5.5.2.2 End-User Management

The end-user management allows end users to customize their services within the bounds set by their subscriber. It involves the following information objects:

- *End-user profile* represents all relevant information of the given end user, which includes registration, session description, usage context, and end-user service profile.
- *End-user service profile* describes a particular service and its relation to the end user, including customizing information. This is similar to a subscriber service profile except that it relates to a single end user.

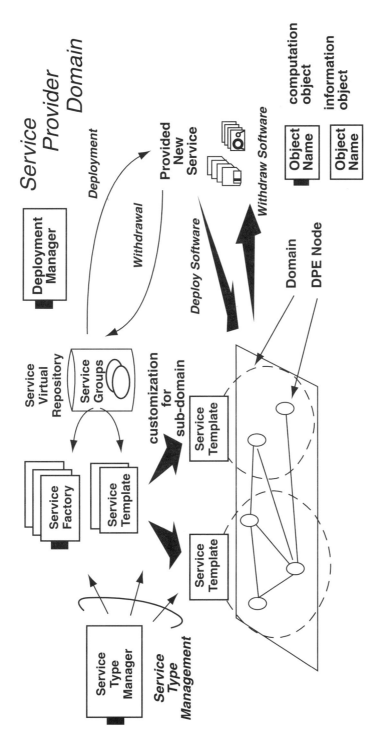

FIGURE 3.5.11 Service deployment and type management fragments. (From Hamada, T., Kamata, H., and Hogg, S., An Overview of the TINA Management Architecture, *Journal of Network and Systems Managements*, 5, 411–435, 1997. Used with permission of Plenum Publishing Corporation.)

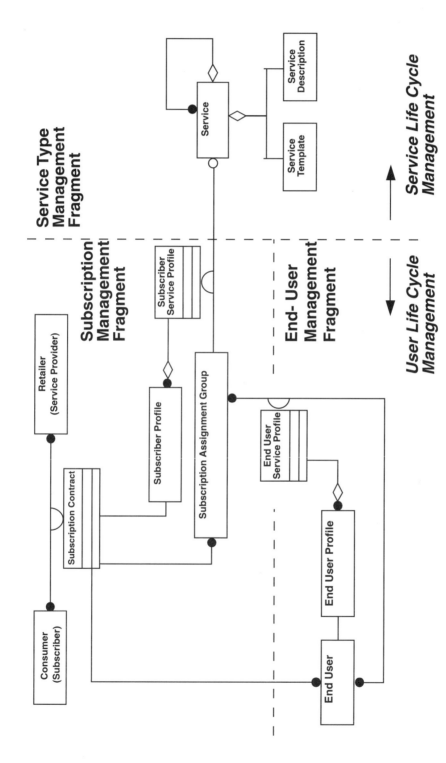

FIGURE 3.5.12 Life cycle management information model fragments. (From Hamada, T., Kamata, H., and Hogg, S., An Overview of the TINA Management Architecture, *Journal of Network and Systems Managements*, 5, 411–435, 1997. Used with permission of Plenum Publishing Corporation.)

3.5.6 Summary and Further Work

TINA imposes a high degree of user control and flexibility in the service layer. It also emphasizes service requirements in a multiprovider environment, and their impact on the resource layer. TINA uses domains to support multiple providers and introduces management policy as a means of controlling domain management. To meet service and functional requirements, the concepts such as management context and service transactions have been devised. The result is a flexible management framework, aimed at supporting consistent management across multiple domains.

TINA also considers service life cycle management and user life cycle management. Service grouping concepts enhance flexibility in deployment and withdrawal of services, and they allow providers to manage the complexity of service life cycle management, particularly when services are interrelated and service components are reused across multiple services. Subscription and subscriber control allows an entity (organization or household) to place controls on user access to services, while users can customize their services, allowing the services to meet individual needs and preferences. It provides retailers with a framework for supporting consumer relations at an enterprise level, which may span multiple services and end users.

Although the current framework presented in this chapter can offer considerable management support to TINA services, there are still many important research issues demanding further study. For example, we consider that service configuration management should be studied as an integrated part of service life cycle management. To derive a practical architecture, however, more use case analyses based on real-world examples are necessary. The service session graph [38] was originally designed as a generic tool for session control, but it can also be used for management support. A security management framework in TINA has yet to be fully developed. Some progress has been made in this direction [28] by using role-based control in conjunction with service session graph.

Acronyms

3Pty	3rd party service (reference point)
CMIP	Common management information protocol
ConS	Connectivity service (reference point)
CORBA	Common object request broker architecture
CSM	Communication session manager
DPE	Distributed processing environment
EFD	Event forwarding discriminator
EML	Element management layer
FCAPS	Fault, configuration, accounting, performance, security
GCCM	General context configuration manager
GDMO	Guideline for the definition of managed object
JIDM	Joint inter-domain management
LNFed	Layer network federation (reference point)
MgmtCtxt	Management context
NEL	Network element layer
NMF	Network management forum
NML	Network management layer
NRA	Network resource architecture
ODL	Object definition language
ODMA	Open distributed management architecture
ODP	Open distributed processing
OMG	Object management group

OSI	Open systems interconnection
PA	Provider agent
QoS	Quality of service
QoP	Quality of performance
Ret	Retailer (reference point)
SAG	Subscription assignment group
SF	Service factory
SGP	Service group
SILC	Service instance life cycle
SLC	Service life cycle
SML	Service management layer
SS	Service session
SSM	Service session manager
TINA [-C]	Telecommunications Information Networking Architecture [Consortium]
TMN	Telecommunications management network
UA	User agent
UAP	User application
USM	User service session manager

References

[1] Barr, W. J., Boyd, T., and Inoue, Y., The TINA Initiative, *IEEE Communication Magazine*, March, 70, 1993.

[2] ISO/IEC 10746-1, *Information Technology — Basic Reference Model of Open Distributed Processing — Part 1: Overview*, 1995.

[3] TINA-C, *Information Modelling Concepts, Ver. 2.0*, April 1995.

[4] TINA-C, *Computational Modelling Concepts, Ver. 3.2*, May 1996.

[5] TINA-C, *TINA Object Definition Language Manual, Ver. 2.3*, July 1996.

[6] TINA-C, *Service Architecture, Ver. 5.0*, June 1997.

[7] TINA-C, *Service Component Specification — Computational Model and Dynamics*, January 1998.

[8] ITU-T, *Principles for a Telecommunication Management Network*, ITU-T Recommendation M.3010, 1992.

[9] ITU-T, *Open Distributed Management Architecture*, ITU-T Recommendation X.703, 1997.

[10] Network Management Forum, *OMNIPoint Technical Architecture, Delivering a Management System Framework to Enable Service Management Solutions*, 1994.

[11] TINA-C, *Network Resource Information Model Specification, Ver. 2.2*, November 1997.

[12] ITU-T, *Architectures of Transport Networks Based on the Synchronous Digital Hierarchy (SDH)*, ITU-T Recommendation G.803, June 1992.

[13] ITU-T, *Architectures of Transport Networks*, ITU-T Recommendation G.805, June 1995.

[14] Pavon, J., Tomas, J., Bardout, Y., and Hauw, L., CORBA for Network and Service Management in the TINA Framework, *IEEE Communications Magazine*, March, 72, 1998.

[15] Eurescom Project P508, *Annex 1 Migration Paths Towards TINA*, ref P508.BT.PL.006, December 1996.

[16] Eurescom Project P508, *CORBA as an Enabling Factor for Migration from IN to TINA*, ref P508.FT.INCWP.4.1, December 1996.

[17] Sidor, D., TMN Standards: Satisfying Today's Needs While Preparing for Tomorrow, *IEEE Communications Magazine*, March, 54, 1998.

[18] TINA-C, *TINA Reference Points, Ver. 3.1*, June 1996.

[19] Network Management Forum, *Network Management Detailed Operations Map — Detailed Process Models to Support the Telecommunications Operations Map*, NMF GB908 Draft 0.9, March 1998.

[20] Network Management Forum, *NMF Technology Map,* NMF GB909 Issue 1.0, March 1998.

[21] Network Management Forum, *NMF Telecom Operations Map,* NMF GB910 Draft 0.2b, April 1998.

[22] Parhar, A. and Handegard T., TINA Object Groups: Patterns in Chaos, *Proc. of TINA'96,* 13, Heidelberg, Germany, September 1996.

[23] ITU-T, *Terms and Definitions Related to the Quality of Telecommunication Services,* ITU-T Recommendation E.801, October 1996.

[24] Hamada, T., TINA Accounting Management Architecture, *Proc. of TINA'96,* 193, Heidelberg, Germany, September 1996.

[25] Eckardt, T. et al., Personal Communications Support in the TINA Service Architecture — A new TINA-C Auxiliary Project, *Proc. of TINA'96,* pp. 55-64, Heidelberg, Germany, September 1996.

[26] Staamann, S., Wilhelm, U. et al., Security in the Telecommunication Information Networking Architecture — the CrysTINA Approach, *Proc. of TINA'97,* Santiago de Chile, November 1997.

[27] Hamada, T., Hoshiai, T., Yates, M., A Perspective on TINA Service Security Architecture, *Proc. of 5th Workshops on Enabling Technologies (WET ICE'96),* Stanford, California, 74, June 1996.

[28] Hamada, T., Dynamic Role Creation from Role Hierarchy — Security Management of Service Session in Dynamic Service Environment, *Proc. of TINA'97,* Santiago de Chile, November 1997.

[29] OMG, *Common Object Request Broker: Architecture and Specification, Revision 2.0,* July 1995.

[30] Appleby S. and Steward S., Mobile software agents for control in telecommunications networks, *BT Technology Journal,* Vol. 12, No. 2, 104, April 1994.

[31] NMF, *NMF Component Sets: CORBA/CMIP/SNMP Interworking,* Draft, CS342, 1996.

[32] X/Open, *Inter-Domain Management Specifications: Specification Translation (JIDM),* 1995.

[33] OMG, *The Common Object Request Broker: Common Object Service Specifications, Revision 1.0,* OMG, 94-1-1, January 1994.

[34] X/Open, *Generic Security Service,* X/Open Publication Set T402, 1994.

[35] OMG, *The Common Object Request Broker: TINA Notification Service Description,* telecom/96-07-02, July 1996.

[36] OMG, *The Common Object Request Broker: Notification Service Request for Proposal,* Draft 4, telecom/96-12-01, December 1996.

[37] TINA-C, *Network Resource Architecture Ver. 3.0,* February 1997.

[38] TINA-C, *Service Architecture Annex, Ver. 5.0,* June 1997.

[39] TINA-C, *Engineering Modelling Concepts (DPE Architecture), Ver. 2.0,* December 1994.

[40] Berndt, H., Darmois, E., Dupuy, F. et al., *The TINA Book — Cooperative Solution for Competitive World,* Prentice-Hall, 1998.

[41] OMG, *Trading Object Service,* OMG RFP 5 Submission, OMG Document orbos/96-07-08, July 1996.

[42] Dupuy, F., Nilsson, G., and Inoue, Y., The TINA Consortium: Toward Networking Telecommunications Information Services, *Proc. of International Switching Symposium (ISS),* Berlin, Germany, 1995.

[43] Rubin, H. and Natarajan, N., A Distributed Software Architecture for Telecommunications Applications, *IEEE Network Magazine,* Vol. 8, No. 1, January/February 1994.

Further Information

All public architectural documents and interface specifications are available at the TINA-C home page (http://www.tinac.com). Other useful information on TINA-C such as FAQ, tutorials from the past TINA conferences, and updates on current activities are also available from the TINA-C home page. Since 1998, TINA-C has been in phase II, and the latest updates and revisions are performed by its working groups. Latest activities of the working groups are also available from the home page. The TINA Book [40] is written by the foremost experts on TINA, and the book stands to be the standard source of reference.

International TINA conferences, which have been held annually since 1990, are excellent sources of the foremost research papers on TINA. For the historical development of TINA and TINA-C, [1] and [42] give good information. For Bellcore Information Networking Architecture (INA) project, a predecessor of TINA, [43] is recommended.

Due to limitation of space, several important topics of TINA had to be omitted from this chapter. For distributed processing environment (DPE), the DPE architecture document [39] is the most reliable source of information. Since a major part of DPE is made compliant with OMG CORBA, the difference between the two distributed architectures is practically minimal. There still remain some differences, however, as influences from ODP are more apparent in TINA DPE. In particular, ODP concepts such as capsule, cluster, stream interface, and multiple interface objects are not yet fully supported in the current version of CORBA [36]. The latest updates on OMG activities are available from its web site (http://www.omg.org). For TINA-related activities at ITU-T, TINA ODL [5] is currently on standardization track at ITU-T SG 10. TINA is also being studied as Long Term Architecture (LTA) at ITU-T SG 11.

Connection management is another important topic in TINA. Network Resource Information Model (NRIM) [11] and Network Resource Architecture (NRA) [37] are the two most reliable sources of information. NRIM describes the information viewpoint, whereas NRA describes the computational viewpoint of network resources.

Many TINA-related researchers have been supported as part of the European ACTS program (http://www.uk.infowin.org/ACTS) and Eurescom (http://www.eurescom.de). Project reports from Eurescom are available (http://www.eurescom.de/public/deliverables/dfp.htm).

For TINA service architecture, the service architecture (SA) document [6] and its companion document, the service component specification (SCS) [7] are the most reliable sources. SA defines basic concepts of TINA service architecture, and SCS describes computational models, interfaces, and event sequences of service components.

3.6   Telecommunications Support Processes

Kornel Terplan

This contribution is based on the SMART TMN Telecom Operations Map by the TeleManagement (TM) Forum (TELE98). The TM Forum is an international, non-profit organization serving the telecommunications industry. Its mission is to help service providers and network operators automate their business processes in cost- and time-effective ways. Specifically, the work of the TM Forum includes:

- Establishing operational guidance on the shape of business processes
- Agreeing on information that needs to flow from one function to another
- Identifying a realistic systems environment to support the interconnection of operational support systems
- Enabling the development of a market and of real products for automating telecom operations processes
- TM Forum makes use of international and regional standards when available, and provides input to standards bodies whenever new technical work is done

The members of the TM Forum include service providers, network operators, and suppliers of equipment and software to the telecommunications industry. With that combination of buyers and suppliers of operational support systems, TM Forum is able to achieve results in a pragmatic way that leads to product offerings as well as to specifications of business processes and support systems. Members meet regularly; several working groups are in action, targeting real challenges in the telecommunications industry. This documentation is extremely valuable to old and new providers who are in the process of re-engineering their support processes. The editor and author thanks the TM Forum for the assistance with this contribution.

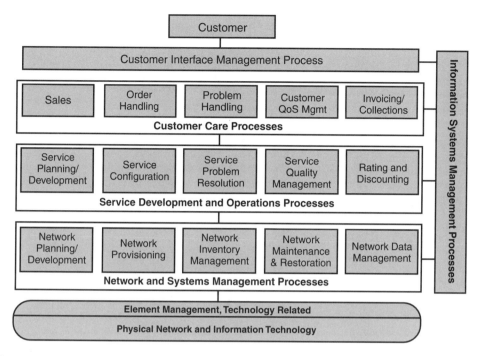

FIGURE 3.6.1 Business model of telecommunications support processes.

A telecommunications service provider efficiently and effectively conducts business by managing its essential business and support processes. These processes can be aggregated to deliver the major requirements common to any service-oriented business, which include:

- Service fulfillment (timely delivery of what the customer has ordered)
- Service assurance (maintaining the service — timely response and resolution of customer and network triggered problems, managing and reporting performance for all aspects of a service)
- Billing for the service (timely and accurate bills, including invoicing, timely adjustment handling and payment collections)

Figure 3.6.1 shows the business process model as recommended by the TM Forum (ADAM96).

3.6.1 High-level Breakdown of Support Processes

Figure 3.6.2 shows a broad breakdown of the business model into the three customer-focused activities identified earlier. The purpose is to show the second dimension in more detail and highlight the predominant processes that need to be involved in the end-to-end tasks of supporting customer services. The network inventory management and network data management processes are split to display their significant role, more than an interface, in both overarching processes; e.g., network inventory management is important to both the fulfillment and assurance processes.

Each of the principal areas are to be described with simplified flow charts. The TM Forum expects to develop robust process flows both at a generic level and in some more specific areas. These process flows are representative, and need customization, but they are an excellent basis for re-engineering processes.

Fulfillment Example

This example of a fulfillment process shows a possible sequence of activities to support a customer inquiry, subsequent order for service, the configuration of the service, and the installation and completion of the request (Figure 3.6.3). Depending on the service provider process, orders can be placed through the sales process and/or directly through the order management process. For a specific service provider, some

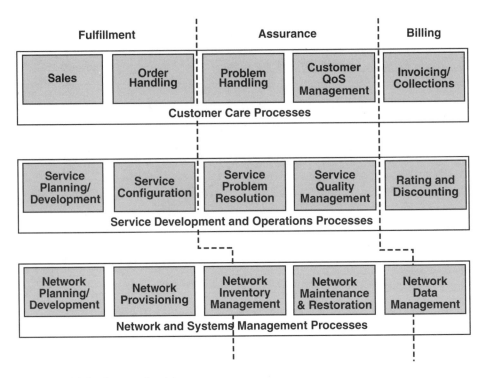

FIGURE 3.6.2   High-level process breakdown.

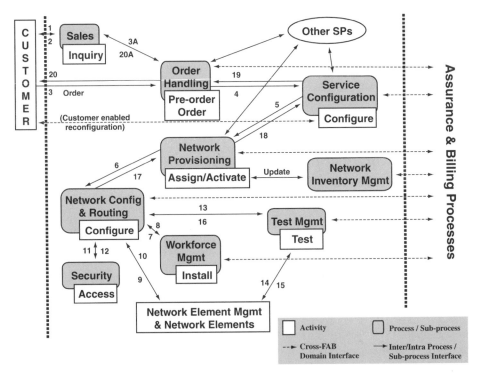

FIGURE 3.6.3   Fulfillment process example.

customers may be supported by a specific sales team that places some or all orders for the customer and tracks them to completion. These dual-trigger process interfaces and follow-ups are shown as 3/3A and 20/20A.

Information exchanges are numbered as follows:

  1 Selling
  2 Selling
  3 Order
 3A Order status and completion
  4 Service request
  5 Assignment request
  6 Network configuration request
  7 Installation request and completion
  8 Installation request and completion
  9 Element configuration complete
10 Element configuration complete
11 Network access check and complete
12 Network access check and complete
13 Test request
14 Perform test and test data
15 Perform test and test data
16 Test complete
17 Network configuration complete
18 Assignment complete
19 Service complete
20 Order status and completion
20A Order status and completion

The complete fulfillment flow-through may not actually be required every time for some simple services, which have preassigned service capacity. For example, the flow for an instance of a service set-up could be bypassed at network provisioning, when configured and tested facilities have been preprovisioned. This depends upon a particular provider's operational process and policy. It will also impact the timing of interactions with network inventory management, hence the interface sequence number has been omitted. Interfaces may be required with other service providers or network operators when the service offered to a customer is one of many different kinds of joint service arrangements.

Service Assurance Example
Most service providers are driving their service assurance processes to become proactive, meaning triggered by automation rather than triggered by the customer. This is important for improving service quality, customer perception of service, and for lowering overall costs. Customer care processes have been basically reactive in the past. The extreme pressure on cost, customer demand for more control, and customer demand for more proactive service support are driving a major shift to proactive support through automation. With the advent of internet access, the goal for processes and automation is now interactive support, including giving the customer the ability to see and act on service performance.

Service assurance processes are interlinked to each other. Figure 3.6.4 shows a possible sequence of activities in response to a network-detected problem. The figure shows two ways a potential service-affecting problem could be identified, e.g., by either an "alarm event" or by synthesis of network data through Network Data Management. Neither is exclusive. Network data management logically collects and processes both performance and traffic data as well as usage data. The usage data is used as logical part of the billing process.

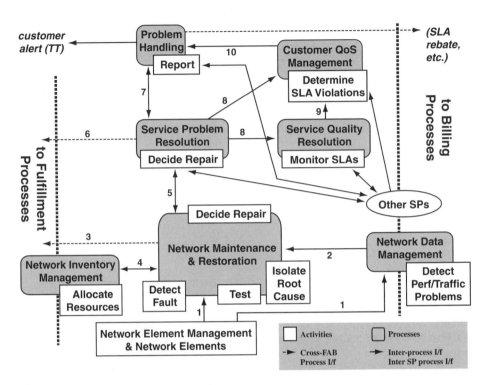

FIGURE 3.6.4   Service assurance process example.

Information exchanges are numbered as follows:

1 Network data, alarm data, events
2 Report degradation
3 Network reconfiguration or change
4 Work order
5 Notify problem/fix
6 Service reconfiguration or change
7 Trouble report or ticket
8 Report problem data
9 SLA impact
10 Service impact

Billing Example

Figure 3.6.5 shows a typical sequence of activities to generate a bill that has flat rate elements (one-time installation, monthly recurring), usage charges, and possible SLA adjustments. Service providers may also choose to apply discounts or rebates for outages and/or SLA breaches to a specific customer's bill, according to service type, promotion, customer relationship, and/or according to its policy or customer contract.

When a service is provided by a combination of different service providers, usage and/or other billing data may be aggregated by the main service provider from input by other secondary service providers, and one bill presented to the customer. This is a trend, but it depends on service provider billing strategy, customer wishes, the actual service arrangement provided, and/or service provider process capability and policy.

Information exchanges are numbered as follows:

1 Network usage data
2 Aggregated usage data

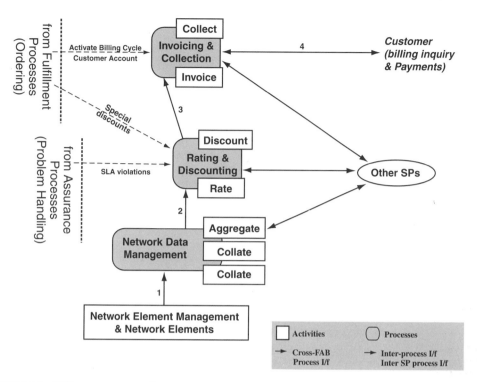

FIGURE 3.6.5 Billing process example.

3 Summarized bill content
4 Generate bills

In the next chapters, each process will be described by its tasks, input, and output connections with internal and external processes and activities.

3.6.2   Customer Care Processes

These processes involve direct interaction with an end customer to provide, maintain, and bill for network services. The end customer is the ultimate buyer of a network service.

3.6.2.1   Customer Interface Management Process

This process may be distinct, or may be performed as part of the individual customer-care processes on an individual service or cross-service basis. These are the processes of directly interacting with customers and translating customer requests and inquiries into appropriate events, such as the creation of an order or trouble ticket or the adjustment of a bill. This process logs customer contacts, directs inquiries to the appropriate party, and tracks the status to completion. In those cases where customers are given direct access to service management systems, this process assures consistency of image across systems, and security to prevent a customer from harming their network or those of other customers. The aim is to provide meaningful and timely customer contact experiences as frequently as the customer requires.

Figure 3.6.6 shows the customer interface management process. Principal functions include:

- Receive and record contacts
- Direct inquiries to appropriate processes
- Monitor and control status of inquiries, and escalate
- Ensure a consistent image and secure use of systems

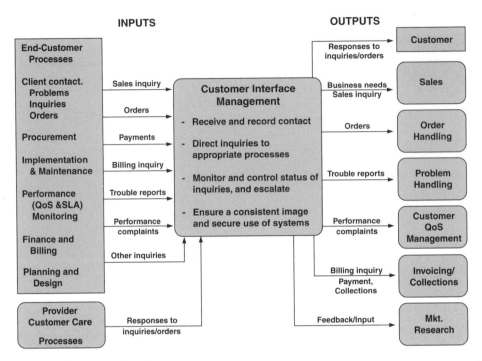

FIGURE 3.6.6   The customer interface management process.

3.6.2.2   Sales Process

This process encompasses learning about the needs of each customer, and educating the customer about the communications services that are available to meet those needs. It includes working to create a match between the customer's expectations and the service provider's ability to deliver. Depending on the service provider process it can be pure selling or can include various levels of support. The sales process may include preorder work and interfaces. The aim is to sell the correct service to suit the customer's need and to set appropriate expectations with the customer. SLA negotiation, Request for Proposal (RFP) management, and negotiation are led from this process.

Figure 3.6.7 shows the sales process. Principal functions include:

- Learn about customer needs
- Educate customer on services
- Match expectations to offerings and products
- Arrange for appropriate options
- Forecast service demand
- Manage SLA and RFP negotiations

3.6.2.3   Ordering Process

The ordering process includes all the functions of accepting a customer's order for service, tracking the progress of the order, and notifying the customer when the order is complete. Orders can include new, change, and disconnect orders for all or part of a customer's service, as well as cancellations and modifications to orders. Preorder activity that can be tracked is included in this process. The development of an order plan may be necessary when service installation is to be phased in, and the need for preliminary feasibility requests and/or pricing estimates may be part of this process when certain services are ordered. The aim is to order the service the customer requested, support changes when necessary and to keep the customer informed with meaningful progress of the order, including its successful completion.

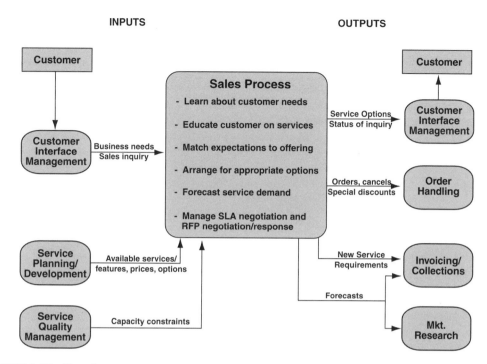

FIGURE 3.6.7   The sales process.

Figure 3.6.8 shows the ordering process. Principal functions of this process include:

- Accept orders
- Determine preorder feasibility
- Prepare price estimates and SLA terms
- Develop order plan
- Perform credit check
- Request customer deposit
- Initiate service installation
- Reserve resources
- Issue orders, and track status
- Complete orders, notify customers
- Initiate billing process

Service orders to other providers can be generated by the ordering process or by the service configuration process, depending on the nature of the service ordered by the customer. They can also be generated from network and systems management processes when part of a network infrastructure.

3.6.2.4   Problem-Handling Process

This process is responsible for receiving service complaints from customers, resolving them to the customer's satisfaction and providing meaningful status on repair or restoration activity. This process is also responsible to be aware of any service-affecting problems, including:

- Notifying customers in the event of a disruption — whether reported by the customer or not
- Resolving the problem to the customer's satisfaction
- Providing meaningful status on repair or restoration activity

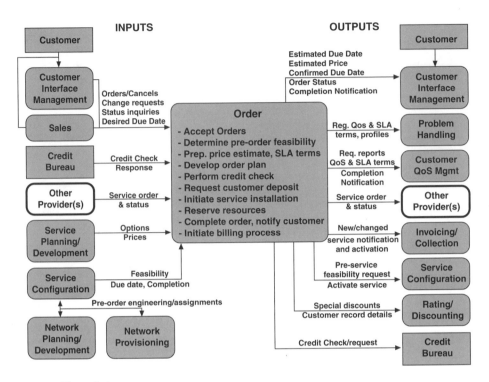

FIGURE 3.6.8   The ordering process.

This proactive management also includes planned maintenance outages. The aim is to have the largest percentage of problems proactively identified and communicated to the customer, to provide meaningful status, and to resolve in the shortest timeframe.

Figure 3.6.9 shows the problem-handling process. Principal tasks include:

Receive trouble notification
Determine cause, resolve, or refer
Track progress of resolution
Initiate action to reconfigure if needed
Generate trouble tickets to suppliers
Confirm trouble cleared, notify customer
Schedule with and notify customer of planned maintenance

When trouble is reported by the customer, a trouble report may be sent to service problem resolution for correction. When a trouble is identified by service problem resolution, then problem handling is notified in order to inform the customer of the problem.

3.6.2.5   Customer Quality of Service (QoS) Management Process

This process encompasses monitoring, managing, and reporting of quality of service (QoS) as defined in service descriptions, service level agreements, and other service-related documents. It includes network performance, but also performance across all service parameters, e.g., orders completed on time. Outputs of this process are standard (predefined) and exception reports including, but not limited to, dashboards, performance of a service against a SLA, reports of any developing capacity problems, reports of customer usage patterns, etc. In addition, this process responds to performance inquiries from the customer. For SLA violations, the process supports notifying problem handling, and for QoS violations, notifying service quality management. The aim is to provide effective monitoring. Monitoring and reporting must provide

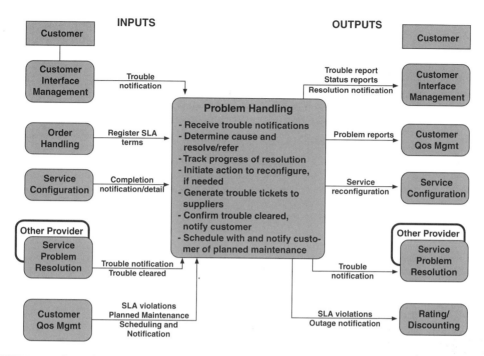

FIGURE 3.6.9 The problem-handling process.

SP management and customers with meaningful and timely performance information across the parameters of the services provided. The aim is also to manage service levels that meet specific SLA and standard service committments.

Figure 3.6.10 shows the customer QoS process. Principal tasks include:

- Schedule customer reports
- Receive performance data
- Establish reports to be generated
- Compile and deliver customer reports
- Manage SLA performance
- Determine and deliver QoS and SLA violation information

3.6.2.6 Invoicing and Collection Process

This process encompasses sending invoices to customers, processing their payments, and performing payment collections. In addition, this process handles customer inquiries about bills, and is responsible for resolving billing problems to the customer's satisfaction. The aim is to provide a correct bill and, if there is a billing problem, resolve it quickly with appropriate status to the customer. An additional aim is to collect monies due the service provider in a professional and customer-supportive manner.

Some providers allow invoicing and collections functions for other providers as a service.

Figure 3.6.11 shows the invoicing and collection process. The principal functions include:

- Create and distribute invoices
- Collect payments
- Handle customer account inquiries
- Manage debt
- Bill on behalf of other providers

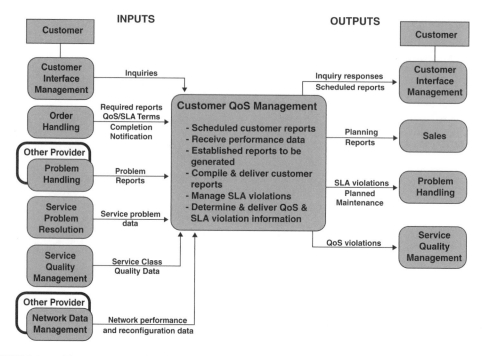

FIGURE 3.6.10   The customer quality of service (QoS) management process.

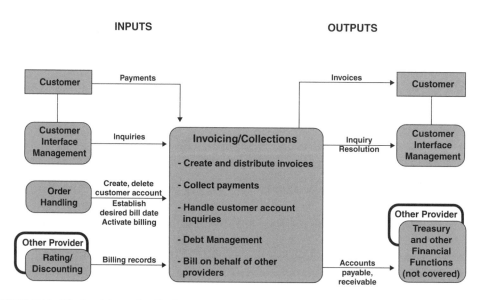

FIGURE 3.6.11   The invoicing and collection process.

3.6.3   Service Development and Operations Processes

These processes are generally one step removed from day-to-day customer interaction. Focus is on service delivery and management as opposed to the management of the underlying network and information technology. Some of these functions are done on a one-time basis, such as designing and delivering a new service or feature. Other functions involve the application of a service design to specific customers or managing service improvement initiatives, and are closely connected with the day-to-day customer experience.

3.6.3.1 Service Planning and Development Process

This process encompasses the following functional areas:

- Designing technical capability to meet specified market need at desired cost
- Ensuring that the service (product) can be properly installed, monitored, controlled, and billed
- Initiating appropriate process and methods modifications, as well as initiating changes to levels of operations personnel and training required
- Initiating any modifications to the underlying network or information systems to support the requirements
- Performing preservice testing to confirm the technical capability works and that the operational support process and systems function properly
- Ensuring that sufficient capacity is available to meet forecasted sales

Figure 3.6.12 shows the service planning and development process. Principal functions are:

- Develop and implement technical solutions
- Develop and implement procedures
- Define and implement systems changes
- Develop and implement training
- Develop customer documentation
- Plan rollout, test, start service, and project management
- Set product/service pricing

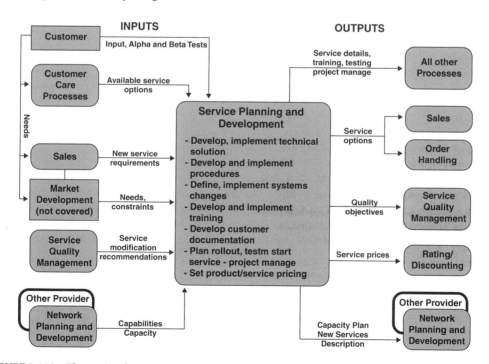

FIGURE 3.6.12 The service planning and development process.

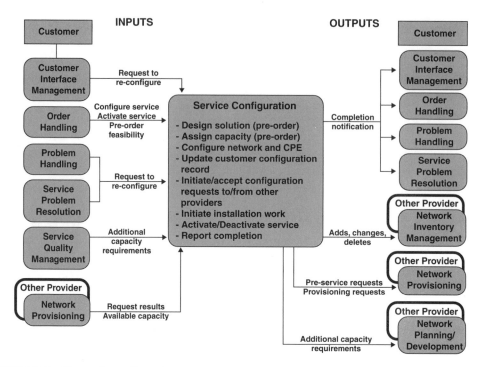

FIGURE 3.6.13   The service configuration process.

3.6.3.2   Service Configuration Process

This process encompasses the installation and/or configuration of services for specific customers, including the installation/configuration of customer premises equipment (CPE). It also supports the reconfiguration of service (either due to customer demand or problem resolution) after the initial service installation. The aim is to correctly provide service configuration within the timeframe required to meet ever-decreasing intervals.

Figure 3.6.13 shows the service configuration process. Principal functions include:

Design solution (preorder)
Assign capacity (preorder)
Configure network and CPE
Update customer configuration record
Initiate/accept configuration requests to/from other providers
Initiate installation work
Activate/deactivate service
Report completion

3.6.3.3   Service Problem Resolution Process

This process encompasses isolating the root cause of service affecting and non-service affecting failures and acting to resolve them. Typically, failures reported to this process impact multiple customers. Actions may include immediate reconfiguration or other corrective actions. Longer-term modifications to the service design or to the network components associated with the service may also be required. The aim is to understand the causes impacting service performance and to implement immediate fixes or initiate quality improvement efforts.

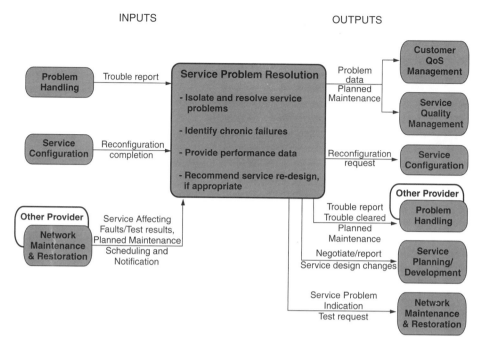

FIGURE 3.6.14 The service problem resolution process.

Figure 3.6.14 shows the service problem resolution process. The principal tasks include:

- Isolate and resolve service problems
- Identify chronic failures
- Provide performance data
- Recommend service redesign, if appropriate

When multiple troubles are reported by customers of a service, a report may be sent from problem handling to service problem resolution for correction. When trouble is identified by service problem resolution, Problem Handling is notified.

3.6.3.4 Service Quality Management Process

This process supports monitoring service or product quality on a service class basis in order to determine whether:

- Service levels are being met consistently
- There are any general problems with the service or product
- The sale and use of the service is tracking to forecasts

This process also encompasses taking appropriate action to keep service levels within agreed targets for each service class and to either keep ahead of demand or alert the sales process to slow sales. The aim is to provide effective service-specific monitoring, to provide management and customers with meaningful and timely performance information across the parameters of the specific service. The aim is also to manage service levels to meet SLA commitments and standard commitments for the specific service.

Figure 3.6.15 shows the service quality management process. The principal tasks include:

- Manage life cycle of service/product portfolios
- Monitor overall delivered quality of a service class
- Monitor available capacity/usage against forecasted sales

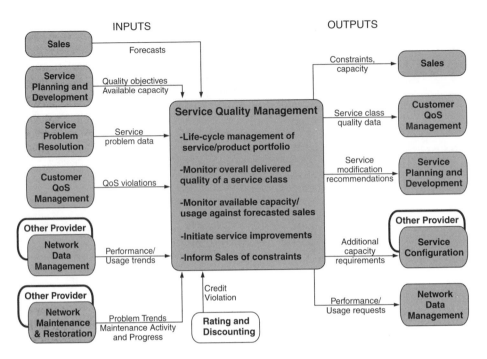

FIGURE 3.6.15   The service quality management process.

- Initiate service improvements
- Inform sales on constraints

3.6.3.5   Rating and Discounting Process

The rating and discounting process encompasses the following functional areas:

- Applying the correct rating rules to usage data on a customer-by-customer basis, as required
- Applying any discounts agreed to as part of the ordering process
- Applying promotional discounts and charges
- Applying outage credits
- Applying rebates due because SLA were not met
- Resolving unidentified usage

The aim is to correctly rate usage and to correctly apply discounts, promotions, and credits. Figure 3.6.16 shows the rating and discounting process. Principal functions are:

- Apply service rates to usage
- Apply negotiated discounts
- Apply rebates

3.6.4   Network and Systems Management Processes

These processes are responsible for ensuring that the network infrastructure supports the end-to-end delivery of the required services. Network Management is a key integration layer between the element management layer and the service management layer. Its basic function is to assemble information from element management systems, and then integrate, correlate, and in many cases, summarize that data to pass on the relevant information to service management systems.

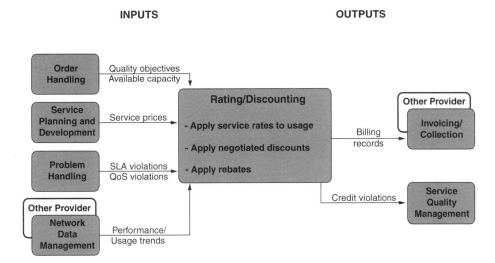

FIGURE 3.6.16   The rating and discounting process.

Network Management is more than just a mediator between the element and service management layers. It has its own responsibilities, e.g., network provisioning and network fault management. The key issue is that management responsibility will be placed at a level where adequate information is present, instead of shifting all responsibilities to service management. At this point, systems management is not explicitly addressed, but would logically need to be integrated into the architecture at this level.

The network and systems management processes manage the complete service provider network and subnetwork. Much of the interface is through element management.

3.6.4.1   Network Planning and Development Process

This process encompasses development and acceptance of strategy, description of standard network configurations for operational use, definition of rules for network planning, installation, and maintenance.

This process also deals with designing the network capability to meet a specified service need at the desired cost and for ensuring that the network can be properly installed, monitored, controlled, and billed. The process is also responsible for ensuring that enough network capacity will be available to meet the forecasted demand and support cases of unforecasted demand. Based on the required network capacity, orders are issued to suppliers or other network operators and site preparation and installation orders are issued to the network inventory management or a third-party network constructor. A design of the logical network configuration is provided to network provisioning.

Figure 3.6.17 shows the network planning and development process. Principal functions are:

- Develop and implement procedures
- Set up framework agreements
- Develop new methods and architectures
- Plan required network capacity
- Plan the mutation of network capacity
- Issues orders to suppliers and other network operators
- Plan the logical network configuration

3.6.4.2   Network Provisioning Process

This process encompasses the configuration of the network, to ensure that network capacity is ready for provisioning of services. It carries out network provisioning as required, to fulfill specific service requests and configuration changes to address network problems. The process must assign and administer identifiers

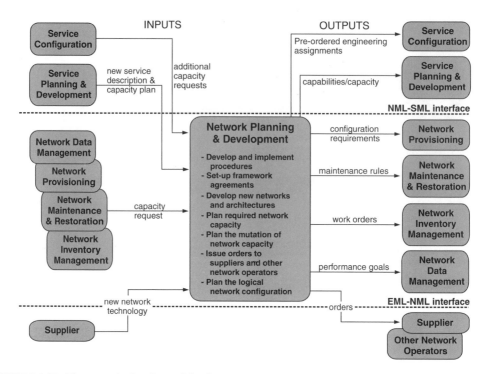

FIGURE 3.6.17 The network planning and development process.

for provisioned resources and make them available to other processes. Note that the routine provisioning of specific instances of a customer service — in particular, simple services such as plain old telephone service (POTS) — may not normally involve network provisioning, but may be handled directly by service provisioning from a preconfigured set.

Figure 3.6.18 shows the network provisioning process. Principal functions of this process are:

- (Re)configuration of the network; installation of initial configuration and reconfiguration due to capacity problems
- Administration of the logical network, so that it is ready for service
- Connection management
- Network testing

3.6.4.3 Network Inventory Management Process

This process encompasses anything to do with physical equipment and the administration of this equipment. The process is involved in the installation and acceptance of equipment, with the physical configuration of the network, but also with handling of spare parts and the repair process. Software upgrades are also a responsibility of this process.

Figure 3.6.19 shows the network inventory management process. The principal tasks of this process are:

- Install and administer the physical network
- Perform work in the network
- Manage the repair activities
- Align inventory with network
- Manage spare parts
- Manage faulty parts

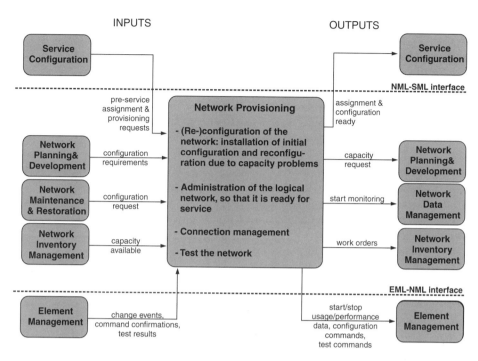

FIGURE 3.6.18 The network provisioning process.

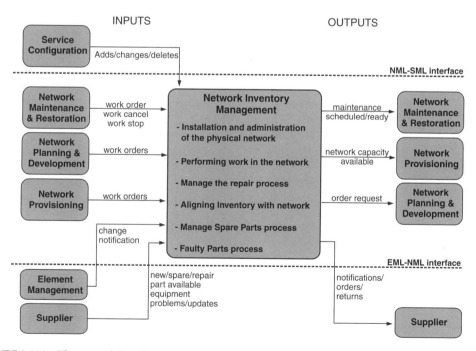

FIGURE 3.6.19 The network inventory management process.

3.6.4.4 Network Maintenance and Restoration Process

This process encompasses maintaining the operational quality of the network, in accordance with required network performance goals. Network maintenance activities can be preventative — such as scheduled routine maintenance — or corrective. Corrective maintenance can be in response to faults or to indications

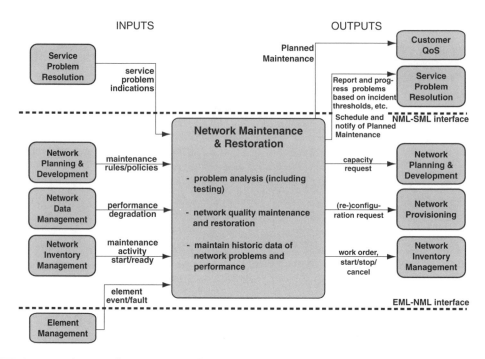

FIGURE 3.6.20   The network maintenance and restoration process.

that problems may be developing (proactive). This process responds to problems, initiates tests, does analysis to determine the cause and impact of problems, and notifies Service Management of possible effects on quality. The process issues requests for corrective actions.

Figure 3.6.20 shows the network maintenance and restoration process. Principal tasks include:

- Problem analysis, including testing
- Network quality maintenance and restoration
- Historic data maintenance of network problems and performance

3.6.4.5   Network Data Management Process

This process encompasses the collection of usage data and events for the purpose of network performance and traffic analysis. This data may also be an input to billing (rating and discounting) processes at the service management layer, depending on the service and its architecture.

The process must provide sufficient and relevant information to verify compliance/noncompliance to SLAs. The SLAs themselves are not known at the Network Management Layer. The process must provide sufficient usage information for rating and billing.

This process must ensure that the network performance goals are tracked, and that notification is provided when they are not met (threshold exceeded, performance degradation). This also includes thresholds and specific requirements for billing. This includes information on capacity, utilization, traffic, and usage collection. In some cases, changes in traffic conditions may trigger changes to the network for the purpose of traffic control. Reduced levels of network capacity can result in requests to network planning for more resources.

Figure 3.6.21 shows the network data management process. Principal tasks include:

- Collect, correlate, and format usage
- Determine performance of capacity and utilization
- Provide notification on performance
- Initiate traffic metrics collection function

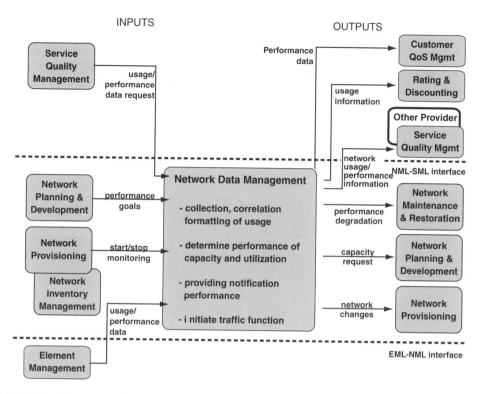

FIGURE 3.6.21 The network data management process.

3.6.5 Summary and Trends

The telecommunication industry has embraced the TMN model as a way to think logically about how the business of a service provider is managed. The model itself is simple, although its implementation is complex. The sheer number of standards now available that address the various interfaces between management systems sometimes makes it difficult to see and appreciate the big picture. These ITU standards are mainly concentrated in the Element Management and Network Management Layers. They have been developed from the bottom up, making it difficult to apply the standards as part of a business case. It is also difficult to have a customer-centric focus.

Smart TMN™ is a holistic, business-driven approach to implementing TMN. It provides investment direction as well as development specifications needed to produce management systems. It starts with the layered TMN model, but goes much further, to address concrete business problems in a pragmatic way. Specifically, Smart TMN™ brings three key elements to bear on the TMN layers:

- A business process-driven approach: Smart TMN™ starts from the premise that service providers and network operators need to automate their business processes, which means information needs to flow from end-to-end across many different systems. Only when a process is understood and the linkages are clear is it possible to apply standards in a way that delivers business value. Unless that is known, a great deal of money can be spent implementing standards that simply don't contribute to the overall business objective.

- Technology-independent agreements: Business agreements about what information will flow between processes must be kept independent of the protocols used to implement those agreements. Technology will continue to change, becoming cheaper and easier to use, and delivering more power. Smart TMN™ applies the right technology for the job instead of forcing a single technology to serve every need. Further, the Smart TMN™ approach documents all agreements in technology-neutral form, so that the same agreement can be implemented in multiple technologies as they continue to evolve.

- Products, not just paper: A main premise of Smart TMN™ is that paper standards are not sufficient to solve business problems. Products, not paper, are the end goal, provided documentation is produced to support the replications of industry agreements across multiple vendors' products.

This contribution has put emphasis on a telecom operations map which is a common model for telecommunications operations processes and a guide for re-engineering such processes.

References

ADAM96   Adams, E., Willets, K.: *The Lean Telecommunications Provider,* McGraw-Hill, New York, 1996.
TELE98   Telemanagement Forum: Smart TMN — Telecom Operations Map, TM Forum, October 1998, Morristown, NJ.

3.7   Management Frameworks and Applications

Kornel Terplan

Management frameworks consist of an application platform and management applications. The application platform consists of management services, computing hardware, and operating system. Management frameworks show unique features that differentiate them from management systems, particularly from proprietary management solutions. This contribution will introduce these differentiating features first. This step is followed by showing typical framework examples for both telecommunications service providers and enterprise users. NetExpert from Objective Systems Integrators, TeMIP from Compaq, TNP from Network Programs, TNG from Computer Associates, OpenView from Hewlett Packard, and FrontLine Manager from Manage.Com are briefly discussed. Management applications can be categorized as device-dependent — provided by vendors of networking equipment, and device-independent — usually provided by independent software vendors (ISV). For the integration between the application platforms and applications, APIs are provided by the framework suppliers assuming that these APIs are going to be supported and used by application providers.

3.7.1   Evolving Management Frameworks

The enrichment of application platforms by generic and specific management applications makes them more powerful. Figure 3.7.1 illustrates this evolutionary process. When application platforms are introduced to the market, they usually provide support for a few management applications in addition to core services that are part of the framework. If the framework is well accepted by users, the number of vertical management applications grows. Step by step, they will be partially or fully integrated into the framework.

3.7.2   Features and Attributes of Management Frameworks

Actually, there is no scientific definition of a management framework. In order to decide whether certain products qualify as a framework, an elaborated list of attributes is going to be addressed first. When products are able to support the basic framework attributes, they are qualified as frameworks. The advance attributes may then serve as differentiators between frameworks. Figure 3.7.2 shows the basic architecture of management frameworks. It consists of a runtime environment, development runtime tools, and application programming interfaces (APIs). The runtime environment is subdivided into management applications, management services, and the basic infrastructure (GHET97). Management services can be further subdivided into basic and advanced services, differentiating management frameworks from each other. Between the runtime environment, development tools, and the implementation, there are APIs to interconnect pieces with each other.

3.7.2.1   Basic Infrastructure

Basic infrastructure concentrates on the hardware and software features of the management frameworks. The most important attributes are listed below.

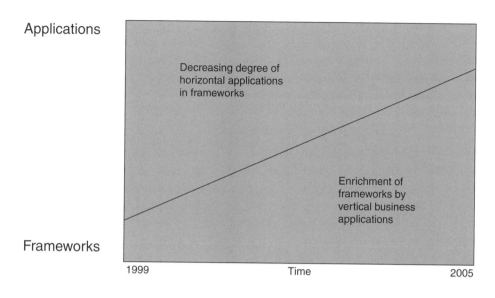

FIGURE 3.7.1 Evolving management frameworks.

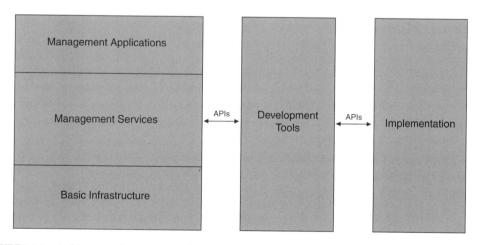

FIGURE 3.7.2 Architecture of management frameworks.

3.7.2.1.1 Hardware Platform

Hardware platform of the product involves a wide variety of items, including Intel 386/486, Pentium, HP 9000, RS/6000, Sun Sparc, Tandem, Alpha, System/88, and eventually others. Backup support should be addressed here as well.

3.7.2.1.2 Operating Systems

The industry expects a certain streamlining or, in other words, a shakeout of operating systems supporting the management platform. At present, the operating systems to be considered include: AIX, DOS/Windows, OS/2, SunOS, Ultrix, Sinix, Unix, Windows, Windows NT, and eventually others. Future solutions will concentrate around Unix versions and Windows NT.

3.7.2.1.3 Directory Services

Management frameworks deal with a great variety of entities and a great number of resources. The service of allocation of human-readable names to each managed resource (object) is the goal of directory services. Directory services are based on commonly agreed standards that model the naming paradigm, provide naming notations, allocate identifiable names to managed resources, translate names into physical addresses of resources, and ultimately provide location transparency of the resource in the system. All these considerations are valid for network and systems management frameworks. Framework management services and management applications use naming information from the directory services in order to perform their functions in relationship with managed resources and other management frameworks. The principal directory service requirements are (GHET97):

- Global information directory service and universal access to directory information
- Separation between the names of managed objects and the underlying physical networks
- Translation capabilities between various directory systems
- Translation between logical addresses and network addresses or routing addresses
- Storage of directory information and access to directory information, including metadata
- APIs in order to easily incorporate directory services into applications

Considering directory service capabilities, the following components are used as examples (GHET97):

- Directory services users: People, management applications, electronic messaging, routers/servers, other management framework services
- Resources requiring naming: People, organizations, computers, processes, files, mail boxes, network devices, printers, object class abstractions, object instances, management applications, management services, management agents
- Directory system types: Centralized, distributed, standard, proprietary; interpersonal communications directory (human), intersystem communications directory (computers and software systems)
- Directory service generic operations: Query (read, list, search, compare), modification (add/remove entry, modify entry, modify naming space, quit), binding/unbinding (security authentication)

Major directory systems include:

- OSI Directory Services (X.500)
- Internet Domain Name System (DNS)
- OSF Cell Directory Service (CDS)
- DEC Naming Service (DECdns)
- Novell Netware Directory Service (NDS)
- AppleTalk Name Binding Protocol (NBP)
- Windows NT Directory service
- Lightweight Directory Access Protocol (LDAP)
- Netscape Directory Server
- Banyan StreetTalk

3.7.2.1.4 Time Services

In distributed systems and network environments, where processes and applications are running on different machines, it might happen that time differences occur between systems. The time difference becomes critical when correct time stamps determine sequencing of events, job scheduling, measurements timing, and reporting intervals. A consistent use of time services is imperative when dealing with global networks, which span multiple time zones. Time service general requirements can be summarized as follows (GHET97):

- Use of an absolute, universal, coordinated time reference source
- Consistent synchronization services across hardware and software components
- Translation of universal time to local time for networks spanning multiple time zones
- Automatic resynchronization of manager and agent platforms after service interruptions
- Ability to operate in a heterogeneous computer and network environment
- Ability to keep the service running in case of major network instabilities
- Ability to provide both clock corrections and time source synchronization

In most cases, Internet Network Time Protocol (NTP) is used. Its primary reference time source is the absolute Universal Time Clock (UTC) or sources directly derived from UTC.

3.7.2.1.5 Software Distribution

In complex environments, management systems are usually distributed. They consist of servers, clients, and the communications paths between them. In order to keep them in synch, software versions or releases running on servers and clients must be compatible with each other. Manual software distribution is too slow and not reliable enough. Electronic software distribution offers two popular alternatives: push and pull. Distributing software by push offers easier scheduling, better automation, and does not require the physical presence of administrators. But, receiver servers and clients should be prepared for the distribution. Distributing by pull offers better control by administrators, and changes during distribution at a price of low automation. Scheduling is flexible and depends solely on human decisions. At this time, pull is privileged in web environments.

3.7.2.1.6 Security Services

Open distributed network environments consist of an increasing number of interconnected computing resources, networks, and users. The networks are no longer closed networks but mixtures of private and public networks. The networks include heterogeneous components, which has bearing on security services as well. Security of a network depends on the security of adjacent networks or other trusted partners. Frequent changes, like adding new resources and new users, bring additional concerns regarding security. Security can be seen as the security management functional feature built into certain management applications, namely security management applications. Since management frameworks control resources, security becomes an issue, as other framework services have to operate securely in order to make the whole system secure. Security is often embedded in framework services such as communications, database management, and object manipulation sevices, which perform management oprtaions.

Basic security requirements include (GHET97):

- Support for basic security features such as authentication, access control, and data integrity
- Ability to protect the system against potential intrusions
- Security features should be included in the whole life cycle of software development
- Distinctions in user security access profiles according to their role in the network
- Ability to group resources and users and apply common security policies to them
- Need to test security features and services against possible violations
- Protection of passwords and encryption keys by storing them in protected, encrypted files
- Mechanism to provide automatic clearing of disabled user accounts, user I.D.s, and passwords
- Capacity to communicate security data in a secured fashion

In typical management environments, management frameworks are protected by using a combination of the following security services:

- Identification of users: Identification is a unique representation of a user, computer, application, or remote system in order to provide accountability and to record action of the identified entity.
- Access control: Allows the requesting party to actually access the system and networking resources if the party was authorized to do so. It is supported by login functions where passwords are built in.

- Authentication: The verification of the entity prior to accessing the system and networking resources. In this case, the entity should prove its identity by using various techniques, such as personal attributes, digital signatures, and others.
- Data privacy: An encryption mechanism using trusted third-party secret keys. Through a combination of private and public keys, the encrypted information can be verified for integrity and accessed for processing.
- Data integrity: A security feature which allows verification of cryptographic data checksums. The correctness of this verification is a proof that the data was not tampered with or corrupted through network transmission.
- Security auditing: Allows the generation of audit logs. These logs should be encrypted and protected against unauthorized access attempts.

3.7.2.2 Management Services

Management services address more specific items toward management applications. The most important features are listed below.

3.7.2.2.1 Communication Services

Network architectures: The targeted networks to be managed are very different. Many products are expected to manage legacy networks and more open networks at the same time. The most widely used protocols supporting network architectures include DECnet, IPX/SPX, OSI, SNA, DSA, APPN, TCP/IP, Guardian, and eventually others. Capabilities of managing SNA and TCP/IP are the highest priority.

Network management protocols: The products are expected to support at least SNMP. It is an additional advantage when they can do more. SNMP support may include the capabilities of working with proxy agents that are capable of converting non-SNMP into SNMP. Protocols to be supported include CMIP, CMOT, LNMP, NMVT, RMON, SNMPv1, SNMPv2, and eventually DMI to manage desktops.

The management platform provides SNMP support in several ways. First and foremost is the ability to poll SNMP devices and receive SNMP Traps, as described previously. However, in order to configure polls on MIB variables of various devices, one must first know what those variables are. Management platforms provide MIB "browsers" for this purpose. A MIB browser queries user-selected SNMP network devices and displays their MIB values. In addition, most platforms can display line or bar graphs of those MIB values, provided they are in numeric form (counters, etc.).

MIB browsers are crude tools at best, displaying raw and often cryptic, low-level device information. For this reason, platforms also provide MIB application builders that allow users to quickly create applications for displaying information on MIB objects in a more meaningful way. MIB applications may include graphing real-time information on selected network nodes. However, even MIB applications builders are limited in supporting the high-level analysis more openly provided by third-party applications.

MIB compilers allow users to bring in third-party, device-specific MIBs (also called "private" or "extended" MIBs) and register them with the management platform. While most platforms ship with a number of third-party MIBs, they do not include all possible MIBs from all vendors. A MIB compiler is necessary for adding support for third parties whose MIBs are not shipped as part of the standard platform.

Some MIB compilers are more robust than others. Some MIB compilers will fail or abort processing if there is an error in the MIB being compiled. Unfortunately, errors in third-party MIBs are not rare. Therefore, it is desirable to have a MIB compiler that can flag errors and recover, rather than stop dead.

3.7.2.2.2 Core Management Services

The management framework is expected to offer core services and interfaces to other applications. The basic management applications to be provided are discovery/mapping, alarm management, and platform protection.

Discovery and Mapping

Device discovery/network mapping discovery refers to the network management system's ability to automatically learn the identity and type of devices currently active on the network. At minimum, a management platform should be capable of discovery-active IP devices by retrieving data from a router's IP tables and Address Resolution Protocol (ARP) tables.

However, even this capability does not guarantee that all IP devices on a given network will be detected. For example, relying solely on routing tables is inadequate in purely bridged networks where there are no routers. Thus, a more comprehensive discovery facility should also include other mechanisms such as broadcast messages (PING and others) that can reach out to any IP device and retrieve its address and other identifying information.

On the other hand, discovery mechanisms that rely completely on broadcasting (e.g., PING) will incur a tremendous amount of overhead in finding devices out on the network. Ideally, a management platform should support a combination of ARP data retrieval and broadcasting.

Furthermore, a complete network discovery facility should be capable of detecting legacy system nodes, such as DECnet and SNA. Currently, most platforms rely on third-party applications or traffic monitoring applications to supply discovery data on non-TCP/IP devices.

Another desirable feature is the ability to run automatic or scheduled "dynamic discovery" processes after the initial discovery, to discern any changes made to the network after the initial discovery took place. In large networks especially, overhead and consumed bandwidth for running a dynamic discovery process continually in background mode may be too great; therefore, the ability to schedule discovery at off-peak hours is important.

It is also important for the user to have the ability to set limits on the initial network discovery. Many corporate networks are now linked to the Internet, and without predefined limits, a discovery application may cross corporate boundaries and begin discovering everything on the global Internet. Some management platforms allow users to run discovery on a segment-by-segment basis. This can help the discovery process from getting out of hand too fast.

Many management platforms are capable of automatically producing a topological map from the data collected during device discovery. However, these automatically generated maps rarely result in a useful graphical representation. When there are hundreds of devices, the resulting map can look very cluttered enough to be of little use.

Even when the discovery process operates on a limited or segment-by-segment basis, there is eventually going to come a time when the operator must edit the automatically generated network map to create a visual picture that is easier for human beings to relate to. Therefore, the ability to group objects on the map, and move them around in groups or perform other types of collective actions, can be a real time-saving feature.

Alarm Capabilities

Management platforms act as a clearinghouse for critical status messages obtained from various devices and applications across the network. Messages arrive in the form of SNMP traps, alerts, or event reports when polling results indicate that thresholds have been exceeded.

The management platform supports setting of thresholds on any SNMP MIB variable. Typically, management platforms poll for device status by sending SNMP requests to devices with SNMP agents, or Internet Control Message Protocol (ICMP) echo requests ("pings") to any TCP/IP device.

The process of setting thresholds may be supported by third-party applications or by the management platform. Some platforms allow operators to configure polls on classes of devices; most require operators to configure a poll for each device individually.

Most platforms support some degree of alarm filtering. Rudimentary filtering allows operators to assign classifications to individual alarms, such as informational, warning, or critical, triggered when thresholds are exceeded. Once classifications are assigned, the user can specify that only critical alarms are displayed on the screen, while all other alarms are logged, for example.

More sophisticated alarm facilities support conditional alarms. An example of a conditional threshold may be "errors on incoming packets from device B > 800 for more than 5 times in 25 minutes." Conditional alarms can account for periodic spikes in traffic or daily busy periods, for example.

Finally, the platform should support the ability to automatically trigger scripts when specific alarms are received.

3.7.2.2.3 User Interface Services

GUI's basic job is to provide color-coded display of management information, multiple windows into different core or management applications, and an iconic or menu-driven user interface. By providing a standardized interface between the user and the underlying tools, the GUI simplifies what a user needs to learn and provides a standard tool for application developers.

Most management operations are available from a menu bar; others from contex menus. Point-and-click operations are standard features, as is context-sensitive help. Most platforms allow some degree of customization of maps and icons.

While most platform GUIs are the same, there can be a few subtle differences. Some GUIs have larger icons than others. While this makes it easier to read information on the icon and distinguish status changes more quickly, a screen can quickly become cluttered with just a few large icons. Icon size is strictly a matter of user preference. The most widely used GUIs are Motif, OpenLook, OS/2 Presentation Manager, and Windows.

3.7.2.2.4 Database Services

The database is the focal point for key data created and used by the management applications. They include MIB data, inventories, trouble tickets, configuration files, and performance data.

Most platforms maintain event logs in flat-file ASCII format for performance reasons. However, this format limits the network manager's ability to search for information and manipulate the data. Therefore, links to relational database management systems (RDBMSs) are now important aspects of the framework architecture.

A RDBMS is essential for manipulating raw data and turning it into useful information. Users can obtain information from a RDBMS by writing requests, or queries, in Structured Query Language (SQL), a universally standard language for relational database communication.

While most management platforms also supply report writer facilities, these tools are generally not top-notch. However, most higher quality third-party reporting applications can extract data from a RDBMS using SQL.

3.7.2.2.5 Object Manipulation Services

Object-oriented and object-based technologies are helpful in relation to user interfaces, protocols, and databases. The use of object request brokers (ORB) and CORBA provides a glue needed to accomplish interoperability between heterogeneous systems. These services provide support for information exchange between objects as abstractions of physical and logical resources ranging from network devices computing systems resources to applications and management services. It includes operations on MIBs, object support services providing location transparency for objects exchanging requests and responses, persistent storage for MIBs, and support for object-oriented applications development.

3.7.2.2.6 Network Modeling Services

Network modeling is an artificial intelligence capability that can assist in automated fault isolation and diagnosis as well as performance and configuration management. Modeling allows a management system to infer status of one object from the status of other objects.

Network modeling is facilitated by object-oriented programming techniques and languages such as C++. The goal of modeling is to simplify the representation of complex networks, creating a layer of abstraction that shields management applications from underlying details.

The building block of this technology is the model, which describes a network element such as a router. A model consists of data (attributes) describing the element as well as its relationships with other elements. Abstract elements such as organizations and protocols can also be modeled, as can nonintelligent devices such as cables. A model may use information from other models to determine its own state; modeling can reduce the complexity of management data and highlight the most important information. In this way, fault isolation and diagnosis can be automated. In addition, models can be used to depict traffic patterns, trends, topologies, or distributions to assist in performance and configuration management.

3.7.2.3 Application Programming Interfaces (APIs) and Development Toolkits

API and developer's toolkit platform vendors encourage third-party applications by providing published APIs, toolkits that include libraries of software routines, and documentation to assist applications developers. Another aspect to this effort is the "partners programs" — the marketing angle of encouraging third-party applications development.

An API shields applications developers from the details of the management platform's underlying data implementation and functional architecture. Management platform vendors generally include in their developer's kits several coded examples of how APIs can be used, as well as the APIs themselves.

In most cases, when an application takes advantage of platform APIs, it must be recompiled with the platform code, resulting in a tightly integrated end product. Many ISVs and other third-party developers lack resources necessary to pursue this level of integration. Or, perhaps a more accurate way of stating this is that ISVs aren't convinced that putting out the extra effort to fully integrate their applications with all leading management platforms will result in a proportionally larger revenue stream. ISVs and other third-party developers face a choice: tightly integrate their products with one management platform vendor, or loosely integrate them with all leading platform providers. Most third parties have chosen the latter route, as they are unwilling to turn off prospective customers who may have chosen a different platform vendor as their strategic management provider.

As a result, at least 80% of the third-party applications available today are only loosely integrated with the underlying management platform — at the menu bar — and completely ignore APIs and other environment libraries. This is expected to change as the market matures, and as platform vendors begin to offer high-level APIs which make porting applications from one management platform to another into an almost trivial exercise.

In summary, published APIs and libraries make it possible for lSVs and other third parties to write applications that take advantage of other basic services provided by the management platform. To date, few third parties have taken full advantage of platform APIs, although this is expected to change over the next several years.

3.7.2.4 Management Operations Support Services

Any management framework consists of framework services and management applications. The services are implemented as a set of related processes, databases, and file sets. The basic thrust of management implies collection and processing of management-related information. The coordination of all the framework processes, including those which are part of the development environment, is done through additional framework components commonly called management operations support services. These services are also responsible for application integration with framework services, and for multiple national language systems support.

Management frameworks are basically a set of interconnected software programs which run on one or more computing platforms. Management operations support services provide supervision, coordination, maintenance, and management of processes, applications, and databases which are part of the management framework. The requirements of management operations support services are the following (GHET97):

- Facilitating interactions between framework services
- Allowing overall coordination and supervision of background processes
- Supporting integration between management services
- Allowing configuration and customization of framework services and associated processes
- Supporting registration of management applications which run on management platforms
- Providing easy integration of management applications with framework services
- Supporting multiple national language systems
- Facilitating incorporation of management information models into frameworks

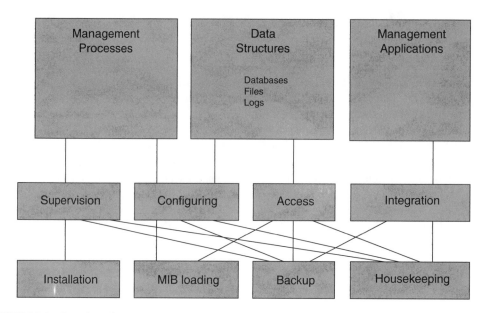

FIGURE 3.7.3   Overview of management operations support services.

- Supporting installation of framework services and management applications
- Supporting MIB loading, backup, and clean-up facilities
- Supporting distribution of management frameworks services and associated databases

This list of requirements indicates that management operations support services play a critical role in monitoring, administration, and management of the management framework itself.

The structure of management operations support services is characterized by a layered architecture. The upper layer consists of management processes, data structures, and management applications (GHET97). The middle layer presents important support functions, such as supervision and synchronization of management processes, configuring processes and databases, access to databases and files, and integration between framework services and management applications. The lowest layer consists of tools, supporting installation, MIB-loading, backups, and other usual housekeeping functions. Figure 3.7.3 shows these layers.

3.7.3   Management Framework Examples for Telecommunications Providers

These examples show very powerful and scalable frameworks with a number of capabilities for both wireline and wireless services. In all cases, third-party management applications can be integrated into the frameworks.

3.7.3.1   TeMIP from Compaq

At the highest level, TeMIP consists of a management information repository (MIR) for storage of data stuctures, functions, and management information, an executive kernel responsible for supporting all the interactions beween components, and a set of interfaces to all the management modules belonging to the framework. Figure 3.7.4 shows the TeMIP architecture.

Three types of management modules interface the kernel:

- Access modules which provide access to various agents attached to real management entities such as physical network elements or systems logic resources
- Presentation modules provide the user interfaces
- Functional modules provide the actual management services such as event management, object manipulation, and management operation support services

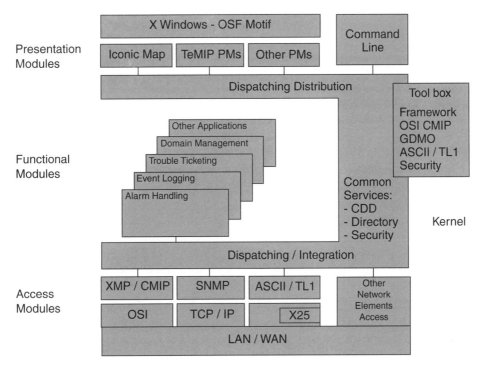

FIGURE 3.7.4 TeMIP architecture from Compaq.

These management modules are a set of cooperative processes rather than independent ones. Compaq/Digital has been adding access and functional modules over the last couple of years, such as SNMP, OSI CMIP, ASCII/TL1, and TMN support. A more detailed view of the framework can be seen in Figure 3.7.5 (GHET97).

An important emphasis is placed on the TeMIP distributed framework which allows any of the constituent modules to run on physically distributed systems. Each of these systems is considered a peer director. Among directors, some play the role of servers, others play the role of clients. Direct communications and management information exchange is provided only between director servers. The implementation of the TeMIP architecture can range from a standalone centralized management system to hierarchical or a cooperative network of manager topologies.

The TeMIP GUI is based on OSF/Motif and XWindows Systems and provides a common view of all the managed resources. The Icon Map PM provides presentation and language localization to the alarm handling, event logging, and trouble ticket FMs. The icon map provides map windows, a navigation box, graph windows, and a toolbox for customization. Forms and command line interfaces are also available.

The platform provides many generic functional modules. The alarm handling and the event logging FMs are based on ISO standards. A log panel window allows the user to customize the logging environment. The trouble ticket FM is based on the recommendations of the Telemanagement Forum. The performance analyzer FM provides normalized and statistical data for TCP/IP hosts, RMON probes, and DECnet nodes. It collects information about DECnet/OSI end systems, data links, intermediate systems, routing ports and routing circuits, circuits, nodes, and protocols. The statistics collected include throughput rates, counts, averages, overhead percents, and utilization metrics.

The information manager is the platform's object request broker and is similar but not compatible with CORBA from OMG. It receives requests from clients along with their arguments. Then, acting as a client, the information manager connects through a RPC binding to the appropriate server. Location transparency is achieved through Distributed Name Services, which provides a global directory service.

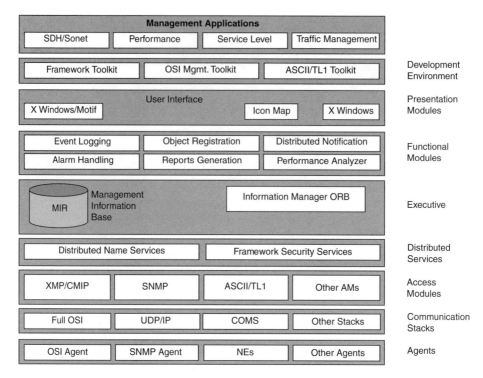

FIGURE 3.7.5 TeMIP framework in detail.

Security services consist of access control (access control filters, user profiles, access control management), logging of operator commands (storage of prefiltered commands entered by users), and a security development toolkit.

The TeMIP framework provides access to managed resources through access modules. All of the relevant network protocols are supported by the framework. The SNMP AM supports the MIB II management information base. In addition, a MIB compiler is provided to check the Concise MIB syntax and to support loading of the MIB into MIR. The SNMP AM allows get and set operations on the agents and can test reachability of an object at the IP level by using the ICMP ping protocol.

The TeMIP applications map is shown in Figure 3.7.6. This map includes three major groupings for external management applications:

- Network management
- Telecommunications management
- Unix systems management

Strengths and weaknesses of the TeMIP framework are summarized in Table 3.7.1.

3.7.3.2 NetExpert from Objective Systems Integrators

The NetExpert framework consists of a series of coordinated modules that fall into three general groups:

- External network element and non-NetExpert subsystem gateways
- Object persistence and behavior servers
- User/operator workstations and web interfaces

NetExpert is a robust, scalable, and distributable archietcture that supports a high degree of configuration flexibility while maintaining individual component independence. Easy to use, modify, and initiate, it is quick to roll out and integrate with existing platforms and management applications.

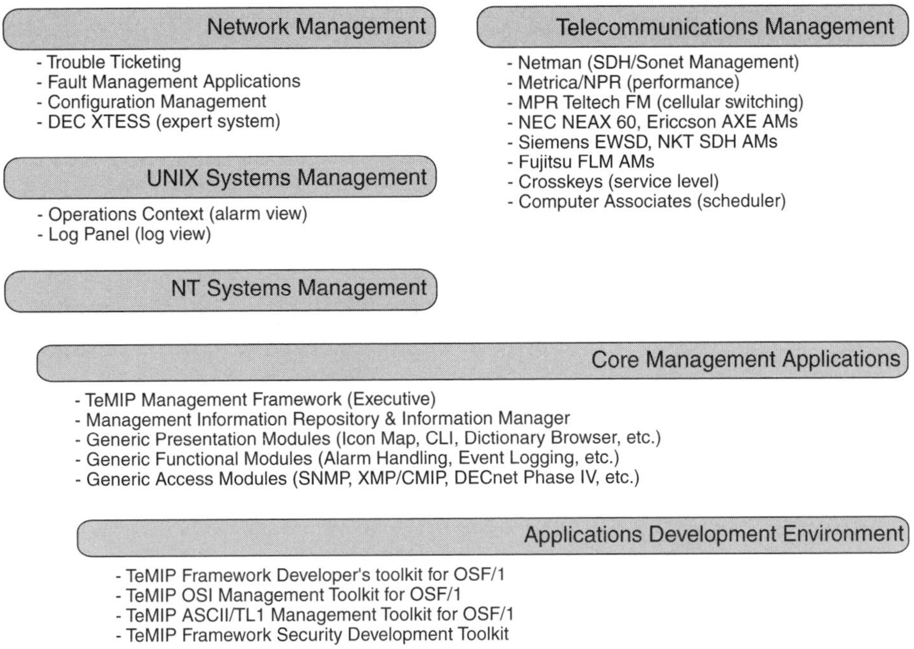

FIGURE 3.7.6 Applications map of TeMIP.

TABLE 3.7.1 Strengths and Weaknesses of TeMIP

Strengths

Modular and functionally distributed architecture
Direct communication capability between TeMIP director servers
Framework functional modules based on OSI management standards
Incorporation of an object request broker
Home-grown management framework design
Distributed notification mechanism
Distributed name and security services
Policy-based management domains selection capability
Partnership with telecommunication service providers and manufacturers
Self-management capabilities

Weaknesses

Very complex architecture and development environment
The Information Manager is not CORBA compliant
Complex documentation
Long learning curve
Small market share
Proprietary internal database and database API
Limited set of systems management application availability
Limited choices of hardware and operation systems platforms
No scaled-down TeMIP management platform alternative

Figure 3.7.7 provides a high-level description of how NetExpert receives "from" and "send" messages to external network elements and operations support systems.

The main attributes of principal subsystems are the following:

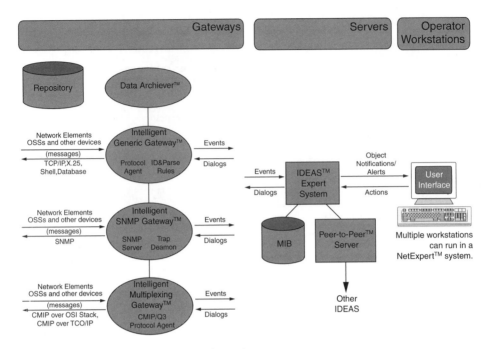

FIGURE 3.7.7   NetExpert framework operational overview.

Gateways:

- Receive raw data from network elements
- Identify important messages, parse relevant data into attributes, and package them into events
- Perform analysis
- Forward events to the IDEAS Expert System Server
- Generate and send dialogs and polls to devices and receive responses
- Forward messages to DataArchiver

Servers:

- Receive events from gateways
- Perform analysis and execute rules
- Generate alerts and send to operator workstations
- Initiate dialogs and polls and send to gateways
- Modify MIB values
- Forward notification to peer systems

Operator workstations include the following modules:

- Gateway control
- Visual agent client windows
- Alert display
- Command and response system
- Managed objects configuration system
- Trouble ticket
- Report maker
- Data browser
- Interface to pagers

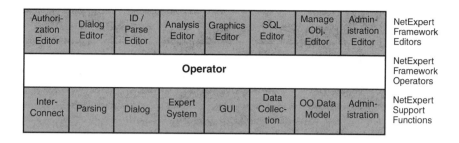

FIGURE 3.7.8   NetExpert functional framework.

The NetExpert framework is a set of modules covering the basic functions that a distributed frameworks needs, including gateways to the system, a way to send messages and events, the intelligence to act on those events, and a consistent operator interface. A customer can distribute these modules across a network, gaining the foundation required to monitor continuous and large volumes of events and traffic. The framework is controlled by rules that replace complex programming languages and enable network analysts to model desired system behaviors. Rules are written with the product's implementation tools. Existing rule sets, called application components, eliminate the complex traditional development process, which entails writing requirements and building a complete solution from scratch. Rule writing is estimated as 10–15 times more productive than traditional development methods.

The functional framework consists of editors, operators, support functions, and other framework enablers. These are shown in Figure 3.7.8.

NetExpert's modifiable application packages provide a comprehensive subset of functions. These can be further tailored to individual customer requirements. This is how the framework accomodates configuration-specific solutions and the demands of the customer's business model. Because they are object oriented, these rule packages can deliver a large number of services; manage any number of tangible elements, such as switches or routers; and model intangible elements, such as knowledge of subject matter experts. Rules make it possible for the same NetExpert framework to manage diverse networks, such as digital cellular, traditional telephony, high-speed data, or hybrid fiber/coax.

Application rules ride on top of the NetExpert framework. They are categorized as point, domain, or corporate level application packages. The differences between each depend on the business focus they are designed to address. Point applications define the native messages required by a network element during, for example, the provisioning process. Domain applications group higher level commands into those associated with, for example, all switch or transport network devices constituting a service provider's network. Corporate applications perform, manage, and control functions associated with the domain- and point-level applications. The layering of corporate, domain, and point applications is illustrated in Figure 3.7.9.

The framework running in concert with NetExpert applications enables users to generate revenue by quickly delivering new services. However, getting to market first is not enough. With NetExpert, users protect past investments, increase the life span of aging equipment, incorporate new elements, and integrate disparate management systems and software rule set packages. Another advantage OSI users have is their ability to deploy network management systems and OSSs in formerly uncharted niches, and integrate these with existing infrastructures. Users are closer than before to automating their business models because OSI delivers the tools they need to translate key processes across systems that support entire telecommunications operations.

Table 3.7.2 lists strengths and weaknesses of NetExpert.

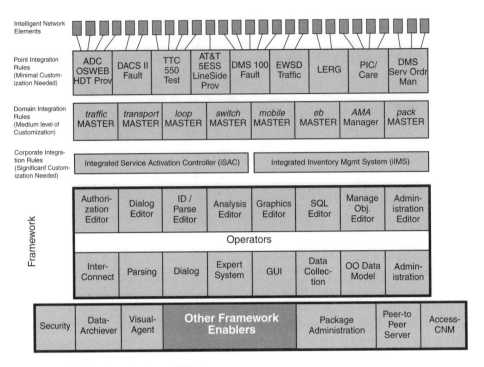

FIGURE 3.7.9   Rules-based applications of NetExpert.

TABLE 3.7.2   Strengths and Weaknesses of NetExpert

Strengths

Support of accelerated design procedures
Flexible integration of new network elements and OSSs
Cross-vendor functions and domain correlation
Common maintenance and operations procedures
Integrated problem resolution capability
Substantial reductions in software costs for development and operations
Support of multiple hardware platforms
Support of CORBA for interprocess communications
Heavy use of Java to support presentation services
Support of Web-based front ends
Large market share with wireless operators

Weaknesses

Portability of rules sets is limited to non-Unix environments
Extensive training is required
Learning curve for subject matter experts is relatively long
Lack of third-party support for integrating management applications
Rule sets are heavily fault management-oriented
No scaled-down alternative for smaller operators
No presence in enterprise environments

3.7.4   Management Framework Examples for Enterprise Users

These examples represent flexible and scalable frameworks with a number of integration capabilities. Some management applications provided by ISVs are the same or at least similar to those provided for the telecommunications industry.

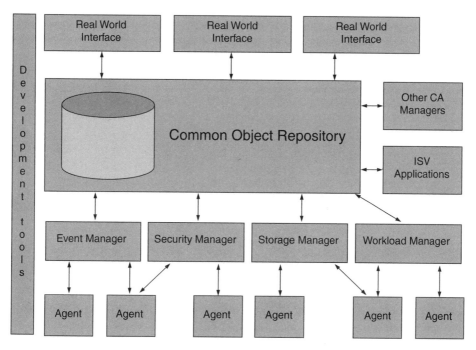

FIGURE 3.7.10 Simplified view of the TNG architecture.

3.7.4.1 TNG from Computer Associates

TNG features, as an absolute novelty for a management platform, a 3-D, animated, graphical user interface called the Real World Interface. The core platform is bundled under the Common Object Repository component which hides the object manipulation and object storage processes. Objects representing the abstraction of actual managed resources and objects created by the platform services are stored in the platform object database. Query and search capabilities allow core management functions and applications to access the management information. Figure 3.7.10 shows the simplified view of the TNG architecture.

The availability of a Java browser, based on either a 2-D or VRLM 3-D interface, provides an alternative graphic environment. This Web interface, in particular, delivers on the framework's promise of managing everything from anywhere.

TNG includes many functional modules, which provide event management, security management, storage management, workload and performance management, as well as backup and recovery functions. Management of distributed resources is based on a manager–agent infrastructure which relies on a mix of proprietary and standard agents. TNG allows scalable, multilevel, hierarchical build-up of manager–agent structures as required by managing large enterprise networks.

The following application packages are running on top of the TNG platform:

- Software delivery
- Advanced help desk
- Open Storage Manager
- Single sign-on
- Internet Commerce Enabler

The company is acquiring products, but more than in the past, pays a lot of attention to integration. Table 3.7.3 summarizes strengths and weaknesses of the TNG architecture.

TABLE 3.7.3   Strengths and Weaknesses of Unicenter TNG

Strengths

Unicenter TNG applications are running on multiple platforms
Extensive experience with mainframe-based management applications
Integration of systems and network management
Interoperability capabilities between various CA-products
Use of advanced 3-D techniques
Multiple alliances
Use of neural technology to deal with large data volumes
Support of Web technology
Promotion of a developer partner program
Filling functionality gaps by acquiring the right best-of-breed products

Weaknesses

Limited experience in network management
Proprietary agent implementations and information exchange
Customer support could be better
No telecom industry-specific management applications are available
Limited application development tools
Support for open, standard systems and APIs is not readily seen in real products
Quality of documentations is not always good
No low entry, PC-based framework version available

3.7.4.2   OpenView from Hewlett-Packard (HP)

The OpenView family provides an integrated network and systems management solution. It consists of a number of products from HP and also from Solution Partners. The most important components of the OpenView family are:

- Network Node Manager (NNM) — meets the requirements for a powerful SNMP core solution.
- IT/Operations — an advanced integrated operations and problem management solution for networks and systems.
- IT/Administration — an integrated solution for change management. It also includes inventory, asset, software, and user management.
- PerfView, NetMetrix, and MeasureWare — performance management solutions for networks and systems; may be considered as the foundation for service level management.
- OmniBack and OmniStorage — typical systems management solutions for powerful backup and storage management.

Figure 3.7.11 shows OpenView with its principal components.

HP is targeting OpenView Network Node Manager at managing the Internet/Intranet infrastructure rather than Web servers and services. HP promotes a three-tier OpenView strategy for managing and leveraging the Internet:

- Manage the corporate Intranet infrastructure, including network infrastructure, servers and Internet applications, and security
- Manage the infrastructure of Internet service providers
- Leverage Internet technologies in OpenView solutions.

The enhancements in NNM have significantly increased the product's scalability, making OpenView-based management of corporate Internet/Intranet infrastructures possible. Management of Web servers and applications is largely provided by the generic server and application management capabilities of IT/Operations. HP is targeting management of Internet service providers infrastructure through its HP OpenView DM offering, and HP OpenView Event Correlation Services. Internet service providers may

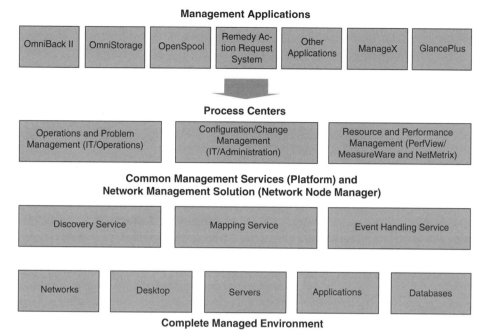

FIGURE 3.7.11   Principal components of OpenView.

include carriers, cable companies, value-added networks, and others. Finally, HP is exploring and prototyping Web technology extentions to OpenView products, including:

- Web access to OpenView event repositories for problem management support
- Web access to the OpenView map
- Internet as a software transport vehicle

Using IT/Operations for Internet Management

IT/Operations is capable of managing processes and applications running on any computer for which HP provides an IT/Operations agent. Supported systems include HP-UX, Solaris, AIX, SCO, and Windows NT, among others. IT/Operations agents are capable of intercepting SNMP traps, Unix logfile messages, and events generated when IT/Operations agents detect threshold crossings. Using these attributes, Netscape Commerce Servers can also be managed. They can run under HP/UX and support secure electronic commerce and communications on the Internet and TCP/IP Intranets. The server permits corporations to publish HTML-formatted documents (Web pages) and deliver them using HTTP. To ensure data security, the Netscape Commerce Server provides server authentication, data encryption, and user authorization. Communications support also includes the Common Gateway Interface (CGI) and the Secure Socket Layer (SSL) protocol.

To support manageability, the Netscape Commerce Server records several kinds of errors, all of which can be collected by an IT/Operations agent reading the logfile of the server. These errors include:

- Unauthorized — occurs when users attempt to access protected server documents without proper permission
- Forbidden — occurs when the server lacks file system permissions needed to execute a read or to follow symbolic links
- Not found — occurs when the server cannot find a document or has been instructed to deny a document's existence
- Server error — occurs when the server has been misconfigured or affected by a core dump, out of memory, or other catastrophic error

HP provides an IT/Operations template for handling these errors. Users can derive proper responses, including forwarding events to the appropriate IT/Operations or database operators, or triggering a script for deleting hypertext links to documents that no longer exist.

Each error type described above can be associated with error codes.

IT/Operations agents are capable of collecting these error messages and forwarding user-specified events to the IT/Operations console for operator attention and problem resolution. For example, in the case of server error, possible causes of the problem may include the following:

- CGI is not enabled on the Web server, preventing electronic commerce application from running permissions that have not been specified properly
- CGI script is not specifying a shell or other program to be run
- Syntax error in the script

Syntax errors are typically resolved by tweaking application scripts, which may be written in CGI, Practical Extraction and Report Language (PERL), or Tool Command Language (TCL). Many Web server applications for electronic commerce are written in C language and implemented with CGI, PERL, or TCL scripts.

Monitoring Web Page Availability with IT/Operations

IT/Operations can be deployed to monitor Web page status as well as Web Server status. Specific functions supported include:

- Monitoring Web access logfiles and error logfiles
- Monitoring the HTTP domain
- Viewing server access statistics
- Integrating the native Netscape administration and configuration tools into IT/Operations
- Starting up and shutting down the Web server and administrative interface
- Modifying access configuration

HP has developed a script that can be used by the IT/Operations agent monitor to check the availability of the Web server system, the HTTP port, and the Web page. The script uses the Korn shell, one of four major Unix command and script interpreters in use today. This script, designed primarily for Netscape Commerce Server, can theoretically be modified and extended to monitor other Web servers as well.

In order to meet the needs of today's IT organizations, a new Java-based user interface has been added to IT/Operations. The features and benefits of such an interface can be summarized as follows:

Ease of use:
- The Java-based interface combines the familiar concepts of IT/Operations with Windows-like concepts (similar to ExplorerView), to minimize training time and reduce the operator's learning curve.
- Most functions available in the IT/Operations Motif Operator user interface are supported in the Java version. The characteristics of Java also add functionality that is not available in the standard user interface, such as sorting and shifting columns in the IT/Operations' message browsers.
- It is available on the Windows NT operating system. The Java user interface allows the management of large heterogenenous environments from PCs running the Windows NT operating system.
- Through a special application-bank entry, local Windows NT applications can be tightly integrated with the Java user interface, resulting in a more powerful, integrated Windows NT operator workstation.

Scalability and distribution:

- The number of operators that can concurrently access IT/Operations from a Java user interface is greatly increased.
- This addresses the needs of customers with large environments. The size of environments is continually increasing.
- The Java user interface minimizes network traffic, enabling it to work over low-bandwidth lines.
- It is not common to have a LAN connection, for example, at home or in a remote office, yet management is available whether at home or on the road.
- The Java user interface runs on any machine with a Java-compliant browser or as a stand-alone application on HP-UX or the Windows NT operating system.

Lower total cost:

- Previously, the operators had to have a Unix workstation or special tools on NT PCs to obtain the same functionality.
- No additional IT/Operations software or hardware needs to be installed and maintained on the client systems, other than a Web browser.

The Java-based interface is designed to take up minimal resources on the client.
Table 3.7.4 summarizes the strengths and weaknesses of the OpenView architecture.

TABLE 3.7.4   Strengths and Weaknesses of OpenView

Strengths

Front runner in the implementation of the framework concept
Modular design with the SNMP and distributed manager framework
Good coverage in HP-developed management applications and products
Extensive coverage in Unix-based applications developed by ISVs
Leader in management framework source code licenses and OEM partnerships
HO provides a testing and certification program for partner applications
Distributed management consoles and GUI services
Partnerships to serve the telecommunications market
Manager-to-manager capabilities for integrated IT/Operations and IT/Administration
Support of application management standardization
Strong performance management applications
Web-based management front ends

Weaknesses

No built-in middleware object request broker
The communication between HP managers contains proprietary elements
Too many processes, too many API calls, no built-in security features
CMIP OSI stack support is environment dependent
Telecom industry support in management applications is still limited
Contains proprietary components and extentions
The application development tools are not yet mature
Delay in delivery of a common management repository
Insufficient customer support for platform implementation and tuning
Scalability is limited for large enterprise networks and systems
Delays in supporting the NT platform

3.7.4.3   FrontLine Manager from Manage.Com

FrontLine Manager unifies the management of rapidly prolferating intranet computing resources across network devices, systems, services, and applications. Designed for systems and network administrators, help desk agents and others who staff the frontline, responding to and solving user support calls. FrontLine Manager uses Web technologies to simplify day-to-day operational tasks and thus lower the cost of intranet management. It begins managing out-of-the-box by discovering and identifying resources, while creating a unified management view of the entire intranet environment. It goes beyond passive monitoring to identifying and diagnosing problems proactively. Embedded software intelligence determines the ideal operating state of each resource and notifies support staff when healthy operating conditions are exceeded. Rapid installation and ease of use are combined with low administrative overhead to maximize the productivity. It is a typical first-tier support tool. All management functionality and the unified management view are accessed via a standard Web browser. The components of FrontLine Manager are:

- FrontLine Manager Server: Each server manages a typical LAN or LANs supporting up to 255 network devices and systems, along with a base set of intranet services and applications. The FrontLine Manager server incorporates a Web server and a scalable object database. As a result, a distributed group of servers can manage up to 1,000,000 nodes.
- Web browser: All functionality and the unified management view are accessed via a standard Web browser. The browser interface simplifies the presentation of complex management information, while giving frontline managers the freedom to manage securely from anywhere, locally or remotely. An active window displays the most recent information for each managed resource.
- Managed Agents: Each managed resource has an associated SNMP or Java agent. Agents transmit management data to the server and can also conduct management tasks. A base set of SNMP and Java agents developed by Manage.Com are included with FrontLine Manager. Third-party SNMP agents already installed on network devices and systems can also be used.

FrontLine Manager is prebuilt with key management features needed to manage the majority of intranet comounting environments. No time is wasted on complex installation, customization, or integration. Quickstart installation automatically discovers all resources and begins monitoring them so that productive management can begin immediately. It begins by proactively discovering and classifying resources during installation, a process that completes within a few hours. It then associates an ideal operating state with each resource and monitors accordingly. If abnormalities are discovered, FrontLine Manager immediately begins to diagnose and isolate the causes. For maximum efficiency, it helps to identify and resolve problems before users report them. The intelligence to identify the healthy operating state of specific resources is built into the product. As a result, it is able to take samples of the intranet continually and determine its overall health. It also launches automatic analysis to diagnose and segment operating problems, often before they are reported by users.

Figure 3.7.12 shows the principal management functions.

This Web-based solution differentiates itself from individual device- or application-dependent products, because it integrates the management of network devices, systems, services, and applications.

Table 3.7.5 summarizes the strengths and weaknesses of FrontLine Manager.

3.7.5   Management Applications

Application platforms are powerless without management applications. They are provided by equipment vendors or by ISVs, and serve various purposes.

3.7.5.1   Device Dependent Applications

Equipment vendors develop and deploy management applications in order to promote sales of their equipment. Today, it's not possible anymore to sell networking gear without element management systems — in other words, without management applications. These applications are offered and sold at reasonable prices. Equipment vendors don't make much revenue with these element management systems

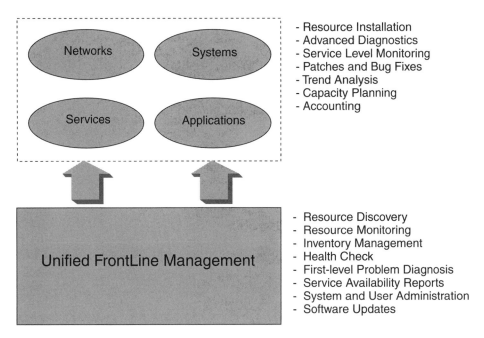

FIGURE 3.7.12 Architecture of FrontLine Manager.

TABLE 3.7.5 Strengths and Weaknesses of FrontLine Manager

Strengths

Unified management
FrontLine management
Intelligent management
Changes are dynamically executable
Distributed architecture
Scalable
Flexible deployment
Extensible and customizable
Use of Internet technologies

Weaknesses

Targeted for small and medium-sized businesses
Not yet widely used
Support of third-party management applications is limited
No compliance to Object Request Broker standards

because they must support multiple frameworks. Web-based management will bring relief by offering an unified interface to management applications. This interface is expected to be supported by all framework vendors.

3.7.5.2 Device-independent Applications

They are designed, developed, and deployed to work in different environments. Usually, they address the following management areas:

- Trouble ticketing
- Performance analysis and reporting

- Security management
- Modeling

Also, these management applications can be integrated into frameworks using Web-based technology. The big benefit is that management applications can be loosely coupled with the framework and with each other.

3.7.6 Summary

Management frameworks are the key for successfully managing communication infrastructures. The frameworks of the future will show very strong core components, and a rich set of management applications. Management applications will be provided by independent software vendors and will address key management process areas of telecommunications services suppliers and of enterprise users. Integration depth is different; the telecommunications are most likely deeper than in the enterprise environment. Some of the management applications are the same for both areas.

References

BALL94   Ball, L. L.: *Network Management with Smart Systems,* McGraw-Hill Series on Computer Communications, New York, 1994.

DORF93   Dorf, C.R.: *Handbook — Electrical Engineering,* CRC Press, Boca Raton, 1993.

GARE95   Gareis, R. and Heywood, P.: Tomorrow's Networks Today, *Data Communications,* September 1995, p. 55-65, McGraw-Hill, New York, 1995.

GHET97   Ghetie, I. G.: Networks and Systems Management — Platforms, Analysis and Evaluations, Kluwer Academic, Norwell, USA, 1997.

NMF95   *Network Management Forum: Discovering OMNIPoint 1 and OMNIPoint 2 — A Common Approach to the Integrated Management of Networked Information Systems,* Prentice-Hall, Englewood Cliffs, USA, 1995.

STAL96   Stalling, W.: *SNMP, SNMP2 and RMON — The practical guide to network management standards,* Addison-Wesley Publishing Company, Reading, MA, 1996.

TERP92   Terplan, K.: *Communication Networks Management,* Second Edition, Prentice-Hall, Englewood Cliffs, USA, 1992.

TOWL95   Towle, T. T.: TMN as Applied to the GSM Network, *IEEE Communications Magazine,* March 1995, p. 68-73.

YAMA95   Yamagishi, K. and co.: An Implementation of a TMN-Based SDH Management System in Japan, *IEEE Communications Magazine,* March 1995, p. 80-88.

3.8   Customer Network Management

Kornel Terplan

3.8.1 Definitions

Customer network management lets corporate users of communication services view and alter their segments of a provider's network. Once such a standardized and open interface is in use, both the service provider and corporate users benefit.

Service providers offer these advantages:

- Keep network loads to a minimum, despite the inexact nature of traffic prediction
- Provide customers with safe access to pertinent OSS and network data, from port assignments to billing and account details
- Isolate individual customer domains without revealing details of the carrier network configuration

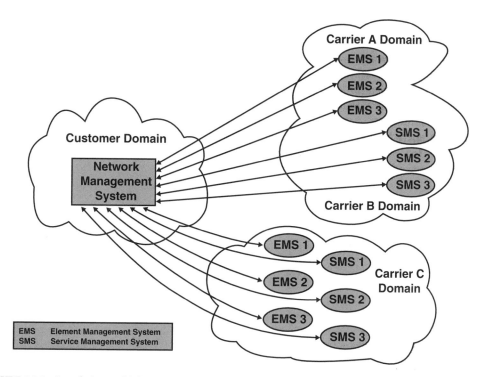

FIGURE 3.8.1   Interfacing multiple services offered by multiple providers.

- Accomplish even the most complex mapping by gathering values from across the network or among OSSs
- Establish customer network domains with full assurance that customers can make only authorized changes

The advantages of corporate users are:

- Alter data network configurations without the delays of paperwork or telephone calls
- Produce any level of report, from performance on a single switch to comprehensive management overviews of account histories
- Manage faults dynamically, reducing the need for carrier intervention
- Streamline troubleshooting with easily generated reports and automatic fixes, even in multicarrier environments
- Integrate to the carrier network whether or not current end-user management systems are robust

The way to a standardized and open interface is long. Today, there are many interfaces and manual information exchange is typical. The customer has to support multiple interfaces that are usually different for each of the providers. Figure 3.8.1 shows this typical case.

This solution can be characterized as follows:

- Support of many proprietary element management systems; most of them are legacy-type systems. They address PBX management, multiplexer management, modem management, management of packet switching nodes, frame relay management, ATM management, wireless management, etc.
- Support of many proprietary service management systems; they are evolving without any core management functions. They address service provisioning, bandwidth management, service assurance, etc.
- Lack of well-understood management protocols
- No easy way of exchanging management information because database and MIB structures are very different.

3.8.2   Concerns of Customers

Customer network management (CNM) has been a long time coming for several reasons. First, it is difficult to measure how much CNM benefits the bottom line. CNM is generally on the cost-saving side which is very hard to sell to management. Selling is easier when the CNM user is actually a value-added provider, which makes CNM a critical component of the value-added service.

Second, when it comes to CNM, many network managers simply do not know what services to require from the supplier. Third, some network managers have serious security concerns about letting an outsider get a detailed look at mission-critical data. Others are afraid that CNM is an attempt by the carrier to lock the customer into a long-term relationship. And some others worry that CNM is the first step toward outsourcing.

Fourth, CNM is very complex to implement. Enabling customers to perform both read and write operations on the internal operations support systems of the telecommunications providers places a considerable stress on those OSSs. Most OSSs are not designed for extra transaction handling and security imposed by CNM. Further, integrating a CNM interface with the network management system of the customer is a difficult task.

Prior to the decision making about implementing CNM functions and features, corporate users should complete a diligency phase consisting of the following tasks (HOLL95):

- Which services come with CNM?
- Are different services integrated in some way?
- What software, hardware, and management platforms do CNM applications run on (Unix, Windows, Sun, HP, IBM, etc.)?
- Can they be easily ported to the company's current network management platform?
- What facilities are furnished to help integrate CNM functions into the existing corporate management infrastucture (CPE, management applications, accounting systems, databases, documentation systems, workflow solutions, etc.)?
- What is the end-user interface to the CNM applications (Windows, Openlook, Motif, etc.)?
- Have provisions been made for training users and technical staff on the CNM system?
- How is corporate data protected against unauthorized access and use?
- What is the cost of the CNM system on a component-by-component basis, including access charges, transport of CNM data to customers, initial installation, integration, and ongoing support?
- How is the CNM system supported?
- What services are offered to help integrate CNM with other management systems?
- What are the procedures in case of significant changes of the OSS?
- What are the impacts on the CNM interfaces and gateways?

After completing this diligency phase, the corporate network manager is well informed about what management functions, databases, and applications can be integrated into the corporate network management systems.

3.8.3   Basic Structures and Core Components

Corporate networks must be able to perform various management tasks. In particular, the following tasks should be supported (HOLL95):

- Fault management, including fault detection, analysis and reporting, tracking and resolution
- Perfomance and quality of service management
- Configuration management, including inventory management, service control, service ordering, and tracking

- Security management, including the protection of the network and its management from both outside and within
- Accounting management, including invoicing, maintaining user and usage profiles, scenario analysis, trend reporting, and exceptional reporting

Most of these tasks must be duplicated for equipment and services of the providers. This redundancy can lead to serious inconsistencies between the provider, corporation, and reality due to the lack of synchronization between inventory files and databases. Moreover, without near real-time information about the provider's network, it is difficult to establish and maintain coherent, end-to-end views of the network, its services, and its performance.

Corporations that buy services from multiple providers find that their problems multiply as the number of interfaces to the service provider rises: operational, fault reporting, inventory, service modification, accounting, and so on. However, even when these interfaces were unified into a one-stop-shopping concept at the provider end, integration with the corporation's internal management systems remains a problem. There are a number of issues to be resolved:

Accounting management: If a customer wants to receive billing information from the provider in near real time (end of shift or end of day) to update an accounting system, some form of electronic interface between customer and provider is needed. Alternatives like e-mail or sending a tape via courier are not the best solutions.

Bandwidth management: Without integrated CNM and enterprise network management, customers who want to change the bandwidth of a service or add more channels to voice and data have to contact the provider through its interface. After confimation, which may take long, customers can start to reconfigure their routers and other network devices. Ideally, using a CNM system, it would require a single application that would accept the request for additional bandwidth. A component of this application would wait for notification that the change has been made and then initiate reconfiguration of the customer network.

Quality of service (QoS): Many customers use their own network management systems to verify that the provider is meeting contracted QoS committments. Doing this properly involves a significant amount of resources. The provider on the other end is probably collecting the same data for the same purpose. It would be best if both parties were working from the same view of the service.

Fault management: Customers will likely perform initial detection and diagnosis of fault using their own network management and monitoring systems. Without CNM, they then must relay this data via phone or fax to the provider and track the progress of fault rectification using the same medium. Assuming the high level of sophistication on both ends, this is not the most efficient way to solve problems.

Table 3.8.1 (HOLL95) summarizes the core CNM components and high priority tasks.

In order to avoid redundancy and inconsistencies, state-of-the-art CNM solutions request a very tight connection between the management architectures and products of the provider and the corporation.

There are a number of ways in which a CNM system could integrate or fail to integrate with customer systems. The first alternative is no integration at all. The provider's CNM system could continue as an independent stand-alone system that provides a convenient point of access to services such as PBX management. Beyond that, the provider could supply customers with a standard interface that encapsulates a particular combination of protocols, information models, and behaviors such as a CNM agent and MIB. This will be the integration point for the management applications at the customer promises. But this still will cause problems if different providers define different interfaces with various information/object models for similar services. At the next level, integration could be achieved via a common graphical interface at the user interface level. The provider would supply a Windows or Motif CNM application that runs alongside the customer's management application on its net management platform. In some cases, the provider would furnish applications as part of the CNM system that uses a private provider CNM agent MIB on the provider side. This is really an extension of the previous approach: the provider offers more of the application functionality to the customer.

TABLE 3.8.1   Core Components of Customer Network Management (CNM)

Fault Management	Configuration Management	Accounting Management	Performance Management	Security Management
Reporting, tracking, and resolution of faults	View inventory of telco-provided customer premises equipment and services	Expenditure tracking on services in near real time	Monitoring of service quality (throughput, delay, and availability)	Access authentication and authorization
Interface to customer trouble ticket or workflow systems	Order new services	Interface to customer accounting system	Ability to generate reports and verify against service contract	Separation of customer data
Fault domain identification	Reconfigure services and network	Extract of histories and usage profiles by customer cost center; cost comparison of rival telco services (ISDN, leased line, xDSL, etc.)	Performance comparison of rival telco services	Separation of telco and customer data

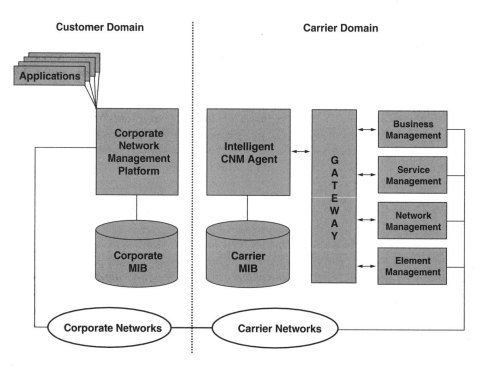

FIGURE 3.8.2   Customer network management.

In order to enhance efficiency and simplification at the same time, the network management platform of the corporation should be connected to a very intelligent "agent" on behalf of the provider. This agent unifies and coordinates the work of multiple managers who are responsible for business applications, along with service, network, and element management. It is also responsible to synchronize data files, databases, and MIBs.

Figure 3.8.2 shows an integrated structure. There are two connections between the systems: one at the physical level and one at the network management level.

This high level of integration is expected to be reached in multiple phases. Telecommunications providers are on the move to select, customize, and deploy powerful management frameworks that will play the role of the intelligent agent.

3.8.4   AccessCNM from Objective Systems Integrators

The functions possible with CNM technology must address the individual customer's need for network management data. They also must meet the provider's need to maintain absolute and integrated control and visibility of their entire network, all OSSs that support it, and each domain contained in it.

Each of these perspectives must be considered if the overall service offering developed is to provide acceptable and competitive features to customers while it answers the nonnegotiable operations and business concerns of service providers.

This seemingly self-conflicting goal defines the challenge that Objective Systems Integrators (OSI) meets with AccessCNM. The company brings to the CNM market an expert awareness from both perspectives and the resources to develop, support, and enhance a comprehensive solution.

3.8.4.1   Basic Functions of AccessCNM

CNM services allow a customer access to segments of a public network that is shared by many users and services. The individual customer's need to know must be balanced against the privacy concerns of the entire customer base and the need for network security and integrity. Therefore, individual access must be controlled. AccessCNM accomplishes this by segmenting the view of network elements and securing them against unauthorized use through three basic functions:

- User authentication, which enforces established access authority
- Flow control to avoid overloading network elements
- View mapping to translate how objects are represented between customer and carrier network views

3.8.4.2   AccessCNM Architecture

The customer access network primarily transports data from the customer site to AccessCNM. It also performs rudimentary flow control and serves in the authentication process, which is a feature of the underlying network service. Access is part of any customer configuration with or without CNM. However, OSI can consult in the design of the access portion to optimize the authentication process and traffic regulation for use with AccessCNM.

The Access Regulation Module (ARM) is the gateway to AccessCNM and acts as firewall. Essentially an IP relay engine, the ARM regulates the load from the server to a module that translates and filters SNMP messages. ARM functions include applying customer-specific flow control; discarding non-SNMP, malformed or errored packets, and those from nonregistered users; and hiding the rest of the carrier network. Traffic from the reverse direction is also regulated.

The deployment of AccessCNM is shown in Figure 3.8.3.

The AccessCNM Core is the system's request processing engine. This is the only module that interprets SNMP messages after the initial phase of the ARM, thus freeing the rest of the system to handle generic database and flow control functions.

The AccessCNM Core performs mapping in three stages; first, the customer view presented in the packets is transformed into the view appropriate to the carrier network. Second, the request is transmitted to one of many potential internal handlers. Third, the handler processes the customer request and, if required, formulates a response.

Handler functions may be customized and their actions extended depending on the desired extent of the resulting CNM features. Some handlers may use SNMP to contact various network elements. Others may obtain database information either locally or from various OSSs. AccessCNM Core is shown in Figure 3.8.4.

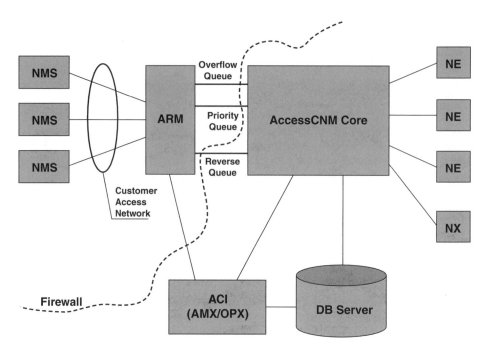

FIGURE 3.8.3   AccessCNM architecture.

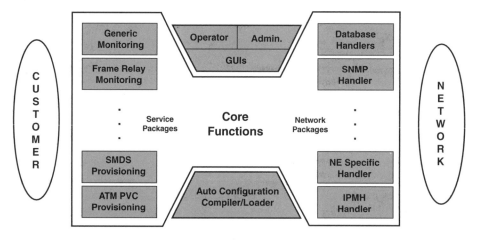

FIGURE 3.8.4   Core of AccessCNM.

Still other handlers and NetExpert interfaces may trigger more complex processes such as provisioning additional services or allocating higher bandwidth, updating customer account files, and confirming that actions have been completed. As the technology evolves, many of these functions will be available off the shelf as feature-specific CNM packages. The AccessCNM Core also is an engine containing the logic and data required to manage customer information. It provides the basis for the MIB traversals, identification of view translator and handler functions, as well as the tables of parameters used in handler processes.

Because AccessCNM associates the actual network view to the segmented and limited customer views, knowing how its processes are executed is important to network security. Information from AccessCNM Core will not flow to the customer unless a corresponding mapping entry is found in the ARM. The AccessCNM Core also includes caching features that keep database lookups to a minimum by storing certain previously accessed data in memory.

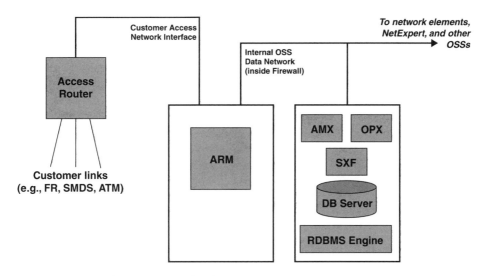

FIGURE 3.8.5 Structure of AccessCNM workstations.

3.8.4.3 Provider User Interfaces

AccessCNM includes operator interfaces for routine modifications, such as provisioning, and maintenance interfaces for privileged commands used for maintenance, monitoring, initialization, and reporting. Interface design may be customized and later revised to match organizational structures and operator
skill levels. Initally, these interfaces will allow these basic functions:

- Assigning interfaces to customers
- Modifying flow control parameters
- Generating configuration reports

An important AccessCNM feature is the ability to perform additions and deletions of user data and
handler functions dynamically, without system restart. AccessCNM offers an open architecture with
editors and compilers (Figure 3.8.5).

3.8.5 Summary

The rising dependence of commercial businesses on internetworking and data communications presents
an excellent chance for service providers to gain the loyalty and partnership of the largest spenders in
their markets. Even in markets where competition has not yet occurred, providers are finding that
customer relations and the efficient implementation of requested services directly impact profit margins
and the provider's general image. If competition is expected soon, the edge may be retained by providing
CNM solutions today and expanding the features offered as the market opens.

Offering comprehensive CNM services complements the most basic reason for being in business:
getting and keeping customers so that revenue generation is assured. End users, demanding the power
to view and alter their segments of the carrier network, gladly pay for the added value and, increasingly,
are comparing CNM functions when deciding on service providers. The right CNM technology gives
providers confidence that their networks and OSSs are secure and that customers have access only to
their own domains. Risks are minimal and profit goes up with AccessCNM because its functions can be
enhanced on demand.

Empowering the customer by implementing a comprehensive CNM solution brings a hidden bonus
to providers: end-user self-government decreases operator intervention and lessens trouble calls, thereby
reducing the cost of doing business. The characteristics of AccessCNM — standards-based, open, and
object-oriented — are the very features required of OSSs as entities. They also are the traits of systems

most likely to return investment quickly because such systems are ultimately less costly. AccessCNM is flexible enough to integrate with existing OSSs and is ready for anticipated expansion.

References

HOLL95   Holliman, G., Cook, N.: Get ready for real Customer Network Management, *Data Communications*, McGraw-Hill, September 1995, p. 67-72.
TERP98   Terplan, K.: *Telecom Operations Management Solutions with NetExpert*, CRC Press, Boca Raton, 1998.

3.9   Aspects of Managing Outsourcing Solutions: Aiming for Success

Carel Marsman

3.9.1   Introduction

The aspects of management of outsourcing solutions and the aim for success — this chapter will handle this broad topic from the customer focus perspective. What are the related processes, what matters from this perspective, and which factors can make a difference? People, as in most service delivery organizations, are key. They can be the competitive edge, if organized and facilitated in an adequate way.

This chapter will discuss how an outsourced solution could be managed in order to deliver world-class services and support, with an emphasis on *could be managed* because it's merely the view of the author based on his experience in the field. Various related topics will be addressed such as the partnership approach of the solutions, service level performance indicators and reporting, integration with vendors, management of and integration with customer operations, decisions to expand or exit, and other related concepts.

These topics will be discussed through the following themes:

- Customer problems and needs in the marketplace
- Strategic outsourcing alliance
- Managing the strategic supplier relationship
- Business processes: What's to be managed?

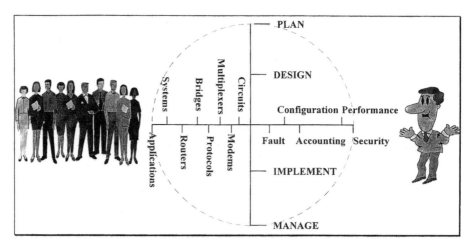

FIGURE 3.9.1

- The partnership approach of service management
- Organizing for success

The aim is not to cover all topics and concepts in the area of management of outsourcing solutions, but to give a comprehensive view of the most important aspects in this exciting and demanding "game." At the end of the day, a customer is looking for a partner to rely upon and to be trusted in handling the company's telecommunications infrastructure. Such a partner offers the right people, processes, and tools for the job, along with the business knowledge, experience, and relationships in the industry — in other words, the core business of the IT outsourcing vendor.

This chapter is written from the customer focus perspective and, for that reason, it emphasizes more the processes, partnership, and people side of the outsourcing business, and not so much the tools and technique side. However, in the references and bibliography section, you will find authors who have contributed in more detail on the technical issues of outsourcing.

3.9.2 Outsourcing — the Evolution

Outsourcing is the product of a long-existing tradition of the make-vs.-buy decision concept. In order to understand the evolution of this product, an overview of the characteristics along the years is given below, including the categories as we still know them:

Decade	Computer Services 1970–1980	Facilities Management 1980–1990	Outsourcing 1990–
Technique	Batch programs on large mainframes	More complex, downsizing of hardware	Highly complex networks, downsizing, telecommunications
Equipment	High purchase costs	Costs decline	Costs decline more
Applications	Costs relatively low to equipment	Costs incline	High costs
Know-how	Low costs	Costs incline	High costs
Customer	Small	Small	Small and large
Presence	National	National/international	International/national
Product	Large amounts of "bread and butter" programs	Standard offers for managing data centers	Managing (parts) of the IT environment based on specific customer requirements
Economic motivation	Shared usage of expensive main frame	Flat costs, higher service level	Saving costs, increased flexibility, better results

In the following sections and examples you will get a better picture of outsourcing. A typical recent end-to-end managed outsourced solution is presented in Figure 3.9.2 in order to create an image of what could be part of the scope of the service delivery.

3.9.2.1 Setting the Stage — Customer Problems in the Marketplace

Before we start setting the stage, we should have a common definition of what outsourcing really is. A definition — not the only one — is the following: *Outsourcing is the transfer of part or all of an organization's existing data processing hardware, software, communications network, and systems personnel to third party* (Due, 1992).

Much of the recent activity in the outsourcing market has concerned vendors developing more desktop services outsourcing capability. These services have been developed to address the challenges that organizations are facing in managing new technologies in the form of distributed LANs. However, changing technology and business requirements are also beginning to have an even more major impact on the ability of organizations to manage their WANs.

In this context, in order to understand what matters from the perspective of the customer, it is important to understand the scope of the customer's problems. A customer's need for integrating and managing multinational infrastructures is hampered by:

Outsourcing Solution Service Delivery

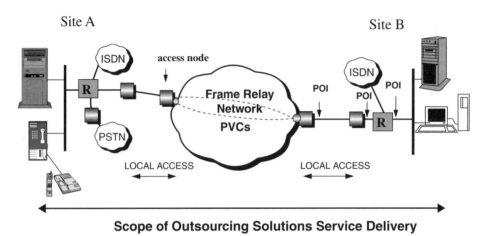

FIGURE 3.9.2   Typical end-to-end managed outsource solution.

- Incompatible and inflexible networks and local infrastructures (LAN)
- Multiplicity of suppliers and products, e.g., multiple telecommunications carriers and vendors
- Difficulty of managing systems and equipment conforming to different standards
- Shortages of affordable skilled resources to manage existing networks and infrastructures
- Lack of global networking infrastructure
- No adequate capabilities to combine voice, data, and image
- Regional barriers — time zones, cultures, languages, lead times, holiday schedules, work shifts — resulting in poor service processes

3.9.2.2   Drivers to Outsourcing — Customer Needs

Altogether, the potential characteristics of outsourcing vendors of perceived importance to customers can be listed in short, i.e., the most common drivers, as follows:

1. Cost savings
2. Focus on core business
3. Avoid headcount increase
4. Availability of new services (voice/data)
5. Staffing/skill problems
6. Headcount reduction
7. Better local support required
8. Better cost control and allocation required
9. Desire for uniform international services
10. One-stop shopping/maintenance
11. Inflexibility of private network
12. Bill consolidation
13. New application requirements
14. Better reliability/resilience required
15. Poor utilization of private network

16. Other companies are doing it
17. Poor network management
18. Poor performance of private network

3.9.2.3 The Importance of the Strategic Outsourcing Alliance

Given these challenges — fast-changing business requirements, technologies, and scarce resources — it is important to understand the concept of the strategic alliance.

Strategic alliances are combinations between firms, designed to support or shape the competitive strategy of one or more of the allies, all for the benefit of the customers served via the alliance. As stated before, companies are not just looking for vendors but also for partnerships. An example:

> Lasher, Ives, and Järvenpää (1991) describe the example of strategic alliances. The objective of the strategic alliance between the United Services Automobile Association (USAA) and IBM was to build a large-scale image processing system for USAA in order to introduce the paperless office. IBM invested heavily in the solution. The requirements of USAA were comparable to those of IBM — a general image solution for all of the application areas of USAA. The advantage for IBM obviously was to integrate this solution in their standard offering. USAA is operating in a niche of the market aiming for financial services to military personnel. The company, in comparison to others, is not worried by IBM's interest to market the product in the insurance industry.

> In the end, the product supported both USAA and IBM in order to achieve important organizational targets and reach a sustainable competitive edge within their own market areas.

3.9.3 Managing the Strategic Relationship — Supplier Management

Given the concept of such strategic relationships: what is important in managing this relationship from the customer's perspective?

In other words — yes, the customer is looking for a partnership but in the same respect wants, exactly for that reason, to get the best "deal" around — one that creates a win–win situation for both parties, as in the case of USAA and IBM. You could say that in outsourcing deals, the offering company has to be some sort of "super-supplier." It has to bring the right cards to the table on strategic, tactical, and operational levels in order to offer the ultimate synergy.

Bearing that thought in mind, let's take a look at what a super-supplier should bring to the table, looking through the customer's eyes. Knowing what they are looking for and putting yourself — as a potential partner — in the position of the customer will give the potential outsourcing partner a head start on the competition.

In an average outsourcing deal, the customer will look for partners performing strong on:

- New technology development
- Taking total cost of ownership
- Globalization and localization issues

For the customer, managing their supplier successfully means looking at what the potential partner's views are on quality management aspects and the integrated business management system. Organizations like Xerox, Honda, and Ricoh[1] are known to be leading practitioners in implementing quality management principles for managing their suppliers.

The right partner should be organized and have:

- Shared values that govern decisions, behavior, and relationships
- Leadership and management systems that provide clear direction and develop, empower, and recognize people
- Organizational structures that provide efficient and effective roles and responsibilities

---

[1]Masaaki, Imai, KAIZEN The Key to Japan's Competitive Success, 1986.

- Processes that create value for their customers
- Information Systems that provide the facts upon which decisions and actions are based
- Their procurement, engineering, product, and quality assurance teams and customers working together early on as one team, in planning and implementing these elements to achieve mutual goals

The underlying principles such a super-supplier should be committed to are:

- Supply and technology strategies driven by business objectives
- The knowledge that allies can achieve more than adversaries
- A willingness to work together to achieve mutual business advantage from the relationship — being open and honest with each other
- Clear objectives, goals, and requirements are always established
- Performance is measured and reviewed on a regular basis
- Strive for continuous improvement to deliver maximum value to the customer and their end customer/end user of the product or service

This means that customers will be looking for the tangible and intangible — after all, it's a people business — proof of all of these high-quality organizational building blocks.

Knowing that the customer is looking for all the above qualities makes it clear that the strategic alliance is very much about building relationships. The more complex the relationship, the more its effectiveness depends on those same factors that are important to personal relationships: trust, mutual respect and dependence, and a shared vision. The joint intent to achieve the spirit, in addition to the letter of a contract, characterizes the most effective relationships.

So, an important step in establishing this relationship is to first understand the characteristics of the customer organization. What are the behavior drivers, the unwritten performance standards, the unspoken generally accepted truths? The customer will be looking for relationships with partners — super-suppliers — that have similar or complementary characteristics. These characteristics will consist of values, beliefs, and norms. Clear, consistent, and *well-communicated* values, beliefs, and norms support both sides of the strategic relationship.

When the partnership is established, the real effort is to implement all of these mindsets on strategic, tactical, and operational levels. Really sharing these values and business senses, living them in the day-to-day business, is what the partnership should be about.

Last but not least, as in any healthy relationship, it should be possible to expand or exit the relationship. The customer could prefer to test the relationship on an operational level through, for instance, a pilot program. Most regular supplier relationships are on a tactical level. Trusting each other to start a long-term strategic relationship should in all cases also be based on clear performance indicators and service levels combined with well-defined contractual exit points. If the relationship starts on an operational level, expansion to a strategic relationship is a logical consequence of the customer's requirements and the partner's capabilities.

3.9.4 Business Processes and Outsourcing — What's to be Managed?

3.9.4.1 Introduction

We now have an impression of the challenges, requirements, and strategic needs of the outsourcing customer.

For a customer, outsourcing (part of) their IT infrastructure means being able to focus on their core business processes and the remaining supporting processes such as finance and human resource management.

This section will provide an overview of the entities within, and the scope of, outsourced solutions. A model in which the to-be-managed processes (core processes for the outsourcing partner!), functions, and objects are displayed serves to build the thesis of this article — how an outsourced solution should be managed in order to deliver world-class service and support.

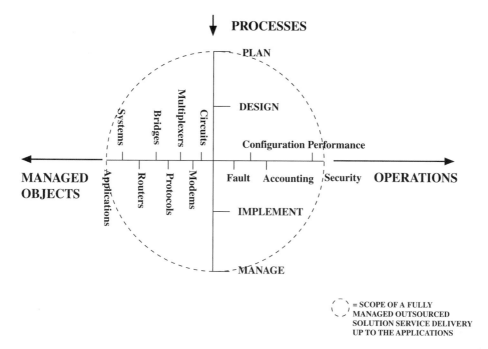

FIGURE 3.9.3 The relationship needed for high-quality services and solution management.

The planning, design, implementation, and management processes will be discussed, as well as the day-to-day management functions. An additional model will offer comprehensive insight into the management functions and related information streams. Furthermore, the various dimensions and definitions of what can be outsourced will be addressed in relation with the to-be-managed objects.

3.9.4.2 The Life Cycle of Processes, Functions, and Managed Objects

In order to be able to present a clear picture to the customer and create a mutual understanding of what the important processes, functions, and elements are, the next model (Figure 3.9.3) depicts the relationship between the different elements needed for high-quality services and solution management.

This model will serve as the basis for explaining how outsourced solutions should be managed in order to create world-class service.

For that reason, the essence of the model will be discussed. It should be noted that any other model could work as well. It's not about this specific model, but more the thought that a picture paints a thousand words. That's exactly the strength of using clear models and pictures in this complex business, and one of the primary added values of a high-quality service provider. In other words: making the provided services and the level of services crisp and clear is really for the benefit of both the customer and the outsourcing partner. After all, as in most relationships, trust is very much about expectations being in sync. This model can be used to aim the combination of processes, daily operations, and managed objects toward the customer requirements.

Horizontal Axis Left: Outsourced Solutions Elements

Every outsourced solution (or part thereof) consists of a number of elements that, working together, establish the service provision to the end user. This combination of hardware and software needs to be managed.

Vertical Axis: Process Elements

Once the decision has been made to implement the solution or to implement changes to an existing solution, a number of processes have to be followed:

- *Analyses.* It is important to do a thorough analysis of the specific requirements that the business processes and the end-user wish to demand from the network. These requirements need to be documented during the analysis phase. This has sometimes already been partly done by the customer and will be complemented by the outsourcing partner in collaboration with the customer.
- *Solution planning/design.* Based on the technical and functional requirements obtained, the solution or the changes on the network can be planned and designed. This is accomplished by outsourcing partner specialists who will regularly update the customer. The functional requirements should be documented in a service level agreement (SLA). The SLA will function as a benchmark and a means for controlling and checking the quality and the performance level of the solution and the solution management service delivery.
- *Implementation.* After the solution design has been approved and accepted by the customer, the implementation will take place. For this, the outsourcing partner should use project management and quality control processes that ensure a correct and timely implementation. After the solution has been implemented, a technical and functional acceptance test should be performed. These tests will be coproduced by the customer and the outsourcing partner.
- *Management.* Once the solution has been approved and accepted by the customer, the management phase will commence. The management service elements are mentioned on the horizontal axis and are described below.

Horizontal Axis Right: Service Elements for Outsourcing Management
The outsourced solution elements need to be maintained and managed on a daily basis, the so-called logical and physical solution management. Logical solution management entails the management of the network protocols and the software for the solution components such as bridges and routers. The logical solution management often can be done remotely. Physical solution management is the management of, for instance, cabling and hardware components. The physical outsourcing management must often be done locally — for example, some of the desktop services.

Outsourced solution management services should be able to perform both the physical and logical solution management for the customer.

To enable this, the following outsourced solution management service elements should be delivered on a daily basis:

- *Fault Management:* This comprises the proactive monitoring of the solution in order to detect possible causes for future problems and prevent them from happening. Next to that, it also should offer help desk services to the customer's user representatives in order to be able to receive user calls for help and have specialists solve the problems. Concerning logical solution management, the outsourcing partner staff will diagnose and solve the problems. In case the problems are caused by the physical solution, the outsourcing partner staff should be able to dispatch a specialist to the faulty component in order to fix the problem.
- *Configuration Management:* The outsourcing partner should be able to perform both physical as well as logical solution configuration management. This means that logical adjustments should be made in case of changes to the information stream requirements, and physical adjustments to the solution should be made when user units change location. Also, the administration of the solution configuration will be performed and reports should be regularly submitted to the customer. The outsourcing partner should control the administration of, for instance, the customer network addresses on the WAN within the points of interconnection (POI). The customer could make the choice to be in control of the addressing on the LANs and the network number registration, depending on the scope of the agreement of the outsourced solution.
- *Performance Management:* The outsourcing partner should constantly monitor and evaluate the solution on the consistent compliance with the performance requirements — indicators — as documented in the functional design. Proactive measurements should be undertaken if the performance level tends

to degrade. The customer representative should regularly receive reports concerning the solution performance.

- *Accounting Management:* Depending on the customer's specific accounting management requirements and the available technology, the outsourcing partner should categorize, quantify, and report the usage of the network.
- *Security Management:* Depending on the customer's specific security management requirements and the available technology, the outsourcing partner should perform the security management of the solution. This could include, for example, access management, password security, and encryption.

3.9.5 The Partnership Approach of Service Management

3.9.5.1 Introduction

In this section, you will learn what service management is, what it offers to the customer, how this should be performed, and the required interaction with the operations in order to deliver a seamless service that surpasses the agreed-upon SLAs.

The "how" will cover the aspects of partnering or relationships with other suppliers/vendors being a truly single point-of-contact for the customer; management around the clock, especially the challenge of the time zones, integration with customer operations; managing customer perception; and building the service level agreement and reporting based on the SLA (service performance indicators, etc.).

Between these points, some customer examples (no names will be mentioned) will be highlighted, along with market research findings.

3.9.5.2 Service Management

The outsourcing partner is a service provider company. What does that mean?

To begin with, services are a mix of tangible and intangible aspects. Tangible aspects are the service levels, the reports, etc. Intangible aspects are the perceived qualities of the service received by the customer, and the way the solutions provider handles the customer. As you will see, there is a strong connection with the essentials of building a strategic partnership.

When we look at a quality service organization, it should distinguish itself through a market-oriented approach. The enterprise of the outsourcing partner should have the following fundamental features related to the marketing concept:

- **Attitude of mind**: the customer is the basic reason for the existence of the enterprise
- **Organizing the enterprise**: all organizational design should stem from the customer and ensure that the customer is "created," won over, and kept by the enterprise
- **Range of activities**: activities necessary to ensure the serving of customer needs emerge as a matter of course (creation, production, delivery of services)
- **Techniques and tools**: enabling the organization to operate as efficiently and effectively as possible in the customer's interest (motivation research, linear programming, discounted cash flows, etc.).

Knowing that services are partly tangible and partly intangible, substantial attention should be paid to the marketing aspects.

Furthermore, the four standard Ps (product, price, place, and promotion) of the marketing mix are expanded with three more Ps (people, physical evidence, and process) in the marketing mix for Services.[2] The three additional Ps should be incorporated in the marketing mix as input for the design of a service organization. Because of their importance to the design, the three extra Ps will be discussed individually. By the way, this is my opportunity to introduce maybe the most important — most forgotten — P of a service organization: pleasure (or fun)! The amount of energy this "P" generates should never be underrated by the management of any service providing company! A happy employee radiates pleasure toward the customer.

---

[2]Cowel, D. The Marketing of Services, 1984.

3.9.5.3 People and Service Organizations

Although the building, equipment, and financial assets are also resources required by organizations, employees — the human resources — are particularly important. People provide the creative spark in any organization. They design the service, control quality, market the service, allocate financial resources, etc. With respect to personnel, close attention should be paid to training, discretion, commitment, incentives, appearance, and interpersonal behavior. With respect to other customers, close attention should be paid to their attitudes, behavior, degree of involvement, and customer/customer contact.

People are the most important assets of a service providing company!

3.9.5.4 Physical Evidence and Service Organizations

In the physical design, it is important to pay attention to the internal environment in terms of:

- Environment: furnishing, color, layout, noise level
- Facilitating goods: presentation equipment, audio-visuals
- Tangible clues: packaging, manuals

3.9.5.5 Process and Tools in Service Organizations

The behavior of people in service organizations is critical. So too is the process — the what and how — of the service delivery. Cheerful, attentive, and concerned staff can help alleviate the customer's problems of having to queue for service or soften the blow of the breakdown of technology. They cannot, however, compensate entirely for such problems. How the overall system operates — the policies and procedures adopted, the degree of mechanization (or high tech) used in the service provision, the amount of discretion employees have, the flow of information and service, etc. — these are operational management concerns that need attention when designing service organizations.

More details of how to successfully organize the service organization will be presented in the Section titled 3.9.6.

3.9.5.6 What Does Service Management Mean in Practice?

Service management in practice means making sure that the agreed service levels are reached and, if possible, surpassed. Consequently, this also means service management is about the art of managing the customer's expectations. No matter how great the quality of service, there always will be dips in the service level. Implementation projects will always run into unforeseen, unexpected, or new hurdles.

The real art lies in really knowing your customer's requirements, the service delivery and operations processes, your supplier requirements, and — last but not least — your customer processes (Figure 3.9.4). A solutions provider should be very aware of this principle. The strongest example is perhaps the role of the (inter) national carriers in global solutions.[3] If a national carrier — a supplier of a major link in the end-to-end service provisioning — closes down the help desk on the weekend, one can tell the impact on the service level if a leased line goes down on Friday at 5:30 p.m.

Knowing this and having the right tools and procedures in place will enable the outsourcing solution partner to proactively inform the customer of any interruptions to the service. Always try to place yourself in the position of the customer—know that they have to face *their* end users.

So knowing the customer requirements — your output, and looking at what that means for the service delivery processes; your input, and its requirements toward the suppliers, e.g., carriers — means having measurements in place, providing and receiving feedback, and communicating. Improving these customer focused business processes[4] is the key to creating a sustainable competitive advantage as a solution provider.

A short and simple example on informing the customer of any interruptions in the service, anytime, anywhere:

---

[3]*Datacommunications International,* May 1997, Rating the World's Best Carriers.

[4]In the References and Bibliography section, a couple of authors are mentioned on this and related topics.

Service Delivery Processes

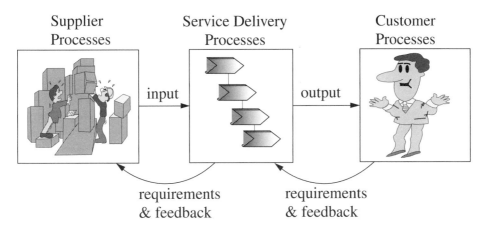

Supplier
Processes

Service Delivery
Processes

Customer
Processes

input

output

requirements
& feedback

requirements
& feedback

FIGURE 3.9.4   Essential areas of awareness for solutions providers.

The service manager of an outsourcing company was instantly informed by the operations manager of a severe problem with the international backbone of a customer. Because of the proactive Management of the managed objectives, the service manager could contact his counterpart within the customer organization. It appeared his counterpart could not be reached that day for family reasons. The service manager, in close cooperation with the customer organization, was able to quickly track down the MIS manager — the counterpart of the service manager — and get him out of an important meeting with one of his end users, an organization. The service manager could now briefly inform the MIS manager of the problem, the corrective actions taken, and the problem analysis so far. The MIS manager was informed from first-hand knowledge and could get the message to his customer — the end user — instead of being confronted with the problem by his customer.

This example may seem like an open door; however, in talking to customers one will find that even the simple and obvious things are not always common practice in the service they receive from their outsourcing partners.

An extra dimension in international service delivery is, of course, the difference in cultures. Minor problems will probably not even be reported from, for instance, the Nordics, but the same problem could sound huge when reported by one of the countries in the south of Europe. The nature of people and customs should always be taken into account. This means taking the time to get to know what these cultural aspects are, and being open minded.

3.9.5.7   Service Reporting

As stated before, the services, in opposition to products, are part tangible and intangible. Therefore, service reporting takes an important place in managing outsourcing solutions. Clear and correct information on the service levels and other relevant information will help both the customer and the partner to control the managed solution, and plan ahead.

After all, the process of analysis, plan and design, implementation, and managing the solution is a never-ending cycle because of changes in business planning and focus, closing/opening of new subsidiaries, short life cycles of products, emerging technologies, and so on.

For instance, monthly service reporting is input for this never-ending cycle and enables the combined customer and outsourcing team to investigate trends, bottlenecks, and other interesting topics emerging from the reporting.

Service reporting should roughly have the following objectives.

- Provide means for customer control of the solution process elements.
- Ensure that the level of solution services remains in line with the customer's business objectives.
- Provide information to authorized customer personnel with an interest in the solution infrastructure.

The contents of the report should be built along the previously discussed elements of the management model. An example of the added value of customer reporting follows:

A customer (a managed outsourced solution much like the typical 1990s managed solution picture mentioned earlier) was confronted on a regular basis with complaints from his end users about the network's performance. After closer examination and further investigation of the customer and outsourcing service provider jointly, as partners, the discovery was made that the degradation in service was the result of heavy file transfers. These file transfers were not part of the regular business processes of the customer's decentralized international user organization. However, the detailed WAN traffic reports showed these bursts in traffic from the central host to the requesting end user.

The customer was able to quickly trace the root cause of the problem that was influencing the service for all the end users — the customer.

By jointly analyzing and discussing the service reports, the outsourcing service manager and the service manager of the customer were able to fix the problem and advise the specific end user's organization to locally request the information. As it turned out, the end user was unaware of some new local data retrieval capabilities.

3.9.5.8 Service Level Performance Indicators and Reporting

Reports should reflect the performance against the agreed service levels. Performance indicators will be agreed upon in SLAs, bearing in mind the process between customer solutions providers/suppliers. A couple examples of indicators important to the customer from a Fault Management perspective (knowing that they also have their customers — end users — to respond to) are given below. Most operational SLAs will have indicators as well on change, accounting, performance, and security management.

- Number of faults closed in the month
- Total number of faults over entire managed solution[5]
- Faults whose outage exceeded the customer-agreed threshold
- Chronic faults (x or more faults per month caused by the same problem)
- Mean time to repair (MTTR) for faults by, for instance, link and total solution
- All equipment failures whose outage exceeded the customer-specified threshold
- Per vendor number of faults (committed MTTR and actual MTTR)
- etc.

Again, the real art is to extract the needed *data* from the managed objects and convert this data into useful *information* to present to the customer. In other words, add value as a solution provider, and not just simply deliver the agreed service levels. Be proactive, look at and discuss trends, indicate opportunities and threats — act as a Partner.

As indicated in the introduction, distributed LANs — and, linked to that, distributed network management — is a major topic and challenge. Quality reporting and indicators on LAN traffic — resulting from adequate solution management — is, for a solution provider, a competetive edge.[6]

---

[5]Up to the agreed point of interconnection, e.g., the boundary of the management responsibility.

[6]Read, for instance, *Network and Distributed Systems Management* by Morris Sloman and *Lan Traffic Management* by Peter Phaal.

3.9.6 Organizing for Success — A People Business

3.9.6.1 Introduction

In this section, the most important aspects of organizing a team in order to deliver world-class services and support will be presented.

What does it mean to create and maintain a team of top specialists, and how do you manage them? What makes or breaks a team? How can you create a learning team? The relevance? The people managing the solution are the competitive edge of the IT outsourcing vendor. They are the human capital of the company and can make the difference in winning or losing a deal or an existing customer.

3.9.6.2 Structuring the Service Organization

Organizing people means some sort of structure should be put in place enabling these people to work together in order to accomplish the company targets. An outsourcing service providing company will mainly have highly educated and dedicated professionals on the payroll. Professionals are people that know their job and — most of the time — are not too fond of too many rules and procedures. However, processes, procedures, and rules are necessary in order to deliver seamless high-quality (inter) national services. The structure of the service organization should take all of these aspects into consideration. Next, a short overview of some of the essentials.

In essence, the *structure* of the service organization should be built around the service delivery processes — the management processes (FCAPS) — in the model. The related subprocesses, activities, and functions should be derived from the major processes. The organization should be able to adequately support the processes. This should be the starting point.

The *systems* — tools and techniques — play an important role in enabling the organization to effectively and efficiently service several customers. After all, the economies of scale aspect is one of the major reasons for an outsourcing company to be able to be cost effective. This goes for the management systems but also for the internal information systems. These last ones should be aimed at the crucial data (customers, orders, design rules, etc.) and the functionality should be fast adaptable.

The classical *line and staff* — advising functions — should have a balanced mix, thus creating "think overhead" sense in the line functions. In other words, at all times it should be avoided that the line functions are caught in turbulence of just managing the day-to-day business. Time and resources should be allocated to think things over, analyze trends, and implement and monitor quality and security policies — thus enabling the organization to learn. Consequently, the modern service organization should strive for horizontal and vertical job specialization together with handing over more responsibilities and authorizations to the work floor. By cutting down on management levels — less coordination and confusion — and by adding value and importance to the tasks performed by the people working close to the customer, the Organization improves its learning curve and self-steering ability of the teams, thereby creating a more *flexible* Service Enterprise.

The management *style* is a very important part of the success of the service organization. Having *shared values* in place and having the management actively practicing these — walking the talk — is rather essential. These values should be the common bond between the employees of a company. Having a solid bond in place enables people to have a basis they can rely upon and which enables them to go that extra mile and be flexible service-minded persons.

The *skills* of the people are probably the most important, because these skills will enable them to participate in multifunctional teams. Organizing skill pools and focusing on key skills will enable the service organization to run an efficient organization. From the people's perspective, they will be able to learn a great deal and have challenging projects to participate in. In this turbulent and dynamic environment, it is not possible and desirable to centrally manage the working force. The outsourcing company employees skill-set should be a mix of soft skills — communication, presentation, teamwork, and hard-skills — management platform technology, specific customer solution technology skills, etc.

The *overall aim* for the structure of the service organization should be to create a learning organization. As stated before, mostly the people, the employees working with the customer every day, will be the competitive edge. Coaching, or enabling, them to learn and grow, be committed, and have fun and pride in their job, will always create value for and to the customer.

3.9.6.3 Teamwork

Last but not least, a couple of words on teamwork. A lot of the success of outsourcing companies managing outsourcing solutions is based on teamwork. This starts during the engagement/sales process, continues during the solution implementation process and, you could actually say, begins during the management process of the solution. After all, it is ten times as expensive to win back a customer that walks away than it is to keep the customer in the first place.

To have professionals work as a team requires special skills from the "coach." The ability to create a team with skill sets that complement each other, stress the team targets, and still reward/penalize the individual for (non) performance are not given to all managers (as it's also a challenge for professionals to be team players!).

Therefore, teams should have clear objectives — both as a team and on an individual basis. Complementing skills sets, thinkers and doers should be committed, and have a sense of urgency. Research has proven that successful teams should have a strong desire to perform — encouraged, for instance, by the company's competitor.

3.9.7   Conclusions

In short, the winners of the outsourcing game are the companies — as far as managing outsourcing solutions — that have a thorough and well-defined plan for their people, processes, and tools. Furthermore, managing the solution is not only having a good plan — it is about the smart planning around the customer of all these three!

Finally, I believe that the outsourcing trend will continue to grow. Reading, for instance, through Tapscott's *The Digital Economy* and Aidarous and Plevyak's *Telecommunications Network Management into the 21st Century*, I see enough indications that lead to the conclusion that companies will seek strategic relationships with suppliers — outsourcing solution providers — that will provide them with the competitive edge of new technologies without having to worry about how to manage these wonders of mankind.

References and Bibliography

Aidarou, S., Plevyak, T., *Telecommunications Network Management into the 21st Century*, 1994.

Brelin H.K., Davenport K.S., Jennings, L.P., Murphy, P.F., *Focused Quality Managing for Results*, 1994.

Cowel, D., *The Marketing of Services*, 1984.

Datacommunications International, May 1997, Rating the World's Best Carriers.

Due, R.T., The real costs of Outsourcing, *Information Systems Management*, winter 1992.

Harrington, H. James, *Business Process Improvement*, 1991.

Hoogeveen, D., *Outsourcing*, 1994.

Katzenbach, J.R., Smith, D.K., *The Wisdom of Teams Creating the High-Performance Organization*, 1993.

Khandpur, N.K., Laub, L., *Delivering World-Class Technical Support*, 1997.

Lasher, D.R., Ives, B., and Järvenpää, S.L., USAA — IBM partnerships in information technology: Managing the image project, *MIS Quarterly*, December 1991.

Masaaki, Imai, KAIZEN The Key to Japan's Competitive Success, 1986.

Minoli, D., *Analyzing Outsourcing Reengineering Information and Communication Systems*, 1995.

Phaal, Peter, *Lan Traffic Management*, 1994.

Sloman, Morris (edited by) *Network and Distributed Systems Management*, 1994.

Tapscott, D., *The Digital Economy: Promise and Peril in the Age of Networked Intelligence*, 1995.

3.10 Support Systems for Telecommunication Providers

Kornel Terplan

The telecommunications industry shows both evolutionary and revolutionary signs. Evolution is seen with incumbent carriers; revolutionary attributes are visible with new entrants. The technology itself shows a mixture of wireline and wireless services, supporting all major telecommunication forms, such as voice, data, and video.

3.10.1 Status, Definitions, and Markets of Operations, Business and Marketing Support Systems

OSSs (operations support systems), BSSs (business support systems), and MSSs (marketing support systems) represent a very complex but increasingly significant segment of the communication industry. All three types of support systems together will be called 3SS. OSS, BSS, and MSS software enables the support, administration, and management from day-to-day operations to traffic trending, capacity planning, and forecasting of communication network services providers. Customer care, billing, provisioning, order processing, and networks operational management are all functions implemented via OSSs, BSSs, and MSSs. Until recently, there was little opportunity for direct investments in this important telecommunications segment. Increasingly, however, both incumbent and new service providers have come to view these systems as critical service differentiating assets. As a result, there is a growing number of public and private companies expected to benefit from the strategic importance of these support systems.

Financial estimates are on the basis of 15% CAGR (compound annual growth rate) over the next few years, approximately until 2002–2005. Unlike the average annual growth rate (AAGR) method, CAGR takes into account the changes from year to year, not only in revenues but also in revenue growth rate. CAGR is the rate at which the amount in the final year represents the future value of the amount in the first year after a specific interval. This CAGR percentage is an average over all market segments, such as customer care and billing, provisioning and order processing, and network operations management.

Industry issues of OSSs, BSSs, and MSSs are:

- Upgrade cycles in support systems: As a result of global deregulation, carrier competition is driving the demand for new, more efficient back-office solutions. In addition to reducing operating expenses, advanced 3SSs improve time to market and often facilitate the introduction of new, revenue-producing solutions.

- Product-based vendor driven solutions: Carriers increasingly demand solutions, rather than raw technology and development kits for custom-developed 3SS solutions. The advent of technology standards encourages the use of best-of-breed vendor solutions.

- Emergence of complex, multiplatform environments: reliability and scalability of large centralized systems remain excellent. Service providers incorporate a multiplatform strategy augmenting existing investments in legacy solutions with newer technologies targeted at profitable customer market sectors.

- Emphasis on telecom systems integration: complex multiplatform, multivendor telecom networks require substantial systems integration for interoperability. With multiple client–server and legacy 3SSs in place, integration capabilities of vendors are in high demand.

- 3SS growth is tied to share-shift among telecom end-markets and carriers: the strongest near-term growth has been achieved by vendors targeting the fast growing telecom end markets, emerging LECs, and wireless carriers.

- Developing 3SS markets: 3SS growth is dominated by new carrier adoptions and incumbent upgrades. Developing markets, such as data solutions, local number portability, and carrier interconnection are likely to justify the next wave of 3SS spending.

- Convergence and telecom consolidation: this accelerates the use of advanced 3SSs. Consolidation of carriers across multiple end markets creates advantages for 3SSs targeting multiple end markets. It increases the complexity of telecom networks and demands for 3SS integration.

- Outsourcing: ongoing structural changes in the telecom industry will place new requirements on 3SSs. In order to concentrate on customer management, some back-office functions may be outsourced to service bureaus. These service bureaus might use 3SSs from the same vendors, but they use them in a shared fashion among multiple service providers.

3.10.2   Market Drivers for 3SSs

The market is changing very rapidly. 3SSs should be positioned well, and should meet telco expectations in a timely fashion. Principal market drivers are addressed in this segment.

3.10.2.1   Growth of the Global Telecommunications Market

Explosive telecom expansion driven by internal growth and acquisition is forcing telecommunications providers to assess the productivity of their current support systems. Growth and acquisition mean that the number of subscribers grow for existing services; new services are provisioned on existing infrastructures and completely new services on new infrastructures are deployed or acquired. Several 3SS vendors have striven to capitalize on this opportunity with solutions that reduce complexity. These 3SS vendors do not usually replace existing systems, but add functionality to accomodate new services, such as:

- Internet, intranet, and extranet
- Special data services on top of voice networks
- Wireless services
- Cable and video services
- Voice services on top of IP

Adding functionality and interoperational features with each other opens new business opportunities for 3SS vendors.

3.10.2.2   Increasing Network Complexity

As a result of customer expectations, the time-to-market of new services is extremely short. Incumbent and new telecommunications services providers do not have the time to build new, but to combine existing and new infrastructures, such as copper, fiber, and wireless. They are deploying emerging services on the basis of a mixture of infrastructures as an overlay. Emerging services use emerged and emerging technologies, such as:

- Emerged technologies (voice networks, ISDN, circuit switching, packet switching, message switching, frame relay, Fast Ethernet, Fast Token Ring, and FDDI/CDDI)
- Emerging technologies (ATM, mobile and wireless, SMDS, Sonet/SDH, cable, xDSL and B-ISDN)

Each of these technologies has its own support system solutions. The only elements in public switched telephone networks (PSTNs) that should be managed are the switches themselves. On average, the ratio of managed elements to subscriber lines is around 1:10,000. The advent of distributed, software-based switching and transmission created a large number of additional managed elements, about one for each 500 subscriber lines. Moreover, multiple elements per subscriber in digital loop carrier systems, digital cellular networks, or hybrid fiber/coax systems may cause an explosion in terms of managed elements. As a result, the size of configuration databases and event messages generated by more intelligent network elements have grown exponentially over the last 20 years.

Growth in the number of network elements has been accompanied by an increase in the complexity of items to be managed. Sonet/SDH, ATM, and digital wireless are highly complex, with a high degree of interdependence among network elements. This in turn makes service activation and fault isolation a challange, especially as the number of service providers increases. As networks shift from lower-speed, dedicated-rate, and inflexible services to mobile, fully configurable, bandwidth-on-demand and high-speed services, 3SSs must adapt to this new situation.

When services are offered in combination, 3SSs should be modified, re-engineered, and connected to each other. This opens new business opportunities for 3SS vendors.

The introduction of standards for support systems is accelerating the demand for third-party 3SSs. Legacy systems are primarily proprietary systems not integrated across functional areas. Service providers depend upon custom development by internal development staff and outside integrators to connect various support systems. The introduction of technology standards such as telecommuication management network (TMN), Distributed Communication Object Model (DCOM), Common Object Request Broker Architecture (CORBA), Telecommunications Information Networking Architecture (TINA), and Web-based Enterprise Management (WBEM) have begun to gain critical support by new 3SS vendors.

TMN is a special network in its own right that is implemented to help manage the telecommunication network of the service provider. As such, it interfaces to one or more individual networks at several points in order to exchange information. It is logically separate from the networks it manages, and may be physically separate as well. However, TMN may use parts of the telecommunication networks for its own communications.

TMN is an extension of the OSI standardization process. It attempts to standardize some of the functionality and many of the interfaces of the managed networks. When fully implemented, the result will be a higher level of integration. TMN is usually described by three architectures:

- The functional architecture describes the appropriate distribution of functionaliy within TMN, appropriate in the sense of allowing for the creation of function blocks from which a TMN of any complexity can be implemented. The definition of function blocks and reference points between them leads to the requirements for the TMN-recommended interface specifications.
- The information architecture, based on an object-oriented approach, gives the rationale for the application of OSI systems management principles to the TMN principles. The OSI systems management principles are mapped onto the TMN principles and, where necessary, are expanded to fit the TMN environment.
- The physical architecture describes interfaces that can actually be implemented together with examples of physical components that make up the TMN.

TMN distributes management responsibilities into several layers, such as business management layer (BML), service management layer (SML), network management layer (NML), element management layer (EML), and into the actual network elements layer (NEL).

DCOM is the heart of Microsoft's ActiveOSS product suite. Basically, DCOM is an integration infrastructure designed to facilitate communication between software components operating on the same host or with DCOM on multiple networked hosts. It was originally developed to create interoperability between components. It is the most widely deployed component object model. Active OSS acts as a centralized management and translation point for an OSS network. Conceptually, applications ride on top of the framework, but communicate through it. DCOM abstracts various application interfaces into objects, basically mapping the functions of the application into a common model that can be stored in a database. The common model allows the various applications to communicate in a uniform manner within the framework or across multiple networked frameworks.

By abstracting interfaces into software objects, applications theoretically can be upgraded and/or changed without affecting surrounding systems because integration is based upon independent software components that communicate, not applications that are heavily modified to fit together one-to-one. In this sense, upgrading an application means mapping a new interface into the framework, or modifying an existing one. The frameworks need to work with the interface, but do not need to affect details of the application. The framework is intended to create uniformity among application services without any modifications to source code. Application services are built into and managed by the framework. The overall architecture also incorporates Smart TMN business process model and related work by TINA.

CORBA is a generic communication framework to connect various network management applications. The object request broker is the coordinator between distributed objects. The broker receives messages, inquiries, and results from objects, and routes them to the right destination. If the objects are in a heterogeneous environment, multiple brokers are required. They will talk to each other in the future by a new protocol based on TCP/IP. There is no information model available; no operations are predefined for objects. But an object does exist containing all the necessary interfaces to the object request broker. For the description, the Interface Definition Language (IDL) is being used. There are no detailed MIBs for objects because OMA is not management specific.

The functional model consists of the Object Services Architecture. It delivers the framework for defining objects, services, and functions. Examples for services are instanciation, naming, storing objects' attributes and the distribution/receipt of events and notification. CORBA services and facilities represent more generic services; they are expected to occur in multiple applications or they are used in specific applications. The driving force beyond designing common facilities for systems management is the X/Open Systems Management Working Group. The Managed Set Service, defined by this group, encourages grouping of objects in accordance to their management needs, with the result of easier administration. In the future, more services are expected to be defined; the next is an Event Management Service that expands the present Object Event Service by a flexible mechanism of event filtering.

Telecommunications Information Networking Architecture (TINA) is based on the concept that call processing in networks, and its control and management are separated from each other. TINA is actually a concept-integrator from IN, TMN, and Open Distributed Processing (ODP) from ISO and CORBA from OMG. The core is OSI-based network management, expanded by the layered structure of TMN. The emphasis with TINA is not on the management of network elements, but on the network and services layers. TINA is going to be standardized by a consortium consisting of telecommunications suppliers, as well as computer and software vendors.

WBEM is a joint initiative of many manufacturers, led by Compaq, Microsoft, and Cisco. The initial announcement called for defining the following specifications:

- HyperMedia Management Schema (HMMS): an extensible data description for representing the managed environment that was to be further defined by the Desktop Management Task Force (DMTF).

- HyperMedia Object Manager (HMOM): data model consolidating management data from different sources; a C++ reference implementation and specification, defined by Microsoft and Compaq, to be placed in the public domain.

- HyperMedia Management Protocol (HMMP): a communication protocol embodying HMMS, running over HTTP and with interfaces to SNMP and DMI.

- Common Information Model (CIM): basis of the information exchange between various management applications.

WBEM is helpful to unify and simplify network management.

The implementation of standard gateways enables interaction between newer client/server solutions with existing legacy systems and eases interoperability among all 3SS systems. In particular, TMN may help to streamline 3SS processes and to position support systems.

3.10.2.3 Deregulation and Privatization

Telecommunications service competition began in the 1980s in the U.S., led by MCI with 3SSs playing a key role. The AT&T divesture in 1984 marked a major breakthrough. The second significant milestone was the Telecom Act of 1996. As telecom deregulation continues, with RBOCs actively pursuing the long distance market and long distance carriers moving into local services, major 3SS re-engineering efforts are expected.

Under the pressure of the European Commission (EC), Europe is in the process of deregulation and privatization. It is a much slower process than in the U.S., because multiple countries are involved, each with their own agenda. Interoperability of 3SSs is more difficult than in the U.S.; but at the same time, it offers opportunities for 3SS vendors.

It is assumed that Asia/Pacific, South America, Eastern Europe, and Africa will follow these deregulation and privatization trends.

Competition is everywhere; long distance, local exchange, ISP, cable, and wireless. In many cases, 3SSs are the differentiators. The best 3SS opportunities are seen with CLECs. 3SS requirements vary substantially from carrier to carrier. As a result, CLEC-3SS-strategies are ranging from internal development to outsourcing to systems integrators and to third-party software/service providers. CLECs could be small or mid-size, they may own facilities, or are facilityless. In all cases, they must interoperate with ILECs by opening the 3SS to permit access by CLECs in various phases of provisioning and order processing, and service activation. Key issues are:

- Local number portability (LNP): it allows customers to retain their telephone numbers even if they change service providers. It is not only the telephone number that is important, customers also typically want to retain access to advanced features they have come to expect from an intelligent network.
- Extranets connecting 3SSs of ILECs and CLECs: ILECs are required to provide access to information on five classes of 3SSs: preordering, ordering, provisioning, repair, and maintenance.
- Directory services: real-time service processing requires additional customer-related data. The expanded directory role includes end-user authorization and authentication. It also includes the real-time allocation of network resources according to a user's class of service and other policy-based variables. Directory Enabled Networks (DENs) promise to increase momentum for directory services by bringing physical infrastructure under the directory umbrella and tackling the standardization of directory information.

Incumbent service providers have turned to advanced 3SSs to differentiate their long distance or local exchange services from each other. After a substantial investment in custom systems over the last few years, many incumbents have begun to focus on upgrading select 3SS systems with best-of-breed technologies. Many of them try to augment older systems to add more flexibility while sustaining traditional levels of performance and reliability. This creates additional complexity and requires that new management solutions designed for advanced equipment also work with older technologies.

As a result, umbrella-types of 3SSs are in demand, opening new opportunities for 3SS vendors with integration capabilities. To remain competitive, incumbent carriers need to deliver an increasingly larger number of new products and services. This has created a mixture of equipment, software, and services within many carriers.

Innovation and re-engineering on behalf of the incumbent carriers show:

- Better customer care: based on call detail record (CDRs) and other resource utilization-related data, unsophisticated customer analysis can be accomplished. It includes discovering trends in customer behavior, traffic patterns, reasons for frauds, and also service-related items.
- Convergent billing: the customer may expect to receive one bill for all services, such as voice, data, video, and Internet. The minimal requirement is to receive multiple bills with electronic staples.
- Rapid provisioning new services: based on additional 3SSs, provisioning can be expedited by better interfaces and more accurate data.
- Service differentiation: still using the same infrastructures, new services can be created and deployed. By carefully defining the value-added nature, it may be considered by customers as differentiators.
- Offering new services: incumbent service providers are expected to react rapidly to new communication needs, including offering Internet access for reasonable money, the deployment of xDSL, VPNS and VoIP.

In each of these cases, either the deployment of new 3SSs or the customization of existing 3SSs are required. In both cases, additional market opportunities open for 3SS vendors.

3.10.2.4 Communication Convergence

Advanced technology, coupled with deregulation, is driving communications convergence. Customers prefer to get all types of services, such as long distance and local voice, data/Internet, cable/video, and wireless access from the same service provider. Voice is expected to support both local and long distance, requiring to play a LEC and IEX role at the same time. Data is gaining importance for both local and long distance, and does usually include Internet access. Data is supposed to reach voice volumes within 5 years, requesting the total rebuilding of circuit switching technology. Cable is expected to accommodate voice and data in addition to video. Wireless does include all kinds of mobile services and satellites supporting voice, video, and data.

Deregulation is meant to encourage competition through the proliferation of new entrants. Looking to gain share, carriers are entering each other's market, blurring traditional lines between services, geographic coverage, and communication platforms. Aggressive new carriers have moved rapidly to establish nationwide service networks, consolidating local, long distance, Internet, wireless, and cable services under one umbrella. Incumbent carriers are trailing this way of convergence. The U.S. shows an excellent example of this convergence, the "big eight" convergence carriers

	Local	Long Distance	Data/Internet	Wireless	Cable/Video
AT&T	Teleport TCI	AT&T	AT&T WorldNet	AWS/McCaw	TCI
Bell Atlantic	Bell Atlantic Nynex		BA New Media	Bell Atlantic	
Bell South	Bell South		BS New Media	Bell South Mobile	Wireless Cable
GTE	GTE	GTE	BBN		
SBC	SBC Pactel Ameritech			Ameritec Wireless	Ameritech in-region cable TV
Sprint	Sprint ION	Sprint	Parenet/Earthlink	Sprint PCS	
Qwest/US West	US West		Internal Enterprise Networking Division (Cisco)		
WorldCom	Brooks MFS	MCI Worldcom	MCI CompuServe/ ANS UUNet		

cover most end markets. But they still leave room for hundreds of point products, mostly best-of-breed telco products and services. Communication convergence necessitates the deployment of next generation 3SSs. Relying upon advanced technologies, client/server or Web-based 3SSs enable convergence carriers to offer their customers higher total value through new, innovative products and services, superior customer service, and customized pricing and billing. At the same time, 3SSs guarantee profitability by increasing effectivity of processes by automation of all routine processes and by supervising quality of services metrics.

3.10.2.5 Customer Orientation

Competition is driving telco service providers to emphasize customer management. Driven by global competition, carriers are likely to focus on improving the total value of their services — quality, support, and price — as means to retain customers. Many of these improvements will come from advanced 3SSs. Besides improving the customer interface — e.g., offering Web-access — granular data available with new 3SSs can be utilized to retain key customers and reduce the number of customer churn. Over a longer range, further differentiation is expected. High-margin customers may receive special treatment, average customers just average services — similar to other industries.

Customer network management (CNM) incorporates a class of 3SSs that enable end users to securely view, troubleshoot, reconfigure, and generate reports on their subscribed telecommunication services. CNM provides strategic links to the customer and allows service providers to further differentiate their offerings. 3SS vendors are expected to offer the following:

- Performance: extraction of the information from the network without slowing overall network operations
- Customization: packaging information so that customers can receive an appropriate level of detail, in a way they can understand
- Security: delivery of the information to the customer in a cost-effective and secure manner, so that customers see only relevant information about their portion of the network

It is expected that Web technology will primarily be used to deliver this service. CNM represents a modest source of incremental growth for 3SS suppliers.

Certain 3SS services can also be outsourced. The customers may not be aware where the 3SS services come from. Today's outsourced solutions are service bureaus. They may outsource all or part of the carrier's support systems. In the latter case, the vendor relies upon remote access to the carrier's existing solution to deliver incremental functionality. For most emerging carriers, the benefits of outsourcing outweigh the negatives.

3.10.3   Strategic Benefits of Advanced 3SSs

Once deployed, advanced 3SSs offer the following strategic benefits:

- Improved operating efficiencies in data, inventory, and network management: it is expected that the management of various objects, such as equipment, applications, databases, etc. is more integrated, and requesting less human resources to manage.
- Reduced support and maintenance costs associated with legacy systems: due to more automation and interconnection, the support and maintenance expenses are decreasing.
- Shorter product development cycles: products and services can be created, tested, and deployed faster due to advanced technology used in 3SSs.
- Speedier deployment of new services and pricing schemes: processes are connected to each other. Rapid service provisioning in combination with pricing guarantee rapid deployment.
- Flexibility to modifying pricing and marketing schemes: due to interconnected processes, changes can be deployed very quickly. Even modeling and simulating resource utilization scenarios is easy to implement.
- New synergistic products and convergent services: products' bonding is very helpful to support convergent services. This bonding integrates OSSs, BSSs, and MSSs.
- Strategic marketing to target and acquire profitable business customers: due to rich information on customers and their traffic generation patterns, marketing strategies can be customized.
- Superior customer management to establish customer loyalty: The significant improvement of customer care will help to avoid customer churn and to sell value-added communication services to loyal customers.

The three principal 3SS process segments are:

- Customer care and billing
- Provisioning and order processing
- Network operational management

Figure 3.10.1 shows the high-level flow between these three process segments. It is important to observe that the corporate database or repository or data warehouse is shared between the principal process groups.

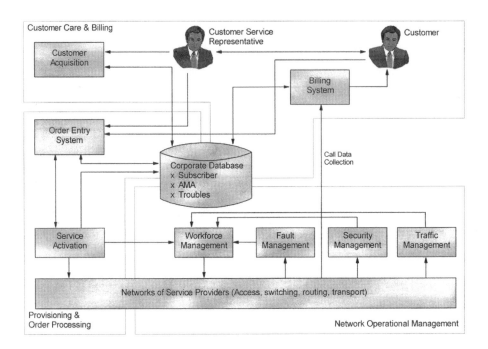

FIGURE 3.10.1   Principal 3SS process segments.

Table 3.10.1 shows the principal processes and functions for each of these segments. The present estimated market shares by segments are:

Customer care and billing   42%
Provisioning and order processing   38%
Network operational management   20%

The ratio of this market share will not change substantially during the next 3–5 years. Support tools are available for each segment offered by approximately 250 to 300 vendors, but it is rare that one product or one vendor can offer solutions for each market segment.

3.10.4   Providers of Operational, Business, and Marketing Support Systems

There are approximately 250 to 300 companies that are successfully competing in the 3SS arena. The categorization of providers is the following:

Software framework vendors (e.g., Telcordia, OSI, TCSI, Micromuse)
Consulting companies (e.g., Andersen Consulting, American Management Systems)
Computer vendors (e.g., IBM, Compaq/DEC, HP, Bull)
Telco equipment vendors (e.g., Ericsson, Siemens, Nortel, Nokia)
Application vendors (e.g., Versant, Vertel, Saville, Kenan, Metrica, Concord, NetScout)
Outsourcers (e.g., EDS, Perot Systems)

The present market shares can be characterized as follows:

• Telcordia and Lucent are equally strong and take up to 20% of the total 3SS market.

• Another 18 companies that are well known in the 3SS branch take approximately 10%

TABLE 3.10.1　　Principal 3SS Processes and Functions

3SS Processes and Functions	Definition
Customer Care and Billing	
Data analysis and mining	Process of analyzing call data details collected from switches and transmits
Mediation	An intermediate step for pre-processing and analyzing CDRs; fraudulent calls can be removed, data input from different switches in multiple formats can be converted into a format appropriate for bill processing. Also, pricing schemes can be inserted here, rather than by the call rating module. Call data can be selected and then transmitted to individual billing platforms, such as for voice, data, wireless, Internet, etc. Mediation is increasingly used for convergent and real-time billing.
Call rating and discounting	Prices call data according to current plan; it does include threshold plans currently popular among wireless carriers. Also, discounts are considered with this function.
Bill compilation and processing	Aggregates the rated call detail records and adds data for multiple services, handles advanced charges and payments.
Bill presentment	Customizes bill formats on a customer or service provider basis, may consolidate multiple statements; delivers bills via mail, online, e-mail, tape, or Internet.
Revenue assurance	Factoring, finding of receivables, credit checks, remittance processing, and customer deposit management.
Collection and credit analysis	Collecting outstanding debt, usually by the help of third-party collectors.
Customer care	Evaluates historical customer requirements, traffic patterns, expectations; reports and solves technical and billing problems.
Customer analysis and acquisition	Billing platforms tend to maintain the most complex picture of telecom customers in terms of resource usage, habits, and traffic patterns. Using these data intelligently, customer churn can be avoided and new services can be sold to customers.
Data warehousing	Call detail records, and additional data sources can be transmitted into warehouses. Data mining and other applications help to determine customers and end-product profitability.
Provisioning and Order Processing	
Local number portability	Allows customers to retain their telephone numbers with multiple service providers. Also, access to value-added services can be retained.
Inventory management	Allows maintaining first of all technical inventory data about equipment and circuits for the geographical reach of the service providers. Both CAD/CAM and GIS solutions may be implemented. Connections to the data warehouse are obvious.
Service creation	Process of creating and testing new or advanced services on the basis of the existing infrastructure of the service providers.
Service activation, provisioning, and assignment	Process of allocating equipment, assigning numbers, and activating circuits or ports at switches and activating customer services.
Service order processing	Based upon customer requests submitted to customer service representatives, creation and activation of services to customers.
Handling service change requests	Based upon customer service change requests submitted to customer service representatives, changing and reactivation of services to customers.
Service assurance	Allows continuous supervision of service level agreements on service indicators, such as availability, throughput, call congestion, packet losses, CDR losses, and others. For violations, the billing module is informed to initiate discounts or reimbursements.
Capacity management	Process of periodic surveillance of capacity in equipment and circuits. If capacity thresholds are exceeded, capacity extensions are initiated automatically.
Network Operational Management	
Call data collection	Collects call detail records (CDRs) from switches and transmits them to a billing database or mediation device. State-of-the-art solutions use complete automation.
Reactive fault management	Process of determining, diagnosing, and resolving faults, detected and reported by customers or by fault monitoring devices.
Proactive fault management	In order to detect problems early, allows the continuous supervision of fault indicators, the identification of causes for chronic troubles, and the evaluation of vendor performance.

TABLE 3.10.1 (continued)   Principal 3SS Processes and Functions

3SS Processes and Functions	Definition
Preventive fault management	Allows evaluation of usage statistics, the causes of performance threshold violations and the impact of additional payload on equipment and circuits.
Performance monitoring	In order to further support preventive fault management, equipment and facilities (circuits) are monitored continuously. In addition, performance metrics are maintained in a repository, which can be part of the data warehouse.
Error repair and maintenance	Allows repair of chronic faults and deployment of preventive maintenance techniques to equipment and to facilities.
Installation and inspection	Allows as part of the provisioning process the physical deployment of equipment and facilities on the basis of provisioning and service order change requests of customers.
Security management	Process of identifying security risks in equipment and facilities, deploying security procedures and tools, creating and evaluating security logs, and protecting operations, business, and marketing support systems.
Workforce management	Allows the central, policy-based, dispatch of workforce to monitor, test, maintain, inspect, and install equipment and facilities.
Testing	Process of testing equipment and facilities prior to deployment or as a part of the error repair process.
Design and planning	Allows, as a result of capacity bottlenecks, initiation of design processes that may include the deployment of new technology to equipment and facilities.
Traffic management	Process of observing typical traffic patterns by customers, customer groups, geographical areas, equipment, and facilities types. As a result, parameters and controls can be changed in equipment and facilities.
Network systems administration	May be considered as part of the maintenance process, limited, however, to version control, backup, archiving, and distribution of software to equipment.

- The remaining 70% is distributed between literally hundreds of companies that are eager to emerge as dominant suppliers of 3SS solutions.

The requirements for being a winner are tough. The profile may look like the following:

- Solutions are scalable: given the large and growing number of network devices, services, and subscribers, 3SSs must grow with the service provider. While there may be a low-end market for small-scale "telco-in-the-box" solutions, it is not expected that solutions that do not scale well will capture significant market segments. Prepackaged functionality will help to reduce the demand of customization required to match 3SSs to a particular service provider's business objectives.

- Domain knowledge (best-of-breed): the implementation and deployment of 3SSs require a sound knowledge of service provider operational procedures. This domain knowledge is not always available, but without it, successful work in the domain is not possible.

- Integration capabilities (best-of-suite): it is absolutely necessary to connect existing point products using electronic bonding or extranets. Standards are emerging to facilitate this work. First implementation results are seen with CORBA and DCOM.

- Supporting multiple products and services: the ability to manage traditional, enhanced, wireless, data, and video products and services in a unified convergent manner is widely viewed as critical to the success of advanced service providers.

- Willingness for partnerships: it is not possible to exhibit open multivendor support without effective partnerships. It is highly likely that leading 3SS vendors will establish partnerships with other industry stakeholders, such as equipment vendors, system integrators, customers, and other 3SS vendors.

- Strong references: perhaps the best selling argument for a 3SS is its existing customer list. Service providers, recognizing the high cost associated with maintaining and enhancing a 3SS platform, view a strong customer base as a way to share development costs for basic 3SS functionality and reduce risk.

3.10.5 Positioning and Evaluating Products

3SS products can be positioned, evaluated, and compared to each other using the following three dimensions:

- Compliance to TMN layers, such as BML, SML, NML, EML, and NEL
- Support of principal 3SS processes and functions (see Table 3.10.1 for details)
- Support of various end markets, such as long distance, local, data/Internet, wireless, and cable

In order to help position, evaluate, and compare support tools, the following tables are recommended:

TABLE 3.10.2 Principal 3SS Process Groups and TMN Layers

Process Areas	Customer Care and Billing	Provisioning and Order Management	Network Operational Management
TMN Layers			
BML	x		
SML	x	x	
NML		x	x
EML			x
NEL			x

The allocation between principal 3SS process groups and TMN layers shows a very clear trend (Table 3.10.2):

- Customer care and billing supports upper TMN layers
- Provisioning and order processing supports middle TMN layers
- Network operational management supports lower TMN layers

This allocation remains relatively stable even when the TMN standard changes over time.

TABLE 3.10.3 Principal 3SS Processes and Functions and TMN Layers

TMN Layers	BML	SML	NML	EML	NEL
OSS Process Areas					
Billing and Customer Care					
Data analysis and mining	x	x	x	x	x
Mediation	x	x	x		
Call rating and discounting	x				
Bill compilation and processing	x				
Bill presentment	x				
Collections and credit analysis	x				
Revenue assurance	x				
Customer care	x	x			
Customer analysis and acquisition	x				
Data warehousing	x				
Provisioning and Order Management					
Local Number Portability	x				
Inventory management	x	x	x	x	
Service creation	x				
Service activation, provisioning and assignment	x	x	x	x	
Service order processing	x	x	x		
Handling service change requests	x	x	x		
Service assurance (SLA)	x				
Capacity management		x		x	

TABLE 3.10.3 (continued)   Principal 3SS Processes and Functions and TMN Layers

TMN Layers	BML	SML	NML	EML	NEL
Network Operational Management					
Call data collection		x	x	x	x
Reactive fault management			x	x	x
Proactive fault management			x	x	x
Preventive fault management			x	x	x
Performance monitoring		x	x	x	x
Error repair and maintenance			x	x	x
Security management	x	x	x	x	x
Installation and inspection			x	x	x
Workforce management			x	x	x
Testing		x	x	x	x
Design and planning			x		x
Traffic management		x	x		x
Network systems administration			x	x	

Breaking down the principal 3SS process groups (Table 3.10.3) does not change the allocation to TMN layers significantly. But for certain processes, exceptions can be observed. In other words, the allocation is not unique. The result is that TMN needs additional work to clarify responsibilities of layers and their functions more accurately.

TABLE 3.10.4   Support of 3SS Processes and Functions by Individual Support Tools

Companies	Functions	Products
Customer Care and Billing:		
Acc*Comm	Billing data collection	TIBS, NetPlus, DCMS, DCMS/NEDS, ANMS TREX*COM NetPlus Pro*Vision
Alltel Information Services	Billing	Virtuoso II
Amdocs	Billing applications	Ensemble
American Management System	Customer care Billing	Mobile 2000, Spectrum 2000 Tieline UP, Tapestry
Andersen Consulting	Billing applications Customer care	IABS (Integrated Access Billing System), Flexcab
Axiom	Billing data collection Fraud management	Sterling Billing Data Collection Sterling Real-Time Fraud Management
Beechwood	Carrier-to-carrier OSS interconnection New carrier systems integration Post-merger IT integration IP network OSS implementation	
Billing Concepts	Service bureau billing	Modular Business Applications (MBA)
Cable Data	Billing	Intelecable
CBIS	Billing software and services	Wireline: Precedent 2000 Wireless: Advantage
Clarify	Customer care applications	
Commsoft	Billing	CommVergence
Corsair	Fraud detection systems	
CSG Systems	Large number of functions, services, and products	CCS with ACSR
Daleen Technologies	Billing	BillPlex
EDS	Service bureau billing	BSM (Billing Services Management), IXPlus, CMIS, Empower

TABLE 3.10.4 (continued)   Support of 3SS Processes and Functions by Individual Support Tools

Companies	Functions	Products
Ericsson	Customer service and billing	TIMS, BIP (Billing Information Processor)
		BMP (Billing Mediation Platform), Progressor
IBM	Customer care and billing	TFS/ICMS
		CARTS (Centralized AMA Records Transfer System)
InfoDirections	Billing	CostGuard
Infozech	Billing	eBill
Intertech Management Group	Billing applications	Network Strategies
Kenan Systems	Billing applications	Arbor/BP, Acumate ES, Arbor Strategist, EC/Arbor
LHS Group	Billing applications	Business Support and Control System (BSCS)
Lightbridge	Customer care applications	Telesto
	Fraud detection systems	
Lucent Technologies	Billing data collection	BILLDATS
Metapath	Billing data collection	
Objective Systems Integrators	Billing data collection	NetExpert, AMA Gateway
Portal Software	Internet billing applications	Infranet
	Customer care applications	
Saville Systems	Billing applications	Convergent Billing Platform (CBP)
Sema Group	Customer care and billing	CABS 2000 Mobile
USCS	Cable billing	Cable Data
	International billing	IBS (Intern. Billing System)

Provisioning and Order Processing

Amdocs	Order management	
Applied Digital Access	Service activation, provisioning	Provisioner
American Management Systems	Service order management	Tieline SOMS
Architel	Service activation	FAMIS, ASAP, OMS
Atlantec	xDSL service activation	
Beechwood	Workflow for provisioning	Flow-through
Bellcore/SAIC	Provisioning	TransportEMS
	Switch administration	NetMemory
	Workforce management	FORCE
	Service order management	Delivery
	Legacy 3SS maintenance	
Call Technologies	Provisioning and enhanced services	Call Profiler, Call Activate, Call Verify, Call Notify, Call Courier, Call Builder, Call Codence, Call Care, Call Plus
CBIS	Configuration management	Switch Manager
CommTech	xDSL and Centrex service activation	BECAS Facility Management
Crosskeys	Service management	Resolve
DSET	Service order administration center	LSOA
	Local service management system	LSMS
EDS	Workforce management	FMS (Force Management (System)
	Inventory and service provisioning	EMAC (Enhanced Mechanized and Control System)
Evolving Systems	Workflow management	
	Event management	
FirsTTel	Service activation for phone systems	
Hewlett-Packard	Service management	OpenView ITA (Admin)
Illuminet		
Lucent Technologies	Service provisioning	CONNECTVU
	Switch administration and management	NetMinder
	Service activation	ACTIVIEW
MetaSolv	Service activation, repository, and provisioning	TBS (Telecom Business Solution)
Network Programs	Rapid service deployment	TNP (The New Platform)

TABLE 3.10.4 (continued)   Support of 3SS Processes and Functions by Individual Support Tools

Companies	Functions	Products
Nortel	Provisioning	DSS-II
Objective Systems Integrators	Service activation	NetExpert, iSOP, iSAC
Quintessent Communications	3SS interconnections for provisioning	QConnect
SmallWorld	Inventory management	GIS-based 3SS
	Network Operational Management	
ADC Metrica	Wireless performance management	NPR (Network Performance Reporting)
ADC NewNet	SS7 management	AcceeMANAGER
Applied Digital Access	Testing	T3AS Test
	Fault detection and isolation	T3AS Monitoring
Ascend	Managing POPs	Navis Access, Navis Core
Axiom	Traffic management	Manifest
Bellcore/SAIC	Operations management	NMA
	Testing	OcuSpan
	SS7 management	NetPilot
	Legacy 3SS maintenance	
Cisco	Service management	Cisco Service Management System
Clear Communications	Data collection and reporting	ClearView Probable Cause, Early Warning, ReportCard, CircuitView, Legacy Gateway
CommTech	Centrex monitoring	Macstar
Compaq/Digital	SS7 management	DECss7
Concord Communications	Data service level monitoring	Network Health
CrossKeys	Network management	Exchange Performance Management, Exchange Traffic Management
	Testing	Open/Test
DeskTalk	Data service level monitoring	TREND
DSET	NPAC-SMS simulator	Simulator
Ericsson	Network management	TMOS Network Traffic Manager, XM (Exchange Manager)
	SS7 management	TMOS SS7
Hewlett-Packard	Network management	OpenView ITO (Operations)
		OpenView
		MeasureWare
Illuminet	SS7 monitoring	
INET	SS7 monitoring	
ISR Global Telecom	Cable, SDH/Sonet management	ObjectEngine, Mask
Lucent Technologies	Testing	SARTS
	Fault detection, isolation, and reporting	ITM, NMF, NOCI
	ChoiceNet	Dynamic filter management
Micromuse	Alarm correlation and analysis	Netcool suite
NetScout	Data network performance monitoring	RMON Probe
Nortel	Network management	S/DMS, NetWORKS, DFMS (Digital Facility Management System)
Objective Systems Integrators	Fault detection, isolation, and resolution	NetExpert
	Alarm correlation and analysis	NetExpert IDEAS
OpenCom Systems	Integration of element managers	TMS2000 Global EMS Product
Remedy Corporation	Trouble ticket management	AR System
Team Telecom	Fault management applications	
Technically Elite	Data network performance monitoring	MeterWorks
Visual Networks	Data network service level monitoring	

This table (Table 3.10.4) groups 3SS products around three principal process groups, such as

- Customer care and billing
- Provisioning and order processing
- Network operational management

Due to acquisitions and changes in product portfolios, such a table needs frequent updates. In addition to these groups, an additional list (Table 3.10.5) is provided for vendors of frameworks and platforms that offer enabling technologies and integration services.

TABLE 3.10.5   Framework and Platform Vendors, and System Integrators

Companies	Functions	Products
ADC SoftXchange	Interconnection platform	DataXchange
Alcatel	Systems integration	1320NM, TSM, CBCU
AllTel	Remote outsourcing	
Amdocs		
American Management Systems	Consulting and integration	
Andersen Consulting	Consulting and integration	
Beechwood	Carrier-to-carrier integration	
	New carrier system integration	
	Post merger IT consolidation	
Bellcore/SAIC	Consulting, integration, and custom design	
Bull	Management platform	Open/Master
CapGemini	Systems integration	OSS, TTM, BSCS, MARS
Compaq/Digital	Systems integration	TeMIP management platform
DMR Consulting	Consulting and integration	
DSET	Development tools	TMN Agent and TMN Manager
	Electronic bonding	EC-Lite, PIC/CARE
	Intelligent networks	LNP
EDS	System integration	
Euristix	Tools and custom development	Raceman EMSX
	Q-Adapter	
Evolving Systems	Platform and management	

TABLE 3.10.6   Support of End Markets by Individual Support Tools

Company	Product	Local	Long distance	Data/Internet	Wireless	Cable/Video
		Customer Care and Billing				
Acc*Comm	TIBS, NetPlus, DCMS, ANMS, TREXCOM			x		
Amdocs		x	x		x	
AMS						
Andersen						
Axiom						
Beechwood						
Billing Concepts		x				
CBIS	Wireline	x	x		x	x
	Wireless					
Clarify						
Corsair						
CSG Systems				x		x
Daleen		x	x			
Illuminet		x		x		
Intertech		x		x		
EDS						
Ericsson						

TABLE 3.10.6 (continued)   Support of End Markets by Individual Support Tools

Company	Product	Local	Long distance	Data/Internet	Wireless	Cable/Video
IBM						
Kenan	Arbor/BP, Acumate ES, EC/Arbor	x	x	x	x	x
LHS Group					x	
Lightbridge						
Lucent						
Metapath						
Objective Systems						
Portal Software	Internet Bill			x		
Saville Systems	Convergent Billing Platform	x	x	x	x	
USCS	Cable Data, IBS					x

Another dimension is offered in this table by grouping 3SSs in accordance with end markets (Table 3.10.6) they support. This table provides selected examples for customer care and billing only.

TABLE 3.10.7   Support of TMN Layers by Individual Support Tools

Company	Product	BML	SML	NML	EML	NEL
	Customer Care and Billing					
Acc*Comm	TIBS, NetPlus, DCMS, Dcms/Neds, ANMS, Trex*Com	L	L	H	H	H
Amdocs		H	H			
Axiom		L		H	H	H
AMS	Mobile 2000 Spectrum 2000, Tieline UP	H	M			
Andersen Consulting	IABS	M	M			
Beechwood						
Billing Concepts		H				
CBIS		H	H			
Clarify		M	M			
Corsair						
CSG Systems						
Daleen		H	H			
Intertech						
EDS	BSM	H	M			
Ericsson	TIMS, BIP	H	H	M	M	M
IBM	TFS/ICMS CARTS	H	H			
Kenan	Arbor/BP, Acumate ES, Arbor Strat EC/Arbor	H	H			
LHS Group		H	H			
Lightbridge		H	M			
Lucent	BILLDATS	H	H			
Metapath				H	H	
Objective Systems	NetExpert AMA, Gateway	M	M	H	H	
Portal Software		H	H			
Saville Systems	Convergent Billing Platform	H	H			
USCS						
	Provisioning and Order Processing					
Amdocs						
Applied Digital Access	Provisioner			H	H	H

TABLE 3.10.7 (continued)   Support of TMN Layers by Individual Support Tools

Company	Product	BML	SML	NML	EML	NEL
AMS	Tieline SOMS		H	M		
Architel	FAMIS, ASAP, OMS		H	H	M	
Atlantec		M	H	H		
Beechwood	Flow Through	H	H			
Bellcore/SAIC	MediaVantage, NCON, Transport, EMS, FORCE, Delivery, NetMemory		H	H	H	
Call Technologies			H	M	M	H
CBIS	Switch Manager			M	M	
CommTech			L	H	H	H
Crosskeys	Resolve		H	H	H	
DSET			L	M	H	H
Evolving Systems		M	H			
FirsTel			H	M	H	
Hewlett-Packard	OpenView ITA		M	M	H	
Illuminet						
Lucent	ACTIVIEW, OneVision, CONNECTIVU, NetMinder		H	H	H	M
MetaSolv	TBS	M	H	M		
Network Programs	The New Platform					
Nortel	DSS-II			M	M	M
Objective Systems	NetExpert, iSOP, iSAC		H	H	M	M
Quientessent		H	H			
SmallWorld			M	H		

	Network Operational Management					
ADC Metrica	NPR			H	H	H
ADC NewNet	SS7		M	M	M	M
Applied Digital Access	T3AS			H	H	H
Ascend	NavisAccess, NavisCore		M	H	M	
Bellcore/SAIC	NetPilot, NMA, OcuSpan	M	H	H	H	
Cisco	Service Management System		H	M	M	M
Clear Comm	ClearView Probable Cause, Early Warning, ReportCars, CircuitView, Legacy gateway	M	H	H	L	
CommTech			L	H	H	H
Concord	Network Health	M	H	M		
Compaq/Digital	DECss7			H		
Crosskeys	Exchange Performance and Traffic Management Open/Test		H	H	H	
DeskTalk		M	H	M		
Ericsson	TMOS Traffic and Exchange Manager, TMOS SS7		M	H	M	M
Hewlett-Packard	OpenView ITO and Measure Ware		M	H	M	M
Illuminet			H	H		
INET			H	H		L
ISR Global	ObjectEngine Mask		M	H		
Lucent	ITM, NOCI, NMF, NetMinder, SARTS, ChoiceNet		H	H	H	H
Micromuse	NetCool		M	H		
NetScout	RMON Probe	H	H	H	H	
Nortel	S/DMS, NetWORKS, DFMS		M	M	M	M
Objective Systems	NetExpert, NX Ideas		H	H	H	H
Remedy	AR Systems		H	H	M	
Team Telecom				H	M	
Technically Elite	MeterWorks		M	M	M	H
TCSI	SolutionCore		L	H	H	M
Visual Networks		M	H	M	M	M
Vertel			L	M	H	H

Using the same 3SSs as in Table 3.10.4, the support of TMN layers is evaluated (Table 3.10.7). As can be observed, there are many empty cells, indicating gaps in support.

The overall conclusions of 3SS evaluations are:

- Even the most powerful companies in the 3SS business (Bellcore and Lucent) and their products cannot support all principal 3SS process groups.
- There are many best-of-breed, and very few best-of-suite products.
- Support for higher TMN layers (BML and SML) is not sufficient.
- Serious segmentation of product's applicability for various end markets exists.
- Enabling technologies for integration and interconnection are absolutely necessary.

3.10.6   Future of Telecom 3SSs

In order to position 3SSs and their vendors, future trends of support systems should be estimated. With other words, the dynamic of principal market drivers should be analyzed in depth.

In order to match the rich service offerings of new entrants, ILECs have implemented multiple upgrade strategies, including modifications by internal staffs, custom development by external system integrators, and integration of third-party products. Most likely, they won't completely replace their existing 3SSs. Several incumbent carriers are incorporating best-of-breed solutions with their legacy systems. This trend opens great opportunities for point 3SSs and for professional services.

Deregulation of the LEC market has stimulated and still stimulates significant demand for CLEC-3SSs. Most of them start from scratch and invite all types of 3SSs vendors with point and integrated products. Larger CLECs with custom-designed in-house solutions are enhancing these to accommodate new services and technologies; they show some similarities with incumbent providers. Replacement of these 3SSs is not expected soon. But, in particular, back-office operational efficiency is expected to improve. Also, network operational management solutions are in demand.

CLECs may want to outsource 3SS services. They have started to evaluate the benefits of outsourcing their back-office services entirely. Outsourcing would eliminate the need for the carrier to invest scarce research and development dollars in 3SSs, allocating spending to their networks and/or customer management systems. Essentially, it allows CLECs to focus on their core business.

Less well-known CLECs either purchase or license point products from third parties, or take advantage of service bureaus. It is highly unlikely that these CLECs are interested with in-house development and maintenance. 3SS-vendors can sell to these CLECs directly or to service bureaus that may share their products between multiple CLECs.

The international market is not easy for 3SS suppliers. Systems integrators are in good position with PTTs and new entrants; they usually subcontract for point solutions.

Carrier interconnections open excellent opportunities for 3SS vendors. The unbundling of local exchange elements for resale requires that resellers of local exchange services provide electronic links to incumbent carriers for ordering, service activation, troubleshooting, and billing. Present 3SSs do not have these interconnect features. There is a significant opportunity for incremental 3SS sales by emerged and also by new vendors. Specialized vendors for LNP will play a significant role during the next 10 years. The best-of-breed solutions are expected to offer provider portability, location portability, and also service portability.

Telecom industry consolidation creates new 3SS requirements, but the need is situation specific. It is difficult to estimate the timeframes of re-engineering or consolidating 3SSs of the consolidated telecommunication providers. Consolidated carriers are likely to work to fully integrate multiple 3SS platforms (customer care and billing, provisioning and order processing, network operational management), to create synergies in products and markets, and to reduce costs. 3SS vendors with system integration capabilities are in demand. 3SSs will mirror trends in the telecommunication industry; full-service 3SS vendors could emerge to serve convergence carriers.

Table 3.10.8 summarizes the present goals and the 2002 targets from the perspective of incumbent carriers, emerging carriers, customers, equipment vendors, 3SS vendors, and system integrators.

TABLE 3.10.8 Goals and Targets

Stakeholder	Goals	Targets for 2002
Incumbent service providers (ILECs, PTTs, global carriers)	Rapid introduction of new services Address Year 2000 issues Customer retention Multi-vendor management Cost reduction	Less internal development More use of systems integrators More 3SS packaged software Pervasive 3SS interconnection
Emerging service providers (CLECs, ISPs, and wireless)	Build network capacity Customer acquisition Improve service quality Offer new differentiated services	Minimal internal development Fully automated processes More 3SS packaged software Less service bureaus Strong 3SS interconnections
Customers (end users)	Increase service reliability Lower transport costs Faster service provider responsiveness Customer network management	Self-provisioning Custom quality of service reporting (QoS) Flexible billing formats
Equipment vendors	Sell more equipment	Outsource element management systems to 3SS vendors Use of open interfaces
3SS vendors	Sell more software Sell more professional services	Full-line of 3SS offerings Target ILEC legacy 3SS replacement Acquire other 3SS vendors Compete with system integrators
System integrators	Sell consulting, custom programming, and integration services	Acquire 3SS vendors Conduct many projects

Acronyms

3SS	Operations, business, and marketing support systems
ASR	Access service request
ATM	Asynchronous transfer mode
BIP	Billing information processor
BMP	Billing mediation platform
BSM	Billing services management
BML	Business management layer
BSS	Business support system
CAGR	Compound annual growth rate
CARTS	Centralized AMA records transfer system
CBIS	Cincinnati Bell Information Systems, Inc.
CDR	Call detail record
CLEC	Competitive local exchange carrier
CNM	Customer network management
CBP	Convergent billing platform
DEN	Directory enabled networks
DFMS	Digital facility management system
DSL	Digital subscriber line
EML	Element management layer
EMS	Element management system
IABS	Integrated access billing system
IEX	Interconnect exchange carrier
ILEC	Incumbent local exchange carrier

ISDN   Integrated services digital networks
LEC   Local exchange carrier
LNP   Local number portability
LSR   Local service request
MSS   Marketing support system
NEL   Network element layer
NML   Network management layer
NPR   Network performance reporting
OSS   Operations support system
PSTN   Public switched telephone network
RBOC   Regional Bell Operating Company
SML   Service management layer
SDH   Synchronous data hierarchy
TBS   Telecom business solution
TNP   The new platform
TSM   Transmission status monitor
VoIP   Voice over IP
VPN   Virtual private network

References

ARL197   Arlitt, M., Williamson, C.: Internet Web Servers: Workload Characterization and Performance Implications, *IEEE/ACM Transactions on Networking,* Vol. 5, No. 5, October 1997.
BULO98   Bulow, D.: Dynamic Compute services, *Datacom,* 7/1998, p. 58–61, Bergheim, Germany.
CASE97   Case, J.: Finding the Right Job, www.nwfusion.com, April 21, 1997.
DETE95   DeTeBerkom: Intelligent Agents: Concepts, Architectures, and Applications, Part 2: Impact of IA Concepts on the telecommunications Environment, June 1995.
FORB97   Forbath, T.: Web-based Management: A recipe for success, *Network World,* May 5, 1997.
GARE97   Gareiss, R.: Casting the Web Over ATM, *Data Communications,* June 1997, p. 35–36.
GHET97   Ghetie, I.G.: *Networks and Systems Management — Platforms, Analysis and Evaluation,* Kluwer Academic Publishers, Boston, 1997.
HERM97   Herman, J.: Web-Based Net Management Is Coming, *Data Communications,* October 1997, p. 139–141.
HEYW97   Heywood, P.: An Impartial Interpreter of Service-Level Agreements, *Data Communications,* November 1997, p. 32–34.
HUNT96   Huntington-Lee, J., Terplan, K., Gibson, J.: *HP OpenView, McGraw-Hill Series on Computer Communications,* New York, 1996.
JAND96   Jander, M.: Distributed Net Management — In Search of Solutions, *Data Communications,* February 1996, p. 101–112.
LARS96   Larsen, A.K.: Mastering Distributed Domains via the Web, *Data Communications,* May 21, 1996, p. 36–38.
LEMA95   Lemay, L.L.: *Web publishing with HTML,* SAMS Publishing, Indianapolis, 1995.
MEGA97   Megandanz, T., Rohermel, K., Krause, S.: *Intelligent Agents: An Emerging Technology for Next Generation Telecommunications,* Research Paper with GMD Fokus, Berlin, Germany, 1997.
NAIR96   Nair, R., Hunt, D., Malis, A.: Robust Flow Control for Legacy Applications over Integrated Services ATM Networks, *Proceedings of Global Information Infrastructure, Evolution Internetworking Issues, Nara, Japan,* IOS Press, Amsterdam, 1996, p. 312–321.
POWE97A   Powell, T.: The Power of the DOM, *InternetWeek,* p. 61–74, September 29, 1997.
POWE97B   Powell, T.: An XML Primer, *InternetWeek,* p. 47–49, November 24, 1997.
REAR98   Reardon, M.: Need Management That Fast? *Data Communications,* 1998, p. 30–31.

ROBE98   Roberts, E.: Load balancing: On a different track, *Data Communications,* May 1998, p. 119–126.

RUBI98   Rubinson, T., Terplan, K.: CRC Press, Boca Raton, 1998.

SANT97   Santalesa, R.: Weaving the Web Fantastic — Review of authoring tools, *InternetWeek,* November 17, 1997, p. 73–87.

SPER98   Spero, S.: Analysis of HTTP Performance Problems, www.w3.org/Protocols/HTTP-NG/http-prob.html.

STEV94   Stevens, W.R.: *TCP/IP Illustrated,* Addison-Wesley, 1994.

TATE97   Tate, D.: Picking Through Piles of Web Pages, *LanTimes,* February 17, 1997, p. 30.

TERP94   Terplan, K.: *Benchmarking for effective network management,* McGraw-Hill, New York, 1994.

TERP96   Terplan, K.: *Effective Management of Local Area Networks,* McGraw-Hill Series on Computer Communications, New York, 1996.

TERP98   Terplan, K.: *Telecom Operations Management Solutions with NetExpert,* CRC Press, Boca Raton, 1998.

THAL98   Thaler, D.; Ravishankar, C.: Using Name-Based Mappings to Increase Hit Rates, *IEEE/ACM Transactions on Networking,* Vol. 6, No. 6, February 1998.

3.11 Performance Management of Intranets

Kornel Terplan

Abstract

After outlining generic and specific challenges of managing intranets, this presentation focuses on emerging new measurement and management tools, such as log file analyzers, traffic monitors, Web server managers, load distributors, and traffic shapers. The presentation ends with a discussion of integration opportunities of these new tools with existing management platforms and applications.

3.11.1 Introduction — Internet, Intranets, and Extranets

Intranet management means deploying and coordinating resources in order to design, plan, administer, analyze, operate, and expand intranets to meet service-level objectives at all times, at reasonable cost, and with optimal capacity of resources. Intranet management can utilize all the experiences collected over the last 25 years with managing data networks. Existing management concepts are still valid. Critical success factors are applicable as well. In managing intranets, those critical success factors include:

- Management processes that can be grouped around fault, configuration, performance, security, and accounting management
- Management tools that are responsible to support management processes and are usually assigned to human resources
- Human resources of the management team, with their skills and network management experiences

Intranet management instrumentation shows similarities with the management of other networks. The architecture is shown in Figure 3.11.1. The management framework is the center and is in charge of consolidating, processing, displaying, and distributing information to authorized persons. The framework is expected to be equipped with Web capabilities meeting the expectations of the majority of users. It means that views and reports are converted into HTML pages and are accessible from universal browsers. Management applications are a mix of well-known ones, such as trouble ticketing, asset management, and change management; and brand-new ones, dealing with log file analysis, load balancing, packet shaping, content authoring, and Web-server management.

The remaining part of this chapter addresses specific challenges of intranet management toward management processes.

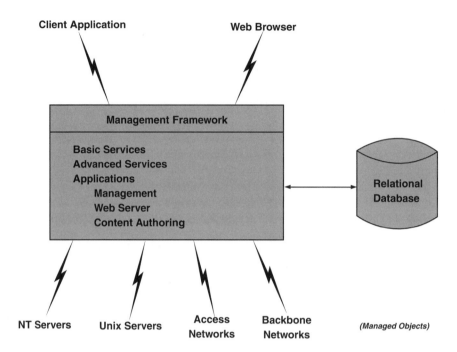

FIGURE 3.11.1　Intranet management framework.

The Internet is an existing network used by millions of people every single day. At the same time, the Internet is a generic term for a bundle of technologies available under the Internet umbrella. The Internet shows a number of similarities with the global phone system. Whoever is a subscriber can be reached by dialing the right country code, area code, and the actual phone number. In the case of the Internet, visitors type in the right universal resource locator (URL) to access the necessary information. Even the billing process shows similarities; the longer the talk or surfing, the higher the bill.

The ownership is not so clear with the Internet as with public phone systems. There are multiple owners of the Internet physical backbone, but they are hidden from users. Administration and management are getting more important as the number of subscribers is growing very fast. Just one administration issue — address management — causes a lot of headaches. Country institutions are coordinated by an independent U.S.-based company. Basically, the Internet can support multiple communication forms, such as voice, data, and video. The predominant use is still data.

This standardization is a threat to proprietary networking architectures, such as SNA from IBM. In order to support both, gateways are being deployed to interconnect both types with each other. It is very tempting to consider the Internet as the central switching point of corporate networking. But performance and security considerations drive corporate network managers to use privately owned Internet-like networking segments, called intranets. Intranet examples are shown in Figure 3.11.2; A, B, C, and D are communicating parties that use the intranet(s) offered by their own company.

Intranets are company-internal networks that are using Internet technology. In particular, Web technology is used for information distribution (e.g., company documentation will be unified this way, internal hiring procedures made visible, etc.) and Web protocols are used for internal information transfer. The backbone of intranets is based in IP on Layer 3. If interconnection is required to other networks, e.g., SNA or to other companies, then firewalls are deployed to protect the company-owned intranet. Firewalls are actually filters; certain packets without the necessary authorization code cannot pass the firewall.

If partnerships are the targets, networking equipment of partnering companies can be connected to each other. In such a case, the connected intranets are called extranets. Doing so, requirements toward firewalls are much lighter. Typical application cases are: car manufacturers and their suppliers of parts;

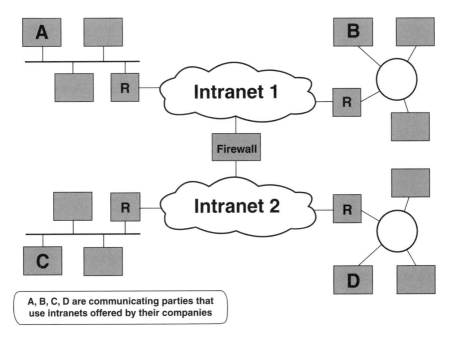

FIGURE 3.11.2 Use of intranets.

airlines in alliance; airlines and travel agencies; telcos with each other to complement local, long distance, and international services; and service providers and customers.

The Internet can still be utilized as part of intranets and extranets. Virtual private networks (VPN) are offering this by just securing channels that are part of Internet, to be used by communicating parties in intra- and extranets. There are a couple of technical solutions that are based either on Layer 2 or Layer 3 technologies.

3.11.2 Generic Intranet Management Challenges

This segment investigates how management functions can be reimplemented in intranets. Challanges will be highlighted in each functional area as well.

3.11.2.1 Performance Management Challenges

Feasible network architectures for intra- and extranets are shown in Figures 3.11.4 and 3.11.5. The components of the intranets are similar to other types of networks. Principal components include:

- Web servers that maintain home pages
- Web browsers that directly support users to view, download, and upload information to/from Web servers
- Backbone offering broader bandwidth for high data volumes
- Access network offering narrower bandwidth for lower data volumes
- Networking components including routers, switches, traffic shapers, and firewalls
- Communication protocols such as IP for the backbone, and higher layer protocols such as HTTP, SNMP, and FTP to support management applications

Figure 3.11.3 shows a typical arrangement in a simplified form. From the management perspective, all these components are managed objects. One additional managed object type must be considered; this object type is the application running in Web servers.

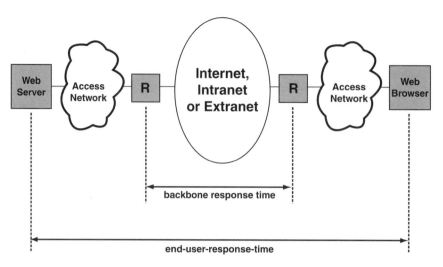

FIGURE 3.11.3   Principal structure of systems and networking components.

By early 1999, approximately 37 million people were accessing the Internet every single day. Altogether, approximately 830 million Web pages were being accessed every day.

Due to these special patterns, performance metrics are extremely important. From the technical viewpoint, everything can be measured. From the practical point of view, however, a few indicators are of prime interest. In particular, two of them are considered in every enterprise: response time and resource utilization.

For the response time, not only the resource-level, but also the user-level response time should be measured. Now, there are several types of tools to choose from: some of them measure throughput rates, some simulate network traffic and tally the results, some gauge performance by running within the applications themselves, and some rely on a combination of those techniques. Altogether, there are four approaches:

- Monitors or packet analyzers
- Synthetic workload tools
- Application agents
- Application Response Measurement (ARM) MIBs

End-user-level response time is helpful for service level agreements. Performance optimization needs more details about the contributors, such as the networks, systems, and applications. When segments of the response time are known, resource optimization by proper capacity planning is possible.

The utilization of resources has a direct impact on the response time. The payload is always an issue with resource utilization. Operating systems put load on the servers; control characters of protocols mean additional bytes to be transferred. Both represent overhead, but they cannot be avoided completely. The same is true with monitors and the transfer of monitored data for further processing. But overhead can be controlled, and then productive operations are not impacted. Further details on performance-related metrics in intranets are shown in other chapters.

In summary, tuning and optimizing intranets may be very different than traditional networks. User behavior, application performance, unusual traffic patterns, asynchronous resource demand, and additional protocols cause unique challanges to performance management of intranets.

3.11.2.2   Security Management Challenges

Due to opening networks, connecting partners, and using a public domain, such as the Internet, security risks increase considerably. VPNs are a possible answer to combine existing infrastructure with acceptable protection. Security expectations may be different in various industries, but the generic security management procedures are identical or at least very similar. Security management enables intranet managers to protect sensitive information by:

- Limiting access to Web servers and network devices by users both inside and outside of enterprises
- Notifying the security manager of attempted or actual violations of security

Security management of intranets consists of:

- Identifying the sensitive information to be protected
- Finding the vulnerable access points to sensitive information
- Securing these access points
- Maintaining the secure access points

Identifying sensitive information means the classification of information. Most organizations have well-defined policies regarding what information qualifies as sensitive; often it includes financial, accounting, engineering, and employee information. But, in addition, there are environments that can have sensitive information unique to them. The main purpose of intranets is to improve the internal documentation and communication within enterprises. Web servers are the focal point of information maintenance. Evidently, not everything is for everyone. Depending on the individual responsibilities, access rights to information sources can be relatively easily structured and implemented. In summary, sensitive information are the home pages with particular content residing on Web servers.

Once the webmaster and network managers know what information is sensitive and where it is located, it must be found out how users can access it. This often time-consuming process will usually require that webmasters and network managers examine each piece of hardware and software offering a service to users. In this respect, intranets are not different from any other complex networks. Generic sensitive access points are (Figure 3.11.4):

- End-user devices, such as browsers
- Access and backbone networks
- Web servers maintaining sensitive information

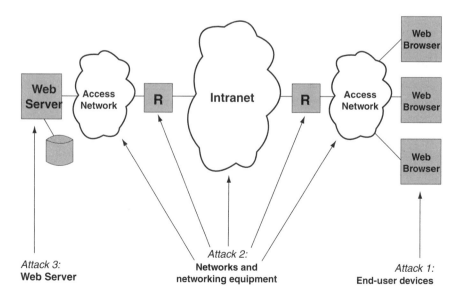

FIGURE 3.11.4 Access points with security risks.

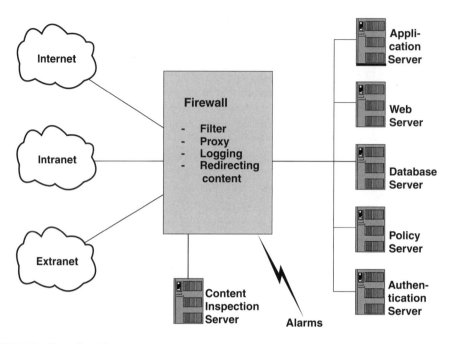

FIGURE 3.11.5   Firewall architecture.

The next step in security management is to apply the necessary security techniques. The sensitive access points dictate how the protection should be deployed using a combination of policies, procedures, and tools. In this respect, the following levels of security techniques must be considered:

- End-user devices, such as universal browsers (use of chip cards or chip keys)
- Access and backbone networks (use of encryption, authentication, and firewalls)
- Web servers (use of server protection, operating systems protection, special tools, and virus protection)

The last step in effectively securing access points in intranets is maintenance. The key to maintenance is locating potential or actual security breaches. It requires an ongoing effort of stress testing intranets, assigning tasks to outside professional security companies, reviewing case studies of security violations, and evaluating new security management techniques and tools.

Firewalls play a significant role in security management of intranets. A firewall (Figure 3.11.5) is a device that controls the flow of communication between internal and external networks, such as the Internet. A firewall serves several functions. First, it acts as a filter for inbound Internet traffic to the servers of enterprises. As a filter, the firewall prevents unnecessary network packets from reaching Web and application servers. Second, it provides proxy outbound connections to the Internet, maintaining authentication of the internal Internet users. Third, the firewall also logs traffic, providing an audit trail for usage reports and various planning purposes.

Firewalls are not without risks. Many companies assume that once they have installed a firewall, they have reduced all their network security risks. Typically, firewalls are difficult to penetrate, but when they are broken, the internal network is practically open to the intruder.

Furthermore, a firewall does not address internal network compromise. Approximately 70% of all network security breaches occur from within the corporation, that is, by persons already past a firewall. A modem dial-up established by the company or by an engineer for remote access is one easy way past a firewall. Also, misconfigured firewalls may cause problems. Firewalls are highly susceptible to human error. In a dynamically changing environment, system managers routinely reconfigure firewalls without regard to security implications. Access control lists on a firewall can be numerous and confusing. Intranet managers should be sure that firewalls have been set up correctly and that they are performing well.

For intranets, a network-based intrusion detection system is required to protect the perimeter network from hacker attack. Network-based intrusion detection systems may be deployed as probes or agents running on servers. Probes are the most effective method at providing network-based intrusion detection. This probe minimizes the impact to existing systems because it is a passive listener reporting back to a centralized console without interruption. Intrusion detection will perform the following functions at the network device level:

- Inspection of data streams as they pass through the network, and identification/action on the signatures of unauthorized activity
- Activation of an alarm immediately upon detection of the event
- Notification of the appropriate security personnel, and triggering of an automated response to several issues to be considered.

In addition to intrusion detection, a TCP proxy aggregator may be considered. This will tighten security through the firewall by limiting the exposed ports. It also provides an offload for session/connection management and a more robust technical implementation in terms of port permutations supported.

Tunneling and encryption are used to deploy networks needing to appear point-to-point, but in fact consisting of various routes to an endpoint, providing data integrity and confidentiality. Usually, tunneling protocols, such as Layer 2 Tunneling Protocol (L2TP), Point-to-Point Tunelling Protocol (PPTP) and Internet Protocol Security (IPSec) and encryption standards such as DES, MD5, Triple DES, and others are used.

Mobile-code programs, such as Java and ActiveX, pose an increasing security threat. Content inspection software should:

- Provide full control over Java, ActiveX, and other mobile code activity in the corporation
- Prevent undetected, costly mobile code attacks, such as industrial espionage and data modification
- Enable safe Internet/intranet/extranet surfing while taking full advantage of Java and ActiveX technologies

A content inspection server will accept mobile contents redirected from a firewall in order to scan for attack signatures. If the scan detects a vulnerability, the contents will be blocked and the client prevented from downloading the mobile code. This denial will alert an appropriate administrator and notify the requesting client. If the scan does not detect any vulnerability, the mobile code is redirected to the firewall for routing to the client.

In summary, security management challenges increase in intranets due to many access points in the network. New techniques and new tools are required in combination.

3.11.2.3 Accounting Management Challenges

As far as the components of intranets are concerned, there are no differences to other types of networks and systems. But there are fundamental differences in terms of traffic patterns that may impact the right accounting strategies. Accounting management involves collecting data on resource usage in order to establish metrics, check thresholds, and finally bill users. Billing is a management decision, but usage measurements are a must in intranets. Principal steps of accounting are:

- Gathering data about the utilization of Web servers, the access and backbone networks
- Setting usage quotas as part of service level agreements with users
- Billing users for their use of resources

In order to gather data on usage, proper instrumentation is necessary. Standalone monitors, log file analyzers, and built-in accounting agents are most commonly used. Accounting management requires continuous measurements, but the amount of collected data is usually not critical in terms of overhead.

Service level agreements may include an expected level of resource utilization by single users or user groups. Either time duration or byte volumes may be agreed upon. Exceeding the agreed data volumes quota, the service and/or the price may change. The agreements and their continuous surveillance help to plan for the right amount of capacity.

Billing for intranet services is a new area, not yet well understood. Users are often billed based on one of the following:

- One-time installation fee and monthly fees
- Fees based on the amount of resources used

The first case is very straightforward. The user is billed for the installation of the intranet access and then a standard fee for each month of use. Using this method, accounting management is not necessary for billing. Although this is the easiest system to implement, it becomes difficult to justify why users with very different traffic patterns and volumes are billed for the same amount.

The second case is more difficult, and requires more engineering. Again, there are more alternatives, such as

- Billing is based on the total number of visits
- Billing is based on the total number of packets sent or received
- Billing is based on the total number of bytes sent or received

The accounting and billing cases are more complicated when multiple suppliers are present in intranets. If so, they must use a clearing house to gather usage data, allocate them to each other, and then generate convergent bills to the users. It is expected that the user receives just one bill for the intranet service.

In summary, the accounting management process can be fundamentally different in intranets in comparison to WANs and LANs of private enterprise networks. In particular, usage-based data collection and convergent billing are the real challenges to accounting management.

3.11.2.4 Configuration Management Challenges

Configuration management is the process of identifying systems and network components and of using that data to maintain the setup of all intranet resources.

Configuration management consists of the following steps:

- Identification of the current intranet environment
- Modifying that environment by moves, adds, and changes
- Maintaining an up-to-date inventory of the components, and generating various reports

Identification of the Current Intranet Environment

This process can be done manually by engineers or automatically by intranet management systems. Intranets don't require special treatment. This discovery and mapping step is identical with other networks and systems.

SNMP-oriented platforms offer configuration and topology services in two different ways; the discovery function identifies all managed objects with valid IP addresses on the LAN or across LANs; the mapping function goes one step further and displays the actual topology of the LAN or across LANs. Both functions can be successfully used for intranets. Managed objects without IP addresses are not discovered. The discovery and mapping processes need time and may impact production. Careful selection of the periodicity is required. Many companies deploy intranet visualization tools, instead of or in addition to discovery and mapping. Usually, they are very user friendly and easy to use, but they are independent from the actual network. Without synchronization of the tool's database with the actual network, these visualization tools are useless. But a combination of the discovery feature of the management platform with a visualization application can be very successful.

Modifying the Configuration Environment by Moves, Adds, and Changes

The intranet environment shows an over-average moves, adds, and changes (MAC) rate. Moves, adds, and changes are due to a user's move, restructuring buildings and infrastructures, deployment of new applications and the usual equipment changes. In order to offer service to mobile users, the change rate is not even predictable. Modification would probably be manual if the data collection method were manual, and automatic, if the data collection method were automatic. It requires stable and well-implementable procedures. Intranets become a very important part of the IT-infrastructure, requiring high availability and good performance. The MAC window is narrowing with the requirements that MACs must be prepared very carefully. The requester is expected to fill in forms, detailing the nature of changes, their impacts on other managed objects, fall-back procedures, desired dates, its priority, and human resources requirements. Also, the MAC process should be carefully monitored. When problems occur, fallback procedures are expected to be triggered. After successfully completing the MAC process, all related files and databases must be updated accordingly.

Maintenance of the Configuration

Asset and inventory management is one of the critical success factors of intranet management. Usually, relational databases are used to store and maintain technical and financial data on systems and network components. Access is usually via SQL; reporting is supported by standard or additional third-party reporting tools.

Asset management is expected to work together with other management tools that are implemented in other management areas. In particular, the following links are obvious in managing intranets:

- Trouble ticketing and asset management
- Performance tuning and asset management
- Security violation traces and asset management
- Accounting details and asset management

In summary, managing the configurations of intranets does not introduce additional challenges to configuration management.

3.11.2.5   Fault Management Challenges

Fault management is the process of detecting, locating, isolating, diagnosing and correcting problems occurring in intranets. Fault management consists of the following steps (Figure 3.11.6):

- Detecting and locating problems: Intranet components generate a number of messages, events, and alarms. Meaningful filtering, combined with user input helps to detect abnormal operations. Management platforms and their management applications are usually able to determine the location of faults. This phase indicates that something is wrong.
- Determining the cause of the problem: Based upon information generated by element managers or correlation results provided by management platforms, the cause of the problem is being determined. This phase indicates what is wrong.
- Diagnosing the root cause of the problem: In-depth measurements, tests, and further correlating messages, events, and alarms will help to determine the root cause of problems. This phase indicates why the problem happened.
- Correcting the problem: Using various hardware and software techniques, managed objects are being repaired or replaced, and operations can return to normal. This phase indicates that the problem has been resolved.

In summary, managing faults in intranets does not introduce additional challenges to fault management.

3.11.3   Specific Challenges to Intranet Performance Management

The emergence of intranets is dramatically altering the way information is accessed within and outside the enterprise. Components of intranets, such as servers, networks, and browsers are known, and are individually well manageable. But their integrated management, as an intranet, generates several challenges to IT managers. Content, server, networks, and browser management are all critical success factors.

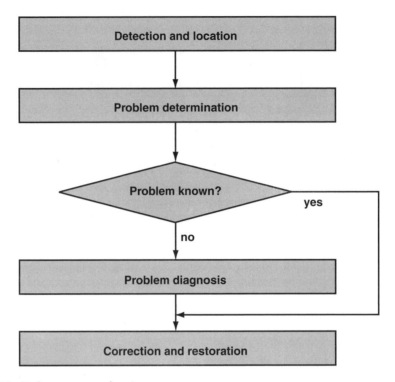

FIGURE 3.11.6   Fault management functions.

Not giving enough attention to any of them will cause IT managers to fail their intranets. Figure 3.11.7 shows the components of intranets.

The emergence of Web computing is dramatically altering the way information is accessed. The heavy use and popularity of the World Wide Web (WWW) is the most dramatic evidence. Looking at the enterprise, there is evidence that the Web browser has become the window of choice into corporate documentation and information. There are several important implications of this trend:

- All information can be viewed as Web content, accessible directly through a Web browser, a browser plug-in, or a dynamic piece of code, (e.g., Java) which is downloaded automatically to the client. This content can be as varied as a static Web page, a CGI script front-ending an existing database application, or new media such as streaming audio or video.
- The information access model has changed from one in which client-specific configuration is required in order to access information to one in which access is always available unless policies are explicitly defined to prevent it.
- Flash crowds, where certain content in the intranet generates significant unexpected traffic, are frequent observations, making traditional network design techniques based on measuring peak and average loads obsolete.
- Information accessed on or through Web servers comprise the bulk of traffic on the intranet (around 80%). Therefore, effective management of Web resources, bandwidth, and traffic is critical if acceptable quality of service (QoS) is required for Web-based computing.

3.11.3.1 Content Management

All information can be viewed as content. Structuring and arranging the content will finally decide about success and failure. Depending on the content for targeted visitors, page layouts may differ considerably. Not only the content of single pages, but also their links to each other have a great impact on visitor satisfaction. Individual visitors expect:

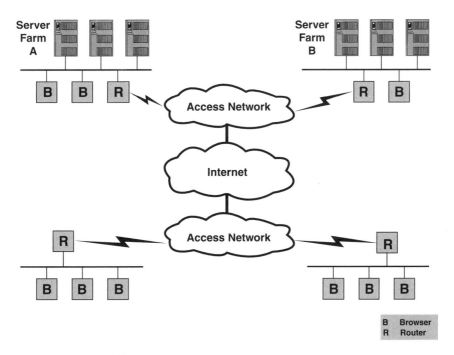

FIGURE 3.11.7 Components of intranets.

- Easy to read layout combining text and graphics
- Easy navigation between pages
- Easy return to the home page
- Rapid painting of pages
- Efficient links to interactive services
- Up-to-date status of pages
- Visualization of the site structure
- Site-wide change management of pages
- Easy ways of selecting pages to print or download

Goals and interests of companies offering information in home pages include:

- Rationalize information distribution to internal customers
- Fully meet content expectations of external visitors
- Manage intranet resources effectively
- Meet performance expectations of external visitors
- Meet business goals by using intranet technologies
- Provide the opportunity of deploying extranets to link business partners
- Meet high security standards
- Monitor visitor's behavior in order to make rapid changes to increase user satisfaction

Improvements in content management will have a great positive impact on overall performance. While Web server performance improvements are part of the performance optimization solution, they must be accompanied by improvements in network and content management technology to have a true impact on WWW scaling and performance. Specifically, developments in the following three areas are critically important:

- Content distribution and replication — By pushing content closer to the access points where users are located, backbone bandwidth requirements can be reduced and response time to the user can be improved. Content can be proactively replicated in the network under operator control or dynamically replicated by network elements. Caching servers are examples of network elements that can facilitate the dynamic replication of content. Other devices and models are likely to emerge over time.
- Content request distribution — When multiple instances of content exist in a network, the network elements must cooperate to direct a content request to the "best fit" server at any moment. This requires an increasing level of "content intelligence" in the network elements themselves.
- Content-driven Web farm resource measurement — A server or cache in a server farm ultimately services a specific content request. Local server, switching, and uplink bandwidth are precious resources which need to be carefully managed to provide appropriate service levels for Web traffic.

3.11.3.2 Web Server Management

Web traffic poses a significant number of challenges to existing Internet and intranet infrastructures. Most Web sessions are short-lived. As such, they have fewer TCP packets compared to batch mode operations such as file transfer. In addition, HTTP traffic tends to spike and fall radically. This creates instant demand for hot content that in turn causes network and server congestions. Web site traffic is highly mobile in that a unique event on a particular Web site could trigger a significantly high hit rate within a very short period of time. This would be typical in cases with periodic management report distribution and major systems and network outages.

Web traffic behavior is significantly different from today's client/server paradigm. It has the following unique characteristics:

- The amount of data sent from a server is significantly larger (5:1) than the amount of data sent from a client to a server. This suggests that optimization of server to client traffic has more significant impact to the intranet and that client request redirection to the best-fit server could have significant performance advantages for Web traffic flows.
- The median transfer size for Web server documents is small (e.g., 5 KB). This implies that Web flows are mostly short-lived flows. They are more likely to create instantaneous congestion due to their bursty nature. This suggests a resource management model must deal appropriately with short-lived flows. Even though HTTP supports persistent connections, due to interoperability issues with existing network caches, it is unclear how widespread deployment will be, or how soon.
- The top 10% of Web server files are accessed 90% of the time and are accountable for 90% of the bytes transferred. This suggests that Web server selection, caching, and content replication schemes that focus on this top 10% will yield the greatest gain.
- A significant percentage (e.g., 15–40%) of the files and bytes accessed are accessed only once. That is, some small number of large files often consumes a disproportionate amount of total server and network bandwidth. In addition, servers suffer performance degradation when subject to significant job size variation. This is due primarily to memory fragmentation, which occurs when buffering variable size data in fixed length blocks. Furthermore, subjecting servers to workloads consisting of both hot and one-time requests will result in lower performance due to frequent cache invalidation of the hot objects. Therefore, a server selection strategy that takes into account content, job size, and server cache coherency can significantly improve network and server resource allocation and performance. In addition, requests for large files may be good candidates for redirection to a server that has a shorter round-trip time to the client.
- Hosts on many networks access Web servers, but 10% of the networks are responsible for over 75% of this usage. This suggests that resource management strategies that focus on specific client populations may yield positive results in some cases.

Real-time traffic is becoming an increasingly significant proportion of Web traffic. Web site resource management strategies must take into account an increasing demand for support of real-time applications such as voice, distance learning, and streaming media. To deal with both legacy and Web traffic as well as real-time Web traffic, these strategies will need to include admission control as well as bandwidth and buffer allocation components.

The hardware of Web servers is practically the same seen with other servers. The software is divided in most cases between Unix and NT; industry analysts expect a clear shift toward NT for price reasons in the future. Web server sizing should follow generic guidelines, and also criteria specified by analyzing Web traffic patterns. If resource demand is higher than server capacity, multiple servers can be put together into server farms. This solution may satisfy the resource demand criteria, but requires careful attention of allocation and flow control.

3.11.3.2.1 *Content Smart Quality of Service (QoS) and Resource Management*

In a typical Web site, the top 10% of Web server files are accessed 90% of the time and are accountable for 90% of the bytes transferred. Therefore, techniques that optimize performance for these files will have the most significant impact on total Web site performance. This requires that the network itself be aware of which content is hot and which servers can provide it. Since content can be hot one instant and cold the next, content-smart switches must learn about hot content by tracking content access history as it processes content requests and responses.

To effectively manage Web site servers, network, and bandwidth resources, something must also be known about the content size and quality of service requirements. These content attributes can be gleaned through the processing of active flows, through proactively probing servers, or through administrative definitions. In addition, it is important to track server performance relative to specific pieces of content. All of this information can be maintained in a content database that provides an analogous function to a routing table in a router or switch. Content-smart switches make a content routing decision based on the information contained in the database, connecting a client to a best fit server in either a local or remote server farm. This enables the emergence of a business model based on replicating content in distributed data centers, with overflow content delivery capacity and backup in the case of a partial communications failure. Additionally, overflow content capacity intelligence minimizes the need to build out to handle flash crowds for highly requested content.

3.11.3.2.2 *Content Smart Flow Admission Control*

Two factors often contribute to congestion in a server farm. One is that servers are not up to the task of handling the amount of incoming traffic. The other is that the link bandwidth from servers to the Internet is overwhelmed by the combination of inbound and outbound traffic; this is complicated by the fact that the amount of outbound traffic from servers is on average about 5 times that of the inbound. As a result, a TCP/HTTP connection could be made successfully only to find out that the server could not be allocated the necessary bandwidth to deliver the requested content. To make matters worse, some server implementations come to a grinding halt when presented with an excessive number of TCP/HTTP connections — sometimes requiring a hard reboot.

3.11.3.3 Load Distribution and Balancing

In order to satisfy the high performance expectations of site visitors, bandwidth in backbone and in access networks should be managed effectively. Usually, servers are consolidated into server farms that are using the infrastructure of LANs. It is very unlikely that the LAN causes any bottlenecks. Larger enterprises may use multiple server farms deployed at various locations. In order to optimize content allocations, traffic and page references should be monitored and evaluated. At different locations in the network, hardware and software are expected to be installed that intelligently analyze the requests and direct the traffic to the right destination. The right destination could be threefold:

- Server farm destination with the requested content
- Server farm destination with the lightest load
- Server farm destination with the closest location to the visitor

There cannot be any compromise on item 1, but there could be a trade-off between 2 and 3, depending on the networking traffic.

The emergence of Web computing and Web traffic over the Internet or intranets has created some unique new problems. It is estimated that over 80% of Internet traffic is related to Web-based HTTP traffic. Even applications such as FTP and RealAudio, which run over TCP and UDP, respectively, typically use HTTP to set up the transfer. Since HTTP is an application protocol that runs over TCP, LAN switches and routers, which run Layers 2, 3, and 4, have very limited ability to influence Web traffic behavior. This burden is left to Web servers, which take on the function of TCP/HTTP connection management and, in some cases, the responsibility to distribute HTTP requests to servers within a server farm. This creates inevitable scaling problems as Web sites grow.

The current Internet can be described by using a model where local bandwidth is plentiful in the premise LAN located at the edge of the Internet. However, the uplink from LAN or remote user to the Internet is often severely bandwidth constrained by orders of magnitude. Although congestion can occur anywhere in the Internet path between a client and a server, the most frequent culprits are the WAN connection between the client and the Internet and the WAN connection between the Web farm and the Internet. Actions taken to ensure that this bandwidth is not overcommitted will help improve end-to-end performance.

Instantaneous bandwidth mismatches can occur for a network device that functions as the demarcation point between the public Internet and the Web farm. Examples are:

- The incoming link of the traffic is a faster media type (e.g., fast Ethernet) and the outgoing link is a slower type (e.g., T1 or T3).
- The instantaneous fan-in, i.e., the number of flows being sent at the same time to the same output port, can vary dynamically from one instant to the next.
- A number of traffic sources (e.g., outbound server traffic) may be sharing the bandwidth of a 45 Mbps T3 pipe in a bursty manner over a very high-speed switching fabric (e.g., 10 Gbps). This creates a need to regulate flow admission into a slower pipe from multiple higher speed traffic sources.

Information about the use of Web pages, their users, the frequency of access, resource utilization, and traffic volumes can also be collected in the network or at the interfaces of the network. In many cases, the borders between tools and techniques in the server and networking segments are not clear. Tools are different from each other; the differentiators are data collection technologies, performance metrics used, and reports offered.

In the Internet and intranet area, effective bandwidth management is a critical success factor. The role of network planners is going to be redefined. Real-time and near-real-time bandwidth allocation definitions are needed. Network managers agree that load balancers are needed.

There is little progress in standardizing on load distribution performance metrics. But the following few metrics can be successfully used:

- Number of referrals to server farms
- Number of lost requests due to load situations
- Number of requests with unacceptable response time
- Number of broken connections due to network problems

3.11.3.3.1   *Content Smart Link Management*

This technique can ensure that more flows are not admitted than can be handled through the switch or on the uplinks on average. It is still critical, however, to deal appropriately with traffic bursts and temporary congestion on these links to ensure that Web flows get the appropriate quality of service. Priority queuing provides a way to prioritize requests based on their type precedence. Fair queuing and weighted queuing methods improve over the priority queuing scheme by addressing the low priority traffic starvation problem with a scheme that separates traffic into well-identified flows so that each receives a "fair" or "weighted fair" share of transmission bandwidth.

Class based queuing (CBQ) was developed by the Network Research Group at Lawrence Berkeley Laboratory, as an improvement upon these existing bandwidth management techniques. It proposes a model which traffic is categorized in hierarchical classes. Flows inherit their flow characteristics from their parent flow class tree and can have local characteristics of their own. Flows are identified based on the IP address and the inner attributes within the IP header and payload. CBQ provides more granular control of transmission bandwidth and distributes it to member flow classes in accordance with their allocation policies. The model itself is independent of the scheduling techniques that run underneath it, therefore implementation details will vary based on the target architecture.

Content smart link management borrows concepts from CBQ. However, where CBQ is a model which operates on a packet-by-packet basis based on Layer 3 and 4 classification techniques, content smart link management classifies flows at admission time based upon the content requested, its attributes, and configured policies. These policies support the enterprise and service provider service models described in an earlier section of this chapter. This facilitates the classification of flows in a two-level hierarchy which includes owners (or customers) and content. Actual scheduling of flows is managed by a hardware-based flow scheduler which supports guaranteed bandwidth flows, prioritized/weighted flows, and best effort flows. Hardware-based scheduling is critical in order to scale the Web farm.

3.11.3.3.2   *Content Smart Load Balancing*

Simple load balancing techniques such as round robin, weighted round robin, and least connections are inadequate for Web traffic. For example, Web traffic load balancers must support "sticky" connections, which allow a particular server to be selected regardless of server load due to content locality or transaction integrity. Because of the disproportionate ratio of hot content files to total content (1:10), it is highly desirable to support a content replication model that does not require that content be equally and fully mirrored among servers in a server farm. This means a load balancing technique must be intelligent enough to recognize if content is available on a particular server before making the selection decision.

Content smart load balancing takes into account several factors that have a significant impact on the overall performance and cost of a Web server farm:

- Server cache hit rate — By directing requests for hot content to a server that has recently serviced that content, this technique ensures that cache hit rate, reducing disk access latency for the most frequently accessed content. Since a significant percentage (15–40%) of the files are accessed only once and 90% of the files are accessed only once or not accessed at all, it is important to keep those infrequently accessed files from thrashing a server cache. That is, an infrequently accessed file should be invalidated in server cache promptly to increase the chances that a more frequently accessed file can remain in cache.

- Burst distribution — Short-lived, bursty flows can best be handled by distributing them among eligible servers so long as the servers have been performing below a defined threshold for a period of time.

- Web flow duration — Most Web flows are short lived. However, a relatively small number of infrequent, long-lived flows have a far significant impact on overall bandwidth and server resource consumption. For that reason, long-lived flows should be separated from short-lived flows from a load balancing perspective and short-lived flows of similar QoS requirements should be aggregated to increase TCP flow intensity and reduce per flow resource allocation overheads.

- Content biased server performance measurement — Current server loading can best be measured by examining the request/response time interval of a server as it handles requests. This measurement is most accurate when connection between the switch and the server is direct. In addition, server performance is not uniform across all content. For example, computer intensive applications may perform better on one server than another. Other servers may perform better for other types of content. Server performance information needs to be qualified by content.

In the Internet and intranet area, effective bandwidth management is a critical success factor. The role of network planners is going to be redefined. Real-time and near-real-time bandwidth allocation definitions are needed. Network managers agree that load balancers are needed. The decisions are if:

- Hardware or software-based load balancers are better
- Embedded or standalone solutions should be preferred
- Use of the combination of both

In the first case, considering high traffic volumes, hardware solutions should be preferred. Software solutions in critical load situations may slow down processes, and risk performance. At this time, there are no accurate guidelines for tolerable workload, but a range up to 5% seems to be reasonable.

Switches, routers, and firewalls are almost everywhere in Internet access networks and in intranets. To embed traffic control and sharing functions would save extra components, but would — as stated earlier — generate additional load and may impair the principal functions. The embedded solution may also include the use of RMON capabilities for real-time load profiling. The standalone solution is sensitive against single point of failure, but would offer an overhead-free traffic and load management. The following attributes may play an important role when evaluating alternatives:

Use of Load Balancing Switches
Benefits:

- Load balancing is performed in a device that is needed anyway in the network
- Centralized management
- Good opportunity to control and guarantee QoS

Disadvantages:

- Performance may be impacted by management functions
- Single point of failure for both switch and management functions

Use of Load-balancing Firewall
Benefits:

- Load balancing is performed in a device that is needed anyway in most networks
- Centralized management
- Includes special functions and services, such as traffic management and application-based load balancing

Disadvantages:

- Switches are still needed
- Single point of failure for both firewall and management functions
- Performance depends on hardware and operating system configuration

Use of Load-balancing Traffic Shapers (Figure 3.11.8)
Benefits:

- Load balancing is performed by a device most likely present in the networks anyway
- Centralized management
- Offers traffic shaping and balancing for Internet or intranet access in addition to server access

Disadvantages:

- In most cases, switches and firewalls are needed in addition to these devices
- Single point of failure for both traffic shaping and load balancing
- Little experience yet with performance and scalability

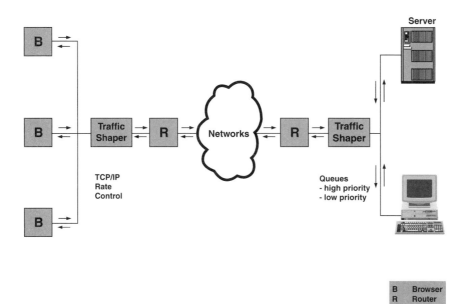

FIGURE 3.11.8 Load balancing packet shapers.

3.11.3.4 Technologies of Access Networks

There are many alternatives of designing and deploying access networks. The basis technology chosen has a significant impact on overall performance. Table 3.11.1 evaluates the most important technological choices according to criteria, such as suitability, maturity, scalability, distance limitations, and costs.

The right choice of access network technology must be seen in connection with content smart control of the bandwidth provided in the access networks.

TABLE 3.11.1 Comparison of Technologies for Access Networks

Criteria	T	ISDN	Frame	ATM	Cable	xDSL
Suitability	Medium	Good	Good	Excellent	Excellent	Good
Maturity	High	High	High	Medium	Low	Low
Scalability	Good	Medium	Mefium	Excellent	Medium	Good
Distance limitations	None	None	None	None	Some	High
Costs	High	Low	Medium	High	Low	Low

3.11.4 Content Management

Authoring tools present a standalone environment in which to build pages. While this requires learning a new program specifically for HTML/XML creation, these tools allow users to make the most of HTML/XML, using features that traditional word processors do not support.

Currently, there are two distinct kinds of tools Web authors can use to bring their words to the Web. Tag-based tools automate HTML/XML syntax, allowing users to see and tweak tags without having to enter their syntax manually. In contrast, WYSIWYG tools hide HTML/XML from the user, generating it in the background instead. If these tools do not support a specific feature of HTML/XML, that feature must be added manually after the document's underlying code is visible, usually in a text editor. Some products use dialog boxes or palettes to accept information before displaying it as HTML/XML code in the body of the document. Since these tools generate HTML/XML for users, they minimize the learning curve for new Web authors and can produce syntactically perfect HTML/XML. Many of the publicly available tools have both standard and professional features, the latter being available only in the registered or commercial version.

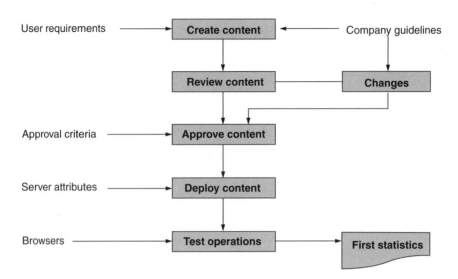

FIGURE 3.11.9   Process of content authoring and management.

3.11.4.1 Design of Home Pages — Content Authoring and Deployment

Most users are challenged by the task of information creation, management, and dissemilation. These activities are time consuming and difficult to control. The Internet and intranets alone cannot solve information management problems unless specific intranet solutions are implemented that directly address the need for document management. The new discipline, called content authoring and deploying, includes the following tasks:

- Creating content
- Reviewing content
- Approving content
- Changing content
- Deploying content

Figure 3.11.9 shows the process of creating, reviewing, changing, enhancing, approving, and deploying home pages.

The prerequisites to successfully execute these tasks are:

- Users must be able to easily add and update content on a periodic basis.
- Users must be able to protect their page contents from changes by other users.
- A content approval process should be defined and in place. This process should emcompass ways to manage and control document revisions, especially changes to shared documents.

As policies and procedures relating to content management are formulated, it is important to designate responsibilities to specific individuals to ensure that they are properly implemented and followed. An internal style guide should be developed that provides page layout, design elements, and HTML/XML code guidelines. Usually, case tools are also involved. The style guide will help the users to maintain a consistent look and feel throughout the Web pages. Sometimes television-like techniques are helpful in this respect. The style guide should contain information on where to obtain standard icons, buttons, video, and graphics, as well as guidelines on page dimensions and how to link the pages to each other. As part of the style guide, it is helpful to create Web page templates. These templates consist of HTML/XML files, and are used to provide a starting point for anyone interested in developing Web pages or content for the intranet. Although it is very easy to create a working Web page and to publish for mass viewing, the real challenges are:

- To maintain the page
- To size the Web server
- To configure the access network

3.11.4.1.1   *Site Design Considerations*

Content authoring includes a number of tasks. The most important tasks are:
  Determine the right page layout, including:

- How to structure a Web site
- How to lay out a Web page
- Ideas for improving usability
- Technical hints to increase display speed
- Collection of examples of well-designed sites that can be used as models
- Consideration of new Web technologies for site design

One of the principal factors in the design of a good Web page is knowledge and understanding of the motivations and goals of the target user as well as the technical platform on which they operate. Given the varying levels of user knowledge and the infinite number of ways a Web page can be constructed, this understanding is essential to creating a usable, effective Web site. Before beginning the design, a user and task-centered analysis should be completed to gain knowledge about the target users and their goals. Important questions are:

- To whom will the page be available?
- What are the business drivers for the site — e.g., to provide information, to collect data, to market products?
- Who are the users — e.g., professional "knowledge workers" or casual intranet users?
- How will a typical user access the page — e.g., fast connection or dial-up?
- What browser will they use?
- What are the most frequent tasks that users perform?

Answers to these questions will provide the necessary background information for the navigational structure of the site. During site design, designers should keep in mind that if the user cannot quickly find what they are looking for and are not engaged by the layout and information contained within the site, they are likely to move on.

Site Registration

The purpose of site registration is to establish content ownership and to facilitate navigation. Through the site registration process, sites are added to the intranet directory and become accessible via the intranet-wide search facility. The Web site is defined as a collection of related Web pages. Typically, a Web site is an administrative unit.

Site Navigation

There are two points to consider when constructing the navigation layout for a Web site — namely, the structure of the information and how access to that information will be provided. First, the layout of the site is usually the most difficult part of the site design process, particularly if a lot of information will be accessible from the site. Adequate time must be put into designing the structure of the information to allow easy access for all users. Second, navigation tools must be clear and easy to use as well as functional within all types of browsers that will be used by the target audience. Navigational design must consider all of the same factors as many other GUI interfaces. Since movement within a Web site is typically non-linear, navigational menus should be planned to allow users quick access to any part of the site.

Content Organization by Menus

A user's ability to move through a Web site and find the information or functions they are searching for plays an important role in determining a site's success. Menus and sub-menus are powerful tools in the design of a Web site. In the same way that menus are used in traditional Windows-based design, HTML menus can be used to subdivide and group relevant content to allow the user to be guided to their topic of interest gradually. The use of more than four levels of menus forces a user to work too hard to find the information they are looking for. Using too few levels may be equally difficult to navigate, in particular, when the information volumes grow. Three to four levels should generally provide appropriate depth and guidance for the user. However, because of the varying content of sites, this is a flexible guideline. It is important to know that the menu structure for the site should be continually evaluated and improved as the site grows.

Interaction Models

There are many ways to organize information contained within a Web site. The term "interaction model" refers to the structure that is implemented to allow the user access to the various pages within a site. The type of model best suited to a particular page will depend on the content and complexity of the information that the page presents. There are a number of interaction models in use. These models may be used independently or in combination throughout a site. These models are:

- Table of contents — This approach is taken from printed books. Users can easily find the headings they are looking for, and then hyperlink directly to that page. This type of access is useful for sites that provide textual or encyclopedic information.
- Image maps — They are graphics that use an embedded linkage map that relates hot spots on the graphic to URLs within the Web site. In this way, the user can view the graphic and point and click to move to different locations on the site.
- Graphic menus — They provide the same visual approach to site navigation as image maps, without incurring the disadvantages of employing one single large graphic, mapped with links. They employ smaller, simpler graphics, strategically placed to provide visual impact.
- Search — Web site searches provide a useful means of allowing a user to access information contained on a particular Web site. Some form of search facility is usually a requirement for larger sites.
- Indexing — It provides functionality similar to book indices. It allows a user to rapidly locate information pertaining to a specific keyword or topic. It may be used in combination with search.

3.11.4.1.2   *Page Design Considerations*

The actual layout of a Web page is highly dependent on the type of information that is being presented. This segment provides some fundamentals of good page design.

Header — The Header provides a user with access to commonly used functions within the company-wide intranet and clearly differentiates intranet content from Internet content. The standard header provides links for navigation to common functions via the following graphics:

- Company logo — Links to the company's home page
- Directory — Links to the company's intranet directory Web site
- Services — Links to the company's intranet service page
- Search — Links to the company's search Web site
- Help — Links to the company's intranet help Web site

Preimaged mapped versions of the company's header are available on the intranet development and support site.

Footer — The footer gives the user important information about the page and provides consistency within the company's intranet. The standard footer usually contains the following:

- A standard horizontal rule as a separator
- Copyright statement
- Statement regarding content ownership with an optional e-mail link to the designated page maintainer; not supposed to be a name of an individual
- Date of the last revision

Page size — Page size must be designed with the actual usable space of the browser window in mind. Typically, this would be the lowest amount of useable space for the standard browser configuration in a 640 × 480 video monitor resolution. When designing a Web page, designers want to limit horizontal scrolling as much as possible. Keeping the width of Web sites less than 600 pixels (using tables) makes it much easier for users to navigate information. In some cases, horizontal scrolling is normal and acceptable.

The acceptable size for an intranet page is 100,000 bytes or less. This limit includes all of the images that are embedded on this page. This size will keep performance within acceptable limits for both LAN, WAN, and dial-up users with 28.8 Kbps modems.

Home page — The layout and design of the "home page" of any Web site is extremely important. Besides being the first thing a user sees upon entering a site, it defines the organizational structure and tone for the entire site. Some essential elements for every home page include:

- Visually appealing design
- Overview of site content
- Links to contents of site
- Company/organization identifying information

Page layout — HTML does not provide graphic designers the flexibility they are accustomed to in existing page layout and editing programs (e.g., MS Word, Adobe PageMaker). However, this does not mean that complex and functional applications cannot be created using HTML. Rather, one must realize that, when used inconsistently, the graphic and typographic controls of HTML can result in inconsistent designs. To avoid the haphazard look of documents, designers should take care in how graphics are placed and organized. A consistent style will also allow for a consistent conversion from non-HTML documents. It is better to use simple icons and images, instead of complex ones. Navigation should be kept in a consistent place.

Text style — Text needs to be short and to the point. Text should be organized in sections of a paragraph. When browsing, visitors tend to scan rathen than read. They are usually searching for information and appreciate when sections are arranged in logical order. Similar ideas or facts should be presented in a consistent way, with the same components presented in the same way in the same order. Consistency is a very important consideration in Web design.

Graphics — Graphics images should be used where appropriate to help the user navigate and find information more quickly. Graphics also provide a "look" to the site that will help the user to identify where they are. Graphics should not be overused for internal publishing applications. Whereas external marketing Web sites often are graphically intense to catch attention, use of graphics in internal Web sites should be based on ease of navigation and usage. The type, sizing, and location of graphics throughout a site should be presented in a consistent manner, items of similar importance should have the same size and type of graphic. If a larger-than-normal graphic is used, the user is likely to assume that there is some additional significance. Often, the visibility and intended use of the site will dictate the level of graphics required for the site. Graphic images should be designed for a 256 color envisonment. A common mistake that professional graphic designers make is designing with higher resolutions and greater color depth than the deployment environment. The color scheme that was designed in 16-bit color may look bad in 256 color or even worse in 16 color environments. Design should follow the requirements of the target environment. Most images are between 10 and 30 K. The exception would be image maps on navigational pages or photographic images, which should be around 50 K. One of the drawbacks with using images on a network is the time it takes to download very large files. Images must be kept as small as possible and fit within the size of the browser's viewable space. For image formats, file formats are the best to be used. GIF and JPEG are both compressed formats. GIF format is better for smaller graphic or line art images.

Local navigation elements — Each Web site should include a sitemap, showing a detailed layout of this site with links to all possible sections and documents. Each page within a Web site should include a link to the sitemap page. Users may link to a Web site or Web page from a number of different places (navigation page, search results page, hyperlinks, etc.). The sitemap page gives the user a quick and easy way to locate the information they need. On long pages, the user may want to quickly go to the top of the page to view the table of contents or other introductory information. The "top of the page" icon helps users more quickly navigate to the top of the current page.

Links — While many Web sites incorporate graphics to support navigation, text links still play an important role in ensuring the usability of a site. Working with text in HTML is easy. In general, because it is easy to create links and change font types, there are several mistakes commonly made. Several guidelines that aid in ensuring a site's readability and usability are listed below:

- Design for scanner, not for readers
- Explain the page's benefit above the "fold"
- Bold typeface will draw attention to a particular section
- Avoid typing in all caps — it is more difficult to read
- Links must be underlined in addition to being colored to assist users who may be colorblind or using black and white monitors
- Avoid blinking text because it is difficult to read and annoying to users

A typical Web page provides both informational text and links to more specific information. Most people are looking for visual clues to whether a page is useful or interesting enough to be worth reading. If they don't find what they want quickly, they will move to another site. One of the difficulties in using text for navigational purposes is the wording of the links. Proper wording of the text allows the user to jump to a new topic or continue reading without losing their place. All links to default pages should be set with a trailing "l." This eliminates the problem of DNS names turning into IP addresses. By default, the Web browser converts any hyperlink that does not include a Web page (such as a link to a home page) to the default page for the server. However, depending on the browser, this may convert the DNS name into the physical IP address of the hosting server. If the DNS name is converted to an IP address and the user adds the page to their favorites, the URL will be stored with the IP address. If the IP address of the site changes, the bookmark will no longer work. To eliminate this problem, simply include a trailing "/" on any link that does not include a page file name. Abstracts and summaries are very helpful for large pages or large graphics. Whenever possible, users should have the opportunity of linking to further information if desired. Very large files or files which are not in a useable browser format (e.g., ZIP files, BMP files, etc.) should have a link which allows the user to download the file to their local PC.

Other graphic elements — Separators are graphic or possible textual elements that are used to break up or visually divide the contents of a single Web page. Separators can be as simple as a horizontal line to a shadowed line graphic or an actual image file. Their use helps to visually discern varying subject matter on the page. While separators can be effective, it is important to remember that separators should not distract the user from page content; rather, their purpose is to divide the information into logical groupings. HTML provides tags for standard information-gathering controls like radio buttons, drop-down menus, and exit boxes. In general, guidelines created for traditional GUI-based development apply to Web page design. Important remarks are:

- In most countries, the eye moves from left to right when reading, so text literals should be left-aligned.
- Exit boxes should be similarly sized and also left-aligned.
- Tabs should move the user downword through the page.
- Controls should be evenly spaced and aligned when possible.
- A default button should be provided.
- Mixed case text should always be used.

Bullets are used in HTML in the same manner they are used in traditional word processing to define a list of items. While textual bullets are fine for use on Web pages, there are also many available graphic bullets that will add just a touch of color to an ordinary Web page.

Background and text colors — The use of appealing backgrounds and text colors can add an artistic look to Web sites, but the way colors are used also affects the usability of the site. Designers must be wary of improperly using color, as colors may have different meanings to different people and some users may be unable to distinguish some colors clearly.

Some user interface guidelines that are applicable to Web sites include:

- Color is second only to movement in attracting attention
- Three colors are sufficient for a color scheme
- Specific colors should be used carefully
- Shades of red attract attention, while the retina responds to yellow the fastest
- Blue is more difficult to focus on, making unsaturated blue a good choice for backgrounds
- Gaudy, unpleasant colors, and combinations of red/green, blue/yellow, green/blue, and red/blue should be avoided
- If backgrounds are going to be used, they should be either a light-colored pattern or a solid color

Printing — When the nature of a site is documentation, users must have the ability to print individual Web pages or an entire site's content. This can easily be accomplished by adding a link to a printable form of the entire document. Documents may also be provided in multiple formats such as Microsoft Office formats to accomodate the maximum number of users.

3.11.4.2 Issues with Content Authoring

The recommendations for Web page design can be summarized as follows:

- Use of standards for the layout of pages
- Standardize links
- Use one or more interaction models, such as table of contents, image maps, graphic menus, search, indexing
- Use of navigation assistance
- Segment long documents into small ones
- If a site includes a significant amount of pages or data, a local search page should be provided to search the content only
- Design pages for rapid and slow searches alike
- Use text pages for users with narrow bandwidth
- Test HTML pages and links before practical use
- Test content on different browsers
- Use recommended templates from the webmaster to create new pages
- Use abstracts or summaries for larger text pages or large images and give the users the option to link to detailed information if desired
- Provide a link to download a concatenated file of a series of Web pages so that a user can print an entire document rather than printing multiple web pages
- Use backgrounds carefully; make sure that users can easily read the text of a page if a background is used
- GIFs should be used for small graphics where there are a limited number of colors and JPEGs should be used for photographic images
- If users link to another site, the site owner of the link must be informed; this will enable the site owner to notify everybody involved if the link changes

But the users should consider the following facts:

- Too much graphic and animation content slows down operations
- Big pictures slow down loading the pages
- Copyright of graphics should be granted

- Proofreading is always necessary
- Browser compatibility must always be checked
- Avoid one-way-streets in HTML-documents

3.11.4.3 Content Authoring Tools

There are many content authoring tools available. The most important ones are listed in Table 3.11.2. Some of them are combined with analysis tools. FrontPage from Microsoft is becoming part of Site Server, a complex product addressing site development, deployment, search and usage analysis.

TABLE 3.11.2 Content Authoring Tools

Vendor	Product
Adobe	PageMill
Allaire	Homesite
FileMaker	HomePage
Golive	CyberStudio
Micromedia	Dreamweaver
Microsoft	FrontPage
NetObjects	Fusion
Softquad	Hotmetal
Symantec	VisualPage

3.11.5 Log File Analysis

Web site activity reporting involves the analysis of:

- Basic traffic statistics (hits, page views, visits)
- Navigation patterns (referrers, next-click, entrance and exit pages)
- Content requested (top pages, directories, images, downloaded files)
- Visitor information (domains, browsers, platforms)
- Fulfillment of the Web site's objective (purchases, downloads, subscriptions)

Clearly, this last characteristic is the reason that Web site activity analysis has become an enterprise-critical priority for organizations investing massive amounts of time and money in their Web presence. How well the Web site is performing relative to its objective is what justifies continued investment. The easiest way to quantify the return on Investment (ROI) is with meaningful Web activity reports.

Reporting is also essential for making decisions about content. Web site activity reports, by providing statistics about the most popular pages or files, give an organization quantifiable measurements as to what type of content appeals to its audience. Without reliable, comprehensive reports, a Web site's content is designed based on an educated guess by the design team or editorial staff.

Similarly, Web site activity analysis reports also tell an organization about their visitors. Where are they coming from, how do they get to the Web site, and what type of browser or platform are they using? When a corporation decides to deploy a Web site, it usually has an idea about who its audience will be. Does the actual audience resemble the predicted one? How does it change over time? What type of content improves visitor retention or session depth?

3.11.5.1 Usage Analysis

Web server monitors and management tools concentrate on how the Web server is utilized and how performance goals can be met. In addition to these tools, other tools are required that are able to continue the analysis by using log files filled by special features of the server operating system. This segment is devoted to log file analyzer tools that are able to give the necessary data for in-depth usage analysis.

Usage analysis is a means of understanding what is happening on an Internet or intranet server such as a Web server. Usage analysis tools piece together data fragments to create a coherent picture of server activity.

Usage analysis can answer the following questions:

- How many individual users visited the site on a particular day?
- What day of the week is the site busiest?
- How many visitors are from a certain country?
- How long do visitors remain on the site?
- How many errors do visitors encounter?
- Where do visitors enter and leave the site?
- How long did it take most visitors to view the home page?
- Which links on other sites send the most visitors to this site?
- Which search engines send the most visitors to this site?

Reports can span any length of time, making it possible to see trends. They can also display any degree of granularity, allowing users to see both broad-ranging reports and detailed reports. Usage analysis is most frequently thought of in terms of Web servers. The reports created by usage analysis tools can be used throughout organizations to help people make informed decisions. Examples are:

- Web developers use these tools to gauge the effects of site design changes. Using this information, they can make further refinements to the design of the site to maximize its effectiveness.
- Marketers use these tools to analyze the effectiveness of marketing programs and online ads.
- Site administrators can spot Web pages that are causing errors, determine future server hardware needs, and track FTP and proxy server activity.
- Salespersons can gather information about prospects including their geographic location, how many pages they viewed, and how they found the site in the first place.
- Executives use the intelligence gathered with log analyzers as a resource when making a broad range of decisions.

Each time a visitor accesses a resource on a Web server — whether it is an image, a HTML file, or a script — the activity is usually recorded as a line in a text file associated with the Web server. This text file is known as the Web server log file. A single line of a typical Web server log file can be interpreted as follows.

Record of the server log file entry:

foo.bar.com --(31/Oct/1998:23:31:44+ 500) "GET home.html HTTP/1.0" 200 1031 http://www.yahoo.com/ "Mozilla/3.0 (Win32;U)"

Interpretation by elements:

Element	Interpretation
foo.bar.com	Hostname of the visitor's computer
31/Oct/1998:23:31:44	Date and time
GET	Method used to request the resource
home.html	Name of the requested resource
HTTP/1.0	Protocol used to request the resource
200	Status code "200" means that the request was successful
1031	Number of bytes transferred to satisfy the request
http://www.logfile.ana.html	Web page that referred the visitor to this page
Mozilla/3.0	Visitor's Web browser and version
Win32	Visitor's operating system

Most Web servers write out log files in the combined log format. It differs from an older common log format in that it contains browser and referral information. Referral information is important to determine what sites are sending the most traffic to the target address and what sites might have out-of-date links pointing to specific user sites. Referral information is also critical for gauging the effectiveness of online ads. Other information that can be included into a log file includes:

Cookie	A persistent identification code assigned to a user which allows the user to be tracked across several visits
Session identifier	Tracks each visitor for the length of the visit only
Amount of time the request took to fulfill	Enables server performance reporting

Basically, there are two types of usage analyzer tools: software-based and on-the-wire-collectors. On the high end of usage, analysis tools are packet sniffers which offer on-the-wire reporting by installing an agent against the kernel of the operating system of the Web server. They run as root in the kernel of the operating system on the Web server. Furthermore, they require that a network runs in promiscous mode in order to expose network traffic to the agent. Usually, there are very few reports packet sniffers can create and log file analyzers cannot. Log file analyzers can create reports on the usage of secure/encrypted communications, while packet sniffers cannot. Packet sniffers are more expensive, offer less reports, and offer just a few report distribution capabilities.

3.11.5.2 Issues of Log File Analysis

When selecting products, there are a number of criteria that must be carefully evaluated. The market is big, addressed by a relatively low number of products. These criteria are also important when the webmaster wants to position log file analysis within their IT administration or when they want to deploy this functionality within their organization.

Architecture of a product answers the question whether the product can support a distributed architecture or not. Distribution means that collecting, processing, reporting, and distributing data can be supported in various processors and at different locations. Figure 3.11.10 shows these functions with a distributed solution.

In Figure 3.11.10, Web servers A, B, and C can be from very different types, such as Netscape Navigator and Microsoft Explorer.

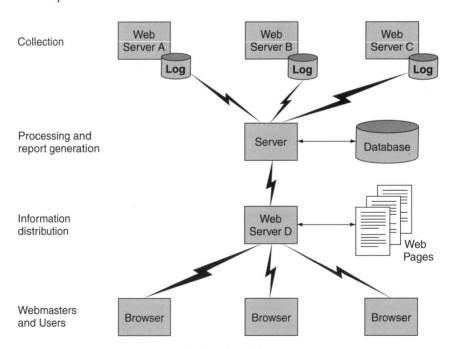

FIGURE 3.11.10 Generic product architecture for log file analysis.

Of course, it is expected that many different Web server types are supported. Also, the hardware and operating system may be a differentiator for products. It is assumed that the Web server hardware has decreasing impact on log file analysis. The role of operating systems is more significant; the product should know exactly how log files are initiated and maintained. No problems are expected with leading Web server solutions, based on Unix and NT.

The data capturing technique is absolutely essential with log file analysis. The first question is where the logs are located. Figure 3.11.10 indicates that they are located in the Web servers. But, more accurate information is required here:

- What memory area is used
- What auxiliary storage area is used
- What is the size of those areas
- What types of log files are supported

If log files are not processed in real time or near real time, it is important to know where they are stored until they are downloaded for processing. Log file analysis is dealing with very large data volumes, and these volumes depend on the visitor's traffic.

Usually, log files are downloaded for processing. It is important to know how downloads are organized and how rapidly they are executed. As indicated in Figure 3.11.10, WANs are involved with sometimes limited bandwidth. The bandwidth is usually shared with other applications, with the result of potential traffic congestion. Bandwidth-on-demand solutions are rare with log file analysis. When transmission is arranged for low traffic periods, the actuality of log file analysis results may suffer. In such cases, local storage requirements increase, and processing, report generation, and information distribution are delayed by several hours or even by days.

Two solutions may help. The first solution is using intelligent profiling at the source of data collection. Redundant data are removed from logs during collection. Data volumes decrease and local storage requirements decrease as well, but processing requirements in Web servers increase considerably. The second solution may use data compression or data compaction with the same results and impacts as with the first solution.

Overhead is a very critical issue with large data volumes. Data capturing is expected to introduce little overhead, when logs are stored away immediately. If local processing is taking place, overhead must be very carefully quantified; if resource demand is high, overall Web server performance may be impacted. Data transmission overhead can be heavy, when everything is transmitted to the site where processing is taking place. WAN bandwidth is still very expensive to be dedicated just to log file analysis. If bandwidth is shared with other applications, priorities must be set higher for business applications than for transmitting raw log file data.

In the case of server farms, a local mediation device could help. The mediation device is connected via LAN; bandwidth is not so critical in LANs in comparison to WANs. Processing and report generation remain at a special server that is consolidating all data from mediation devices.

It is absolutely necessary that all data are captured that are necessary to conduct a detailed Web site analysis of visitors or groups of visitors:

- Who is the visitor?
- What is the purpose of the visit?
- Where is the visitor coming from?
- When has the visit taken place?
- What key words have brought the visitor to the site?
- What search machines helped to access the site?
- How long was the visit?

Data losses cannot be completely avoided. Logging functions of Web servers, storage devices, or components of the transmission may fail; in such cases, there will be gaps in the sequence of events. Backup capabilities may be investigated, but IT budgets won't usually allow too much to spend for backing up large volumes of log file data. In the worst case, certain time windows are missing in reporting and in statistics. Those gaps may be filled with extrapolated data.

Also, the management capabilities are very important. One of the functions here includes automatic log cycling. In order not to lose data, multiple logs are expected to be used. When one of the logs is full, the other log seamlessly takes over. Another function is the translation of domain name service (DNS). Its speed is absolutely important for real-time information distribution. In order to generate more meaningful reports, it is required that results of log file analyzers are correlated with other data sources. These other data may be maintained in other databases. In order to correlate, *ad–hoc* database links should be established and maintained. Management of logs of any log file analyzer can be taken over by the operating system of Web servers. The basic services are supported today; additional services may follow. In the case of server farms or of many individual Web servers, the coordination of log transfers and processing is no trivial task. Event scheduler may help in this respect.

Cookie support is important to speed up work initiated by visitors. It is a logical connection between Web sites and browsers; a persistent identification code is assigned to a user, which allows the user to be tracked across several visits.

Due to considerable data volumes, databases should be under consideration to maintain raw and/or processed log file data. Database managers would then offer a number of built-in features to maintain log files. Clustering visitors may be deployed from various perspectives, such as geography, common applications, common interests on home pages, and data and time of visits. Automatic log cycling can also be supported here by the database managers. Open database connectivity (ODBC) support helps to exchange data between different databases and to correlate data from various databases. Besides log files, other data sources can be maintained in the same data warehouse. Besides routine log files analysis with concrete targeted reports, special analysis may also occasionally be conducted. This special analysis, called data mining, can discover traffic patterns and user/visitor behavior. Both are important to sizing systems and networking resources.

One of the most important questions is how log file analysis performs when data volumes increase. Volume increase can be caused by offering more pages on more Web servers, more visitors, longer visits, and extensive use of page links. In any case, collection and processing capabilities must be estimated prior to deciding for precedures and products.

In order to reduce processing and transmission load of log files, redundant data should be filtered as near as possible to the data capturing locations. Filters can help avoid storing redundant data. Filters can also be very useful in the report generation process. Again, unnecessary data must not be processed for reports. Powerful filters help to streamline reporting.

Not everything can be automated with log file analysis. The user interface is still one of the most important selection criteria for products. Graphical user interfaces are likely, but simple products are still working with textual interfaces. When log file analyzers are integrated with management platforms, this request is automatically met by management platforms.

Reporting is the tool to distribute the results of log file analysis. Predefined reports and report elements as well as templates help to speed up the report design and generation process. Periodic reports can be automatically generated and distributed for both single Web servers and Web server farms. In the cases of many Web servers, report generation must be carefully synchronized and scheduled. Flexible formatting helps to customize reports to special user needs.

Output alternatives of reports are many. The most frequently used solutions include Word, Excel, HTML, and ASCII. Also, the distribution of reports offers multiple choices:

- Reports may be stored on Web servers to be accessed by authorized users who are equipped with universal browsers
- Reports can be uploaded into special servers or even pushed to selected users

- Reports may be distributed as attachments to e-mail messages
- Reports can also be generated at remote sites; this alternative may save bandwidth when preprocessed data instead of completely formatted reports are sent to certain remote locations

Documentation may have various forms. For immediate answers, an integrated on-line manual would be very helpful. Paper-based manuals are still useful for detailed answers and analysis. This role, however, will be taken over by Web-based documentation systems. In critical cases, a hot line can help with operational problems.

Log file analysis is actually another management application. If management platforms are used, this application can be integrated into the management platfrom. There are many ways to integrate; most likely a command line interface (CLI) will be deployed.

3.11.5.3 Drawbacks of Pure Log File Analyzers

Log file analysis can give a good entry-level summary about the activities in and around Web servers. But this technology shows major problems that are analyzed as follows.

The first major problem is traffic volumes. As traffic levels quickly reached exponential growth rates, nightly log file downloads quickly became afternoon-and-evening, and then even hourly downloads, since server disk drives would fill with log file data so quickly. Compounding this problem was the fact that higher-traffic sites needed to load-balance across several servers and physical machines, so that log file downloads needed to be done not only many times a day, but also across several machines each time. The quick fix to this problem was typically an automated script that would download log files on a preset schedule. However, this failed to account for unexpected spikes in traffic and also clogged internal networks with huge log files being transmitted across the network several times a day.

The second major problem is data processing speed. Even if there were an easy way to continuously transfer log file data to a consolidated area, there was still the problem of how to process the gigabytes of log files into database tables in an efficient, continuous, and robust manner. Batch processing of log file data requested a considerable amount of time. In addition, the human resources demand for log file collection, processing support, and report compilation has exceeded the expectations.

The third major problem involved incomplete data. Beside log files, there are significant alternate sources of site activity data which contain more information than even the longest, most complex custom log file format can provide. A log-file only approach cannot guarantee a complete picture for Web activities. A good example of missing data is certain network-level data that the Web server and the server's log file never get to see. For instance, a visitor requests a page that turns out to be too slow to download, and decides to hit the browser STP button, BACK button, or otherwise terminate the request in mid-download. In this case, the network layer will log that action, but it will not notify the Web server about it. Similarly, there is much data that is seen by Web servers, but never written to the log file. Therefore, any measurement approach based solely on log files would occasionally miss critical information about user activity on the Web site.

The fourth major problem with the log file approach is flexibility. As sites become more sophisticated, one of the first obvious enhancements is to add dynamically generated content. Regardless of the type of content management system used, dynamic content typically results in URLs that are very difficult, if not impossible, for a human reader to decipher. Since log files are just transaction records, dump reporting systems simply pass the nonsensical URLs through to the end-user report as the page that was requested, resulting in an unintelligeble report with meaningless page names and URLs. The ideal solution would be to interpose some intelligent classification system between the raw activity data and end-user report. In practice, however, the reality of gigabytes of raw log files often leave an in-house analysis team with few human resources to add even more complexity to an already slow log-based process. The inflexibility of log files to handle the tracking of new technologies has been observed not only with dynamic content but also with personalization applications, applet-based multimedia technologies, and a host of other new capabilities which the log file approach was never designed to handle.

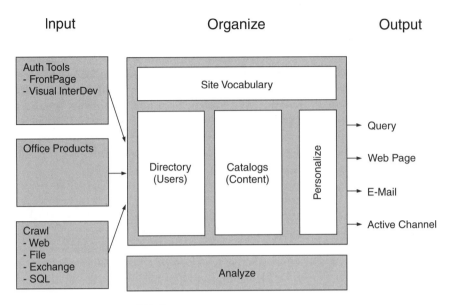

FIGURE 3.11.11 Key components of SiteServer.

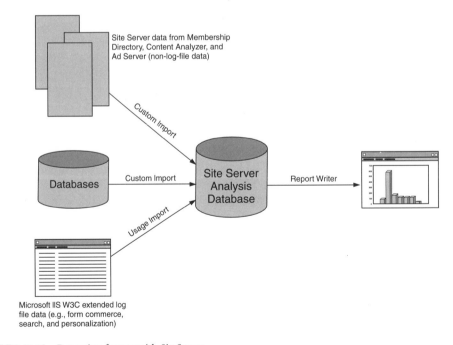

FIGURE 3.11.12 Reporting feature with SiteServer.

Figure 3.11.11 shows the principal components of SiteServer from Microsoft. Figure 3.11.12 displays the reporting process with SiteServer.

Finally, Table 3.11.3 summarizes a general Web server statistic based on log file analysis.

In summary, though log files were a convenient approach to measurement in the early days of using the Web, they rapidly highlighted problems of:

TABLE 3.11.3 General Web Server Statistics

Date and time this report was generated	Friday, December 4, 1998; 07:46:17 a.m.
Time frame	11/01/98 01:03:38–11/30/98 22:05:41
Number of hits for home page	1031
Total number of successful hits	5729
Total number of user sessions	1119
User sessions from the U.S.	0% (not broken down)
User sessions from outside the U.S.	0% (not broken down)
Origin unknown user sessions	100%
Average hits per day	190
Average user sessions per day	37
Average user session length	67:11:49

- Labor intensity
- Slow data processing speeds and turnaround times measured in weeks
- Incomplete data, missing server- and network-level data
- Ineffective tracking of new feature enhancements such as dynamic content, personalization, and applet-based multimedia

In response to these problems, hybrid products have been developed and deployed.

3.11.5.4 Log File Analysis Tools

There are numerous log file analysis tools. Their depth and functionality are very different. Some of them are complex in their nature, and offer more than just log file analysis. The most widely used tools are listed in Table 3.11.4.

TABLE 3.11.4 Log File Analysis Tools

Vendor	Product
met.Genesis	net.Analysis
WebManage	NetIntellect
WebTrend Corporation	WebTrends
Marketware	Hit List
Andromedia	ARIA
Microsoft	SiteServer

3.11.6 Wire Monitors

Log files are not the only source of information for analyzing Web sites. There are other tools that are residing "on-the-wire" or LANs and collecting information on performance and traffic metrics. The information depth and the overhead are significant indicators that may differentiate between log file analyzers and these products. In certain environments, the most effective results can be achieved only when both types of tools are deployed in combination.

Over the past several years, companies have adopted distributed multi-tier network infrastructures, and moved business operations from traditional client/server applications to distributed Web-based applications. However, as more and more users come to depend on Web servers and TCP-based services, IT organizations are discovering that their current infrastructures are unable to offer the performance and availability expected by users; nor do they provide the management and monitoring capabilities required by IT organizations themselves.

3.11.6.1 Changes in Networking Infrastructures

Over the past several years, large corporations have begun re-engineering their enterprise networks and establishing distributed, multi-tier infrastructures. These multi-tier infrastructures typically include three levels:

- At the wide area network (WAN) level to enable communication across multiple points of presence (POPs)
- At the Web level, to support server farms providing a wide range of TCP-based services, including HTTP, FTP, SMTP, and Telnet
- At the application level, to support farms of application servers that offload computation from Web servers to increase overall site performance

IT organizations are deploying newly distributed, Web-based applications to take advantage of this new enterprise infrastructure. In place of fat software clients and centralized application servers, corporations are deploying Web browsers on every desktop, Web servers in departments and divisions, and application servers residing at multiple locations.

The new Web-centric model offers several advantages over the client/server model it replaces. IT departments can deploy Web browsers quickly and affordably to every desktop platform. Basic Web skills can be learned quickly, and are popular with users. If an application need requires modification to reflect changing business practices, IT departments need only modify the application itself, not the complex clients that used to work with the application. Most importantly, distributed, Web-based infrastructures move content and applications closer to users and provide improved reliability and availability. Employees can leverage this new infrastructure to improve internal business practices, communication with partners and suppliers, and services for customers.

While distributed, multi-tier infrastuctures offer considerable advantages over earlier network architectures, they still do not offer the performance and availability expected by end users; nor do they provide the management and monitoring capabilities expected by IT organizations. Multi-tier architectures are physically well connected, but not logically well connected. Standard network equipment enables traffic to flow, but not necessarily to the server best suited to respond. IT departments deploying these networks need traffic management solutions that intelligently direct TCP traffic to optimal resources at each tier of the enterprise infrastructure. An optimal traffic management solution requires communication between tiers. For example, there is little point in a DNS server directing traffic to a local server, if that server is down or overloaded, while another server is available with processing cycles to spare. To perform its job optimally, the DNS server needs availability and load information from the servers to which it directs requests.

The multi-tier model itself, when implemented with the standard software products available today, does not monitor services for system failures or spikes. Nor does it provide other capabilities that IT departments require to manage busy, distributed networks effectively. Specifically, it provides no:

- Policies for scheduling TCP traffic based on specific events centralized
- Remote management reporting integration with standard network management tools

IT organizations need integrated software systems that can be layered on top of the existing infrastructure to provide intelligent scheduling of requests and information.

3.11.6.2 Issues of Data Collection

The targeted metrics are the same as with log file analyzers, but the source of data is different. When selecting products, there are a number of criteria, such as information depth, overhead, and reporting capabilities, that must be carefully evaluated. The market potentials are good, addressed by a few vendors. These criteria are also important when webmasters want to position traffic measurements within their IT administration or when they want to deploy this functionality within their organization.

Architecture of a product answers the question whether or not it can support a distributed architecture. Distribution may mean that collecting, processing, reporting, and distributing data can be supported in various processors and at different locations. Figure 3.11.13 shows these functions with a distributed solution.

The monitors are passively measuring the traffic in the network segments. They are actually micro-computers with ever increasing intelligence. Their operating systems are either proprietary or based on

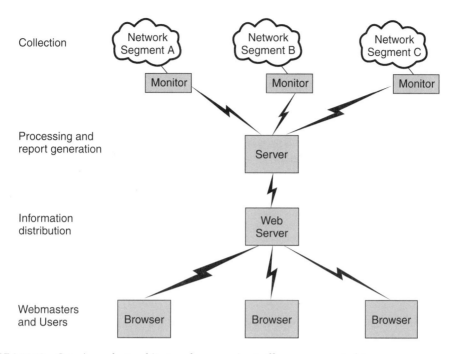

FIGURE 3.11.13 Generic product architecture for processing traffic measurements data.

Unix or more likely on NT. Usually, they are programmed to interpret many protocols. TCP/IP and UDP/IP, and HTTP are high on the priority list of vendors.

The data capturing technique is essential with traffic measurement tools. The measurement probes are attached to the digital interface of the communication channels. They can reside directly on the network (standalone probes) or co-located with networking equipment. In this case, the probe is used as a plug-in. Even software probes can be used and implemented into networking components or into end-user devices. The hardware or software probes usually include event scheduling. It means determining polling cycles and time periods when downloading of measurement data is intended. Transmission should be scheduled for low-traffic periods. Probes are expected to deal with large data volumes. These volumes depend — to a large degree — on visitor's traffic in networking segments. Probes have limited storage capabilities; implementation examples show capabilities up to 24 hours. When this limit is exceeded, measurement data are overwritten by new data. Usually, measurement data are downloaded for further processing. It is important to know how downloads are organized and how rapidly they can be executed. As indicated in Figure 3.11.13, wide area networks are involved that may show bandwidth limitations. The bandwidth is usually shared with other applications with the result of potential traffic congestions. Bandwidth-on-demand-solutions are rare with measurement probes. When transmission is arranged for low traffic periods, the actuality of measurement results may suffer. In such cases, local storage requirements increase, and processing, report generation, and information distribution are delayed by several hours or even by days.

Two solutions may help. The first is using intelligent filtering during and shortly after data collection. Redundant data are removed from captured packets during collection. Data volumes decrease, local storage requirements decrease as well, but processing requirements of the probes increase. The second solution may use data compression or data compaction with the same results and impacts as can be observed with the first solution.

Overhead is a very critical issue with large data volumes. Data capturing is expected not to introduce any overhead in case of hardware-based probes. Overhead is minimal with software-based probes. It is assumed that measurement data are stored away immediately after collection. If local processing is taking place, overhead must be critically quantified. If resource demand is high, probes must be upgraded properly. Data transmission overhead can be heavy, when everything is transmitted to the site where processing takes place. Dedicated bandwidth would be too expensive for measurement and management purposes only. If bandwidth is shared with other applications, priorities must be set higher for business applications than for transmitting measurement data.

It is absolutely necessary that all data are captured that are necessary to conduct a detailed Web site analysis of visitors or groups of visitors.

- Who is the visitor?
- What is the purpose of the visit?
- Where is the visitor coming from?
- When has the visit taken place?
- What key words have brought the visitor to the site?
- What search machines helped to access the site?
- How long was the visit?

Data losses cannot be completely avoided. Probes, monitors, networking devices, user workstations, or transmission equipment may fail; in such cases, there will be gaps in the sequence of events. Backup capabilities may be investigated, but IT budgets won't usually allow too much to spend for backing up large volumes of log file data. In the worst case, certain time windows are missing in reporting and in statistics. Those gaps may be filled with extrapolated data.

Due to considerable data volumes, databases should be under consideration to maintain raw and/or processed data. Database managers would then offer a number of built-in features to maintain data. Clustering visitors may be deployed from various perspectives, such as geography, common applications, common interests on home pages, data, and time of visits. Automatic log cycling can also be supported here by the database managers. Open database connectivity (ODBC) support helps to exchange data between different databases and to correlate data from various databases. Besides measurement data, other data sources can also be maintained in the same data warehouse. Besides routine log file analysis with concrete targeted reports, special analysis may also occasionally be conducted. This special analysis, called data mining, can discover traffic patterns and user/visitor behavior. Both are important in sizing systems and networking resources.

One of the most important issues is how measurement data analysis performs when data volumes increase. Volume increase can be caused by offering more pages on more Web servers, more visitors, longer visits, and extensive use of page links. In any case, collection and processing capabilities must be estimated and quantified prior to deciding procedures and products.

In order to reduce processing and transmission load of measurement data, redundant data should be filtered out as near as possible to the data capturing locations. Filters can help to avoid storing redundant data. Filters can also be very useful in the report generation process. Again, unnecessary data must not be processed for reports. Powerful filters help to streamline reporting.

Not everything can be automated with analyzing measurement data. The user interface is still one of the most important selection criteria for products. Graphical user interfaces are likely, but simple products are still working with textual interfaces. When measurement data are integrated with management platforms, this request is automatically met by management platforms.

Reporting is the tool to distribute the results of log file analysis. Predefined reports, report elements, and templates help to speed up the report design and generation process. Periodic reports can be automatically generated and distributed for both single Web servers and Web server farms. In the cases of many Web servers, report generation must be carefully synchronized and scheduled. Flexible formatting helps to customize reports to special user needs.

Output alternatives of reports are many. The most frequently used solutions include Word, Excel, HTML, and ASCII. Also, the distribution of reports offers multiple choices:

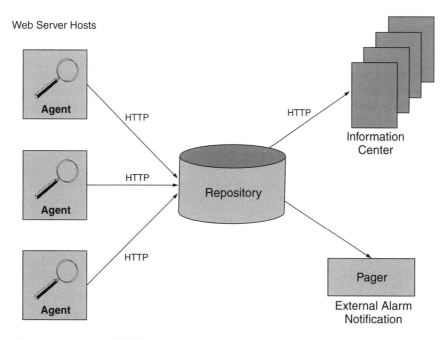

FIGURE 3.11.14   Architecture of WebSniffer.

- Reports may be stored on Web servers to be accessed by authorized users who are equipped with universal browsers
- Reports can be uploaded into special servers or even pushed to selected users
- Reports may be distributed as attachments to e-mail messages
- Reports can also be generated at remote sites; this alternative may save bandwidth when preprocessed data instead of completely formatted reports are sent to certain remote locations

Figure 3.11.14 shows the architecture of WebSniffer, one of the well-known wire monitors. Figure 3.11.15 shows distributed monitoring capabilities with Net.Medic Pro, known for for its rich reporting selection.

Documentation may have various forms. For immediate answers, an integrated on-line manual would be very helpful. Paper-based manuals are still useful for detailed answers and analysis. This role, however, will be taken over by Web-based documentation systems. In critical cases, a hot line can help with operational problems.

Measurement data analysis is actually another management application. If management platforms are used, this application can be integrated into the management platform. There are many ways to integrate; most likely a command line interface (CLI) will be deployed.

3.11.6.3   Traffic Monitoring Tools

There are just a few tools supporting traffic monitoring. Table 3.11.5 displays the list of these tools. The best results are expected in such cases, where these tools are used in combination with load balancers and traffic shapers.

3.11.7   Web Server Management

The content of Web pages is maintained on Web servers. Usually, they are processors running under Unix or NT. They must be flexible and scalable enough to cope with significant workload fluctuations.

Server management is composed of several functions:

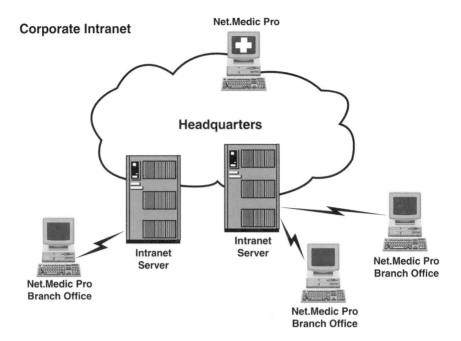

FIGURE 3.11.15   Distributed monitoring with Net.Medic Pro.

TABLE 3.11.5   Traffic Monitoring Tools

Vendor	Product
Network Associates	WebSniffer
Resonate	IntelliFlow
Sane Solutions	NetTracker
Telemate Software	Telemate.Net
Visual Networks	OnRamp
Vital Signs	Net.Medic

- Server monitoring — This function is the base component of server management. This requires someone or something to keep a constant watch on the status of the managed servers. This is a painfully tedious task for human beings, much of which, fortunately, can be automated by management platforms, such as Unicenter TNG or HP OpenView. Monitoring is essential for detecting problems as soon as they occur, and for gathering data for use in performance management.

- Workload management — This function consists of scheduling and tracking the jobs that run across one or more servers in a heterogeneous environment. Workload management takes into account calendar requirements such as time of day, day of week, or holidays. It also considers dependencies between workloads, such as Job A must be finished before Job B should be started, as well as what to do in the case of a failure.

- Server performance management — While monitoring focuses on server availability, the purpose of server performance management is to ensure that servers are working efficiently. The keys to this function are data collection and trend analysis.

- Server capacity planning — While performance management focuses on current effectiveness, capacity planning ensures that servers will work effectively in the future. The keys to this function are historical analysis and forecasting.

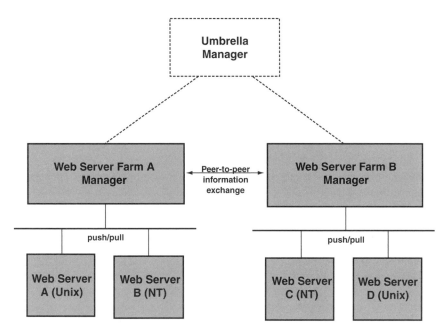

FIGURE 3.11.16 Decentralized Web server management.

The management architecture for Web servers may be central or decentral, or a combination of both. A centralized solution assumes that all Web servers can be managed from one location. When the number of Web servers to be managed exceeds a certain number, this solution could become critical in terms of networking overhead. It is assumed that with the exception of collecting raw data, all processing functions are executed in the manager.

With a decentralized solution, domain managers take over the responsibility of managing a certain number of Web servers (Figure 3.11.16). Each domain is actually a centralized solution on its own. Domain managers may communicate with each other or can even be connected to an umbrella manager. Network overhead can be well controlled and kept to a minimum. Domain managers usually just exchange consolidated data with each other. The result is that the communication overhead can be kept to a minimum.

Practical arrangements usually work with a combination of these two alternatives. If umbrella management is the choice, this manager can also manage other components such as switches, routers, and other components, and can correlate data with server management.

It is important to know when different architectural alternatives are under consideration which operating systems of Web servers can be managed. Web servers are usually deployed on Unix or NT platforms.

In terms of hardware and software of the manager, there are multiple choices. The software is Unix or NT; the hardware is constantly losing importance, because both leading operating systems are working with a number of hardware platforms.

Data capturing techniques are critical for both overhead and performance of the management architecture. Measurement probes or agents are located inside the operating system; they run with relatively high priority. These agents can supervise both hardware and software components of Web servers. Raw data are expected to be stored away immediately. Processing can be done here in the Web server or in the manager. The targeted metrics to be collected include:

- What is the CPU utilization by applications?
- What are physical and logical I/O-rates?
- Can the list of active applications be generated?
- What is the average queue length for CPU?
- What is the average queue length for I/O devices?

- How high is the CPU/I/O overlap?
- Are process wait times measured and displayed?
- How high is the disk utilization?
- How high is the memory utilization?
- Are swap rates measured?
- What are resources that processes are blocked on?
- What reporting is used?
- Can user by application be identified?

Raw data or preprocessed data are stored at the Web servers with the intention of being uploaded for further processing by the manager. Upload may be controlled in two different ways:

- Upload is triggered by events, such as filling percentage of storage spaces, or time, or when critical data are captured, or
- Upload is controlled by polling cycles initiated by the manager

Both alternatives have pros and cons; the selection depends on the actual configuration, data volumes, and the communication protocols in use. Web server management can utilize SNMP for transmitting data, assuming Web server metrics are stored and maintained in MIBs. Another alternative is the use of DMI-like standards for storing and transmissions. The recent alternative is the use of embedded Wbem-agents that are supporting the common information model (CIM) for storing and exchanging data. In this case, HTTP is the protocol of choice.

As for overhead, concerns are similar to those experienced with traffic monitors. Data capturing is expected to introduce little overhead when data are stored away immediately. If local processing is taking place, overhead must be very carefully quantified; if resource demand is high, overall Web server performance may be impacted. WAN bandwidth is still very expensive to be dedicated just to transmitting management data. If bandwidth is shared with other applications, priorities must be set higher for business applications than for transmitting raw log file data.

Here as well, data losses cannot be completely avoided. Data capturing functions in Web servers, storage devices, or components of the transmission may fail; in such cases, there will be gaps in the sequence of events. To protect as much data as possible, we re-emphasize the importance of database use to properly maintain Web server measurement data, and effective processes to filter redundant data and facilitate timely report generation/documentation.

One of the most important questions is how Web server management performs when the number of managed Web servers are maxed out, and as a result of this, data volumes increase. All resources, such as processors, storage devices, I/O-devices within the Web servers, and networking components may become the bottleneck.

Figure 3.11.17 displays information sources for NT management.

Managing Unix and NT servers represents just another management application. If management platforms or umbrella managers are used, these applications can be integrated into the platform. There are many ways to integrate; most likely a Command Line Interface (CLI) solution will be deployed. Integration may even be supported by a management intranet. Every participant is equipped with a universal browser and communicates with management applications residing in managed objects and being equipped with lean Web servers.

The majority of Web-server-implementation is based on Unix or NT. Some of them are on NetWare, but their market share is not significant.

Table 3.11.6 displays examples for Web server management tools. The tools are grouped by operating systems they support, such as Unix and NT.

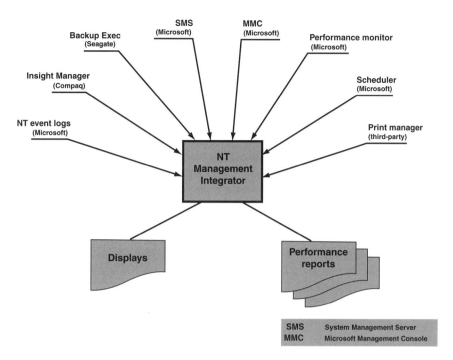

FIGURE 3.11.17 Information sources for NT management.

TABLE 3.11.6 Web Server Management Tools

Vendor	Product
	Unix Management
Computer Associates	Unicenter TNG
BMC	Patrol Knowledge Module for Unix
Hewlett-Packard	PerfView
Hewlett-Packard	GlancePlus
Hewlett-Packard	PerfView RX
	NT Management
BMC	Patrol Knowledge Module for NT
Computer Associates	Unicenter TNG
Heroix	Robomon
Hewlett-Packard	ManageX
NetIQ	AppManager
Seagate	ManageExec

3.11.8 Load Balancing

In order to help IT managers track IP performance and to optimize bandwidth usage across WANs, several new vendors offer hardware- and software-based load balancing products. Load balancers typically reside at the edges of corporate networks and decide about traffic priorities. They apply a policy that defines different traffic types and determine what happens to each. A very simple policy may call for priorities for a specific sender. Other criteria may be TCP port numbers, URLs, and also domain name service (DNS). Traffic shaping may be supported by queueing or via TCP rate control. There are products available for both categories.

Optimization is accomplished by controlling enterprise traffic flows at the boundary between the LAN and the WAN. Because these products give priority to traffic according to application type or even individual users, they will let IT managers take the first steps toward policy-based QoS in their networks. These products are a logical evolution from the passive probes that gave users a certain level of visibility for fault operations monitoring but no actual control over traffic. These products go further, and can manipulate traffic. IT managers expect that this new class of traffic-shaping tools will ease the congestion for bandwidth without forcing purchase of more and larger physical transmission lines.

This segment of the book introduces a couple of innovative solutions provided by start-ups and known flow-control companies.

3.11.8.1 The Needs for Bandwidth, Service Quality, and Granularity

Bandwidth management is rapidly becoming a must for internet service providers (ISPs) as well as corporations running their own global intranets. The reasons for bandwidth management are the following.

The Move to Internet/Intranet-based Business
Corporate networks are rapidly evolving from a classic client/server paradigm toward an intranet-based model, based on information-sharing and Web navigation. Analysts predicted that by the year 2000 there would be over 3 million private intranet sites, compared to approximately 650,000 Internet sites. The result is the demand for significantly more bandwidth. Adding more channels and more bandwidth to each channel will not guarantee availability and performance, where it is needed most. An intranet-based model implies the following factors:

- Changing patterns or network use and unpredictable demands for bandwidth. Global users access the network 24 hours a day, 7 days a week. As information appears and disappears on Web sites, access patterns change and saturation moves around the network.
- Demand for increased amounts of bandwidth. People may stay on the link for extended periods of time and download large amounts of data.
- Demand for guaranteed QoS in terms of bandwidth and minimum delay. Emerging Internet applications are both bandwidth intensive and time sensitive, often requiring support for voice, video, and multimedia applications across the network infrastructure.
- Lack of control by IT staff. Workgroups and departments generally create their Web sites without IT approvals, generating increased traffic without necessarily having the infrastructure to handle it. This often results in excessive traffic at the fringes of the network where Web sites are situated, generating traffic precisely where there is least provision.
- A change in user attitude. Users expect instant access to information without delays or restrictions, especially if that information is critical to their work.

The Need for Guaranteed Bandwidth
Current networking technology has two major limitations:

- The bandwidth available on a link at any given moment cannot be predicted in terms of quantity or quality. Bandwidth management is needed to allow applications which require a specific quality of service, in terms of bandwidth and delay (such as desktop video conferencing), to reserve the bandwidth quality of service they need.
- It is difficult to control which applications or users get a share of the available bandwidth. In some circumstances, an application or a user can take control of all the available bandwidth, preventing other applications or users from accessing the network. To solve this problem, the user can either add extra capacity at additional costs, resulting in an overprovisioned network that still does not guarantee equal access, or the user can introduce bandwidth allocation.

The Need for Service Level Agreements
Virtual private networks (VPN) are a popular value-added Internet service that corporations are increasingly moving toward. Enterprise customers seeking a VPN provider are more likely to sign with an ISP that can offer a contractual service level agreement — one that guarantees quality of service.

While service level agreements (SLA) cannot guarantee end-to-end service across the public Internet, they can be implemented for transport over a single-vendor network or for Internet server hosting. In these areas, a SLA is an important differentiator for an ISP.

Generally, the customer subscribes to a particular class of service, and signs a SLA accordingly. Packet throughput is monitored as part of the agreement. Value-added services were expected to grow at almost 175% in the U.S. up to the year 2000. ISPs that want to get a piece of this additional business clearly need to implement bandwidth management in order to meet SLAs which guarantee QoS toward customers. Only efficient bandwidth management can enable them to tune network behavior so that customers receive the quality of service they are charged for.

The new paradigm is a service-driven network. This is a responsive, reliable, modular infrastructure, based on the latest generation of management technology and built on dynamic, flexible management services. To respond to today's business needs, ISPs and large enterprises must deploy the service-driven network. It delivers innovative services, such as unified roaming, push browsers, multicast, on-line shopping, etc. to customers faster and at a lower cost than ever before.

The Need for Granularity

Bandwidth allocation based simply on filtering by protocol is not sufficient to meet bandwidth management needs. One of the key issues in this area is the extensive and increasing use of HTML/HTTP systems for OLTP. Within the next few years, the volume of HTTP-based OLTP traffic is expected to exceed the volume of traditional OLTP traffic. A fine level of granularity is needed for bandwidth management to take into account more than just the protocol when assesing the relative importance of network traffic. Bandwidth management must base allocation not only on protocol type, but also on the application and users involved.

3.11.8.2 Issues of Deploying Load Balancing Products

Load balancing helps to utilize resources more effectively. At the same time, the end-user response time may be stabilized and improved. This is an emerging area with a number of innovative products that work hardware- or software-based. Even there, a few implement load balancing functions in both hardware and software. The hardware solution is faster; the software offers more flexibilities if changes are required.

The functionality of a load balancer can be deployed in a standalone device or embedded into existing networking components, such as routers, switches, and firewalls. The standalone solution offers broad functionality without impacting any other routing, switching, or firewall functions. But it will add components into the network that must be managed. It may add another vendor that may be managed as well. The embedded solution is just the opposite; easier management at a price of conflicting functions with its host.

Load balancers are only successful when policy profiles can be implemented and used. Policy profiles are most likely based on supporting various transmission priorities. Priorities may be set by applications, by users, or a combination of both. The technology of solution may differ from case to case, and product to product, but most frquently the TCP flow is intercepted.

Load balancers are expected to support a number of services, such as quality control, resource management, flow control, link management, and actual load balancing. Advanced products support all these services in dependency of page content. It requires more work to gather the necessary information about content, but it offers better services for high-priority content.

Functions in a narrower sense include traffic shaping, load balancing, monitoring, and baselining. Baselining means to find the optimal operational conditions for a certain environment. It may be expressed by a few parameters, such as resource utilization, availability, and response time. Load balancers should monitor these few metrics, and act on them. Traffic shaping and load balancing help restore normal conditions by splitting traffic, redirecting traffic to replicated servers, delaying payload transport, etc.

One of the most important questions is how load balancing performs when data volumes increase. Volume increase can be caused by offering more pages on more Web servers, more visitors, longer visits,

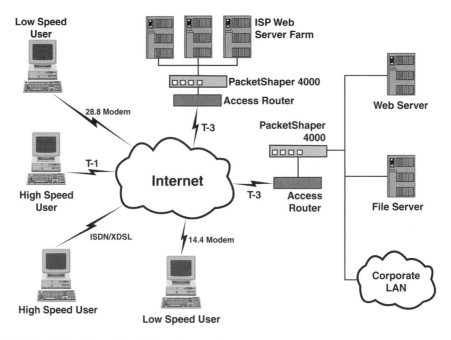

FIGURE 3.11.18 PacketShaper from Packater in operation.

extensive use of page links, etc. In any case, collection and processing capabilities must be estimated prior to deciding for precedures and products.

Load balancing products can be managed by SNMP- or Wbem-agents. They are handled by managers as with any other kind of managed object. As before, various approaches may be taken for documentation and assistance to generate documentation.

Figure 3.11.18 shows PacketShaper in operation.

Managing load balancers out of a management platform offers integration at the management applications level. Baselining and monitoring may even be supported by other applications. In case of using management intranets, universal browsers may be used to view, extract, process, and distribute management information. The only prerequisite is that Wbem agents have been implemented and that CIM is supported for information exchange.

3.11.8.3 Load Balancing Tools

Table 3.11.7 lists the presently available tools supporting load balancing and traffic shaping.

TABLE 3.11.7 Load Balancing Tools

Vendor	Product
Allot Communications	AC200 and AC300
CheckPoint Software Technology	Floodgate-1
Internet Devices	Fort Knox Policy Router
NetGuard	Guidepost
NetReality	WiseMan
Netscreen Technologies	Netscreen 10 and Netscreen 100
Packeter	PacketShaper
RND	Web Server Director
Structured Internetworks	IPath 10 and IPath 100
Sun Microsystems	Bandwidth Allocator
Ukiah Software	Trafficware
Xedia	Access Point

3.11.9   Look-through Measurements

Web application requirements have gone from zero to mission-critical within a very short period of time. The available tools have not kept up with this speed. In a business environment where "Connections failed" means the same thing as "Closed for business," IS/IT professionals are left to struggle with the challenges of building a highly avalable, high-performance server infrastructure.

Many problems interact with each other:

- The majority of Web sites, both Internet and intranet, use single Unix or NT servers. Like mainframe solutions of the past, these centralized servers have become single points of failure. Even minor system upgrades become major service problems for demanding users.
- As the demands of interactivity grow, the cost of WAN bandwidth becomes a major factor. System configurations that force all user access out across the WAN for each request stretch out retrieval times, and raise users' frustration levels.
- The increasing complexity of Web applications add even more overhead; electronic commerce and multi-tier content architectures that build pages on the fly out of applications and databases make high reliability an even more important — and costlier — goal.

The severe problem in addition to all of these is that the Web technology base is narrow. In other words, solutions that can be applied to these problems are expensive and not very effective. Adding WAN bandwidth and a larger server are just the first steps in a never-ending circle. Adding mirrored, distributed servers increases server costs significantly as well as the complexities and costs of content distribution. Hiring more webmasters and Web administrators to reboot downed web applications and servers is not the ultimate solution. And, in a world of increasingly dynamic content and transactions, how effective will server caches and load balancing tools really be?

3.11.9.1   Response Time Measurements

Response time is one of the key metrics in all SLAs. Its definition varies, but most users consider the duration between sending the inquiry until receiving the full answer as response time. There are two alternatives:

- Time up to the first character of the response on the screen of the user
- Time up to the last character of the response on the screen of the user

The second definition is better suited for the working cycle of users. The difference between RT2 and RT1 depends on many factors, such as the throughput of the backbone and access networks, servers in these networks, number of hops, and the hardware/software capabilities of the client's workstation or browser. Present measurement technology offers the following alternatives:

- Monitors and packet analyzers: They filter and interpret packets and draw inferences about application response times based on these results. These monitors are passively listening to the network traffic and calculate the time it takes specific packets to get from source to destination. They can read the content of packages, revealing eventual application errors and inefficiency. But they cannot measure response time end to end.
- Synthetic workload tools: They issue live traffic to get a consistent measurement of response time on a particular connection in the intranet or for a given application. These tools are installed on servers, desktops, or both. They typically send TCP messages or SQL queries to servers and measure the time of the reply. Results from multiple sources are correlated to give a more detailed view about intranet response times. They are very accurate to the end-to-end response time.
- Application agents: They work within or alongside applications, using software that monitors keystrokes and commands to track down how long a specific transaction takes. They can run at both the client and server. They clock specific portions of the application at the server or at the

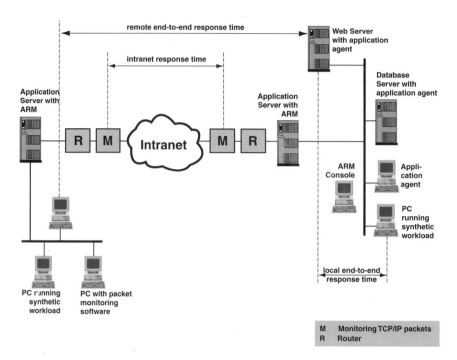

FIGURE 3.11.19   Positioning response time measurement tools.

workstation. The use of agents needs customization and the correlation of many measurements in order to give users a performance estimate about their intranet.

- Use of ARM MIBs: ARM defines APIs that allow programmers to write agents into an application so that network managers and webmasters can monitor it for a range of performance metrics, including response time. It is a complete offer to application management. But it requires rewriting of existing code that many companies are unwilling to do.

Figure 3.11.19 shows the locations of these tools and agents.

When evaluating products, many components must be factored in. These factors are:

- Customization needs
- Maintenance requirements
- Deployment of code
- Overhead of transmitting measurement data
- Load increase due to synthetic workload
- Reporting capabilities
- Capabilities to solve complex performance problems
- Capabilities to conduct root-cause analysis
- Combination with modeling tools
- Price of the tools

3.11.9.2 Highlighting Bottlenecks

End-to-end service level monitoring is getting extremely popular with Web-based applications. Monitoring is targeting availability and response time measurements. Element-centric management platforms "look down" and manage elements. Response time monitoring tools "look through" the infrastructure from one end to the other.

Applications-related measurements can also be done with RMON probes. The way to do this, according to NetScout Systems Inc., is to track an application on its entire path across enterprise. To support that approach, the remote monitoring vendor is able to collect and report traffic statistics IT managers use to measure how quickly an application makes its round-trip run.

NetScout is leading off its application flow management strategy with a new multiport Fast Ethernet probe, a RMON2 agent for NT servers and Web-based reporting software. Applications can be observed and measured as they run using AppScout, a browser-based solution. AppScout monitors SAP R/3, Microsoft Exchange, Lotus Notes, and TCP/IP applications.

Typical look-through products work on the principle of Java applets in combination with C++ scripts. The code is distributed to various selected end points on the network. These agents generate synthetic transactions against targeted applications, such as databases or intranet Web pages. Response time for these scripted transactions — including the response times over each individual "hop" along the route — are logged on a management server, which assembles and organizes the collected data. The data is then available to users through a client-side Java interface.

The new type of network instrumentation closely mimics the end users' actual experience since it measures the end-to-end responsiveness of an application from one or more outlying LAN nodes to the application server and back again. By doing so, it delivers a metric that accurately reflects application and service performance levels on the network. Trying to gauge the end-to-end performance level of an application over the network by monitoring each distinct element along the service delivery path has not proven successful. Element-specific monitoring is still essential for troubleshooting and maintenance, but network managers have to start looking at some new kinds of instrumentation if they want to view the environment from the end-user's point of view.

3.11.9.3 Tools for Look-through Measurements

Table 3.11.8 lists all the tools that may be considered for end-to-end response time measurements.

TABLE 3.11.8   Tools for Look-through Measurements

Vendor	Product
Avesta	Trinity Measurement Software
Freshwater Software	SiteScope
International Network Services	Enterprise Pro
Jyra Research	Service Management Architecture
NextPoint Networks	NextPoint S3
NetScout Systems	Application Management
Proactive Networks	ProntoWatch
Response Networks	VeriServ

3.11.10 Trends of Intranet Performance Management

Intranet management is an emerging area to webmasters and Web administrators. It combines existing processes for fault, performance, configuration, security, and accounting management with new management tools. Performance and security management are the two most challanging areas. Usage pattens, traffic peaks, unbalanced input/output streams from/to Web servers, server overload, and unstable performance mean challenges to webmasters and network capacity planners. Partitioning networking segments properly, selecting and implementing firewalls, stress testing firewalls and the use of the right authentication techniques mean challenges to security officers of all corporations operating intranets.

New intranet-related management tools are content authoring and auditing instruments, log file analyzers, traffic monitors, load balancers, and application monitors. They can be used individually, or in combination with each other. It is expected that they will soon be integrated into systems and network management platforms.

References

ALDR99   Aldrich, S.: Freshwater's Web Application Management, *Patricia Seybold Group e-Bulletin,* January 21, 1999.

BOBR98   Bobrock, C.: Web developers follow old scripts, *Interactive Week,* November 2, 1998, p. 29.

BOCK98   Bock, G.E.: Microsoft Site Server — Organizing and Sharing the Contents of a Corporate Intranet, *Workgroup Computing Report, Patricia Seybold,* August 1998.

BRUN99   Bruno, L.: *IP Balancing Act: Sharing the Load Across Servers, Data Communications,* February 1999, p. 29.

GIBB98   Gibbs, M.: Pinning down network problems, *Network World,* March 2, 1998, p. 43.

HERM98   Herman, J., Forbath, T.: Using Internet Technology to Integrate Management Tools and Information, http://www.cisco,com/warp/public/734/partner/cmc/bmi_wi.htm.

HUNT96   Huntington-Lee, J., Terplan, K., Gibson, J.: *HP OpenView — A Manager's Guide,* McGraw-Hill, New York, 1996.

JAND98   Jander, M.: Clock watchers, *Data Communications,* September 1998, p. 75–80.

JAND99   Jander, M.: Network Management, *Data Communications,* January 1999, p. 75.

KAPO98   Kapoor, A., Ryan, J.: Reassessing networks for an IP architecture, *Telecommunications,* October 1998, p. 48.

LARS97   Larsen, A.K.: All Eyes on IP Traffic, *Data Communications,* March 1997.

LEIN93   Leinwand, A., Fang, K.: *Network Management — A Practical Perspective,* Addison-Wesley Publishing Company, New York, 1993.

POWE97B   Powell, T.: An XML Primer, *InternetWeek,* p. 47–49, November 24, 1997.

REAR98   Reardon, M.: Traffic Shapers: IP in Cruise Control, *Data Communications,* September, 1998, p. 67.

RUBI98   Rubinson, T., Terplan, K.: Network Design — Management and Technical Perspectives, CRC Press, Boca Raton, 1998.

SANT98   Santalesa, R.: Weaving The Web Fantastic — Authoring Tools, *InternetWeek,* November 17, 1997.

SCHU97   Schultz, K.: Two Tools for Monitoring Your Web Site, *InternetWeek,* October 27, 1997, p. 60–61.

STUR98   Sturm, R.: *Working with Unicenter TNG,* QUE Publishing, Indianapolis, 1998.

TAYL96   Taylor, K.: Internet Access: Getting the Whole Picture, *Data Communications,* March 1996, p. 50–52.

TERP96   Terplan, K.: *Effective Management of Local Area Networks,* Second Edition, McGraw-Hill, New York, 1996.

TERP98a   Terplan, K.: Web-based systems and network management, Xephon Briefing, London, October 14, 1998.

TERP98b   Terplan, K.: *Telecom Operations Management Solutions with NetExpert,* CRC Press, Boca Raton, 1998.

TERP99   Terplan, K.: *Web-based Systems and Network Management,* CRC Press, Boca Raton, 1999.

4

Future Telecommunications: Trends and Directions

James Anderson
Alcatel

4.1   Introduction

Imagine for a moment how daily life would be affected if the telecommunications services and applications that we take for granted were to be removed. The daily paper would contain mainly local news and any international stories would be describing events that were weeks or months old. We would spend much of our time during the week traveling from house to house and town to town as we tried to keep in touch with our friends and business associates. We would tend to live close to where we were born and raised otherwise we would risk losing contact with friends and family. Finally, the number of envelopes, paper, and stamps sold would be constantly increasing as people wrote letters in order to have their presence felt in far-off locations without having to travel. The contrast between our everyday life and this example clearly shows just how significant the impact of today's telecommunications services has been on how we communicate. As hard as it is to imagine a day without the communications systems and services that have become such an integral part of our lives, so too will it be impossible for future generations to imagine living in our times with our "primitive" telecommunications infrastructures and applications!

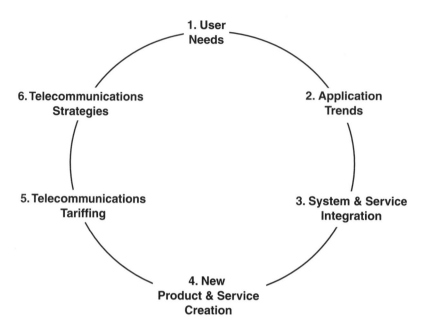

FIGURE 4.1   Telecommunications trend lifecycle model.

In this chapter we will be looking at where the field of telecommunications is evolving to. This type of prediction is not without a great deal of risk: a similar analysis done as recently as 1990 could not have hoped to accurately identify the impact that the Internet now has on the way we communicate today! However, the basic building blocks that will control the evolution of the field of telecommunications, the telecommunications DNA if you will, are reflected in the state-of-the-art services, applications, and equipment available today. We will look at the current trends along with the end-user requirements and competitive market forces that will shape the future of telecommunications.

To help focus the consideration of such a large topic as the future of telecommunications, it is helpful to have a model to frame the discussion. The model that we will use in this chapter to identify future trends in telecommunications is shown in Figure 4.1.

This telecommunications trend lifecycle model that we will be using is intended to provide a high-level view of how the effects of changes "ripple" throughout the telecommunications field. We will be discussing the model in a sequential manner, starting with an analysis of the changing needs of end users. It is important to keep in mind that innovation and change in real life is often chaotic and seems to resist following orderly models. Therefore, as long as we understand that a new telecommunications trend can potentially start at any step of the trend lifecycle model (i.e., a new equipment technology is invented in a research lab and only later is it understood well enough to be used to address end-user needs), then we will be able to correlate this chapter's analysis and the real world.

4.2   User Needs

The modern world is currently undergoing its third major communications transformation. It took 38 years for radio to garner 50 million listeners; likewise, it took 13 years for television to achieve a similar number of viewers. Incredibly, the worldwide computer communications network known as the Internet has required only 4 years to reach that milestone. In the U.S., as of this writing, there are more than 62 million Internet users and another 7 million are estimated to be joining them soon. These users will be joining a worldwide community of over 100 million Internet users. As is to be expected, when more people make use of the Internet, more information needs to be processed by the networks and computers

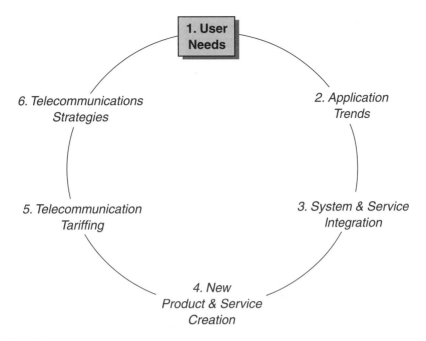

FIGURE 4.2 Trend analysis — user needs.

that make up the Internet. The U.S. Commerce Department (as of April 15, 1998) estimates that the amount of information processed over the Internet is doubling every 100 days. The needs of these users and others like them will form the drivers of telecommunication trends in the future.

In this section we will examine the user needs that will form the basis — and demands of — tomorrow's telecommunications systems and applications (Figure 4.2). We will start by determining exactly what types of users' needs we have to understand. Next, we'll explore the specific problems and challenges that each group of users is currently trying to solve. Finally, we'll identify several general trends in user needs that will have the greatest impact on future telecommunications services.

4.2.1 Types of Users

It can be argued that almost everyone in industrialized countries could be considered to be an end user of telecommunications services and applications. A recent study by the International Telecommunications Union (ITU) standards body reported that in high-income countries (per capita GDP of more than U.S. $8955) there exists a "teledensity" of more than 50 phone lines for every 100 people. This would lead one to conclude that in these countries, telecommunications services and technologies will evolve to meet the needs of the general public. However, in order to identify specific future trends in telecommunications, we need to limit our focus to only those users who either have the financial resources or sheer numbers to generate and sustain a trend in telecommunications. We will also avoid focusing on narrow vertical application segments such as healthcare and banking in order to identify trends because their influences on future applications and services can be safely generalized into broader end-user groups without losing their contribution. In this chapter, we will segment end users into four primary groups for further study. These groups can be characterized in the following ways:

- **Businesses:** This segment of telecommunications end users is defined to be a group working toward a common goal at one or more locations. As a rule, businesses need to interconnect each of their workers on a frequent basis. Depending on the size and type of business, this interconnection requirement can result in the need for large amounts of bandwidth. The business segment is also characterized by its growing need for 7 days per week × 24 hours per day × 365 days per year connectivity in order to support globally distributed operations. Businesses are fairly price resistant — they are willing to pay more for access to applications that they feel will provide enough of a competitive advantage to recover their costs.

- **Mobile Professionals:** These end users generally interact with business segment end users. The difference between these segments is that mobile professionals generally operate either by themselves or as part of small focused teams. Mobile professionals don't have a fixed location connected to telecommunications services; rather, they need to have services find them or permit them to access the services from a wide variety of remote locations. Once again, the mobile professional segment is fairly price insensitive to the price of telecommunications services that have a direct correlation to a competitive advantage.

- **SOHO:** The small office/home office (SOHO) segment is a rapidly growing portion of the market, as larger businesses discover it is more economical to outsource many of the tasks they used to perform internally. Tax incentives from many local and federal governments designed to decrease commuting congestion and pollution have also added to the economic incentive for this segment to experience explosive growth. Telecommunications applications have been crucial to fueling the growth of this segment. Existing applications have permitted home office workers to have access to similar communications resources that centralized workers also enjoy. The SOHO segment is price sensitive; however, their large numbers can often be used to create attractive business cases for both the end users and the service providers.

- **Residential:** This segment of end users wants to have telecommunications services delivered to their homes. The telecommunications applications desired by this segment often are used to communicate with other residential end users, businesses, or for entertainment. This segment is very price sensitive; in order to pay for a telecommunications application or service, something else will have to be given up. Each application is subjected to a tradeoff evaluation by the end user.

4.2.2   Different Users Have Different Needs

Each of the different user groups we have identified is facing a different set of challenges that can be addressed in a variety of ways by telecommunications services. In this section we will explore the environmental and social drivers that have created these end-user needs. In the final section of this chapter we will identify the common drivers that apply to each segment of end users. As you read this section, it is important to keep in mind that although the specific details of how end user problems will be addressed may change over time, the core set of conditions that have created the needs will not change.

4.2.2.1   Business End-User Needs

Businesses exist to earn a profit and they do this by offering some combination of better products, lower prices, or by meeting the specific needs of a particular customer better than any other firm. For the purposes of this discussion, we group together businesses of all sizes from the very small to the very large. Although the specifics of the problems they are trying to solve may differ, all businesses face the same basic set of challenges.

The communications needs of business end users can be divided into two basic groups: internal needs and external needs. A business's internal needs relate to how it communicates the way that it wants to do business to its employees and how those employees communicate status and learned information throughout the firm. The external communication needs of a firm relate to how it exchanges information with members of its business environment. These members include other businesses (trading partners) and customers alike. We will now examine the drivers in each of these different groups of needs in detail.

In the last decade, firms have come to realize that one of their primary sources of competitive advantage can come from how well they exchange information internally. Having used the recent explosion of networking and computer storage technology to collect, store, and distribute large amounts of information, firms are now looking to refine their operations. What businesses have realized is that they have a major challenge of providing everyone in their organization with access to the specific types of information that they require in order to perform their jobs better. A key challenge is that each employee in a firm performs a different task (or performs the same task in a different business context) and therefore needs to have access to different types of information at different times. How to provide such connectivity presents a significant challenge to businesses of all sizes.

One of a business's most valuable resources is its internal knowledge of how problems were identified and solved in the past. A key communications objective for a firm is to find a way to share problem-solving experiences throughout the firm. Meeting this challenge is critical for the firm, otherwise it will face the expense of solving the same problem for the first time over and over again. The solution involves communication solutions that not only provide access to detailed records of past projects, but also include identification and access to the employees who were involved in solving the problems. Only by finding a way to meet this challenge can firms refine their problem-solving processes and become more competitive.

The cost of producing products or services has received a great deal of attention in recent years. Businesses have implemented a wide range of control and monitoring systems that are able to evaluate the operations of different internal processes. Such systems include enterprise resource planning systems that can control the supply chain of a product's production process, quality improvement tracking systems, and just-in-time manufacturing systems. One of the primary purposes of each system is to permit a firm to more effectively use its resources and raw materials — in other words, they help a firm run a "lean operation" in which all of its assets are fully utilized. Such tightly run operations require a business to establish and maintain a wide variety of communications between its internal divisions no matter where they may be located. Additionally, there is a direct correlation between how fully the firm's assets are utilized and how rapid communications between the different parts of the firm are executed. These processes and systems force a business to walk a tightrope between operating at peak efficiency and not having to correct materials to operate at all. Firms must identify what communication is required to support such mission-critical systems and then implement and use their telecommunications solutions to gain a competitive advantage.

Finally, businesses are often thought of as a collection of employees who come together at company-owned locations to perform work. Businesses are now starting to realize that the arrival of relatively inexpensive computing resources, coupled with the availability of numerous communications services, call for rethinking about how they conduct their daily operations. Firms have already realized that many of the noncritical or nonstrategic processes they perform can be effectively outsourced to other firms that are able to perform these processes more efficiently and at a lower cost. Firms are now starting to reexamine how and where their remaining employees work and interact. The popularity of telecommuting and rotating "work from home" days shows how firms are starting to explore these uncharted waters. One of the primary keys to making a widely distributed workforce successful is to identify communications solutions that permit the firm's employees to interact as though they were together in an office, without the actual expense of the office.

Advances in transportation and communication have permitted businesses of all sizes to compete on a global scale. New businesses are able to offer their products to almost any international market, starting on their first day of operation. Existing businesses that have saturated their traditional domestic markets are able to seek new revenue streams in unexplored global markets. One side effect of operating and competing on a global scale is that all of the telecommunication systems that a business established to facilitate internal communications for its domestic operations must now be extended to become both location and distance insensitive. This requirement affects all forms of communication including voice, video, and data. As a clear confirmation of this growing need, the Federal Communications Commission's (FCC's) statistics show that since 1987 the growth of the U.S. long distance market has been propelled by a 14.5% compound annual growth rate (CAGR) in international long-distance revenues. Traditionally, such services have been very distance sensitive, thus making telecommunications expenses a significant expenditure for a globally distributed business. As the number of firms that operate internationally has increased, so too has the number of telecommunications service providers. This increase in service providers has provided businesses with an opportunity to seek out and use those providers who are able to help them minimize their telecommunications costs. Once again, the FCC's statistics show that the composite cost of an international phone call has dropped from U.S. $1.00 in 1992 down to U.S. $0.68 in 1997.

As businesses study how they can maximize their profits, they have realized they can reduce their costs by streamlining interactions with their suppliers. This new understanding has led to the sharing of information, such as current sales results and stocking data between retailers and their many suppliers. The high volume and near real-time characteristics of this information have created a growing need for more sophisticated telecommunications services. Once again, since retailers and suppliers may be located in different areas, the telecommunications systems must be distance insensitive.

Finally, the most important interaction that a business has is with its customers. Customers are demanding that it become easier and quicker to interact with a firm. They want to see updated product lists and information; in some cases they want to be able to custom-design their own solution from a firm's product lines; and they want to be able to review and perhaps pay their bills electronically. This increased level of interaction with customers who are not physically located in a firm's place of business demands an entirely new set of sophisticated telecommunications services.

4.2.2.2   Mobile Professional End-User Needs

As business become more decentralized and at the same time more customer-focused, the ranks of the mobile professionals are swelling. This new breed of employee can no longer be thought of as being only a salesperson; rather, the mobile employee may be part of any one of a number of project teams that have been brought together to solve a specific problem. As more and more employees start to operate away from the firm's offices for longer periods of time, the ability to use communications systems and services to provide information, obtain status updates, and share learned knowledge becomes even more important. Let's take a look at some of the specific needs of this group of end users.

Arguably, the most critical need of a mobile professional end user is his need for up-to-date information. Since a mobile user is operating away from a centralized office environment, his ability to learn about changes in products or company strategy is limited to what information is sent to him — the critical real-world "water cooler" information exchange system is no longer available to him. New means of identifying important information need to be created along with an effective two-way system for distributing that information and getting end user responses and feedback.

Since the mobile end user is often away from the office and in fact may be spending much of the time with a customer, it is impractical to carry all of the product and service reference material that may be required to perform the tasks. Therefore, it's important that the mobile end user be able to quickly access all of the material that may be required to support the current task. Note that the information required may take many forms including text, pictures, animation, and video. Many firms that sell large, complex software systems have changed the way that they now perform product demonstrations. Instead of taking complex computer systems to the customer's site, they use a standard laptop and establish a communications link back to their office, where the application is running on the more complex hardware system. This is one way for the firm to better utilize its expensive resources and better support its mobile users at a lower cost. Such services are only the start of what will be required to support the growing mobile user community.

The type of data that can be accessed by mobile users is another critical issue. Current analog modem links over voice-grade phone lines limit mobile users to a bandwidth between 28.8k bps and 56k bps, which is acceptable for accessing small- to medium-sized text documents. As more and more information is stored in richer formats such as video and integrated multimedia documents, new telecommunications services will have to be created to support mobile users. The need for access to multimedia information is especially critical for mobile users whose firms design, manufacture, or sell complex products. The multimedia information for these products can help the mobile user to shorten the selling cycle by permitting such complex products to be clearly and simply communicated. New telecommunications solutions are required to ensure that mobile users are able to access *all* of the information they require in order to perform their jobs.

A unique requirement of mobile end users is that, unlike stationary users, information must "find" its way to the mobile user. The mobile user is expected to change locations quite often and can't be expected to be reached via an addressing scheme that requires the user to always be at a given geographic location. This applies not only to voice communication but also to all forms of electronic information interchange. This issue has been partially addressed by some of today's current telecommunications solutions; however, such solutions generally work only within a limited geographical area (country or artificially determined service provider territory) and completely different solutions have been designed for voice and data services. Mobile users require solutions that provide seamless integrated voice and data solutions of ubiquitous coverage.

Although many of the needs of a mobile end user relate to ensuring reachability at all times, the opposite is also a concern. One of a mobile user's more valuable resources is time. Giving others the ability to communicate with the mobile user also gives them the ability to appropriate time. The mobile user needs to be able to limit who has what level of access. Additionally, the mobile user needs to be able to decide if and how to respond to each request for valuable time.

A mobile user's toolkit consists of several groups of information to help do the job on a daily basis. These groups of information consist of a variety of phone lists, customer names and addresses, customer lead lists, internal corporate directories, etc. As this collection of data grows in size, so too does it grow in value to both the mobile user and the company as a whole. The telecommunications challenge is how this information can be shared among the wide variety of communication devices used by the mobile user without having to retype the information each time.

The demanding lifestyle that being a mobile user requires often results in the lines between a worker's personal and professional lives being blurred. Since the mobile user may be away from home for long periods of time, it is critical that personal messages from various sources and in varying formats must be able to find their way to where the mobile professional is. Additionally, personal communications must be clearly identified as such and must be easily differentiated from work-related communications. Both mobile workers and the firms that employ them appear to be drivers for this type of requirement — both parties realize that good communications can help a mobile worker strike the correct balance between different roles and responsibilities.

Change and movement are key components of a mobile user's typical day. Because of this, there is no single best way for messages and information to reach the mobile user. Therefore, the mobile user needs to be able to access a message in any one of several different ways: e-mail via the phone, and voicemail via the laptop. It is critical that the information is able to reach the mobile end user as quickly as possible without restricting how the user chooses to retrieve the information.

The era in which groups of the same people worked together for years or even entire careers is quickly coming to a close. Mobile users are at the forefront of this change and represent the new breed of worker: they are part of dynamic teams quickly created to solve specific problems. Once the problem has been solved and a solution implemented, the team is then dissolved and its members go on to join other dynamic teams. From a communications perspective, the mobile user needs to be able to easily exchange and work on the same information with other members of the dynamic team during the time that the team exists. The security associated with such communication is a critical factor. In today's customer-focused markets, employees of the customer may be part of the same dynamic team as the mobile user. In such cases, the ability to filter and restrict a team member's access to sensitive data is required in order to ensure that the internal and external team members are able to work together smoothly.

4.2.2.3 SOHO End User Needs

In contrast to large established firms, employees of small firms have different communications needs. We include in this group those workers, who may work for firms of almost any size, operating out of their homes. Corporate outsourcing and the increasing number of new businesses have caused this small office/home office (SOHO) group of end users to increase in size on a yearly basis. As the telecommunications service marketplace becomes more and more competitive, the SOHO segment of end-users has started to receive the attention of telecommunications service providers. The key to a provider being able to successfully serve this market will be an ability to correctly identify the needs that will motivate the SOHO end-users to purchase telecommunications services.

Unlike either the business or the mobile user, the SOHO end-user is extremely price conscious. Smaller organizations naturally tend to have smaller budgets and therefore will have less to spend on telecommunications services of any kind. However, SOHO end users are generally involved in very competitive market niches and so they feel that it's necessary to their continued survival that they arm themselves with any tools that provide a competitive advantage. The end result of these two conflicting conditions is that the SOHO end user will purchase or subscribe only to those telecommunications services that are priced within budget and which can be clearly demonstrated to give a competitive advantage.

SOHO end users do share some of the same basic needs that mobile end users have. Specifically, those SOHO end users who operate out of their homes will have the need to be able to separate personal messages from business messages. This issue is a little more complex than it was for mobile end users because all of the messages are delivered to a single location — the user's home. An extension to this need is that the at-home SOHO end user, just like the mobile end user, needs to be able to control who can communicate and when. Since all requests for time (phone calls, e-mail, etc.) will come to home, the SOHO end user needs to be supported by telecommunications services that can be told which role the end-user is currently playing — homeowner or worker.

Most SOHO establishments share a desire to one day be bigger then they are now. As a move in that direction, SOHO end users want to be able to start projecting a "big company" facade at all times when dealing with customers. This requirement manifests itself in several different ways: addresses and staffing levels. In the days prior to electronic addresses, small firms could use postal boxes to obscure their less impressive residential or strip mall addresses. As we move into the future of electronically linked businesses and electronic commerce, the importance of an impressive electronic address will take the place of the postal box. Additionally, since SOHO operations are generally staffed at very lean levels (i.e., perhaps a single employee), SOHO-end users are always on the lookout for telecommunications services that can take the place of additional nonexistent staff members and which can be used to provide superior customer contact. An example of such an application would be the "automated attendant" feature on many small business phone systems which automatically provides company information and basic directory services.

For the SOHO end user, the previous requirement can be further extended. It is once again the limited amount of staff available in the SOHO environment that generates the need for additional telecommunications services. These services are needed to permit potential customers to easily show themselves the SOHO firm's products, prequalify themselves, and then get in touch with actual employees. This use of telecommunications services to handle initial customer interest and then using valuable human resources only when the customer has demonstrated that they are a viable potential customer may be one of the most important drivers for SOHO telecommunications requirements. It certainly is one of the easiest to justify spending money!

Like the mobile end user, a SOHO end user must often work with others in order to secure large business orders, due to a SOHO's small size. This can often result in a SOHO establishment being required to *ad-hoc* partner with another business on a per-project basis. The telecommunications requirements that would be driven by this opportunistic type of limited partnering would be to support the exchange among the temporary partners of such information as schedules and project information. Once again, security would be critical; just because partnering is occurring on this project does not exclude the possibility that these partners may be competing against each other in the future.

Unlike the mobile end user, the SOHO end user has a "base of operations" — an office. It will be used to store almost all of the information related to the SOHO operation. This organizational structure produces a telecommunications need to permit the SOHO end user to access the information while away from home. Such access requirements include the ability to retrieve voice messages, electronic data, and any other information or formats that may be required. There is also the need for notifying SOHO end users that new information has arrived at the office in their absence. Note that once again, the information can arrive in a multitude of different formats.

Finally, since a SOHO end user faces the dual dilemma of operating under a tight budget today but believing that the operation will grow larger tomorrow, whatever telecommunications decisions are made today must be able to grow and change with the business. Solutions that must be removed and replaced are unacceptable both in terms of costs and time lost.

4.2.4   Residential End-User Needs

Our final segment of end users is also arguably the largest. In the U.S. there are currently over 120 million homes; it is these residential end users to whom a wide variety of service providers hope to sell additional telecommunications services. The marketing success of standard telephone service and the mixed success of various cable and Internet-related services clearly shows that the residential end-user community is a complex and multi-faceted group. The service providers hoping to capture a significant share of this diverse group must be willing to spend the time to understand what shared needs are currently unsatisfied.

Perhaps the most important factor that must be considered when attempting to understand the needs of the residential end-user is that, unlike the other end user segments that we've studied, the residential end-user has a relatively fixed budget from year to year. The result of this is the simple fact that every purchase is a tradeoff: if a new telecommunications service is to be purchased, then something else must be passed over. In most cases, this means that any service that does not provide a clear return for the residential end-user's investment is certain to fail. A good example of this occurred when the next generation of phone services based on the Integrated Services Digital Network (ISDN) technology were introduced. Despite the technology being sound, one reason that they failed was because residential end users judged them to not provide enough of a benefit to justify their cost.

As communication systems have improved our lives, they have also permitted us to move faster throughout the day and get more done. The result of this has been that the residential end user views the ability to manage time as a critical need. Any product or service that can provide more control over how limited time resources are spent seems attractive. However, as we have previously discussed, other factors such as price and availability will still play a very significant role in determining the residential end user's final acceptance.

As more and more information arrives at a residence, a striking advantage of postal mail over telephone service starts to emerge: information that is delivered via the postal system clearly identifies its intended recipient. On the other hand, a phone call arrives with no attached address and so whoever is first to answer the phone is required to perform a crude routing function in order to ensure proper delivery. This problem will only continue to grow as Internet access requires separate e-mail addresses and cable services permit channel and scheduling selections to be customized on a per-viewer basis. Any services that seek to address these needs of the residential end user must make sure that they are able to handle information that arrives in a variety of formats and that both end-user addresses and information processing preferences are handled by the service.

People are tribal by our very nature — we accomplish our daily activities by interacting with a wide variety of other people in our community, neighborhood, and extended family. Residential end users have a need to stay in touch with their contact group which resides locally as well as their extended families who may not live locally. The specific relationship defines the frequency of this contact and the format where it needs to occur. Today, such contact is mainly limited to text (letters or e-mail) and voice (via the phone). However, the arrival of the Internet and its support for a diverse set of multimedia communication formats has started to acquaint residential users with new options for communicating.

A very important constraint on any new telecommunications service is that it must be easy for the residential end user to use. Since the educational background and technical sophistication of residential users can vary widely, the majority of residential end users require that systems they purchase be easy and intuitive to use. One of the reasons that basic telephone service has been such a success is that the service is intuitive and simple to use. Note that the amount of end-user training time that it takes to learn to use a phone is very short! A key point for service providers to remember when introducing new services is that, in the mind of the residential end user, ease of use is a more important factor than additional bells and whistles.

Residential end users are always on the lookout for bargains whenever they are preparing to make a purchase. This mentality can be seen in the types of retail establishments that dominate the U.S. landscape: Wal-Mart, Kmart, and an almost infinite variety of strip malls. One of the greatest advantages of the Internet as it exists today is it permits skilled users to rapidly perform comparison shopping prior to going out and making a purchase. In the future, telecommunications services that standardize such comparisons and permit product offerings to be compared on multiple criteria including price, features, and availability would meet a need of the residential end user.

As we move into a new millennium, it is becoming evident that the skills required to survive and thrive in the modern world are changing. An example is found in automobile repair. The number of residential end users who service and maintain their car themselves has dropped substantially due to increased complexity in automobile design (anti-lock brakes, turbo-charged engines, etc.) and a decrease in the amount of time available to perform such basic tasks. Interestingly enough, when a car is taken to a repair shop to be worked on, one of the first steps that the mechanics perform is to attach computer input cables to various parts of the car in order to diagnose its operational health. Residential end users understand that this change in required life skills is occurring and they are eager to not be left behind. Therefore, they see access to education and information resources as a critical need and they desire telecommunications products and services that can improve, supplement, or provide greater access to such educational resources.

One of the greatest benefits of modern communications services is allowing people to interact with others who share a common interest. Without such services, perhaps these people would otherwise never know about each other. Residential end users desire services that will permit them to interact with other (potentially) remote end users who share a common interest. Examples would be collectors, fantasy-league sports players, on-line action houses, and support groups. New telecommunications services offer the possibility of permitting such interactions to occur on a global scale.

In the past, if a residential end user wished to gain access to valuable resources such as technical help, a stockbroker, etc., they had few options: schedule an appointment and then travel to meet with the resource provider face-to-face or phone them and either wait on hold or wait for them to call back. Telecommunications services that can streamline access to such valuable and limited resources are desired by all residential end users.

Access also plays a key role when it comes to a residential end user's finances. Better access to financial resources such as loan information, checking/savings account information, and stock portfolios has always been desired but not widely available. Key barriers to such services in the past have been concerns regarding both the security of transactions and the inability to validate the identity of the user, and the lack of appropriate equipment at the end user's residence to support such services. Both of these issues are being dealt with and will not continue to be barriers.

Residential end users seek ways to supplement other activities and thereby produce a richer experience for themselves. Users desire a way to gain more information or to follow up on something else that they have read about or seen. An example would be PBS's *Nova* programs, which display different Web links that point to supplemental material about the portion of the show that is currently being viewed. Additionally, residential end users would like to be able to follow up and obtain more information on advertised products that they see in different media — note that this accounts for the fact that Web addresses have become a standard part of any auto advertisement!

In a fashion similar to both mobile and SOHO end users, residential end users are very concerned about both their privacy and how they spend their valuable time. Residential end users want to be able to control who is able to get access to them and when such access is permitted. Therefore, they are interested in finding solutions that permit them to control who is able to send them information and how they are notified when that information arrives.

Finally, the ultimate benefit of technology is that it permits residential end users to plan events around their schedule rather than the other way around. Residential end users would like to be able to pick what time they want to be entertained instead of having to arrange their lives around external entertainment schedules.

4.2.3   End User Requirements Summary

As we conclude this section, it is important that we review the needs that are facing the four main segments of end users who will be driving the evolution of telecommunications into the future: business, mobile, SOHO, and residential. It is important to note that each of these segments is attempting to accomplish a different set of goals with different sets of available resources. This simple fact becomes quite evident when one looks at the differences in how much each of the different segments is going to be willing to spend on new telecommunications applications and services.

Although there are significant differences between each of the major end-user segments, several common themes have emerged. One of the most fundamental needs that each segment is trying to address is the ability to better control how its time is spent. Telecommunications services have the unique ability to eliminate distances and to permit time to be "shifted" — that is, to allow interaction between different parties to occur when it is most convenient for all of the involved parties. This need is further supported by each segment's desire to be in control of when and how they communicate with someone. The curse of modern technology is that it severely limits our ability to make ourselves unreachable when we so desire. The ability to regain this ability is a need that has been expressed by end users in all segments. Finally, the realization that end users are working harder at their jobs and the fear that this will cause their professional and personal lives to blur into an undistinguishable mass has generated a common set of needs. Users are seeking a way to be able to clearly distinguish communication and information that is associated with one role that they play from their other roles.

The recognition of these common basic end-user needs provides a clear prioritization for the development and deployment of future telecommunications applications and services. At its core, telecommunications is a field that exists to improve lives and solve problems. Advances in telecommunications often appear to be based on the latest "gee-whiz" technologies; however, for a new service or application to be successful, it must address one or more of the basic end-user needs that we have identified.

4.3   Application Trends

Telecommunications applications provide solutions to the problems faced by people who wish to exchange information ("end users"). We define telecommunications applications as the software that provides end users with access to the functions that permit information to be exchanged. A wide variety of telecommunications applications are in use today: the software in telephone switches that provides such services as emergency 911, caller I.D., three-way calling, etc; e-mail and Internet Web browsers; distributed synchronized databases such as Lotus Notes™, etc. Each type of application was developed to solve a specific set of end-user problems. Future telecommunications applications will also be developed to meet the needs of end users.

The types of future telecommunications applications will be directly related to the end-user needs discussed in the previous section. In order to focus our investigation into telecommunications applications, we will use the same segmentation of end users from the previous section. Figure 4.3 shows the stage of the Telecommunications Trend Analysis Model that is covered by this section. Our investigation will consist of two main parts: application functionality and functionality implementation. Looking at the application functionality trends that are occurring will help us to understand how application

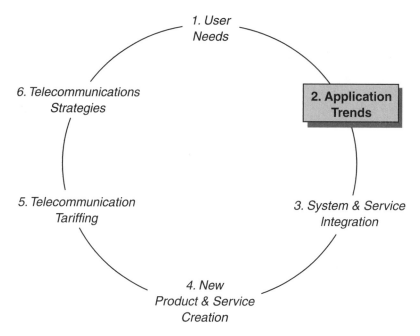

FIGURE 4.3 Trend analysis — application trends.

developers and service providers are working to address end-user communication needs. We will explore how this new functionality will be deployed in the real world when we go one step farther and look at how vendors and service providers are planning to implement the new application functionality.

4.3.1 Application Functionality

All four major categories of end users will require more functionality from their telecommunications applications. Because of the large purchasing power of each segment, competition among service providers has started to increase in the past few years. This trend is most noticeable in the U.S. and England; however, the arrival of the European Union (EU) and a unified currency (the Euro) in western Europe is also helping to make those telecommunications markets competitive.

A result of multiple competing service providers means that, at the very least, all segments of end users will shortly be presented with multiple sources for all existing services. Additionally, the number of services offered to end users will increase more rapidly than in the past due to the need for providers to distinguish their offerings from each other.

The eventual result will be that the telecommunications applications offered to all end user segments will become more customized in order to meet the specific needs of a particular segment. Since end users are best suited to determining their exact needs, the process of subscribing to a telecommunications application will change from the selection of "all-or-nothing" applications in which the end user had little or no choice to participating in a "build-your-own" functionality selection in order to create a customized application.

This ability for end users to design their own applications will be the arrival of true multimedia applications that combine voice, video, and data features into a single customized application. This customization will cause the functionality provided by applications to increase over what is available in today's applications. New functionality will be apparent in the following five areas: Internet services, e-mail, videoconferencing, wireless services, and enhancements to traditional services. We will now look at application functionality improvements we can expect in each of these areas.

4.3.1.1 Internet Applications

The recent explosion in the popularity of the Internet (an unmanaged collection of interconnected computer networks that are all able to "speak" the same communications protocols) has forever changed what telecommunications applications will be expected to do. Studies of Internet usage are difficult to do because of its rapid growth; however, in the early 90s the Internet was used by a handful of researchers and scientists, and studies eventually predicted that the Internet was expected to reach more than 200 million end users and 60 million hosts by 2000. With this kind of growth, it is very conceivable that the usage of the Internet will catch up to, and perhaps surpass, the use of the telephone in the not so distant future.

Today's Internet applications lack the functionality required for end-users to perform e-commerce transactions efficiently. Electronic commerce ("e-commerce"), the use of the Internet to facilitate the buying and selling of goods, is viewed by many as "the next big thing." The Internet offers sellers of goods the ultimate virtual storefront: without having to rent physical space, they can display and demonstrate their products for potential buyers. What is currently missing is the end-user's ability to feel confident making a purchase of the displayed goods directly over the Internet. The reasons for this lack of confidence are varied: lack of a secure environment, lack of an appropriate exchange mechanism, and privacy concerns.

Users are well aware of the fact that as they exchange information with a retailer's Internet application, it is possible for a malicious user to monitor and record their transaction. This could result in the malicious user obtaining credit card or bank account identification information that could then be used to steal funds from the unsuspecting user. Enhancements to functionality are being made to both the retailer's and the end-user's applications. Basic encryption is now available that can be used to secure the transaction information before it is transmitted in order to negate the effect of any interception of the transmission. As this type of functionality is added to end-user's browsers and Internet-aware applications, user confidence in secure Internet transactions will increase and e-commerce can be expected to grow at an explosive rate. In the short term, some service providers are offering guarantees to make good on any losses incurred while using their networks in order to "jump-start" e-commerce activities.

E-commerce is currently complicated by the lack of an agreed upon form of "digital cash." Despite gains in the past decade regarding the increasing use of credit and debit cards, the majority of retail transactions still occur using either paper money or checks. Neither of these two popular forms of exchange translate well to being used in the Internet's all-electronic environment. Once again, several different approaches to this problem are currently being investigated. Recent agreements among many of the major credit card companies have identified the required functional and exchange procedures that will be required to support electronic forms of currency for existing and new Internet applications.

Finally, as more and more of everyday life becomes computerized, consumers are starting to become concerned about how much information retailers are able to obtain regarding personal habits and buying patterns. As the use of the Internet to purchase goods increases, a retailer's ability to track the user's entire buying experience will also be increased. Such information could include a history of goods that the consumer looked at but did not purchase, how often and at what times of day the user visited a specific electronic "store," and all of the products that the customer has ever purchased. Consumers have become alarmed that retailer's applications will be able to "mine" their purchasing history to target other goods for advertising purposes or that retailers will sell their information to other retailers for their use in trying to sell goods to the consumer. As Internet e-commerce applications mature, consumers are going to insist that retailers clearly identify what consumer-related information is being tracked and post their polices regarding use or sale of that data. Internet applications will have their functionality enhanced to support and enforce such privacy policies.

Although the Internet is a worldwide phenomenon, the majority of its content has been created in the English language. The reasons for this are varied; however, the origination of the Internet in the U.S. and the high availability of both computers and Internet access in English-speaking countries has definitely played a major role. Future Internet applications will be required to be able to deal with multiple languages. The tools to make this possible are slowly starting to emerge. Internet-based language translation products are now available that offer translation services for several languages. Whereas the amazing translation devices seen in some popular science-fiction movies may still be a long way off, the ability to translate text found on the Internet into another language or the ability to select a language for the purchasing process are just around the corner.

- Some service providers who are deploying high-speed digital access services are also establishing on-line communities built around high-speed access. These communities provide an opportunity for businesses to set up on-line shops, as well as a place for both residential and business customers to receive e-mail, purchase goods, access applications, and find out current event information for their local areas.
- Service providers are starting to explore the opportunities presented by integrated bills, accepting payment and providing customer care electronically over the Internet. Voice services and Internet services can be consolidated onto a single bill. Additional applications can electronically present the bill to customers and accept payments over the Internet. This type of application can be used with all types of telecommunications services including paging, IP voice, and long distance. An additional benefit of this approach is that it permits targeted marketing of specific customers and offers a better chance of capturing an impulse buying opportunity.
- Many vendors are looking for ways to replace today's ubiquitous fax machines. Some of the more innovative solutions are coming from companies that are trying to reduce their product support costs. One approach to directly provide a user with only specifically requested information uses Internet based "push" technology. This information delivery technique requires a user to log on to the company's server via an Internet connection. Then the company is able to "push" or force the display of specific information. The true power of this approach becomes clear when the user is able to talk with the company at the same time by using a separate line. These hybrid solutions are a cross between e-mail and fax services. Companies have found that this type of solution works best when the company has a great deal of information that the user would otherwise have to work through in order to find what is needed.
- Firms are discovering that an estimated 10% of customers sometimes need assistance when using the firm's Web site. So-called "chat" applications are being added to Web sites to provide customers with the ability to receive real-time one-on-one guidance from employees of the firm.

4.3.1.2 E-mail Applications

E-mail has become such a critical part of how so many people communicate that we choose to treat its functionality separately from that of Internet applications. A 1998 survey by Forester Research revealed that 83% of Internet users send e-mail, making it the most popular on-line activity. Surfing the Web is the second most popular and attracts 81% of users.

- Adding voice and video to e-mail represents the next step in e-mail's evolution. Some service providers are now able to deliver e-mail that contains embedded links to additional voice and video components of the message. The additional e-mail components are then sent to the user through streaming technology that uses a service provider's computers to do the majority of the required processing, and then ships only the resulting images to be displayed on the end user's Web browser application. The challenge is to avoid disappointing the end users with poor application performance that causes them to revert to standard text-only messages.
- Estimates show that up to 40% of users' time on the Internet is spent on e-mail. In 1997, America Online (AOL) had 11 million members and it processed 15 million e-mails per day, which roughly relates to 23% of its members on-line time.
- E-mail is fairly pervasive, fast, and relatively free. One of the next logical steps is to make it secure. Currently, the majority of financial and legal communications occur using either paper or the

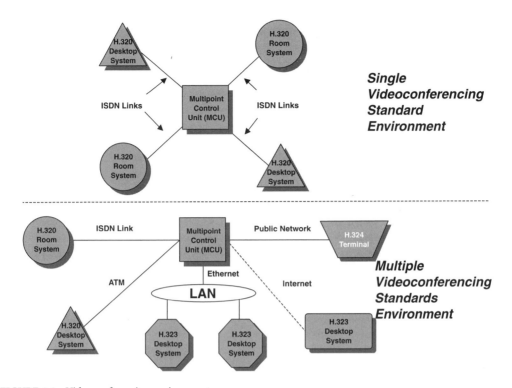

FIGURE 4.4 Videoconferencing environments.

somewhat dated electronic data interchange (EDI) systems. The problem with existing e-mail is that it can be easily faked. Internet security has five key requirements: access control, authentication, privacy, integrity, and non-repudiation.

• Current secure e-mail solutions use a public key infrastructure (PKI). PKI is a set of security services that can be used to provide security over public networks. PKI services consist of encryption, digital signatures, and digital certificates. PKI services require the use of a two-part key: a public key and a private key. Information is sent to a user after having been encrypted using their publicly advertised "public key," and can only be decrypted using the user's secret "private key." Every PKI exchange is monitored and authenticated by a company that provides digital security services.

4.3.1.3 Video Conferencing Applications

• Videoconferencing (Figure 4.4) offers many benefits, including savings in corporate travel and savings in employees' time. The U.S. market for videoconferencing service revenue is projected to top $27 billion by 2002. In 1995, videoconferencing service revenue was $2.5 billion. Important pieces must be in place for videoconferencing to happen: rising demand from multinational corporations, improvements in technology, solidification of key standards, and proliferation of standards-compliant video-enabled products from heavy hitters such as Microsoft and Intel. Key issues for service providers are reliability, quality, and ease of use. Current standards include:

Video Conferencing Standard	Purpose
H.320	Videoconferencing over ISDN
H.323	Videoconferencing over LANs, WANs, intranets, and the Internet
H.324	Videoconferencing over regular dial-up telephone lines

- According to networkMCI Conferencing, about 250,000 videoconferencing-capable devices are currently in place worldwide; by 2000, there will be over 50 million. A big user issue is service complexity: it can take 40 to 50 minutes to set up a call because all endpoints need to be configured to the same line speed, audio rate, frame speed, and resolution rate. How both vendors and service providers have interpreted standards can also affect the service: a mismatch in interpretations can result in dropped calls. Videoconferencing systems that are able to talk to different standards-compliant endpoints are now becoming available (e.g., H.323/H.320 gateways).

- IP muliticasting will be able to provide multipoint H.323 videoconferencing. IP multicasting will save users' bandwidth on packet networks because the information needs to be transmitted only once over a given link, with routers replicating information as required. One challenge associated with multicasting is that it imposes a significant communications load on the processor at each endpoint since each endpoint, must send information to every other endpoint. This means that IP multicasting is not currently scalable for large videoconferences.

4.3.1.4 Wireless Applications

- Wireless data service providers are starting to shift their focus from vertical to horizontal applications. In the past, wireless data applications have been traditionally targeted at the public safety and utility markets. Newer applications target members of the financial community, such as bankers, analysts, and traders, by providing real-time access to stock information. One of the key success factors to entering horizontal business markets will depend on the service provider's ability to create appealing service bundles.

- In the U.S., the future of the mobile data market is based on the cellular digital packet data (CDPD) technology. CDPD is TCP/IP implemented over cellular networks. CDPD is well suited for certain types of transmission, especially short file transfers. CDPD was first specified in 1992; however, it has been slow to be adopted and there are currently fewer than 500,000 data customers on all U.S. cellular networks. Although CDPD may be well suited to supporting Internet-related applications, it is currently limited by two factors. The first is the fact that CDPD-based services are only available in selected markets. The second is that CDPD's bandwidth is currently limited to 19.2k bps and actual connection throughput can drop as low as 2.4k bps when network voice traffic is high. CDPD transmission rates as high as 56k bps have been discussed; however, support for such rates is not currently provided.

4.3.1.5 Enhancements to Traditional Services

- Vendors are starting to work with service providers to create service solutions that meet end user needs. Unified messaging products are the first examples of such services. The service alerts users that they have e-mail via their service provider's Web site and their own voice/fax mailbox. This will be provided as a first step for customers who only want basic service. To be added: products that use text-to-speech technology. Good approach: everybody doesn't need everything. Future services include integrated e-mail, voice, and fax mailbox; non-subscriber voice connect — allows e-mail users to send voice messages to anyone; and consolidated wireless/wireline mailbox with improved phones that contain text display screens.

- The Universal International Freephone Number (UIFN) system allows a single toll-free number to be used around the world. Users apply to the Internatinoal Telecommunications Union (ITU) for an eight-digit number that is accessible by dialing the appropriate international access code, "800," and then the new number.

- Many new telecommunications applications are being developed for call centers. These applications are being designed to help companies gather information about their customers and make sure that the products and services that the company offers are meeting the needs of their customers. This type of application uses computer telephony integration (CTI), automatic call distributors (ACDs), and interactive voice response (IVR) systems.

- One of the primary motivations for firms to use virtual private networks (VPNs) is to avoid the costs of expensive dedicated leased lines. Vendors are now making VPN products that contain combinations of functions, including serving as IP routers, corporate firewalls, and certificate authorities, along with the required VPN functions of encryption and authentication. A key drawback to today's VPN products is that the processing power required to perform VPN functions such as encryption severely limit the throughput of the devices.

4.3.2 Functionality Implementation

The enhanced telecommunications application functionality described in the previous section requires that the way applications are designed must be radically altered. As the computing equipment available to end users continues to improve, the intelligence required to support the application is migrating from within the network to the endpoints. In new and emerging applications, much of an application's functionality may reside in the end-user's equipment. This is dramatically changing how networks are designed. We will investigate these types of changes later in this chapter.

The competitive environment that service providers are starting to operate in will no longer permit deployment of new applications at the current somewhat leisurely rate. End users will demand new applications as soon as they identify problems they need to solve. The service provider who is the first to be able to offer a solution to such end users stands the best chance of capturing the largest share of the market. Past history has clearly shown providers that it is better to be first to market and bring additional functionality later rather than wait until a new application is perfect.

The arrival of networking equipment that is able to provide exponentially larger amounts of bandwidth will aid developers of new telecommunications applications. Table 4.1 identifies several of the network bandwidths now available for use with new applications. The result of greater bandwidth availability is that less development time will have to be spent attempting to minimize the amount of data that telecommunications applications exchange. This reduced development time will result in applications that are richer in functionality, being made available to end-users more rapidly.

TABLE 4.1 Standard Transport Bandwidths

Transport Type	Bandwidth
OC-3	156M bps
OC-12	622M bps
OC-48	2.5G bps
OC-192	10G bps

Recent increases in the amount of bandwidth provided by data networking equipment, coupled with the initial availability of products that can provide voice services over a data network, have fueled a focus on Internet Protocol (IP)-based networks. As competitive service providers build new networks to provide services, they are selecting networking equipment that permits them to build IP-based networks rather than the traditional Class 5 voice switches. These new service providers believe that in the very near future all information (voice, video, and data) transported by a provider will be viewed as data and can be encapsulated in the IP data network protocol.

If current application trends continue as expected, almost all future telecommunications applications will be "Web-Aware." Simply put, this means that such applications will have the ability to obtain information from and provide information to other applications via World Wide Web (WWW) Internet protocols. Although still in its infancy and facing an unsure future, the Java programming language has popularized a highly distributed programming model that will influence the design of such future applications. In this model, the network has the responsibility for advertising what applications it supports and storing the logic required to provide the application. The end-user's customer premises equipment

(CPE) will then download the needed functionality and execute it locally, thus distributing application processing from the network's limited resources.

The use of data networks for telecommunications application interconnection will have the interesting side effect of causing what has been called the "death of distance." Because end users are currently charged for the size of the connection that they use to access the Internet, it no longer matters how far the data travels once it is transmitted. This will result in a greater use of more widely distributed applications. A more detailed discussion of the effects of changes in telecommunications application pricing is provided later in this chapter.

One of the greatest bottlenecks in deploying new telecommunications applications resides in the back office operations of the service providers themselves. After an end user selects a service and negotiates any customizations, there is often a delay (sometimes, significant delay) as the provider processes the order and reconfigures its network to deliver the requested service. The telecommunications applications introduced in a competitive environment must be deployed with minimal support costs and must start to generate revenue as quickly as possible. One of the most promising means of accomplishing both of these goals simultaneously is to automate the telecommunications application service ordering process. Assuming that the obvious security issues can be solved, interfacing the application directly to the provider's operation support systems (OSS) will reduce the support costs for the application while at the same time decreasing the delay between when the service is ordered and when the application is available to the end user.

Finally, the near-panic caused by the so-called Year 2000 (Y2K) bug, which caused some applications to be unable to distinguish between 1900 and 2000 due to historical efforts by software developers to minimize the amount of memory required to execute an application, will forever change how telecommunications applications are developed. Immediately after having experienced the expense and turmoil caused by the hunt for potential Y2K errors in hundreds if not thousands of hardware platforms, operating systems, and applications, end users can be expected to demand protection from future errors. Although complete protection from software errors can never be guaranteed, new telecommunications applications will most certainly contain enhanced testing capabilities that will permit the end user to simulate program execution in an off-line environment in order to determine how it will react to a given set of inputs.

Let's now take a close look at some of the issues surrounding how some of this enhanced application functionality will be implemented in two important segments of telecommunications applications: Internet applications and wireless applications.

4.3.2.1 Internet Functionality Implementation

- A carrier-grade IP telephony gatekeeper that complies with the emerging H.323 standard is now available in evaluation versions. This product can be used to tie together IP and public network gateway systems from other vendors. This product is significant because it represents the first phase of multivendor interoperability. Ericsson plans on using applications to differentiate its gatekeeper product — specifically for applications that are better suited to reside inside the carrier network.

- Hammer Technologies has introduced an IP test system that monitors the quality of voice on IP networks. The system automatically tests voice quality, measures audio quality, and includes a Voice over Internet (VoIP) protocol analysis tool as an IP traffic generator.

- Microsoft, Netscape/AOL, and Sun are all competing to supply commercial Web server and application platforms to public network service providers. Each of these companies has a different vision of what the next generation of telecommunications applications will look like. Microsoft sees applications being built on top of low-cost PC-based Microsoft operating systems. Netscape/AOL sees applications as being distributed and platform independent. Sun sees applications running on open, fault-tolerant systems that use the Java language.

- Some service providers are aggressively deploying advanced high-speed digital subscriber services. Many of these service providers own and operate switch-based networks and feel that a switched network infrastructure routes packets faster and more reliably than a routed one. Such providers are offering Internet access and LAN-like services.

TABLE 4.2   Digital Subscriber Line Bandwidth

Symmetrical Service Rates	Asymmetrical Service Rates
256k bps	4M bps downstream, 1M bps upstream
512k bps	7M bps downstream, 1M bps upstream
768k bps	
1M bps	

- In order to provide access to popular Internet content to users in other countries, creative applications are being developed to distribute the information. Using a combination of satellite links, multicasting software, and local caching, service providers are using public Internet kiosks to permit users to view the most popular web pages. This eliminates long waits for dial tones and conflicts over access to what precious bandwidth exists. This new approach "pushes" content to where the user is instead of requiring the user to pull content off of North American servers.
- Dynamic HTML will allow designers to make richer, multilayered pages. Dynamic HTML will allow designers to create Web pages more efficiently so each link of information doesn't have to be downloaded from the server.

4.3.2.2   Wireless Functionality Implementation

- Microsoft has announced that it is developing a non-standard microbrowser as a part of its goal to enter the wireless data marketplace. The microbrowser will permit wireless users to browse the Web, provision services, and access billing information. The Wireless Application Protocol Forum released the open wireless mark-up language (WML) microbrowser specification.
- Researchers have been able to crack the messaging encryption algorithm used in U.S.-based CDMA and TDMA digital cellular networks. The researchers have broken the Cellular Message Encryption Algorithm (CEMA) code. The CEMA code has been designed to safeguard dialed digits that are sent over the airwaves. Different encryption algorithms are employed for user authentication and voice privacy. The reason that the researchers were able to crack the CEMA code was, in part, due to the fact that the wireless industry has watered down its security algorithms in order to appease the U.S. federal government.

4.4   Systems and Service Integration (Figure 4.5)

4.4.1   Introduction

As telecommunications technologies are able to provide end users with more and more complex services with an increasing number of interrelated features, end users have started to complain. Just as when you go to purchase a car, you don't want to be required to make decisions regarding issues that are relatively unimportant to you. An example of this would be when taking an airline flight, you do care about flight times and where you sit; however, you don't care what altitude you fly at or what movie is shown. End users simply want to be able to use telecommunications services to make their lives easier and to make themselves more productive — they don't want to be telecommunications experts in order to select and use such services.

Service providers and network equipment vendors are responding to these needs by integrating what are currently separate service offerings into new feature-rich services and by consolidating technology-specific networks into single networks that are simultaneously able to handle voice, video, and data information exchanges. In this section, we will explore some of the drivers for service and system integration and identify how these are going to affect the telecommunications services available to end

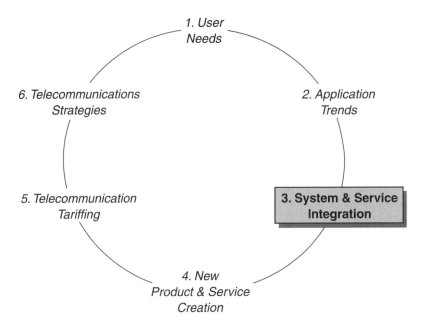

FIGURE 4.5   Trend analysis — system and service integration.

users in the future. We will then take a look at specific trends in service and system integration as they relate to each one of the four classes of end users identified earlier in this chapter.

4.4.2   Drivers for Integration

Integration of services and systems requires both considerable effort and expense. In order to make such an investment worthwhile, there needs to be a future payoff for service providers and application developers who make the integrated solutions. In fact, there are several distinctly different motivations that are in the process of creating integrated solutions. The drivers for integrated services are as follows:

- **Competitive differentiation:** As the number of service providers is increasing, the number of end users in each segment is remaining relatively constant. This means that service providers will only be able to grow by wooing end users away from their current providers. In order to accomplish this, a provider will have to be able offer the end user a compelling reason to switch. Integrated services can be such an enticement, and such tactics are starting to appear in the form of "follow me" offerings where voice, paging, and mobile services are linked to a single service. With such a service, someone trying to contact the end user dials a single number which then attempts to establish a connection with the called party via each different communication method. If the desired end user is not reachable, then a message can be left on a voice mail system that the end user can check via any of the available technologies.
- **Single provider:** Recent surveys of end users have revealed that, all other things (such as price) being equal, users desire to receive all of their services from a single provider. The reasons are simple: a single provider means a single bill and one number to call in the event of any problems with the service.
- **Technology advancements:** The integration of multiple services into a single offering to the end user has its own potential risks. An integrated application requires a significant amount of end-user customization in order to provide the maximum benefit. An example of an integrated service that has suffered from low end-user acceptance due in part to its complex configuration is the Integrated Digital Services Network (ISDN). Advances in network intelligence and equipment processing allow the configuration of new services to be simplified and have allowed much of the

configuration process to be performed automatically by the network equipment itself. Additionally, improved network element processing has permitted multiple elements in different technology domains to exchange the required information to support integrated services.

- **Improved billing systems:** Amazingly enough, one of the greatest limitations on integrating services has been the billing system used by service providers. Such large and complex billing applications were originally designed to support a specific set of services offered via a single technology network.

Likewise, there are several identifiable drivers working together to motivate service providers to create integrated systems. The drivers for integrated systems are:

- **Reduced network deployment costs:** End users are starting to demand services that have voice, video, and data components. Service providers will have the choice of building separate redundant networks to provide such services or of building a single high-speed network to handle all three forms of communication. As you may well imagine, the decision to build a single network becomes very straightforward once the economics of building a single network to deliver all services is considered.

- **Reduced operations costs:** A significant cost of delivering a service to an end user can be attributed to the operational expenses required to keep the network working correctly. The use of an integrated network reduces the number of network elements required to deliver services, simplifying operational requirements and thereby lowering the ongoing cost of offering the service.

- **Bandwidth breakthroughs:** The possibility of using an integrated system to offer services to end users could not be realized until improvements in network equipment and transport technologies occurred. Recent increases in the bandwidth that can be provided by a single network have made it possible to build a single network that can support multiple services.

- **Tariffing:** Existing tariffing of telecommunications systems was designed years ago when the primary offerings to end users were voice services. Data networks are currently free of many of the limitations that restrict what and where services can be offered via traditional voice networks. We will discuss specific tariffing-based motivations later in this chapter.

4.4.3   Integration for Service Providers

- The arrival of wavelength division multiplexing (WDM) systems has caused service providers to reevaluate their existing time division multiplexing (TDM) systems. Network planners currently believe that the two different approaches can be used together to create networks that provide the lowest bandwidth costs.

- As service providers prepare to reshape their circuit-switched networks into IP packet-routed networks, the issue arises about what type of operation support system (OSS) will be needed. Service providers have a range of functions to support: provisioning of new services, service assurance, and network management. Existing service providers will most probably address this problem by reusing part or all of their existing billing or customer care systems. The greatest challenges will come in the areas of network management and provisioning.

- Sprint Corporation announced in June 1998 that they were planning on carrying all of their voice, video, and data traffic over its asynchronous transfer mode (ATM) network. Sprint CEO William T. Esrey predicted that this approach would cut Sprint's cost of delivering a voice call by 70%. Sprint expects to achieve this level of cost reduction because of the much higher performance:cost ratio of data switches vs. conventional circuit-switched voice switches.

- Traditional circuit-switched system service providers are watching the success of facilities-based Internet service providers (ISPs) and their packet-switching and routing forwarding networks. In addition, the large circuit-switched network equipment vendors are also modifying their equipment in order to transition them to work in a packet-switched environment. Many are forgetting

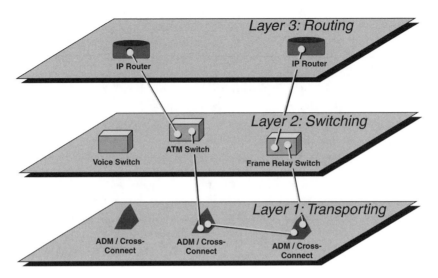

FIGURE 4.6   A new network architecture.

that the existing circuit-switched, connection-oriented public network infrastructure has been built up over decades and includes layers of resiliency and fault tolerance built in. Another key point is that time-slotting information in hardware allows for guaranteed latency and delay parameters that simply cannot be achieved in many packet-switched systems. It is possible that in the future, the circuit-switched network will serve as a mission-critical backup system for the public packet-based system and will only be used for those cases where the "call must go through." A possible casuality of the move to a packet-based public network could be the current computer-telephony integration (CTI) market.

- ISPs are now at the front line of telecommunications equipment design. Today's ISPs are building their own facilities, laying their own optical fiber and installing their own carrier-class switches in points-of-presence (POPs). Traditional circuit-switched service providers have been taking more data and even voice traffic off traditional circuit switches and putting the traffic on packet-switched networks that were formerly considered to be "data-overlay" networks.

- The current public network consists of voice switches interconnected via transport systems. New user demands are causing this network architecture to be reshaped to now support voice, video, and data services. This new network architecture (Figure 4.6) uses a transport infrastructure which supplies the required interconnectivity to create its foundation. The architecture's switching layer provides the call set up and teardown functions throughout the network that are required to deliver services using a variety of protocols. Finally, a routing layer is used to provide the final step in the process of delivering data services to end users.

- Within the telecommunications industry there is still disagreement about whether packet switching, ATM, or traditional circuit switching has the best performance:cost ratio. Peter Sevcik, a senior associate at Northeast Consulting Resources, Inc. has shown that each successful new generation of switching technology cuts the performance:cost doubling time in half. Sevcik says that traditional central office circuit-switched telephone switches double their performance:cost ratio every 80 months; ATM switches do the same every 40 months; packet switches and routers double their ratio every 20 months; frame relay switches double their ratio every 10 months.

- Hewlett-Packard (H-P) is helping ISPs develop an architecture that delivers fast, consistent, and differentiated service over the Internet. The goal is to enable individual businesses to guarantee service levels to end customers over the Internet and to offer predictable and differentiated services.

H-P's product is a bundle of special H-P hardware and software, along with add-on services from Cisco Systems, that will offer ISPs end-to-end, mission-critical service level guarantees and Internet service level agreements. Technically put, this product offering will integrate control, measurement, and management across the servers in the network in order to guarantee delivery of service levels.

- Sprint's announcement of their plan to build an integrated on-demand (IOD) ATM-based network to deliver voice, video, and data sets the standard for future public network developments.

- As service providers start to offer services that use multiple technologies, equipment vendors are modifying their existing equipment to support the providers' new needs. Traditional voice switch vendors are enhancing their wireline switching products to also support wireless services. Some switch architectures are so flexible that providers can mix and match wireline and wireless modules to permit subscribers to connect to a cell site or the public network using the same switch.

- U.S. West reports that the average voice call is approximately 5 minutes. The average data call is about 32 minutes — this is causing congestion in the central office.

- U.S. West is conducting trials with two ISPs to weed out Internet traffic from voice calls. Using distributed SS7 technology, ISP-bound data calls are identified at the user's ingress switch and immediately routed to the ISP over a parallel data network. This differs from the way traditional end-user data calls are set up. Data calls are normally routed to the ISP through the phone network via ISDN or digital switched services such as channelized T1. The data call rides the public network the whole way.

- Service providers are showing a renewed interest in video services. Vendors are demonstrating products that can push 26 Mbps over existing twisted pair wiring for up to 4000 feet. Broadband wireless equipment vendors are mainly focused on data applications; however, some have demonstrated videoconferencing and distance learning applications.

4.4.4 Integration for Business Users

- Some observers suggest that more than 60% of the costs associated with modern data networking lie in the cost of ownership.

TABLE 4.3 Application Driving Network Growth*

Application	Data Types/Sizes	Network Traffic Implication	Network Need
Scientific modeling, engineering	Data files 100s of megabytes to gigabytes	Large files increase bandwidth required	Higher bandwidth for desktops, servers, and backbones
Publications, medical data transfer	Data files 100s of megabytes to gigabytes	Large files increase bandwidth required Low transmission latency High volume of data streams	Higher bandwidth for desktops, servers, and backbones
Internet/Intranet	Data files now Audio now Video is emerging High transaction rate Large files, 1 MB to 100 MB	Large files increase bandwidth required Low transmission latency High volume of data streams	Higher bandwidth for desktops, servers and backbones Low latency
Data warehousing, network backup	Data files Gigabytes to terabytes	Large files increase bandwidth required Transmitted during fixed time period	Higher bandwidth for desktops, servers, and backbones Low latency
Desktop video conferencing, interactive whiteboarding	Constant data stream 1.5 to 3.5 Mbps at the desktop	Class of service reservation High volume of data streams	Higher bandwidth for desktops, servers, and backbones Low latency Predictable latency

*Source: Gigabit Ethernet Alliance

- Ethernet LANs typically offer 10 Mbps of shared bandwidth. As the volume of network traffic increases, however, this amount of bandwidth quickly becomes inadequate to maintain acceptable performance to support demanding applications. These traffic jams are fueling the need for higher-speed networks. Fast Ethernet, or 100BASE-T, has become the leading choice of high-speed LAN technologies. Building on the near-universal acceptance of 10BASE-T Ethernet, Fast Ethernet technology provides a smooth, nondisruptive evolution to 100 Mbps performance. The growing use of 100BASE-T connections to servers and desktops, however, is creating a clear need for an even higher-speed network technology at the backbone and server level. Ideally, this technology should also provide a smooth upgrade path, be cost effective and not require retraining. The most appropriate solution now in development is Gigabit Ethernet. Gigabit Ethernet will provide 1 Gbps bandwidth for campus networks with the simplicity of Ethernet at a lower cost then other technologies of comparable speed. Gigabit Ethernet will be an ideal backbone interconnect technology for use between 10/100BASE-T switches, as a connection to high-performance servers and as an upgrade path for future high-end desktop computers requiring more bandwidth than 100BASE-T can offer.

- Although Gigabit Ethernet is primarily an enterprise LAN technology, several service providers (most of them ISPs) have begun evaluating it for use in local and metropolitan area sections of their networks. Gigabit Ethernet can connect network equipment such as the server, routers, and switches within a service provider's POP, both inexpensively and at high speeds. One of Gigabit Ethernet's biggest selling points is that its cheaper and faster than asynchronous transfer mode (ATM) or Synchronous Optical Network (SONET), which many service providers now use to link gear in their POPs. Gigabit Ethernet's heavy data orientation and distance limitations are red flags, however, for established telcos looking for technologies that can support voice, video, and data.

- Inverse Multiplexing for ATM (IMA) is a specification for provisioning multiple ATM circuits in T1 increments. IMA was created to bridge the bandwidth gap between T1 (1.544M bps) and T3 (45M bps). Using IMA, several low-cost T1 lines can be used to aggregate the bandwidth and distribute ATM traffic across multiple physical circuits.

- Frame relay speed and capacity improvements are being designed in order to keep pace with the needs of the new public network for data services. The two major changes to frame relay are the emerging frame relay over SONET (FROSONET) and multi-link frame relay (MLFR) standards. FROSONET provides specifications for frame relay to run at OC3/STM-1 or OC12/STM-4 speeds. MLFR adds scalability to frame relay networks, thus helping service providers keep pace with growing traffic demands while providing an incremental capacity jump for users who are outgrowing T1 capacity but are not ready for the speed or expense of DS3/E3 lines. The FROSONET specification uses the same high-level data link control (HDLC) over SONET mapping that is being used for Point-To-Point Protocol (PPP) over SONET (PoS). This saves costs by allowing the same hardware to be used for both PPP and frame relay interfaces. MLFR trunks combine multiple physical links between switches in the public network into a single higher-capacity logical facility. Additionally, frame relay's existing quality of service (QoS) functionality permits it to be used by service providers to offer such capabilities as service level agreements (SLAs) and customer network management (CNM) functionality.

- Time division multiplexing (TDM) is used to combine individual connections in order to traverse longer distances. Switched circuits, such as those used in telephone networks, provide dedicated connections between two points. Switched-packet protocols, such as Ethernet, provide good utilization of the backbone but have no provisions for providing the equivalent of a switched circuit over a network. Switched-cell protocols such as ATM provide good utilization of the backbone and have provisions for providing CBR and UBR virtual circuits, but are expensive when compared to the newer switched packet systems such as Gigabit Ethernet.

- Wide area networks (WANs) tend to be rings like FDDI or various star configurations. Generally, the number of entry points into the network tend to be very limited. WANs are designed to transmit data over long distances, tend to be focused on isochronous data, and lean toward circuit switching because they were often devised by telephone companies to carry voice.

- Five leading car and truck manufacturers have banded together to lead the Automotive Network Exchange (ANX) project. This is designed to create a specialized high-end, Internet-like VPN to link North American automakers and their suppliers. This very reliable and secure network may act as the beginnings of a parallel "Business Internet."

4.4.5 Integration for Mobile Professionals

- The CDMA Development Group (CDG) is coordinating location technology trials among member carriers and vendors. Trial focus will be on the three types of technology: global-positioning system-based, network-based, and a combination of the two. The FCC has mandated that carriers are required to be able to locate callers within 125 meters. Some carriers believe that they will ultimately implement multiple location technologies. Network-based solutions are less precise but may meet the FCC mandate. A handset-based solution may locate users more accurately and, if used with a network-based solution, may allow a carrier to offer enhanced services such as location-sensitive billing and concierge services.

- A spokesman for the CDG reports that wireless data represents only 10% of the total wireless airtime in the U.S. New CDMA products have been introduced recently that will increase data rates to 64 kbps.

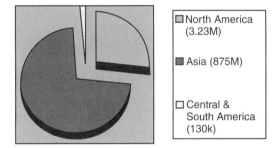

FIGURE 4.7 Worldwide CDMA subscribers (source: CDMA Development Group).

- Service providers and equipment vendors are currently working on developing standards for the third generation of wireless products: "3G." The first generation was analog, the second was digital, and the third will be wireless broadband that will be used to support high-speed mobile data services.

- What standard will be used for the next generation of wireless services is still undecided. Possibilities include W-TDMA which suffers from limitations in growth and W-CDMA which suffers from limitations in power and processing.

4.4.6 Integration for SOHO Users

- In September 1998, the ITU ratified a single standard (V.90) for 56k bps access over the Public Switched Telephone Network (PSTN). V.90 data transmission technology overcomes the theoretical limitations imposed on standard analog modems by using the digital server connections that most Internet and online service providers use at their end connection to the PSTN.

- Community area networks (CANs), as represented by cable modems, have a unique topology that is not served well by existing LAN or WAN topologies. They have a large connection count of shared wire like LANs, but have distances like those of WANs. Current CAN implementations generally utilize a single downstream CATV channel that is shared by all network participants. A separate upstream CATV channel is also shared for transmitting from the home to the cable head end.

- Low and medium earth orbit satellite systems (LEO and MEO).
- Geostationary (GEO) satellite systems are being used to deliver data broadcasts. There are two primary types of GEO services: very small aperture terminals (VSAT) and direct broadcast satellite (DBS). Two DBS services have been announced: an entertainment service and a data service. The data service will serve corporate customers with occasional and regular broadcasts, along with residential customer service. One possible use of the data service is for software distribution. Another company is offering three versions of its DBS Internet access service: direct delivery of text files at 12 Mbps, multimedia at 3 Mbps, and Internet access at 400 kbps. This service is asymmetrical: customers send information requests to the service provider via telephone lines and receive data via the customer's 24-inch antenna. VSATs are used for corporate broadcasts of data including price updates. VSATs operate at speeds between 14 kbps and 64 kbps, with high-speed bidirectional communication
- Home based LANs
- According to U.S. West, DSL services can be up to 250 times faster than a 28.8 kbps modem.
- Virtual private networks (VPNs) provide an alternative to leased-line connections. VPNs provide an inexpensive way to extend the corporate network to telecommuters, home workers, day extenders, remote offices, and business partners.
- VPNs are implemented through tunneling, in which the data to be transmitted and header information are encapsulated inside standard IP packets, usually after encryption and sometimes after compression as well.
- Three VPN tunneling protocols are currently in line to become industry standards: PPTP (Point-to-Point Tunneling Protocol), L2TP, and IPSec (IP Security).
- Security is a critical component of a VPN implementation, especially for those implemented over the public Internet. Encryption delivers the "private" in virtual private networking, but it is very process intensive. Because of this, hardware-based VPN products deliver the best performance.
- Infonetics Research estimates that service providers' share of the VPN market will grow to U.S. $8.8B by the end of 2001.
- Despite severe quality limitations, users have already started using the Internet to deliver video-conferencing. Many of these early systems use 56 kbps links to deliver video images of 160 by 160 pixels at a rate of 12 to 14 frames per second — that translates to a small video box confined to just part of the computer screen, showing a jumpy image. Existing corporate videoconferencing systems deliver full-screen images at 30 frames per second — a much higher quality image. Internet-based videoconferencing services won't be real until new QoS standards are in place such as the Resource Reservation Protocol (RSVP) and IP version 6 (IP6).

4.4.7 Integration for Residential Users

- Researchers are issuing cautions regarding unsolved problems with digital subscriber line (DSL) and cable modem services. Complexities have been identified regarding the mixing of POTS and DSL services without using a splitter. One of the biggest issues concerns what happens when a user picks up a telephone handset in a splitterless service — the result is an immediate change in load on the local loop, resulting in a loss of one to two orders of magnitude in signal amplitude. Additionally, crosstalk can occur when several POTS twisted pairs in the same bundle are used to provide DSL service.
- A new service that is being considered combines Internet and television services so that end users can simultaneously surf the Internet and watch enhanced broadcast television at home. One approach uses cable television systems to deliver downstream data to advanced set-top boxes. Other approaches are more software oriented and don't necessarily need set-top boxes.

- The FCC has mandated that broadcasters must have started offering some high-definition television (HDTV) digital programming in 1999 and complete their transition to digital by 2006.
- MSNBC, the cable broadcaster owned by Microsoft and NBC, has experimented with technology that allows broadcasters to send digital signals embedded within television signals to PCs. MTV, along with Intel, launched Intercast Jam, which broadcasts videos to PCs alongside rock artist information via a webbrowser in the broadcast signal.
- For years, TV stations have been beaming out data in small doses — in the form of closed captioning, test signals, ghost canceling, and messages to affiliates. That data is carried mainly through the vertical blanking interval (VBI). This offers a total of between 150 and 200 kbps of available bandwidth. This bandwidth is now being used by the Intercast consortium to transmit ancillary data streams to PCs via the VBI.
- Vendors are starting to create products that support videophone services. One such product puts a video camera in a set-top box and displays its image on any cable-ready TV. A touch-tone phone provides audio, dialing, and navigation of the system's on-screen controls. The system includes a built-in 10BaseT Ethernet interface to link directly to a cable modem, digital subscriber line modem, or corporate network.

4.5 New Product and Service Creation (Figure 4.8)

4.5.1 Introduction

The increasingly competitive telecommunications environment will require service providers to create and deploy new products and services faster than ever before. Providers were able to create new services at a much more leisurely rate in the past. A provider could wait until the next generation of technology had been deployed into the network before introducing the services that used the new technology.

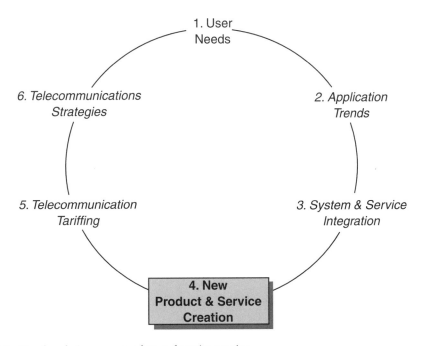

FIGURE 4.8 Trend analysis — new product and service creation.

As the number of service providers increases, it is generally agreed that the ones who will succeed are those who are able to best understand their end user's needs and deploy services that best meet those needs. In order to accomplish this, a provider will need a new approach to designing and deploying future services.

Changes in network equipment and in the types of networks that are being deployed are equipping providers with the essential tools. In this section we will first look at some of the drivers and constraints that providers are facing as they struggle to change how they create services. Next, we'll focus on how services will be created in tomorrow's network. Finally, we'll investigate how network bandwidth affects the types of services that can be created and what is being done to provide more bandwidth for new services.

4.5.2 Drivers and Constraints

- U.S. West is now offering a PCS service that includes mobile dial tone and advanced messaging and routing capabilities. The dial tone service combines a handset-generated and network-generated dial tone. The handset portion allows users to hear dial tone while dialing, and the network-generated portion allows users to hear a dial tone while they are initiating features. Customers have said that they associate dial tone with reliability and quality. The service also includes a same-number feature that routes calls made to a home, office, or PCS number to a PCS phone. It can also route all messages to a single mailbox, notifying users of messages via a light on the handset.

- Two different standards are being considered for using ATM to switch IP traffic: Multiprotocol over ATM (MPOA); ATM Forum; based on LAN emulation, seen as a campus backbone solution. Multiprotocol Label Switching (MPLS); IETF; designed with the large-scale WAN in mind

- Service providers are looking for new ways to rapidly introduce new services to meet growing customer demands. Existing circuit switches can generally only be modified by their vendors, which takes too long and costs too much. Programmable switches (Figure 4.9) consist of three main parts: a programmable switching fabric, controlling software ("host program"), and external media used to provide enhanced-services functions. New functions can be added to programmable switches by simply adding services and features to the host program. Open interfaces and APIs permit third-party developers to create vast libraries of available services.

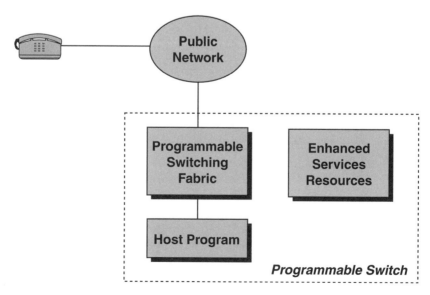

FIGURE 4.9   Programmable switch architecture.

- In order to help ISPs that are not ready to make a full-fledged investment in electronic commerce with an option to start a little smaller, e-commerce software vendors are getting creative. One vendor permits ISPs to operate small on-line stores (less than ten items to be sold) for free. As they increase the size of the "store," they then start paying the vendor for the use of the software.

- As of 1997, there were more than 1.3 billion televisions in the world, compared with 245 million PCs and 741 million telephone lines.

- Both the competing standards for digital television, one promoted by the U.S.'s Advanced Television Systems Committee and the other promoted by Europe's Digital Video Broadcasting Group, offer an almost unlimited potential to broadcast data to end users. In tests, broadcasters have been able to transmit 60 Mbytes of data during a 51-second commercial.

- Some cable operators are now able to offer traditional switched voice services using their cable networks. These services are proving to be very popular and in fact are more popular then the highly touted cable modem services.

- Joint research done by International Data Corporation, Zona Research Inc., and Literature Searches reveals that U.S. corporations spend more than U.S. $14 billion annually for their own expensive but reliable data networks.

- In order to permit Internet traffic to be prioritized, two sets of networking protocols are working their way through the IETF. The first set is called Differentiated Services (DiffServ). It provides routing mechanisms designed to manage various QoS profiles, or performance parameters. The other set of protocols is the multiprotocol label switching protocol. MPLS is a routing mechanism designed to group all packets within an IP session into a single "flow" at the networking layer (Layer 3) and "tag" each session as such for expedited passage through router hops.

- Wireless service providers who use TDM are looking for ways to differentiate their services. Their latest attempt is called Wireless Office Services. These allow users to access PBX features from their wireless phones while they are in the office and when they leave the office. This permits them to use such features as four-digit dialing and call forwarding in all environments.

- AIN platforms and capabilities are ways that new services can be introduced into the public network. However, the change to a packet-based network puts the future of AIN services in some doubt.

- Vendors' research labs are starting to produce products that implement some of the latest advances in speech recognition technology. This type of interface is seen as a major step toward the convergence of telephony and Internet applications. Call center applications are expected to be among the first to benefit from these types of products.

- Vendors are offering service development products for Internet protocol-based voice service providers. Service providers will be able to use these products to add key features such as universal messaging, follow-me services, and paging to their IP/public network gateways. All incoming messages are stored in a single mailbox and can be converted to a variety of formats that the user can then access via Web-browser, e-mail, or telephone.

- The Voice-over-IP (VoIP) Forum recently ratified an implementation agreement that defined an interoperability profile based on the H.323 standard from the ITU. H.323 was designed to be a technology for multipoint-multimedia communications over packet-based networks, which include IP-based networks, such as the Internet. It can be applied in a variety of ways — audio only (IP telephony or VoIP); audio and video (video telephony); audio and data; and audio, video, and data.

- One debate in the communications community is how to successfully deliver QoS and implement service-level agreements (SLAs). QoS, a network-wide performance characteristic, refers to the network's ability to fulfill a traffic contract — the SLA — between the WAN network provider and the subscriber for the minimum service provided by the network.

- ISPs are replicating content across multiple services in order to balance user demand loads. Vendors are now starting to offer products that allow service providers to automatically route end-user requests to the replicated server that has a low enough load to facilitate the request.
- User demand for access to multimedia Internet content has resulted in novel solutions being created by vendors. One approach uses satellite links to bypass the Internet and deliver multimedia content to local ISPs where it can be cached for access by local users. This approach can be further extended to caching of popular websites in order to speed up local access speeds.

4.5.3   New Service Creation

- The emerging consumer vehicle tracking service is called telematics. Telematics systems combine GPS and cellular networks to offer safety and concierge services to consumers in automobiles. The number of users is expected to grow from 58,000 subscribers this year to 1.2 million by 2003, according to the Strategis Group. Most U.S. telematics operate on AMPS because of its near-ubiquitous coverage.
- Smaller ISPs are using audio and video conferencing capabilities to distinguish themselves from competitors. These service providers are starting to investigate using client and server software solutions that permit videoconferencing over the Internet. Initial users include schools that have a need to provide a one-on-one tutoring experience but don't need an elaborate room-based videoconferencing system. The supporting software systems are all H.323 compliant.
- Business travelers want to be able to access the Internet even when they are traveling internationally. This is currently not possible — such travelers must reach their ISPs POP in order to access the Internet. Some service providers are attempting to build international POPs to meet this need. Other smaller ISPs are banding together to create consortia to offer Internet access to their collective customers. An additional service that is being investigated would offer roaming users access to their corporate intranets via secure tunneling.
- Bell Atlantic Mobile is offering utilities the ability to read customer's meters automatically via wireless data transmission using the cellular digital packet data (CDPD) network. The service would allow utilities automatically to read meters and monitor energy flows, among other services, from a central location, skipping the need to send personnel to customer locations. This service offers utility companies an advantage in a deregulated market because they can offer their customers a better picture of their usage patterns and then offer them a special deal to keep them from going to other utility providers.
- As of 1998, the Strategis Group reported that CDPD services had only 17,000 subscribers.
- Service providers are starting to offer enhanced fax services. These services include mailbox, which provides a secure fax mailbox accessible from any location; never-busy fax transparent service stores faxes for later delivery; fax-on-demand lets businesses create a library of faxable documents that customers can access; and fax broadcast delivers a document to as many as 10,000 locations with just one transmission.
- ISPs are starting to roll out Internet protocol voice services to corporate users. Initially, business users can connect their PBXs to the ISPs IP network, thereby cutting costs on internal long-distance calling. The next step is to combine IP voice with extranets. Businesses would then be able to call other businesses at remote locations using five-digit dialing.
- Many PC games now come with multi-player Internet options. Users first connect to the Internet, then select a specific server which "hosts" a gaming session. Then as the end-user plays the game in multi-player mode, the server allows them to exchange information with other players in real time.
- Consumers and businesses will soon have the ability to both view and pay bills via the Internet thanks to various forms of electronic bill presentation and payment (EBPP). This new service will

allow billers to cut paper processing costs and garner customer loyalty and website hits. Financial institutions, bill consolidators, Internet portals, and makers of personal financial manager (PFM) software products look forward to capturing market share.

- Studies show that 40% of U.S. homes have a PC and only 20% of those are plugged into the Internet.

- Electronic commerce is struggling with the issue of how to reach customers who are not connected. Companies that have to use both the telephone and the Internet to reach customers are looking for a way to tie the two systems together — "v-commerce." These firms want to develop new applications that will link voice and data, telephone, and PC to let Internet vendors reach customers who can't reach their Web pages. These new applications will use Motorola's VoxML markup language which simplifies embedding speech into Web pages.

- The Web provides an opportunity for delivering a new type of picture called immersive photography. This technology allows you to use your PC to navigate around a digitized 360-degree photo. This technology is targeted toward Internet retailers who want to give their customers a wrap-around view of their goods, including high-end real estate agents, travel agents, cruise lines, and destination marketers.

- Visual communication services are poised for proliferation as new advances eliminate the final technological and market obstacles. The ideal solution for multimedia services combines the organization and simplicity of the telephone system with the multimedia and open nature of the Internet.

- Telemedicine is a broad term for several facets of medical care. Collaborative videoconferences between sites, on-line access to patient records, medical libraries and databases, and continuing medical education all fall under the term. Most telemedicine programs today are either simple store-and-forward systems or ISDN videoconferencing systems adapted for use in a health care setting.

- Automobiles are being equipped with more and more electronics and telecommunications devices. Many cars now have Global Positioning System (GPS) receivers and computers to help the driver from becoming lost. The U.S. government, state governments, and a variety of industries are considering spending U.S. $200 billion on the Intelligent Transportation System (ITS) initiative. ITS will provide automated cross-border fleet services for North America, enhanced driver navigation, automated accident reporting, and toll collection. Futurists foresee a day in which a car monitors its "health" and can then use wireless communications to identify repair stations in the event that a potential part failure is detected.

- Prepaid wireless services have become a big business in the U.S. The industry may see more than U.S. $650 million in prepaid card service revenues in 2000. Customers generally must pay to have their wireless service activated, then they must purchase a prepaid denomination, often in the form of a card from a retail distributor. The next step is to initialize the prepaid service via an interactive voice response (IVR) service.

4.5.4 Increasing Bandwidth

- Wireless cable operators are starting to offer high-speed Internet access services using multichannel multipoint distribution service (MMDS). Without converting to digital, the most wireless cable operators could offer in video is 33 channels, which can't compete with average landline cable or satellite providers.

- Wavelength Division Multiplexing (WDM) technology is being added to the network in order to increase backbone capacity to handle new high-speed access technologies. Initial WDM systems were only 2 to 4 channels. Recently, 32 channels appeared in dense WDM. Now hyperdense or ultradense WDM (UDWDM) systems with channel densities of 40 and up and capacities of 400 Gbps are becoming available. Providers are upgrading a few fibers on a route and then upgrading the others over a few years. One vendor boasts that its terabit demo could carry the Internet's entire traffic on a single fiber.

Time Period	WDM Capabilities
1980s	• 2 channel • Wideband WDM • 1310, 1550 nm
Early 1990s	• 2–4 channels • 3–5 nm spacing • Passive WDM components/parts
1996	• 16 or more channels • 0.8 nm spacing • DWDM, integrated systems

• GTE is using multichannel multipoint distribution service (MMDS) technology to deliver 68 video channels, 32 music options, and near video-on-demand with 40 channels of pay-per-view. GTE has also rolled out hybrid fiber/coax (HFC) cable-based digital video networks. These systems transmit at 750 MHz downstream and 40 MHz upstream.

• U.S. West has introduced a VDSL platform that provides subscribers with integrated digital TV and high-speed Internet access. Included in U.S. West's bundle are on-screen caller ID, voice messaging, and 120 channels of programming, including pay-per-view. The service operates at 256 kbps.

• Bell Canada now uses its HFC system with 10 Mbps downstream and 1 Mbps upstream to offer a picture-within-a-picture service that allows users to go online and watch television from the same screen simultaneously.

• Cable operators that want to start offering high-speed Internet access services to their subscribers without having to perform expensive upgrades to make their cable network two-way are getting creative. They are using their existing one-way cable networks to deliver content to end users while the end-users use their telephone to send information requests. Although this solution may be well suited for rural cable providers who will never have the funds to make their systems two-way, this one-way approach may not provide the bandwidth required by the growing SOHO market.

• Cable operators are able to offer residential subscribers Internet access at 1.5 Mbps rates using a cable modem. A report from Forester Research concedes the residential market to cable operators over telephone companies: cable operators are predicted to have 13.6 million cable modem customers by 2002, while telecos will have only 2.2 million ADSL users.

• U.S. West markets its DSL services to three types of residential users: consumer/Web browsers (want "always on"), gamers ("entertainment"), and work-at-home users ("looking for bandwidth and the user experience").

• The cable company MediaOne has found, through internal studies, that nearly all cable modem owners use their Internet connections seven to nine times more often than when they had a dial-up connection.

• MediaOne marketing cites a recent study that claimed the average Internet user wastes a total of 50 hours a year waiting to connect to the Internet and waiting for pages to download.

• Telecos, ISPs, and CLECs that are rolling out ADSL services are finding that the earliest adopters of the services are in the small business market. Telephone companies will stress the security of ADSL over cable modem's shared media to small business owners.

• In the U.S., the FCC has auctioned off 1.3 GHz of spectrum in the 28 and 31 GHz ranges for use in local multipoint distribution service (LMDS) two-way services.

- Broadband wireless networks have many benefits: they are fast and easy to deploy; they have minimal infrastructure and real estate requirements; they feature grow-as-you-go network build-out; and they can deliver voice, video, and data services from 64 kbps to 155 Mbps.

- LMDS can be used to offer many services. Business-oriented services include wire speed LAN interconnect and fractional and full T-1. Teleworking at 10 Mbps is virtually as fast as being at the office. Megabit per second Internet access is geared to residential users. Other services include 100 broadcast video channels in competition with cable, and second and third phone lines at home or the office.

- LMDS services compete with DSL and hybrid fiber/coax (HFC) services. LMDS is better than both DSL and HFC at offering high-speed symmetrical services.

- Wireless cable operators have spectrum in the 2.5 GHz range (MMDS).

- The H.323 protocol, used to provide VoIP services, defines ways in which multimedia formats such as phone calls, computer data, pictures, or video can be exchanged and managed seamlessly across packet-switched networks.

- A variety of broadband wireless providers have already introduced services that use multichannel multipoint distribution service (MMDS) and local multipoint distribution service (LMDS). MMDS service providers have been around for awhile, whereas LMDS providers have only recently bought their licenses. MMDS offers a broader coverage reach while LMDS offers greater capacity. Current service offerings use either the public network or a cable modem for the return path.

4.6 Telecommunications Tariffing (Figure 4.10)

4.6.1 Introduction

Perhaps no aspect of telecommunications is as overlooked as how services are priced. All segments of end users have differing amounts of funds available to spend on telecommunications services. Pricing a

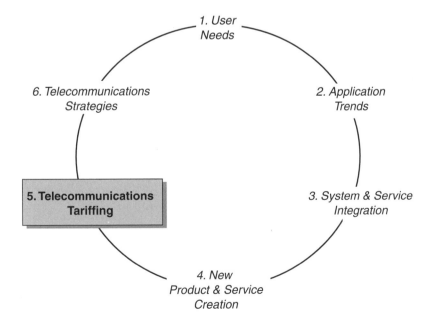

FIGURE 4.10 Trend analysis — telecommunication tariffing.

service too high will cause end users to seek lower price alternatives. Pricing a service too low will result in the service provider missing out on revenues that could have been used to fund the next service.

In the past, service providers have enjoyed monopoly status in both North America and Western Europe. Under this system, prices for services were closely regulated by governments. This is in the process of changing, and in the future service prices will be driven by market factors. This change will require existing service providers to change the metrics used to measure service and the pricing philosophies that have been used to create service rates in the past.

In this section, we will explore trends in tariffing in the leading markets of North America and Western Europe. The effect of competition on service pricing will also be examined. Finally, we'll discuss the impact that new technologies will have on the pricing of future services.

4.6.2   Regulatory Trends

- The FCC is proposing that the Bell companies be permitted to create separate subsidiaries to offer data communication services. These subsidiaries would be less regulated and could set interstate service prices without filing to the FCC. The Bell's regulated units would still be required to sell capacity to competitors but the separate subsidiaries wouldn't.

4.6.3   Service Pricing Trends

- The average long-distance call in the U.S. costs about 13 cents per minute, but the average international price is 89 cents per minute. Telco revenues per minute on international calls are predicted to fall more than 20% annually through 2001.
- Cable & Wireless USA hopes to use pricing and inexpensive long distance to draw residential customers to its Internet service. CWIX will offer customer 150 hours of on-line service, e-mail, and a free Web page for a monthly fee of U.S. $14.95. Some analysts doubt if bundling long distance with Internet access will attract new customers. They point out that the intersection of households that are on-line and use long distance heavily is not large — perhaps 15% of the total.
- In most U.S. telco service areas today, termination fees of up to U.S. $36,000 to break a tariffed service contract are still alive and kicking, despite efforts by competition to eliminate them. These contracts can prevent a customer from purchasing the services offered by a competitive provider because they still have a year or two to go on their current contracts.
- The paging industry grew by 14% in 1997 to a total of 50 million subscribers. However, in 1997 four of the top firms, which together control almost 40% of the market, reported almost a half-billion dollars in losses on combined record revenues of more than $2 billion. Many paging companies are suffering from expensive network buildouts. Paging companies seem to hold a high number of customers who refuse to upgrade beyond basic plans, according to analysts. Price wars and new technologies have driven down the costs of average basic local service from $20 a month a decade ago to less than $10 a month. In some markets, the price has shrunk to less than $5 a month.
- Although extending wireless service to the high percentage of credit-challenged users was a chief driver in the development of prepaid service, wireless carriers are discovering that prepaid strategies may be almost as critical to future growth of their overall customer bases as traditional post-paid service. BellSouth Mobility intends to have prepaid accounts for 30% of its new sign-ups.
- Traditional methods of buying and selling bandwidth are not adequate in today's competitive market. A new Internet-based service permits providers with bandwidth to sell their available bandwidth for bidding purposes. Buyers are then able to see the available bandwidth along with information regarding destination country, size (T1, OC-3, etc.), and the length of the contract. If a qualified registrant posts a bid, then the service puts the bidder in touch with the service provider to see if they can work out a deal.

- An interexchange carrier has entered into a partnership with one of Florida's tourism groups. The carrier will share its profits with hotel property owners when hotel guests make calls from their hotel rooms using the carrier's service.

- Cable operators that provide Internet access services via their cable networks are already dropping the price of their service in order to capture more of the Internet access market. Some cable operators see this as the only way to push their Internet access service beyond the early adopters. These cable providers hope to use their lower prices to attract lighter users and cut into the market share of ISPs.

- U.S. cell phone users pay for all incoming and outgoing calls that use their phone. About 80 to 85% of all cellular calls originate from a wireless phone — this means that cellular subscribers are either not giving out their phone numbers or they are turning off their phones. One way to balance traffic is to upgrade equipment to accommodate calling party pays (CPP) billing. The caller typically pays 35 to 45 cents a minute, an average rate for an outbound call from a cell phone.

- One reason that domestic long-distance services have not switched to an IP network is because circuit-switched voice is already cheap: rates are below $0.05 per minute for corporate customers and below $0.09 for residential customers. The bottom line is that to make the numbers work domestically requires 10,000 minutes a month to a single location to justify the cost of a private IP telephony network.

- The cost to complete an international voice call is much higher. Carriers charge as much as U.S. $4.00 per minute to complete a call to North Korea and other countries where it is hard to find a good termination.

- To make greater wireless penetration and increased billable minutes a reality, carriers must embrace "calling party pays" (CPP) as the prevalent billing model, rather than "wireless party pays" (WPP).

- Juan Fernandez of Frost & Sullivan reports that when CPP was implemented in Argentina, the market grew from 700,000 subscribers to 2.1 million in 11 months.

- Giving a customer the first incoming minute of a call for free is an interim way that service providers are trying to increase the number of billable minutes.

- ISPs jumped *en masse* onto the flat-rate bandwagon in 1996, only to find that "all-you-can-eat" pricing has a way of eating away at the bottom line. Some service providers have found that the flat-rate strategy delivers something that they wanted to get all along: lots and lots of customers.

- Flat-rate pricing can be a nightmare for providers, especially if their costs are largely dependent on usage and that usage is difficult to predict. Frame relay and Internet services fall into this category.

- Providers gain from using flat-rate pricing because they don't have to cover the cost of administrating usage-base pricing. That can be a significant gain considering that these expenses can run as high as 18% of the total cost of the service.

- Usage-based pricing becomes just as attractive as flat-rate pricing if the cost to deliver a service increases substantially as service usage grows.

- Wireless service providers are starting to offer prepaid services in order to address the 20 to 40% of the market that didn't qualify for service because of bad or nonexistent credit histories.

- Prepaid systems have become more attractive in recent years due to several improvements: they lacked a real-time billing engine (couldn't cut off calls in mid-conversation), and they didn't accommodate incoming calls.

- Wireless service providers can use either a switch-based or a handset-based approach to implementing prepaid services. Most providers have selected the switch-based approach because it works with any handset and it is less prone to tampering.

- The initial investment in prepaid infrastructure can be heavy, but payback periods can be quick. Along with expanding the potential customer base, prepaid wireless lowers the cost of acquiring a customer, since it eliminates the need to do a credit check.

- Despite the upfront charges (for a phone), prepaid services aren't necessarily a tough sell to credit-challenged customers. The per minute charges are comparable to those levied under low- and mid-tier pricing plans, and they include taxes and interconnection charges. Prepaid customers aren't charged monthly access fees.

- Carrier consolidation and interconnection, increased competition, service bundling, and new technology introductions all are contributing to the need for more intelligent and flexible customer care and billing systems.

- Convergent billing means using a single billing system to create all bills — it does not necessarily mean sending a customer a single bill!

- New and existing service providers competing against each other are selling telephone services that are roughly the same. Their goal is to avoid a commodity war of attrition.

- In the United Kingdom, there are 150 licensed telecommunications providers contending to supply the country's 30-odd million adults with fixed wireless, data, voice, and video communications.

- Although pricing is becoming increasingly important in telecommunications (especially voice telephony service), customer service, branding, billing, and value-added services are all keys to success.

- Types of carefully constructed rates and calling plans include bundling, demographic profiling, "loss leaders," incentive schemes, "flattened" prices, calling circles, postalized rates, and special rates.

- Service providers seek to bundle multiple telecommunications services in order to provide one-stop shopping for their customers.

- There are concerns that bundling may reduce churn for a company as a whole, but not necessarily for individual lines of business.

- When customers are asked which company they would use for bundled services, customers overwhelmingly prefer local and long-distance carriers.

- WorldCom has announced its International Business Links (IBLs) and end-to-end ATM services within Europe and to the U.S. The announcement is a culmination of its transatlantic Gemini cable project with Cable & Wireless, together with its European fiber-laying activities to link the former islands of MFS' metropolitan networks. WorldCom has made a habit of breaking the traditional telecom mold. This service announcement is no exception. Other ATM services have been slow to emerge on the commercial market. WorldCom will launch constant bit rate (CBR), variable bit rate (VBR), near-real time (NRT), and available bit rate (ABR) services, pegged favorably against existing leased circuit and frame relay tariffs.

4.6.4 Impact of New Technologies

- Networkwide QoS is needed to deliver priority service to higher-paying customers. Service providers want to use QoS as a basis for offering various classes of services to different segments of end users. By doing this, they can create different pricing tiers that correspond to QoS levels. That might be one of the best ways to offer new revenue-generating services in public data networks.

- Smaller ISPs are using centralized functionality to improve their competitive situation. Most ISPs store subscriber information on up to five different servers, thus preventing them from using data mining tools that are essential to customizing services. This opens the door to content-based billing. Software can be used to create something similar to the call detail records used with voice calls, but it will consider a subscriber's profile.

- PCS services have reduced many of the advantages of paging through longer battery life, first minute free, free/bundle voicemail, free caller I.D., prepaid plans for less creditworthy customers, and competitive pricing.

- From a connectivity perspective, the Internet is well suited for telephony because of its global reach. From an engineering perspective, it is efficient — a dedicated T1 can support as many as 130 IP voice calls vs. 24 simultaneous calls as in today's carrier networks.
- When talking about billing for IP services, the two key issues are metering and settlements. Metering is relatively straightforward. Settlements introduce trouble because the number of billing arrangements between carriers grows exponentially with the number of Internet telephony service providers.
- According to Duane Ackerman, chairman and CEO of BellSouth, 17% of new PCS customers in Louisiana recently signed up for "untethered" service as a replacement for wireline.
- Finland has the greatest wireless penetration of all markets: 42% at the end of 1997.
- Some service providers — BellSouth and Pacific Bell among them — are now betting millions of dollars that Web-based electronic billing systems are essential for hooking lucrative but finicky business customers — and eventually even some residential ones — who are interested in fast, responsive billing.
- Many service providers expect less than 5% of all telecom customers to use Internet billing in the near future.
- On-line billing, however, has its challenges. It not only requires Internet access but is also costly and complicated to set up, especially for big service providers with massive billing systems already in place.
- One of the major benefits of electronic billing is that it saves the service provider money. The more that customers opt to pay their bills through a Web site, the lower the cost of running a paper-based billing system. By some estimates, the entire paper trail from stuffing an envelope, mailing the bill, and processing the payment costs a service provider 75 cents to $1.50 per account every billing cycle. BellSouth estimates that it spends 7 cents to send every printed page.
- Initially, the IP did a poor job of tracking and generating the appropriate data to accurately measure usage for customer billing. Changes are being investigated because of the interest in using IP telephony for voice and fax.
- Many different usage-based services are currently being planned: least-cost routing, time of day routing, dynamic bandwidth allocation, volume discount rates, callback, security enhancements, Web hosting, e-mail, chat lines, whiteboards, videoconferencing, work group collaboration and multimedia sessions, software applications distribution, applications rental, and classes of service quality.
- Many technical challenges of IP-based services must be tackled. Foremost are extrapolating and scrubbing down traffic information from routers and switches and matching that against customer account data for bills. This invoices tracking packet volumes, counting bits or bytes and logging origination or destination IP addresses.

4.7 Telecommunications Strategies (Figure 4.11)

4.7.1 Introduction

The brave new world that represents the future of telecommunications will consist of a group of aggressive global service providers who are competing for the same segments of end users. How each of the service providers hopes to succeed at the expense of its competitors is a fundamental part of its long-term strategy.

A provider's strategy for increasing its market share must be in part based on its current situation. In this section we will look at the current situations that describe many of today's up-and-coming service providers as well as some of the well-established players. We will examine their business goals and how they may go about achieving them. Finally, we'll identify some of the possible events that could dramatically change existing strategies.

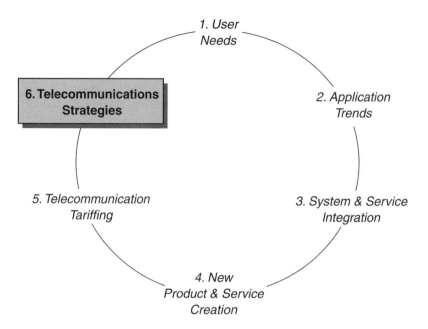

FIGURE 4.11   Trend analysis — telecommunications strategies.

4.7.2 The Players

The value chain of products and services will dictate the positioning of telecommunications service providers. The positioning process usually starts with answering a number of questions, such as:

- What is the perceived quality of my network?
- Is the network keeping pace with the growth of subscribers?
- How much should be invested?
- Where do I need to invest?
- How can I get more revenue out of existing services?
- How can I reduce operating costs?
- How do I know if a problem is just a solitary abnormality or a building problem?
- How can I reduce customer churn?
- How can I predict future capital expenditures?
- How can I get system usage information to improve marketing and sales?

The traditional value chain was very simple. The equipment suppliers — a closed market of monolithic suppliers — have provided hardware with hard- or soft-wired integrated services. This equipment was key for network and service providers who have based their service offers to their customers on the capabilities of this equipment. Change cycles and service creation were extremely long, hardly meeting the customer's expectations.

The actual value chain includes the following principal components (TERP01):

1. Infrastructure
   IT component suppliers
   OSS application suppliers
   Network element suppliers
   System integrators

2. Network Products and Services
   Network operators
   Service providers
3. Hosting and Processing
   Hosting services providers
   Processing services providers
4. Applications and Media
   Applications services providers
   Context, content packaging, and management
   Content services providers
5. Customer

The players are not yet evenly distributed. Most of them are still emerging from the traditional service providers, and can be allocated to Network Products and Services. Examples are:

- ILEC (Incumbent Local Exchange Carrier): Strong provider who owns a considerable amount of telecommunications facilities and doesn't want to give away this position easily. Most likely, number of legacy support systems with little interoperability and integration in use. The result is high operating costs.
- CLEC (Competitive Local Exchange Carrier): Smaller, flexible provider who owns little or no telecommunications facilities (facility-less). By offering excellent customer care and new services, they try to build the support structure step-by-step. Their support systems are state-of-the-art, lightweight, and less expensive to operate. In certain cases, they use service bureaus for billing and provisioning.
- IEX (Inter Exchange Carriers): Primarily responsible for long-distance services with stepwise penetration of the local exchange area. They can be both incumbent and competitive providers with the result of the need for very heterogeneous support systems.
- PTT (Post, Telegraphy, and Telephone): Strong provider who owns a considerable amount of telecommunications facilities and doesn't want to give away this position easily. Most likely, number of legacy support systems with little interoperability and integration in use. The result is high operating costs.
- CAP (Competitive Access Provider): Facilities-based or non-facilities-based; similar to the ILEC, but have carefully selected local loops for high-profit commercial customers.
- NSP (Network Service Provider): Responsible for providing a highly reliable networking infrastructure, consisting of equipment and facilities. Its responsibilities are usually limited to the physical network only, but element management systems are usually included into their offers.

However, integration is important for many customers. Thus, ISPs and ICPs will play an important role as well. The short definitions are:

- ISP (Internet Services Provider): Its main goal is to provide Internet access to business and attract customers. Major challenges include peering to each other and to other carriers, managing quality, and offering acceptable performance.
- ICP (Integration Communications Provider): Emerging provider with integrated services offer, concentrating on next generation, high-speed data and wireless services, in particular for profitable business users. Its acceptance in the marketplace is expected to be high. In terms of support systems, they buy instead of build; occasionally, they use service bureaus for billing and provisioning. They take advantage of the fact that intranet, extranet, virtual private networks, eCommerce, and multimedia applications require more bandwidth than is available over traditional circuit-switched voice networks.

Hosting and processing will be most likely dominated by traditional mainframe and server manufacturers that are flexible enough to make the necessary facelifts to their equipment to meet requirements of load distribution, load balancing, storage management, and security. IBM and Compaq may be mentioned here as examples of providing reasonable services, using server farms with high availability features.

Application and media need many new competitive players. At the beginning, ASPs and ESPs will dominate this market. The short definitions are:

- ASP (Application Services Provider): Emerging service provider, which must combine application, systems, and network management. Service level expectations are extremely high; the whole business of customers may rely on this provider.
- ESP (Enterprise Services Provider): Emerging service provider from the enterprise environment. It offers services for a limited user community with similar attributes to the provider. It uses and customizes its existing support systems that may not scale well.

Hosting and processing enable Web presence and interactivity on the Internet. They are typically provided by ISPs and NextGen service providers which are active in IP services. They mainly include hosting of Web server infrastructures and content and Web-enabled transaction software and hardware which allow the execution of online transactions. It is an infrastructure type of activity, although it is characterized by added value and significant amount of additional services. Service offer alternatives are:

- Web hosting: Keeping content on Web server farms and offering access with good performance
- Value-added Web hosting: In addition, content, and database maintenance and Webmaster services
- Data hosting by offering Storage Area Networks
- Data management services, including search machines
- Public Key Infrastructure services, including trust center functions
- Centralized Web transaction services
- Web community and Internet account management
- Transaction authentication services

The typical customers of these services are businesses. While large businesses previously deployed their own Web infrastructure in-house, they now also realize the efficiencies of lower complexity and economics of scale given by professional service providers. This customer base can easily extend to the future. The key differentiators will be the service level and complexity of services offered to customers. Hosting and value-added hosting emerge as a volume business. Large data centers with server farms, load balancers, and traffic shapers combined with high availability and excellent performance will take business away from smaller service providers with lower availability and limited Internet access capability. Most service providers are inexperienced in this area.

The continuation of the value chain is dominated by innovative services that are to a certain extent IP-based. It means that the traditional circuit-switched architectures are replaced by packet-switched architectures. For the underlying physical architecture, there are many choices, including:

1. IP + ATM + SDH/Sonet = B-ISDN	Traditional approach, which has the most supporting network elements and their element managers
2. ATM transport	Includes both SDH/Sonet-less ATM transport and ATM/SDH/Sonet hybrids
3. Switched routing	ATM/IP hybrids
4. IP over SDH/Sonet	PPP or HDLC-framed IP mapped to SDH/Sonet
5. Optical IP	Transport of PPP or HDLC-framed IP over WDM with fast photonic restoration
6. Use of enhanced frame relay	Substitution of ATM by frame relay in any of the approaches 1, 2 or 3

There is no doubt that the new area of competition is content.

Content delivery management is taking off and service providers are well positioned to earn revenues there. Content delivery management helps content owners to provide seamless and fast website access for customers by

- Large scale caching
- Distribution of Web server farms
- Complex Internet routing services on managed network segments

All these aspects help to deliver reasonable performance. Processing is an increasingly important revenue generation opportunity as traditional infrastructure business shrinks. Transaction, and therefore processing, which is the infrastructure and software enabler of transactions, is believed to grow to become the single most important revenue input of the Internet value chain. Processing by no means is related to the core business of service providers, but it is important for eCommerce service offers. Services can be created by the IT organization of service providers in collaboration with systems integrators.

The Application and Media elements of the value chain create and translate traditional and digital content into Web-ready format and creates the actual interface between the digital product and the customer. This service is targeting an end-to-end process which covers creation, manufacturing, delivery, and presentation of content to customers. This is believed to be the most promising business opportunity of the Internet. It carries the highest growth potential, but at the same time the highest risks, too. Telecommunications service providers, IT companies, media enterprises, retail chains, and several other industries are competing for revenues.

Service offers are:

- Application Services to be provided by service providers, integrating IT, software, system integration, telecommunication, and consulting skills.
- Content authoring, auditing, deployment, and maintenance combined with bandwidth management, server-load management and traffic management, supporting generic, corporate, and specialized niche portals, and B2B and B2C operations.
- Content creation targeting videos, movies, audio, photo archives, encyclopedic articles, analyst reports, financial evaluation, and many others. Music and also written material combine with broadband access to support multimedia to be delivered over the Internet.

These innovative service areas must be seriously investigated by service providers. In other traditional areas, the profit margins are narrowing; in the IP area they have to face other competition. To be successful, innovative minds are required. It means more collaboration with customers, mergers, acquisitions, investment into smaller companies that may be acquired later, and flexibility in service creation, fulfillment, and quality assurance.

Another Internet-based service is Immersive Photography, allowing customers to use PCs to navigate around a digitized 360-degree photo. This technology is targeted toward Internet retailers who want to give customers a wraparound view of their goods, such as high-end real estate agents, travel agents, cruise lines and destination marketers.

Whether retailers are between service providers and customers depends on the marketplace. No general guidelines can be given in this respect.

4.7.3   Goals

The goals are different for each cluster of service providers. Table 4.4 summarizes the most obvious goals and future targets for each cluster of service providers, referenced in segment 4.7.2 (TERP01).

References

TERP01Terplan, K.: *OSS Essentials: Support System Solutions for Service Providers*, John Wiley & Sons, New York, 2001 (in production).

TABLE 4.4   Goals and Future Business Targets for Service Provider Clusters

Service Provider Clusters	Goals	Business Targets
	Infrastructure	
IT component suppliers	Sell more software	Replace legacy solutions
	Sell professional services	Acquire OSS application suppliers
		Integrate legacy and innovative systems
OSS application suppliers	Sell more software	Full-line of offerings of support systems
	Sell more professional services	Target ILEC legacy replacement
		Acquire other vendors of support systems
		Compete with system integrators
Network element suppliers	Sell more equipment via best of breed and best of suite offers	Outsource element management systems to vendors of support systems
		Use of open interfaces
		Develop solutions for eCommerce
System integrators	Sell custom design, development, and deployment	Acquire vendors of support systems
		Conduct many projects
	Sell custom integration	Consolidate products
	Sell consulting	Compete with OSS suppliers
	Network Products and Services	
Network operators (ILECs, PTTs, IEXs, NSPs, CAPs and global carriers)	Rapid introduction of new services	Less internal software development
	Cost reduction	More use of systems integrators
	Customer retention	More packaged software of support systems
	Multi-vendor management	Pervasive interconnection of support systems
	Convergent ordering	Customer relationship management
	Up-to-date asset management	Self-care with support systems for customers
Service providers (CLECs, ISPs, ICPs)	Build network capacity	Minimal internal development
	Customer acquisition	Automated processes
	Improve service quality	More packaged software for support systems
	Add facilities	Less service bureaus
	More carrier interconnection	Integration of support systems
	Support of micropayment and prepaid services	Customer relationship management
		Self-care with support systems for customers
	Hosting and Processing	
Hosting (mainframe manufacturers, server manufacturers)	Use existing storage resources	Penetrate the Web market
	Reengineer business processes	Support of eCommerce
	Use load balancers	Advanced asset management
		Support of Storage Area Networks (SAN)
Processing (mainframe manufacturers, server manufacturers)	Use existing processing resources	Penetrate the Web market
	Reengineer business processes	Support of eCommerce
	Use caching	Advanced asset management
	Applications and Media	
Applications (ASPs, ESPs)	Sell service	Resource integration
	Customer acquisition	Good management of the infrastructure
	Early profitability	Advanced asset management
		Excellent service levels
		Use of packaged software
Context, content packaging & management (ISPs, ESPs)	Real-time rating	Usage-based billing
	Service creation on-the-fly	Multimedia support
	Mid-range profitability	Multicasting for distribution
Content providers (ISPs, ASPs)	Real-time rating	Billing for content value
	Service creation on-the-fly	Web switching technology
	Mid-range profitability	

TABLE 4.4 (continued)   Goals and Future Business Targets for Service Provider Clusters

Service Provider Clusters	Goals	Business Targets
Customer	Increase service reliability Lower transport costs Faster service provider responsiveness Customer network management	Self provisioning via Web Custom quality of service reporting Flexible billing formats Electronic bill presentment and payment Usage-based accounting

Index

J

O